de Gruyter Studies in Mathematics 32

de Gruyter Studies in Mathematics

1 Riemannian Geometry, 2nd rev. ed., *Wilhelm P. A. Klingenberg*
2 Semimartingales, *Michel Métivier*
3 Holomorphic Functions of Several Variables, *Ludger Kaup and Burchard Kaup*
4 Spaces of Measures, *Corneliu Constantinescu*
5 Knots, 2nd rev. and ext. ed., *Gerhard Burde and Heiner Zieschang*
6 Ergodic Theorems, *Ulrich Krengel*
7 Mathematical Theory of Statistics, *Helmut Strasser*
8 Transformation Groups, *Tammo tom Dieck*
9 Gibbs Measures and Phase Transitions, *Hans-Otto Georgii*
10 Analyticity in Infinite Dimensional Spaces, *Michel Hervé*
11 Elementary Geometry in Hyperbolic Space, *Werner Fenchel*
12 Transcendental Numbers, *Andrei B. Shidlovskii*
13 Ordinary Differential Equations, *Herbert Amann*
14 Dirichlet Forms and Analysis on Wiener Space, *Nicolas Bouleau and Francis Hirsch*
15 Nevanlinna Theory and Complex Differential Equations, *Ilpo Laine*
16 Rational Iteration, *Norbert Steinmetz*
17 Korovkin-type Approximation Theory and its Applications, *Francesco Altomare and Michele Campiti*
18 Quantum Invariants of Knots and 3-Manifolds, *Vladimir G. Turaev*
19 Dirichlet Forms and Symmetric Markov Processes, *Masatoshi Fukushima, Yoichi Oshima and Masayoshi Takeda*
20 Harmonic Analysis of Probability Measures on Hypergroups, *Walter R. Bloom and Herbert Heyer*
21 Potential Theory on Infinite-Dimensional Abelian Groups, *Alexander Bendikov*
22 Methods of Noncommutative Analysis, *Vladimir E. Nazaikinskii, Victor E. Shatalov and Boris Yu. Sternin*
23 Probability Theory, *Heinz Bauer*
24 Variational Methods for Potential Operator Equations, *Jan Chabrowski*
25 The Structure of Compact Groups, *Karl H. Hofmann and Sidney A. Morris*
26 Measure and Integration Theory, *Heinz Bauer*
27 Stochastic Finance, 2nd rev. and ext. ed., *Hans Föllmer and Alexander Schied*
28 Painlevé Differential Equations in the Complex Plane, *Valerii I. Gromak, Ilpo Laine and Shun Shimomura*
29 Discontinuous Groups of Isometries in the Hyperbolic Plane, *Werner Fenchel and Jakob Nielsen*
30 The Reidemeister Torsion of 3-Manifolds, *Liviu I. Nicolaescu*
31 Elliptic Curves, *Susanne Schmitt and Horst G. Zimmer*

Andrei V. Pajitnov

Circle-valued Morse Theory

Walter de Gruyter · Berlin · New York

Author
Andrei V. Pajitnov
Laboratoire de mathématiques Jean Leray
UMR 6629 du CNRS
Université de Nantes
2, rue de la Houssinière
44322 Nantes
France

E-mail:
pajitnov@math.univ-nantes.fr
andrei_pajitnov@yahoo.co.uk

Mathematics Subject Classification 2000:
Primary: 58E05, 37C10, Secondary: 57M25, 57R70, 37C27, 37C30

Keywords:
Circle-valued Morse functions, the Morse complex, the Novikov complex, cellular gradients, dynamical zeta functions.

⊗ Printed on acid-free paper which falls within the guidelines of the ANSI to ensure permanence and durability.

Library of Congress – Cataloging-in-Publication Data

A CIP catalogue record for this book is available from the Library of Congress.

ISSN 0179-0986

ISBN-13: 978-3-11-015807-6
ISBN-10: 3-11-015807-8

Bibliographic information published by the Deutsche Nationalbibliothek

The Deutsche Nationalbibliothek lists this publication in the Deutsche Nationalbibliografie; detailed bibliographic data are available in the Internet at http://dnb.d-nb.de.

Printed in Germany.
Cover design: Rudolf Hübler, Berlin.
Typeset using the author's TEX files: Matthias Pfizenmaier.
Printing and binding: Hubert & Co. GmbH & Co. KG, Göttingen.

To the memory of my mother,
Nadejda Vassilievna Pajitnova

Contents

Preface

In the early 1920s M. Morse discovered that the number of critical points of a smooth function on a manifold is closely related to the topology of the manifold. This became a starting point of *the Morse theory* which is now one of the basic parts of differential topology. Reformulated in modern terms, the geometric essence of Morse theory is as follows. For a C^∞ function on a closed manifold having only non-degenerate critical points (*a Morse function*) there is a chain complex $\mathcal{M}_*$ (the *Morse complex*) freely generated by the set of all critical points of f, such that the homology of $\mathcal{M}_*$ is isomorphic to the homology of the manifold. The boundary operators in this complex are related to the geometry of the gradient flow of the function.

It is natural to consider also *circle-valued Morse functions*, that is, C^∞ functions with values in S^1 having only non-degenerate critical points. The study of such functions was initiated by S. P. Novikov in the early 1980s in relation to a problem in hydrodynamics. The formulation of the circle-valued Morse theory as a new branch of topology with its own problems and goals was outlined in Novikov's papers [**102**], [**105**].

At present the Morse-Novikov theory is a large and actively developing domain of differential topology, with applications and connections to many geometrical problems. Without aiming at an exhaustive list, let us mention here applications to the Arnol'd Conjecture in the theory of Lagrangian intersections, fibrations of manifolds over the circle, dynamical zeta functions, and the theory of knots and links in S^3. The aim of the present book is to give a systematic treatment of the geometric foundations of the subject and of some recent research results.

The central geometrical construction of the circle-valued Morse theory is *the Novikov complex*, introduced by Novikov in [**102**]. It is a generalization to the circle-valued case of its classical predecessor — the Morse complex. Our approach to the subject is based on this construction.

We begin with a detailed account of several topics of the classical Morse theory with a special emphasis on the Morse complex. Part 1 is introductory: we discuss Morse functions and their gradients. The contents of the first chapter of Part 2 is the Kupka-Smale Transversality theory; then we define and study the Morse complex.

In Part 3 we discuss the notion of *cellular gradients* of Morse functions, introduced in the author's papers [**113**], [**108**]. To explain the basic idea, we recall that for a Morse function $f : W \to [a, b]$ on a cobordism W the gradient descent determines a map (not everywhere defined) from the upper boundary $f^{-1}(b)$ to the lower boundary $f^{-1}(a)$. It turns out that for a C^0-generic gradient this map can be endowed with a structure closely resembling the structure of a cellular map. We work in this part only with real-valued Morse functions, however the motivation comes from later applications to the circle-valued Morse theory.

In Part 4 we proceed to circle-valued Morse functions. In Chapter 11 we define the Novikov complex. Similarly to the Morse complex of a real-valued function, the Novikov complex of a circle-valued Morse function is a chain complex of free modules generated by the critical points of the function. The difference is that the base ring of the Novikov complex is no longer the ring of integers, but the ring $\overline{L}$ of Laurent series in one variable with integral coefficients and finite negative part. The homology of the Novikov complex can be interpreted as the homology of the underlying manifold with suitable local coefficients.

The boundary operators in the Novikov complex are represented by matrices with coefficients in $\overline{L}$ (the Novikov incidence coefficients). One basic direction of research in the Morse-Novikov theory is to understand the properties of these Laurent series. The Novikov exponential growth conjecture says that these series always have a non-zero radius of convergence. A theorem due to the author (1995) asserts that for a C^0-generic gradient v of a circle-valued Morse function, every Novikov incidence coefficient is the Taylor series of a rational function. This theorem is the basis for the contents of Chapter 12. The reader will note that in general we emphasize *the C^0 topology* in the space of C^∞ vector fields; we believe that it is the natural framework for studying the Morse and Novikov complexes.

These results are then applied in Chapter 13 to the dynamics of the gradient flow of the circle-valued Morse functions. We obtain a formula which expresses the Lefschetz zeta functions of the gradient flow in terms of the homotopy invariants of the Novikov complex and the underlying manifold.

The last chapter of the book contains a survey of some further developments in the circle-valued Morse theory. The exposition here is more rapid and we do not aim at a systematic treatment of the subject. I have chosen several topics which are close to my recent research: the Witten framework for the Morse theory, the theory of fibrations of manifolds over a circle and the circle-valued Morse theory for knots and links.

Brief historical comments can be found in the concluding sections of Parts 2, 3 and 4, and some more remarks are scattered through the text. However I did not aim to present a complete historical overview of the subject, and I apologize for possible oversights.

The book is accessible for 1st year graduate students specializing in geometry and topology. Knowledge of the first chapters of the textbooks of M. Hirsch [**61**] and A. Dold [**29**] is sufficient for understanding most of the book. When we need more material, a brief introduction to the corresponding theory is included. This is the case for the Hadamard-Perron theorem (Chapter 4) and the theory of Whitehead torsion (Chapter 13).

Acknowledgements.

I am greatly indebted to S. P. Novikov for sharing his insight during several years starting from 1985.

I am grateful to many people with whom I discussed the Morse-Novikov theory throughout the years 1985 – 2005, in particular to P. Akhmetev, V. Arnol'd, M. Farber, V. A. Ginzburg, J.-C. Hausmann, A. Hohlov, F. Laudenbach, S. Marchiafava, G. Minervini, M. M. Postnikov, J. Przytycki, A. Ranicki, J.-C. Sikorav, V. Turaev, O. Viro.

I am grateful to D. Millionschikov who read Chapters 1-5 of the manuscript. His comments have led to numerous improvements.

Many thanks to A. Ranicki for his constant support and encouragement. I am indebted to de Gruyter publishing house and in particular to R. Plato for help and patience.

The pictures in the book were created using J. Hobby's Metapost programming language. I am grateful to J. Hobby for his advice. Many thanks to A. Shen and V. Shuvalov for their help with numerous TEX issues.

Moscow, August 2006 Andrei Pajitnov

Introduction

A C^∞ function $f : M \to \mathbf{R}$ on a closed manifold M must have at least two critical points, namely maximum and minimum. This lower bound for the number of critical points is far from exact: the existence of a function on M with precisely two critical points implies a strong restriction on the topology of M. Indeed, let v be the gradient of f with respect to some Riemannian metric, so that

$$\langle v(x), h \rangle = f'(x)(h)$$

for every $x \in M$ and every $h \in T_x M$ (here $\langle\ ,\ \rangle$ denotes the scalar product induced by the Riemannian metric). Assuming that f has only two critical points: the minimum A and the maximum B, the vector field v has only two equilibrium points: A and B, and it is not difficult to see that every non-constant integral curve γ of v has the following property:

$$\lim_{t \to \infty} \gamma(t) = B, \qquad \lim_{t \to -\infty} \gamma(t) = A.$$

Therefore the one-point subset $\{B\}$ is a deformation retract of the subset $M \setminus \{A\}$. The deformation retraction is shown in the next figure:

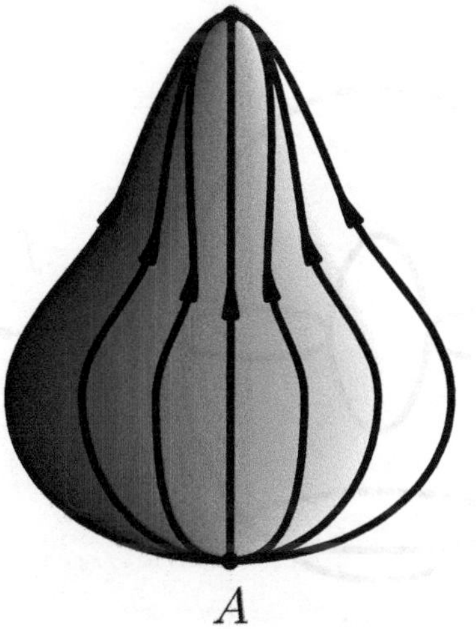

$M \setminus \{A\}$ is deformed onto B along the flow lines of v. In particular $M \setminus \{A\}$ is contractible, and it is not difficult to deduce that M is a homological sphere.

This example suggests that the homology of M can provide efficient lower bounds for the number of critical points of a C^∞ function on a manifold. Such estimates were established by M. Morse in his seminal paper [**98**]. Here is an outline of his discovery. Recall that a critical point p of a function f is called *non-degenerate* if the matrix of the second order partial derivatives of f at p is non-degenerate. The number of the negative eigenvalues of this matrix is called the *index of* p. We shall consider only C^∞ functions whose critical points are all non-degenerate (*Morse functions*). Let $f : M \to \mathbf{R}$ be such a function. Put

$$M_a = \{x \in M \mid f(x) \leqslant a\}.$$

M. Morse shows that if an interval $[a, b]$ contains no critical values of f, then M_a has the same homotopy type as M_b. If $f^{-1}([a, b])$ contains one critical point of f of index k, then M_b has the homotopy type of M_a with one k-cell attached. The classical example below illustrates this principle. Here M is the 2-dimensional torus $\mathbf{T}^2$ embedded in $\mathbf{R}^3$, and f is the height function.

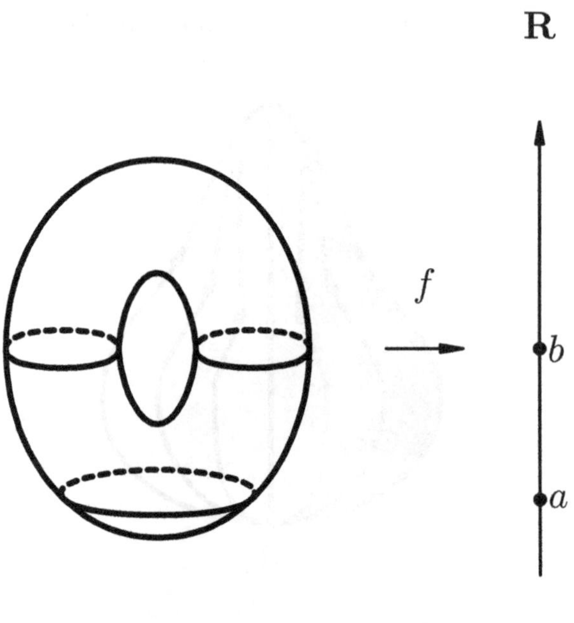

The homotopy type of M_b is clearly the homotopy type of M_a with a one-dimensional cell e_1 attached:

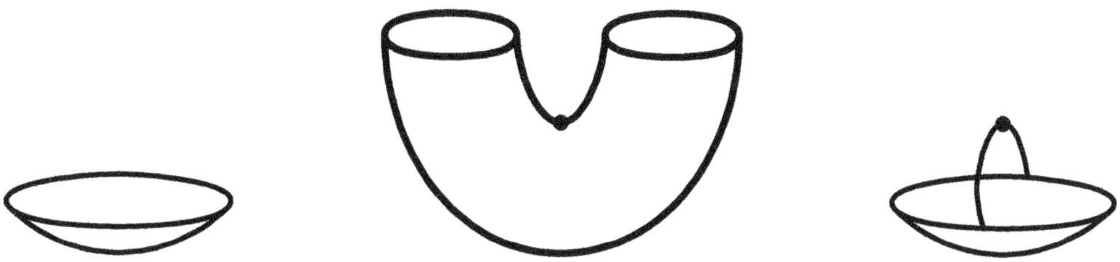

M_a $\qquad$ M_b $\qquad \sim \qquad$ $M_a \cup e_1$

Returning to the general case, it is not difficult to deduce that the manifold M has the homotopy type of a CW complex with the number of k-cells equal to the number $m_k(f)$ of critical points of f of index k. This leads to the *Morse inequalities*:

$$m_k(f) \geqslant b_k(M) + q_k(M) + q_{k-1}(M)$$

where $b_k(M)$ is the rank of $H_k(M)$ and $q_k(M)$ is the torsion number of $H_k(M)$, that is, the minimal possible number of generators of the torsion subgroup of $H_k(M)$. (This version of the Morse inequalities is due to E. Pitcher [**125**]; it is slightly different from Morse's original version.) The applications of these results are too numerous to cite here; we will mention only the classical theorem of M. Morse on the infinite number of geodesics joining two points of a sphere S^n (endowed with an arbitrary Riemannian metric) and the computation by R. Bott of the stable homotopy groups of the unitary groups.

The construction described above can be developed further. Intuitively, it is possible not only to obtain the number of cells of a CW complex X homotopy equivalent to M, but also to compute the boundary operators in the corresponding cellular chain complex. In more precise terms, starting with a Morse function $f : M \to \mathbf{R}$ and an f-gradient v, one can construct a chain complex $\mathcal{M}_*$ such that $\mathcal{M}_k$ is the free abelian group generated by critical points of f of index k and the homology of $\mathcal{M}_*$ is isomorphic to $H_*(M)$.

The explicit geometric construction of $\mathcal{M}_*$ is a result of a long development of the Morse theory (especially in the works of R. Thom [**157**], S. Smale [**150**] [**149**], and E. Witten [**163**]). By definition, $\mathcal{M}_k$ is the free abelian group generated by the set $S_k(f)$ of all critical points of f of index k. The boundary operator $\mathcal{M}_k \to \mathcal{M}_{k-1}$ is defined as follows. Let v be the Riemannian gradient for f with respect to a Riemannian metric on M. For two critical points p, q of f with $\operatorname{ind} p = \operatorname{ind} q + 1$, denote by $\Gamma(p, q; v)$

the set of all flow lines of $(-v)$ from p to q. It turns out that under some natural transversality condition on the gradient flow, this set is finite. The gradients satisfying this condition are called *Kupka-Smale gradients*, they form a dense subset in the space of all gradients of f. One can associate a sign $\varepsilon(\gamma) = \pm 1$ to each flow line γ of $(-v)$ joining p with q (we postpone all the details to Chapters 4 and 6). Summing up the signs we obtain the so-called *incidence coefficient* of p and q:

$$n(p, q; v) = \sum_{\gamma \in \Gamma(p,q;\, v)} \varepsilon(\gamma).$$

Now we define the boundary operator $\partial_k : \mathcal{M}_k \to \mathcal{M}_{k-1}$ as follows:

$$\partial_k(p) = \sum_{q \in S_{k-1}(f)} n(p, q; v) q.$$

One can prove that $\partial_k \circ \partial_{k+1} = 0$ for every k and the homology of the resulting complex is isomorphic to $H_*(M)$. This chain complex is called *the Morse complex.*

Here is a picture which illustrates the 2-torus case, considered above:

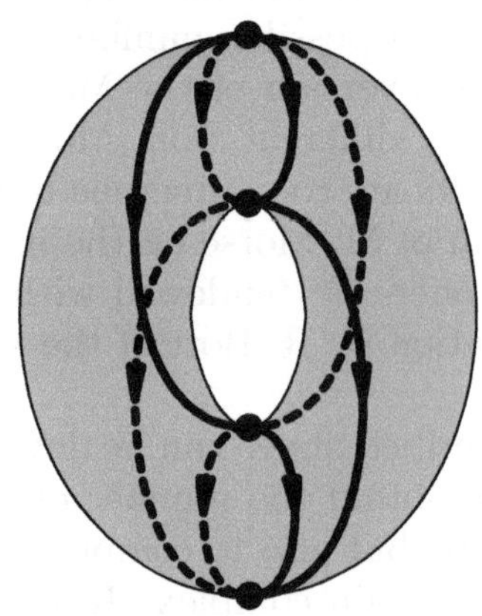

There are four critical points: one of index 0 (the minimum), one of index 2 (the maximum), and two critical points of index 1 (saddle points). There are eight flow lines of $(-v)$ joining the critical points of adjacent indices; they are shown in the figure by curves with arrows. The Morse complex is as follows:

$$0 \longleftarrow \mathbf{Z} \longleftarrow \mathbf{Z}^2 \longleftarrow \mathbf{Z} \longleftarrow 0$$

where all boundary operators are equal to 0.

It is natural to consider also the *circle-valued Morse functions*, that is, C^∞ functions with values in S^1 having only non-degenerate critical points. Identifying the circle with the quotient $\mathbf{R}/\mathbf{Z}$ we can think of circle-valued

Morse functions as *multi-valued real functions*: locally the value of such a function is a real number defined up to an additive integer.

A systematic study of circle-valued Morse functions was initiated by S. P. Novikov in 1980 (see [**102**]).[†] The motivation came from a problem in hydrodynamics, where the application of the variational approach led to a *multi-valued* Lagrangian (see the papers [**101**], [**104**]). The formulation of the circle-valued Morse theory as a new branch of topology with its own problems and goals was outlined in S. P. Novikov's paper [**102**], and in more detail in the survey paper [**105**].

The central geometric construction of the circle-valued Morse theory is the *Novikov complex* which is a generalization to the circle-valued case of its classical predecessor, the Morse complex. To understand the fundamental difference between the constructions of the Morse complex and the Novikov complex let us have a look at the following figure.

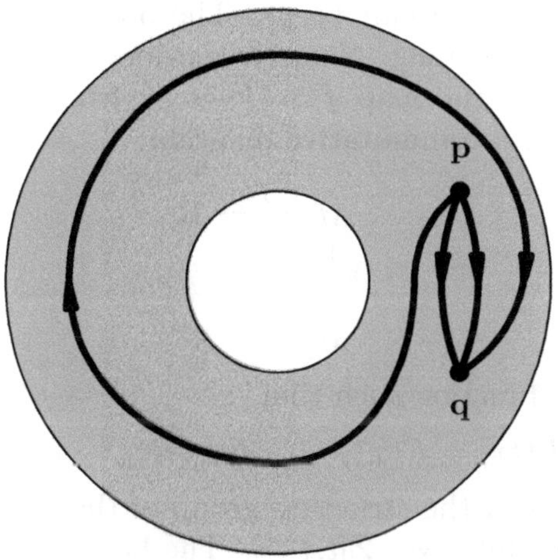

The shaded area depicts the manifold M; p and q are critical points of f. Contrary to the real-valued case a flow line of $(-v)$ can turn around several times; it can well happen that the set of flow lines joining p and q is infinite, and we will not be able to apply the procedure described above to the present situation. A way to overcome this difficulty was suggested by S. P. Novikov: for each positive integer m one counts the flow lines of $(-v)$ joining p and q and intersecting m times a given level surface of f (generically there is only a finite number of such flow lines). We obtain

[†] The first recorded instance of circle-valued Morse theory is most probably the paper [**124**] by Everett Pitcher.

integers $n_m(p,q;v)$ and form a power series in one variable

$$N(p,q;v) = \sum_{m \in \mathbf{N}} n_m(p,q;v)t^m \in \mathbf{Z}[[t]]$$

(the *Novikov incidence coefficient*). For technical reasons it is convenient to consider these series as elements of a larger ring, namely the ring

$$\overline{L} = \mathbf{Z}((t)) = \{\lambda = \sum_{m \geqslant m(\lambda)} a_m t^m \mid a_m \in \mathbf{Z}\}$$

of all Laurent series with integral coefficients and finite negative part. Let $\mathcal{N}_k(f,v)$ be the free $\overline{L}$-module freely generated by critical points of f of index k. Introduce the homomorphism

$$\partial_k : \mathcal{N}_k \to \mathcal{N}_{k-1}, \quad \partial_k p = \sum_{q \in S_{k-1}(f)} N(p,q;v)q.$$

One can prove that $\partial_k \circ \partial_{k+1} = 0$ for every k. We obtain the *Novikov complex* associated to the pair (f,v). The homology of this complex has a natural geometric meaning. Namely, consider the infinite cyclic covering $\bar{M} \to M$ induced by the map $f : M \to S^1$ from the universal covering $\mathbf{R} \to S^1$. We obtain a commutative diagram

$$\begin{array}{ccc} \bar{M} & \xrightarrow{F} & \mathbf{R} \\ {\scriptstyle\pi}\downarrow & & \downarrow \\ M & \xrightarrow{f} & S^1 . \end{array}$$

Here F is a Morse function such that

$$F(tx) = F(x) - 1 \quad \text{for every} \quad x \in \bar{M},$$

and t is a generator of the structure group of the covering π. The homology $H_*(\bar{M})$ is a module over $\mathbf{Z}[t,t^{-1}]$. The basic property of the Novikov complex is the following:

$$(1) \qquad H_*(\mathcal{N}_*(f,v)) \approx H_*(\bar{M}) \underset{L}{\otimes} \overline{L}$$

(where $L = \mathbf{Z}[t,t^{-1}]$, $\overline{L} = \mathbf{Z}((t))$). This theorem was stated in [**102**], see [**110**] for the proof. The ring $\overline{L} = \mathbf{Z}((t))$ is a principal ideal domain, and an easy algebraic argument allows us to deduce from the isomorphism (1) the following *Novikov inequalities* for the number $m_k(f)$ of critical points of f of index k:

$$(2) \qquad m_k(f) \geqslant \widehat{b}_k(M) + \widehat{q}_k(M) + \widehat{q}_{k-1}(M)$$

where $\widehat{b}_k(M)$ is the rank of the $\overline{L}$-module $\widehat{H}_k(\bar{M}) = H_k(\bar{M}) \underset{L}{\otimes} \overline{L}$ and $\widehat{q}_k(M)$ is the torsion number of this module.

One can ask whether these inequalities are optimal, or *exact*: given a manifold M and a cohomology class $\xi \in H^1(M) \approx [M, S^1]$ is there a function $f : M \to S^1$ whose homotopy class equals ξ and all these inequalities are equalities? A theorem of M. Farber [**30**] says that this is the case for any closed manifold M with $\pi_1(M) \approx \mathbf{Z}$, $\dim M \geqslant 6$ and any $\xi \neq 0$. The restriction on the fundamental group is essential for this theorem already in the case when we expect the existence of a function f without critical points, that is, a *fibration* over S^1. Vanishing of the Novikov homology $\widehat{H}_*(\bar{M})$ is in general not sufficient for fibring the manifold. However the construction of the Novikov complex can be generalized further to obtain a more sophisticated version (*the universal Novikov complex*) defined over a certain completion of the group ring $\mathbf{Z}[\pi_1(M)]$. Using these tools it is possible to obtain a necessary and sufficient homotopy-theoretic condition for the existence of a fibration (this was done in the works of J.-C. Sikorav [**147**], the author [**111**], F. Latour [**82**], A. Ranicki [**134**], [**132**]). The problem of existence of a fibration of a manifold over a circle had been intensively studied in the 1960s and the 1970s and a homotopy-theoretic criterium for the existence of a fibration was obtained in the works of W. Browder and J. Levine [**17**], and T. Farrell [**34**]. Another approach was developed by L. Siebenmann [**144**]. Thus the Novikov theory provides yet another necessary and sufficient condition for fibring which at first glance is completely different. A deeper analysis leads to the identification of the two fibring obstructions; this was done by A. Ranicki [**134**], [**132**]. Thus the Browder-Levine-Farrell-Siebenmann obstruction theory is embedded into the Morse-Novikov theory as a particular case of "functions without critical points". See Sections 2 and 3 of Chapter 14 for an introduction to this subject.

Now let us have a closer look at the incidence coefficients $N(p,q;v) = \sum n_k(p,q;v)t^k$ in the Novikov complex. These series contain a lot of information about the gradient flow, and it is natural to ask what are the asymptotic properties of $n_k(p,q;v)$ when $k \to \infty$. S. P. Novikov conjectured that the coefficients $n_k(p,q;v)$ have at most exponential growth when $k \to \infty$. This conjecture remains one of the most challenging problems in the field. See the works of V. I. Arnold [**3**], [**4**], [**5**], and of D. Burghelea and S. Haller [**19**], where different aspects of the problem are discussed.

In 1995 the author proved that C^0-generically the incidence coefficients $N(p,q;v)$ are rational functions of the variable t (see [**112**], [**113**]). Namely, for every Morse function $f : M \to S^1$ there is a subset $G_C(f)$ of the space of all f-gradients which is open and dense in C^0-topology, such that for every $v \in G_C(f)$ the Novikov incidence coefficients $N(p,q;v)$ are Taylor series of rational functions. The elements of $G_C(f)$ are called *cellular gradients.* In Chapter 12 we study cellular gradients and prove this rationality theorem.

Let us return now to the isomorphism (1) between the homology of the Novikov complex and the completed homology of the infinite cyclic covering. One can strengthen this result and construct a natural chain equivalence

$$\mathcal{N}_*(f,v) \xrightarrow{\phi} \Delta_*(\bar{M}) \underset{L}{\otimes} \overline{L},$$

where $\Delta_*(\bar{M})$ is the simplicial chain complex of $\bar{M}$. It turns out that this chain equivalence is an important geometric invariant of the pair (f,v). The source and the target of ϕ are free finitely generated complexes with naturally arising free bases. A standard algebro-topological construction associates to such a chain equivalence its *Whitehead torsion.* In the particular case of the ring $\overline{L}$ this torsion can be considered as an element $w(f,v)$ of the ring $\mathbf{Z}[[t]]$. Define the *Lefschetz zeta function* of the gradient flow by the following formula:

$$\zeta_L(-v) = \exp\Big(\sum_\gamma i_F(\gamma) t^{n(\gamma)}\Big) \in \mathbf{Z}[[t]]$$

where the sum is extended over the set of all closed orbits γ of $(-v)$ and $n(\gamma) \in \mathbf{N}$, $i_F(\gamma) \in \mathbf{Q}$ are certain numerical invariants associated with each closed orbit γ (see Chapter 13 for details). We prove that for a generic f-gradient v we have

$$w(f,v) = \big(\zeta_L(-v)\big)^{-1}.$$

A relation between the torsion invariants of the Novikov complex and the zeta function of the gradient flow was discovered in 1996 by M. Hutchings and Y-J. Lee [**64**]. They treated the case when the Novikov complex becomes acyclic after taking the tensor product with $\mathbf{Q}((t))$. Our approach is based on the techniques of cellular gradients, which allows us to get rid of the acyclicity assumptions and to generalize the formula above to the case of the universal Novikov complexes. We first prove the result for cellular gradients, and then deduce the general case by approximation techniques (Chapter 13).

There is one class of spaces where circle-valued Morse functions appear in a very natural way. Let K be a classical knot, that is, the image of a C^∞ embedding of S^1 to S^3. The knot K is called *fibred,* if there is a C^∞ map $f : S^3 \setminus K \to S^1$ such that f has no critical points and the level surfaces of f form an *open book structure* in a neighbourhood of K, as shown in the next picture (here $L_0, L_{\pi/4}, L_{\pi/2}$ denote the level surfaces corresponding to the values respectively $0, \pi/4, \pi/2 \in S^1$):

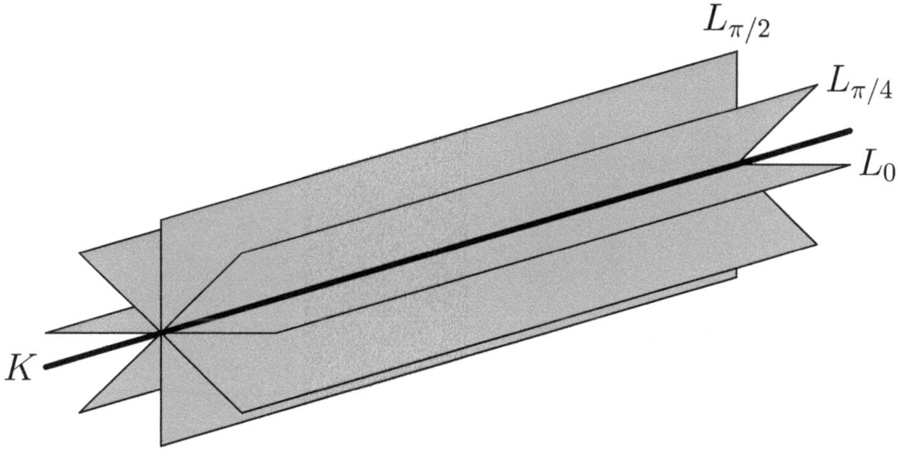

Now let K be an arbitrary knot, non-fibred in general, and $f : S^3 \setminus K \to S^1$ be a Morse function such that its level surfaces form the open book structure in a neighbourhood of K. It is not difficult to show that the number of critical points $m(f)$ of such a Morse function is finite. Put

$$\mathcal{MN}(K) = \min m(f)$$

where the minimum is taken over the set of all such Morse functions f (see [**119**]). The Novikov inequalities (2) in this case are reduced to one single inequality:

$$\mathcal{MN}(K) \geqslant 2q_1(S^3 \setminus K),$$

which is not exact in general. To obtain better lower bounds for $\mathcal{MN}(K)$, the best tool would be the universal Novikov complex. However it is very complicated algebraically and explicit computations with this complex are at present beyond our reach. In a joint work with H. Goda we introduced the *twisted Novikov homology* which is in a sense intermediate between the Novikov homology (1) and the universal Novikov homology. On one hand it reflects the essentially non-abelian structure of the knot groups, and on the other hand it is effectively computable in many cases with the help of modern software. A discussion of these invariants and related topics is the contents of the last section of the book.

Part 1

Morse functions and vector fields on manifolds

CHAPTER 1

Vector fields and C^0 topology

It is well known that if a vector field v is close to a vector field w in C^1 topology, then the integral curves of v are close to the integral curves of w in C^1 topology (see for example [**28**], Ch. 10, §7). Less well known is the fact that the property cited above remains true if we replace C^1 *topology* by C^0 *topology* . A statement of this kind can be found for example in [**13**] §4 "Continuity". This C^0-continuity property is the main topic of the present chapter.

Before we start let us set the terminology. The term *manifold* means C^∞ paracompact manifold without boundary, having a countable base. A *closed manifold* is a compact manifold without boundary. The term *manifold with boundary* or equivalently *∂-manifold* means C^∞ paracompact manifold with possibly non-empty boundary, having a countable base (observe that the boundary of a ∂-manifold may be empty). The term "smooth" is equivalent to "C^∞".

1. Manifolds without boundary

1.1. Basic definitions. Let M be a C^∞ manifold and v a C^1 vector field on M. Any C^1 map $\gamma : I \to M$ defined on some open interval $I \subset \mathbf{R}$ and satisfying the differential equation

$$\gamma'(t) = v(\gamma(t)) \quad \text{for every} \quad t \in I$$

will be called an *integral curve of v.*

An integral curve of v is called *maximal* if there exists no extension of γ to an interval J with $I \subsetneqq J$.

For every $\alpha \in \mathbf{R}$ and every $x \in M$ there is a unique maximal integral curve of v satisfying the *initial condition*

$$\gamma(\alpha) = x$$

(this follows from the standard theorems on the existence and uniqueness of solutions of differential equations).

If M is compact then every maximal integral curve is defined on the whole of $\mathbf{R}$.

The value at t of the maximal integral curve γ satisfying $\gamma(0) = x$ will be denoted by $\gamma(x,t;v)$. For the curve itself we shall use the notation $\gamma(x,\cdot;v)$. For a subset $A \subset \mathbf{R}$ we denote by $\gamma(x,A;v)$ the set of all points $\gamma(x,t;v)$ with $t \in A$.

The restriction of the maximal integral curve $\gamma(x,t;v)$ to the subset of all non-negative $t \in I$ will be called the *trajectory of* v or the v*-trajectory* starting at x.

A subset X of M is called v*-invariant*, if

$$a \in X \Rightarrow \gamma(a,t;v) \in X \quad \text{for every} \quad t \geqslant 0.$$

A subset X of M is called $\pm v$*-invariant*, if it is v-invariant and $(-v)$-invariant. In other words, X is $\pm v$-invariant, if every maximal integral curve passing through a point of X never leaves X.

Two maximal integral curves γ_1, γ_2 of v are called *equivalent* if they can be obtained one from another by reparameterization, that is, there is C such that $\gamma_1(t+C) = \gamma_2(t)$ for every t. The classes of equivalence of maximal integral curves are called *flow lines*.

When every maximal integral curve of v is defined on the whole of $\mathbf{R}$ we shall also use the notation

$$\gamma(x,t;v) = \Phi(v,t)(x)$$

so that for every given t the map $\Phi(v,t)$ is a diffeomorphism of M onto itself.

The set of all C^1 vector fields on M is denoted by $\mathrm{Vect}^1(M)$. Assume now that M is endowed with a Riemannian metric and denote by $||\cdot||$ the induced norm on the tangent spaces. A vector field v on M is called *bounded* if the function $x \mapsto ||v(x)||$ from M to $\mathbf{R}$ is bounded. The set of all bounded vector fields of class C^1 is denoted by $\mathrm{Vect}^1_b(M)$. We endow this space with the C^0-norm

$$||v|| = \sup_{x \in M} ||v(x)||.$$

The topology induced on this space by this norm is called C^0*-topology*. Observe, that the space $\mathrm{Vect}^1_b(M)$ is *not* a Banach space if $\dim M > 0$, $M \neq \varnothing$. The same definition and notation is valid also in the category of ∂-manifolds.

1.2. The evaluation map. In this section M is a Riemannian C^∞ manifold (not necessarily compact). In the set

$$\mathcal{L} = M \times \mathbf{R} \times \mathrm{Vect}^1_b(M)$$

consider the subset $\mathcal{D}$ consisting of all triples (x,t,v), such that the domain of definition of the maximal integral curve of v passing through x at $t = 0$

contains the interval $[0,t]$.† In many cases $\mathcal{D}$ equals $\mathcal{L}$; this occurs for example for any compact manifold M. This is also the case for $M = \mathbf{R}^m$ endowed with the Euclidean metric (see [**13**], Theorem 8 of Chapter 5). For $M =]0,1[$ (endowed with the Euclidean metric) the set $\mathcal{D}$ is a proper subset of $\mathcal{L}$.

We have the natural *evaluation map*

$$\mathcal{E} : \mathcal{D} \to M, \quad \mathcal{E}(x,t,v) = \gamma(x,t;v).$$

We endow the vector space $\mathrm{Vect}_b^1(M)$ with the C^0-topology, and the space $\mathcal{L}$ acquires the product topology.

Theorem 1.1. *The set $\mathcal{D}$ is open and the map $\mathcal{E} : \mathcal{D} \to M$ is continuous.*

Proof. We shall first do the particular case of Euclidean space (Lemma 1.2), and the general case will be reduced to this one.

Lemma 1.2. *Theorem 1.1 is true when M is the space $\mathbf{R}^m$ endowed with the Euclidean metric.*

Proof. The set $\mathcal{D}$ is open since

$$\mathcal{D} = \mathcal{L} = \mathbf{R}^m \times \mathbf{R} \times \mathrm{Vect}_b^1(\mathbf{R}^m).$$

To prove the continuity of the map $\mathcal{E}$ we will apply the standard techniques of *approximate solutions* from the elementary theory of differential equations. Recall that for a vector field v defined in $\mathbf{R}^m$, a map $\phi : I \to \mathbf{R}^m$ defined on some interval I is called an *ϵ-approximate solution* of the equation

$$y'(t) = v(y(t)) \tag{1}$$

if

$$||\phi'(t) - v(\phi(t))|| \leqslant \epsilon \quad \text{for every } t \in I.$$

Here is the basic result about ϵ-approximate solutions.

Theorem 1.3. [[**21**], Ch.2, §1.5] *Let $\phi_1, \phi_2 : I \to \mathbf{R}^m$ be ϵ_1 — respectively ϵ_2 — approximate solutions of the equation (1) (where $I = [0,t_0]$ and $t_0 \geqslant 0$). Let $x_1 = \phi_1(0)$, $x_2 = \phi_2(0)$. Assume that $||v'(x)|| \leqslant k$ for every $x \in \mathbf{R}^m$. Then for every $t \in I$:*

$$||\phi_1(t) - \phi_2(t)|| \leqslant ||x_1 - x_2||e^{kt} + (\epsilon_1 + \epsilon_2)\frac{e^{kt} - 1}{k}. \qquad \square$$

Let us now return to the proof of the continuity of $\mathcal{E}$.

† If $t \leqslant 0$ the symbol $[0,t]$ denotes the interval $[t,0] = \{x \in \mathbf{R} \mid t \leqslant x \leqslant 0\}$.

Lemma 1.4. *Let*

$$(x, t, v) \in \mathbf{R}^m \times \mathbf{R} \times \mathrm{Vect}_b^1(\mathbf{R}^m)$$

be an arbitrary point of $\mathcal{L}$. There is $R > 0$ such that for every $(x', t', w) \in \mathcal{L}$ with

$$|t - t'| \leqslant 1, \; ||w - v|| \leqslant 1, \; ||x - x'|| \leqslant 1$$

we have

$$\gamma(x', t', w) \in B(x, R).$$

Proof. We have

$$||\gamma(x', t'; w) - x|| \leqslant ||\gamma(x', t'; w) - x'|| + ||x - x'|| \leqslant ||\gamma(x', t'; w) - x'|| + 1;$$

$$||\gamma(x', t'; w) - x'|| \leqslant ||w|| \cdot |t'| \leqslant (||v|| + 1) \cdot (|t| + 1).$$

Thus

$$||\gamma(x', t'; w) - x|| \leqslant 1 + (||v|| + 1) \cdot (|t| + 1)$$

and our assertion follows. □

We shall prove that $\mathcal{E}$ is continuous at any point $(x, t, v) \in \mathcal{L}$ with $t \geqslant 0$; the case of negative values of t is similar. Choose R as in the previous lemma. Let $h : \mathbf{R}^m \to \mathbf{R}$ be a C^∞ function such that

$$h(x) = 1 \qquad \text{for} \quad x \in D(x, R),$$

$$h(x) = 0 \qquad \text{for} \quad x \notin D(x, 2R).$$

It follows from the previous lemma that for

$$|t - t'| \leqslant 1, \; ||w - v|| \leqslant 1, \; ||x - x'|| \leqslant 1$$

the values of the integral curve $\gamma(x, \tau; w)$ are in the disk $D(x, R)$ for $\tau \in [0, t']$, so that we have

$$\gamma(x, t; w) = \gamma(x, t; h \cdot w).$$

Let $\epsilon > 0$. If

$$t, t' \in \mathbf{R}_+, \quad ||w|| \leqslant ||v|| + 1, \quad |t' - t| \leqslant \frac{\epsilon}{2(||v|| + 1)},$$

then

$$(2) \qquad ||\gamma(x', t'; w) - \gamma(x', t; w)|| \leqslant |t - t'| \cdot ||w|| \leqslant \frac{\epsilon}{2}.$$

For a vector field $w \in \mathrm{Vect}_b^1(\mathbf{R}^m)$ satisfying $||w - v|| \leqslant \alpha \leqslant 1$, every integral curve

$$y(\tau) = \gamma(x', \tau; w) \quad \text{with} \quad \tau \in [0, t] \quad \text{and} \quad ||x - x'|| \leqslant 1$$

is an α-approximate solution of the equation

$$y'(t) = h(y(t)) \cdot v(y(t)).$$

Indeed,

$$||y'(t) - h(y(t)) \cdot v(y(t))|| = ||w(y(t)) - h(y(t)) \cdot v(y(t))|| \leqslant \alpha$$

since the integral curve lies entirely in the domain where h equals 1. The vector field hv on $\mathbf{R}^m$ has a compact support, its derivative is bounded, and applying Theorem 1.3 we obtain

$$||\gamma(x, t; v) - \gamma(x', t; w)|| \leqslant \frac{\epsilon}{2}$$

if α and $||x - x'||$ are sufficiently small. Combining this inequality with (2) we deduce that for $t' - t,\ v - w,\ x' - x$ sufficiently small we have:

$$||\mathcal{E}(x', t', w) - \mathcal{E}(x, t, v)|| \leqslant \epsilon,$$

and this completes the proof of the lemma. □

Corollary 1.5. *Let $v \in \mathrm{Vect}^1_b(\mathbf{R}^m)$, $a \in \mathbf{R}^m$, $t_0 \geqslant 0$. Let $U \subset \mathbf{R}^m$ be an open neighbourhood of the set $\gamma(a, [0, t_0]; v)$. Then for all triples (x, t, w) sufficiently close to (a, t_0, v), the set $\gamma(x, [0, t]; w)$ is still in U.* †

Proof. The contrary would mean that there are sequences x_n, t_n, v_n converging respectively to a, t_0, v, and a sequence $\tau_n \in [0, t_n]$ with $\gamma(x_n, \tau_n; v_n) \notin U$. Choose any subsequence τ_{n_k} converging to some $\tau_0 \leqslant t_0$. Then $\gamma(x_{n_k}, \tau_{n_k}; v_{n_k}) \to \gamma(a, \tau_0; v) \in U$, so we have a contradiction. □

Now we proceed to the proof of our theorem for the case of an arbitrary manifold M. Let

$$(a, t_0, v) \in \mathcal{L} = M \times \mathbf{R} \times \mathrm{Vect}^1_b(M)$$

be a triple such that $\gamma(a, \cdot; v)$ is defined on $[0, t_0]$. Let V be a neighbourhood of $b = \gamma(a, t_0; v)$. Our task is to prove that for every triple $(y, t, w) \in \mathcal{L}$ sufficiently close to (a, t_0, v) the trajectory $\gamma(y, \cdot; w)$ is defined on $[0, t]$ and $\gamma(y, t; w) \in V$.

Replacing v by $-v$ if necessary, we can assume that $t_0 \geqslant 0$. We are going to reduce the proof to the case of Euclidean space treated above. Subdividing the interval $[0, t_0]$ into smaller ones if necessary, we can assume that the curve $\Gamma = \gamma(a, [0, t_0]; v)$ is contained in the domain of definition of some chart $\Psi : U \xrightarrow{\approx} \mathbf{R}^m$. It suffices to consider the case $V \subset U$.

Choose $R > 0$ sufficiently large so that $\Psi(\Gamma) \subset B(0, R)$. Let

$$\widetilde{v} = \Psi_*(v).$$

The vector field $\widetilde{v}$ is not defined on the whole of $\mathbf{R}^m$, and may be unbounded. We want to apply Lemma 1.2, so we shall modify this vector

† Recall that the space $\mathrm{Vect}^1_b(\mathbf{R}^m)$ is endowed with the C^0-topology.

field so as to obtain a vector field in $\mathrm{Vect}_b^1(\mathbf{R}^m)$. Let $h : \mathbf{R}^m \to [0,1]$ be a C^∞ function such that

$$h(x) = 1 \quad \text{for} \quad x \in B(0,R) \quad \text{and} \quad h(x) = 0 \quad \text{for} \quad x \notin B(0,2R).$$

The vector fields $\widetilde{v}$ and $h \cdot \widetilde{v}$ are equal in a neighbourhood of $\Psi(\Gamma)$, therefore the Ψ-image of Γ is also an integral curve of the vector field $h \cdot \widetilde{v}$ (have a look at Figure 1).

Lemma 1.2 says that there is a neighbourhood $A \subset B(0,R)$ of $\Psi(a)$ and a number $\delta > 0$ such that for any vector field $w \in \mathrm{Vect}_b^1(\mathbf{R}^m)$ the three conditions

$$||w - h\widetilde{v}||_{\mathbf{R}^m} < \delta, \quad x \in A, \quad |t - t_0| < \delta$$

imply that the curve $\gamma(x, [0,t]; w)$ is still in $B(0,R)$ and $\gamma(x,t;w) \in \Psi(V)$.

The initial Riemannian metric on M and the metric induced on U from the Euclidean metric in $\mathbf{R}^m$ are not equal in general. To compare them choose any positive real number C such that

$$||\Psi'(y)(\xi)||_{\mathbf{R}^m} \leqslant C||\xi||_M \quad \text{for every} \quad y \in \Psi^{-1}(D(0,2R)),\ \xi \in T_yM$$

(such a number exists, since $D(0,2R)$ is compact). We have:

$$||u - v||_M < \delta/C \ \Rightarrow \ ||h\widetilde{u} - h\widetilde{v}||_{\mathbf{R}^m} < \delta,$$

so if

$$||u - v||_M < \delta/C, \quad y \in \Psi^{-1}(A), \quad |t - t_0| < \delta,$$

then

$$\gamma\big(\Psi(y), [0,t]; h\widetilde{u}\big) \subset B(0,R), \quad \gamma\big(\Psi(y), t; h\widetilde{u}\big) \in \Psi(V). \tag{3}$$

The function h is equal to 1 in $B(0,R)$, therefore the Ψ^{-1}-image of the curve $\gamma(\Psi(y), \cdot; h\widetilde{u})$ is an integral curve of u starting at y, and defined (at least) on $[0,t]$. The endpoint $\gamma(y,t;u)$ of this curve is in V, and this completes the proof of our theorem. □

Similarly to Corollary 1.5 we deduce the next one.

Corollary 1.6. *Let $(a, t_0, v) \in \mathcal{D}$, where $t_0 \geqslant 0$. Let $U \subset M$ be an open subset such that $\gamma(a, [0,t_0]; v) \subset U$. Then there is a neighbourhood S of (a, t_0, v) in $\mathcal{D}$, such that for every $(x,t,w) \in S$ we have $\gamma(x, [0,t]; w) \subset U$.* □

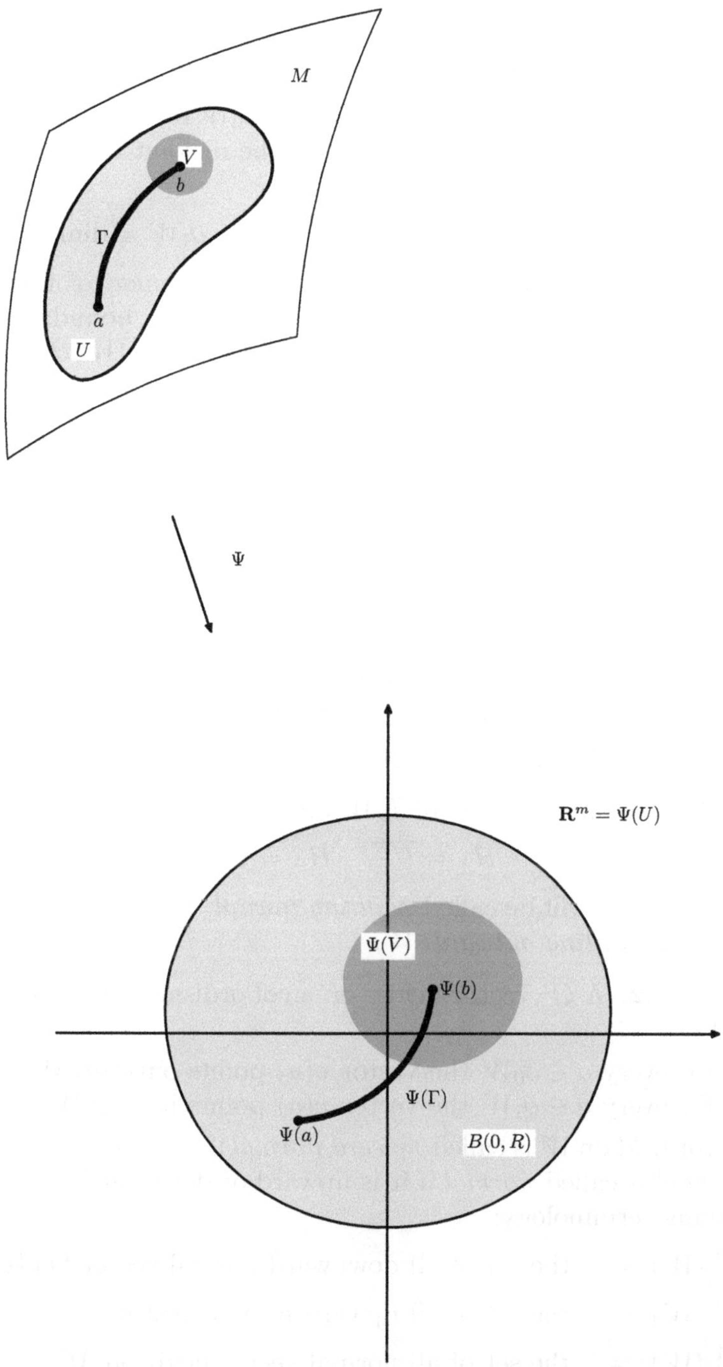

FIGURE 1.

2. Cobordisms

2.1. Basic definitions.

Definition 2.1. A *cobordism* is a triple $(W; \partial_0 W, \partial_1 W)$ where W is a compact ∂-manifold and the boundary ∂W is the disjoint union of two closed submanifolds:

$$\partial W = \partial_0 W \sqcup \partial_1 W \text{ with } \dim \partial_0 W = \dim \partial_1 W = \dim W - 1.$$

The manifold $\partial_0 W$ will be called the *lower component* of the boundary, and $\partial_1 W$ will be called the *upper component* of the boundary.† The set $W \setminus \partial W$ will be also denoted by W°. The cobordisms $(W; \partial_0 W, \partial_1 W)$ and $(W; \partial_1 W, \partial_0 W)$ obtained from each other by interchanging the upper and the lower components will be called *opposite* to each other. We shall abbreviate the notation $(W; \partial_0 W, \partial_1 W)$ to W, when no confusion is possible.

At every point $x \in \partial W$ the tangent space $T_x \partial W$ is naturally identified with a hyperplane in the vector space $T_x W$. This hyperplane divides $T_x W$ into two half-spaces:

$$T_x W = H_+ \cup H_-, \quad H_+ \cap H_- = T_x \partial W,$$

where we denote by H_+ the half-space formed by the tangent vectors $\gamma'(0)$ to all the C^1-curves $\gamma : [0, a] \to W$ with $\gamma(0) = x$, and $a \geqslant 0$. Put

$$U_+ = H_+ \setminus T_x \partial W, \quad U_- = H_- \setminus T_x \partial W,$$

then U_+, U_- are open subsets of $T_x W$ and

$$H_+ = \overline{U_+}, \quad H_- = \overline{U_-}.$$

The vectors in U_+ will be called *pointing inward* W, and the vectors in U_- will be called *pointing outward* W.

Definition 2.2. A C^1 vector field v on a cobordism W is called *downward normal* if

(1) for every $a \in \partial_0 W$ the vector $v(a)$ points outward W;
(2) for every $a \in \partial_1 W$ the vector $v(a)$ points inward W.

A C^1 vector field on W is called *upward normal* if $(-v)$ is downward normal. A vector field is called *normal* if it is upward or downward normal. We use the following terminology:

$$\begin{aligned}
\mathrm{Vect}^1_\perp(W) &= \text{the set of all downward normal vector fields on } W,\\
\mathrm{Vect}^1_\top(W) &= \text{the set of all upward normal vector fields on } W,\\
\mathrm{Vect}^1_N(W) &= \text{the set of all normal vector fields on } W.
\end{aligned}$$

† Note that these sets do not have to be *connected components* of ∂W.

Exercise 2.3. Show that on any cobordism there are downward normal vector fields. (*Indication:* use the partition of unity techniques.)

A downward normal field on a cobordism W is an upward normal field on the opposite cobordism. It is easy to see that the set $\mathrm{Vect}^1_N(W)$ is an open subset of $\mathrm{Vect}^1(W) = \mathrm{Vect}^1_b(W)$, and the subsets $\mathrm{Vect}^1_\top(W)$, $\mathrm{Vect}^1_\perp(W)$ are the connected components of $\mathrm{Vect}^1_N(W)$.

Let v be a downward normal field. The Cauchy-Lipschitz theorem implies existence and uniqueness of a solution of the equation

$$\gamma'(t) = v(\gamma(t)), \quad \gamma(\alpha) = x$$

on an interval of the form

(1) $]\alpha - \epsilon, \alpha + \epsilon[$ with some $\epsilon > 0$ if $x \in W^\circ$,
(2) $[\alpha, \alpha + \epsilon[$ with some $\epsilon > 0$ if $x \in \partial_1 W$,
(3) $]\alpha - \epsilon, \alpha]$ with some $\epsilon > 0$ if $x \in \partial_0 W$.

Such solutions will be called *integral curves* of v.

Every integral curve of a downward normal field can intersect $\partial_0 W$ only once; the same applies for $\partial_1 W$. As in the case of manifolds without boundary, each integral curve can be extended to a maximal integral curve. There are four possibilities for the domain of definition of a maximal integral curve of a downward normal field:

(1) $\mathbf{R}$;
(2) $]-\infty, a]$ with $a \in \mathbf{R}$;
(3) $[b, \infty[$ with $b \in \mathbf{R}$.
(4) $[b, a], \; a, b \in \mathbf{R}, a < b$.

In the first case the curve belongs entirely to W°. In the second case $\gamma(a) \subset \partial_0 W$. The third case corresponds to $\gamma(b) \in \partial_1 W$. In the fourth case we have: $\gamma(b) \in \partial_1 W, \; \gamma(a) \in \partial_0 W$.

A similar analysis can be carried out for the case of upward normal vector fields; we suggest that the reader formulates the corresponding results.

Definition 2.4. Let v be a normal vector field on a cobordism W. A subset $X \subset W$ is called *v-invariant* if, for every $x \in X$ and for every $t \geqslant 0$ such that the interval $[0, t]$ is in the domain of definition of the maximal integral curve $\gamma(x, \cdot; v)$, we have $\gamma(x, t; v) \in X$.

A subset $X \subset W$ is called *$(\pm v)$-invariant*, if it is v-invariant and $(-v)$-invariant.

2.2. Collars. In this subsection we study the structure of normal vector fields on a cobordism in a neighbourhood of the boundary (cf. [**92**], §2, [**61**]), Ch.4, §6).

Definition 2.5. Let v be a downward normal C^1 vector field on W, and $\alpha > 0$. We say that $\partial_1 W$ *has a collar of height α with respect to v* if for every $x \in \partial_1 W$ the v-trajectory $\gamma(x, \cdot; v)$ is defined on the interval $[0, \alpha]$. In this case the map

$$\partial_1 W \times [0, \alpha[\to W, \quad (x, t) \mapsto \gamma(x, t; v)$$

is a diffeomorphism on its image which is called a *collar* of $\partial_1 W$ in W.

We say that $\partial_0 W$ *has a collar of height* α *with respect to* v if for every $x \in \partial_0 W$ the v-trajectory $\gamma(x, \cdot; -v)$ is defined on the interval $[0, \alpha]$. In this case the map

$$\partial_0 W \times [0, \alpha[\to W, \quad (x, t) \mapsto \gamma(x, t; -v)$$

is a diffeomorphism on its image, which is called a *collar* of $\partial_0 W$ in W.

If v is an upward normal vector field on W, then we say that $\partial_1 W$, respectively $\partial_0 W$, *has a collar of height* α *with respect to* v if $\partial_1 W$, respectively $\partial_0 W$, has a collar of height α with respect to $(-v)$.

If v is a normal field on W, then we say that ∂W *has collars of height* α if both $\partial_1 W, \partial_0 W$ have collars of height α with respect to v.

Using compactness of $\partial_1 W$ it is not difficult to show that for a given v there are always collars of some height α.

Definition 2.6. Let v be a downward vector field on W, and $x \in W$. If the interval of definition of the maximal integral curve $\gamma(x, \cdot; v)$ is bounded from above, then $\gamma(x, \cdot; v)$ intersects at some moment $\tau = \tau(x, v) \geqslant 0$ the lower component $\partial_0 W$. We say that γ *quits* W, and $\tau(x, v)$ is called *the moment of exit* of γ. The point $\gamma(x, \tau(x, v); v) \in \partial_0 W$ will be called *the point of exit* of γ and denoted by $E(x, v)$.

For a given vector field v the function $x \mapsto \tau(x, v)$ is not defined everywhere in W in general. The detailed study of the moments of exit is postponed to Subsection 2.4 (in particular we will show that $\tau(x, v)$ is a continuous function of both its arguments); now we are going to do only some preliminary work. Assume that $\partial_0 W$ has a collar of height α with respect to v, and consider the corresponding map

$$\Phi : \partial_0 W \times [0, \alpha[\to W, \quad \Phi(y, t) = \gamma(y, t; -v).$$

The composition of the function $\tau(\cdot, v)$ with Φ is equal to the second projection

$$P : \partial_0 W \times [0, \alpha[\to [0, \alpha[,$$

therefore the function $\tau(\cdot, v)$ is continuous; moreover, if v is of class C^∞, then $\tau(\cdot, v)$ is also of class C^∞.

Lemma 2.7. *Let v be a downward normal field. There is a neighbourhood V of $\partial_0 W$, and a neighbourhood $\mathcal{W}$ of v in $\mathrm{Vect}^1_N(W)$ such that for every $(x, w) \in V \times \mathcal{W}$ the w-trajectory starting at x quits W, and $\tau(x, w) \leqslant 2\tau(x, v)$.*

Proof. Assuming that $\partial_0 W$ has a collar of height α with respect to v, let

$$\Phi : \partial_0 W \times [0, \alpha[\to W, \quad \Phi(y,t) = \gamma(y, t; -v)$$

be the corresponding map, and let $V = \Phi(\partial_0 W \times [0, \alpha/2])$. The vector field $\Phi_*^{-1}(v)$ on $\partial_0 W \times [0, \alpha/2]$ has coordinates $(0, -1)$. Therefore there is a neighbourhood $\mathcal{W}$ of v in $\mathrm{Vect}^1_N(W)$ such that for every $w \in \mathcal{W}$ the second coordinate of the field $\Phi_*^{-1}(w)$ is between -2 and $-1/2$. This implies that the $\Phi_*^{-1}(w)$-trajectory starting at a point

$$y = (a, t) \in \partial_0 W \times [0, \alpha/2]$$

reaches the lower component of boundary

$$\partial_0\big(\partial_0 W \times [0, \alpha/2]\big) = \partial_0 W \times 0$$

and we have the following estimate for the time of exit:

$$\tau\big(y, \Phi_*^{-1}(w)\big) \leqslant 2t = 2\tau(x, \Phi_*^{-1}(v)).$$

The same is true for the vector field w on the cobordism V. □

2.3. The evaluation map. Now we can establish the C^0-continuity property for normal vector fields on cobordisms. Let W be a cobordism. Put

$$\mathcal{M} = W \times \mathbf{R} \times \mathrm{Vect}^1_N(W).$$

The set $\mathcal{M}$ has the natural topology of a direct product (recall that the space $\mathrm{Vect}^1_N(W)$ is endowed with the C^0-*topology*).

Let $\mathcal{D}$ denote the set of all triples $(x, t, v) \in \mathcal{M}$, such that the domain of definition of the maximal integral curve of v starting at x contains the interval $[0, t]$. Define the *evaluation map* $\mathcal{E} : \mathcal{D} \to W$ as follows:

$$\mathcal{E}(x, t, v) = \gamma(x, t; v).$$

Theorem 2.8. *The set $\mathcal{D}$ is a closed subset of $\mathcal{M}$ and the map $\mathcal{E} : \mathcal{D} \to W$ is continuous.*

Proof. Put

$$\mathcal{D}_0 = \mathcal{D}_0(W) = \{(x, t, v) \in \mathcal{D} \mid t \geqslant 0, \ v \in \mathrm{Vect}^1_\perp(W)\}.$$

The plan of the proof is as follows. We prove first that $\mathcal{E}|\mathcal{D}_0$ is continuous, then we show that $\mathcal{D}_0$ is closed, and the theorem will be deduced from these two assertions.

Proposition 2.9. *The map $\mathcal{E}|\mathcal{D}_0$ is continuous.*

Proof. Let (x_n, t_n, v_n) be a sequence of points in $\mathcal{D}_0$ converging to $(x, t, v) \in \mathcal{D}_0$. We have to prove that

$$\gamma(x_n, t_n; v_n) \to \gamma(x, t; v).$$

The argument depends on the position of the points $x, \gamma(x, t; v)$. If the integral curves $\gamma(x_n, \cdot; v_n)$ and $\gamma(x, \cdot, v)$ are contained in $W^\circ = W \setminus \partial W$ the proposition would follow by a direct application of Theorem 1.1. We will consider below the case when $x \notin \partial W$, leaving the general case to the reader as a useful exercise.

During the proof we shall need an estimate of the length of integral curves of v and v_n. Pick any $C > 0$, such that $||v|| \leqslant C$ and $||v_n|| \leqslant C$ for every n. Then

$$(4) \qquad \rho\big(\gamma(x, t; w), \gamma(x, t'; w)\big) \leqslant C|t - t'| \quad \text{for} \quad w = v \quad \text{or} \quad w = v_n.$$

Here ρ stands for the distance on W induced by the Riemannian metric on M, that is $\rho(x, y)$ equals the infimum of lengths of piecewise smooth curves joining x and y.

Let us distinguish two cases:

(A) $t = 0$ so that the curve $\gamma(x, [0, t]; v)$ is reduced to one point.
(B) $t > 0$.

In the case (A) our task is to show that $\gamma(x_n, t_n; v_n) \to x$. This follows immediately from the fact that the distance from x_n to $\gamma(x_n, t_n; v_n)$ is not greater than $Ct_n \to 0$.

Proceed now to the case (B). Choose any $\alpha \in]0, t[$. Since $t_n \to t$ it suffices to do the proof under the assumption that $\alpha < t_n < t$ for every n. Then the interval of definition of the curve $\gamma(x_n, \cdot; v_n)$ contains $[0, t_n - \alpha]$, and

$$\gamma(x_n, [0, t - \alpha]; v) \subset W^\circ \quad \text{for each} \quad n.$$

Applying Theorem 1.1 we deduce that for every $\alpha \in]0, t[$ we have

$$\gamma(x_n, t_n - \alpha; v_n) \xrightarrow[n \to \infty]{} \gamma(x, t - \alpha; v).$$

Furthermore observe that the distances

$$\rho\Big(\gamma(x_n, t_n - \alpha; v_n), \gamma(x_n, t_n; v_n)\Big), \quad \rho\Big(\gamma(x, t - \alpha; v), \gamma(x, t; v)\Big)$$

are both bounded from above by $C\alpha$ (recall that C is any positive constant such that for every n we have $C \geqslant ||v_n||$). Therefore every accumulation point of the sequence $\gamma(x_n, t_n; v_n)$ is at distance at most $2C\alpha$ from $\gamma(x, t; v)$. Since this is true for every $\alpha \in]0, t[$, we conclude that there is only one accumulation point which coincides with $\gamma(x, t; v)$. □

Proposition 2.10. *The subset $\mathcal{D}_0$ is closed in $\mathcal{M}$.*

Proof. Let $(x_n, t_n, v_n) \in \mathcal{D}_0$ be a sequence, converging to a point $(x, t, v) \in \mathcal{M}$. We have to prove that $(x, t, v) \in \mathcal{D}_0$. The contrary would mean that

$$t > \tau(x, v).$$

Then for some $\epsilon > 0$ we have

$$t_n \geqslant \tau(x, v) + \epsilon \quad \text{for every} \quad n \quad \text{sufficiently large.}$$

Put

$$y_n = \gamma(x_n, \tau(x, v); v_n),$$

then $\tau(y_n, v_n) \geqslant \tau_n - \tau(x, v) \geqslant \epsilon$ for every n sufficiently large. On the other hand

$$y_n = \gamma(x_n, \tau(x, v); v_n) \to \gamma(x, \tau(x, v); v) \in \partial_0 W,$$

since the function $\mathcal{E}|\mathcal{D}_0$ is continuous, and we deduce from Lemma 2.7 that $\tau(y_n, v_n) \to 0$, which leads to a contradiction. □

Now we can complete the proof of our theorem. Put

$$\mathcal{D}_1(W) = \{(x, t, v) \in \mathcal{D} \mid t \geqslant 0,\ v \in \mathrm{Vect}^1_{\top}(W)\},$$
$$\mathcal{D}_2(W) = \{(x, t, v) \in \mathcal{D} \mid t \leqslant 0,\ v \in \mathrm{Vect}^1_{\perp}(W)\},$$
$$\mathcal{D}_3(W) = \{(x, t, v) \in \mathcal{D} \mid t \leqslant 0,\ v \in \mathrm{Vect}^1_{\top}(W)\}.$$

Then

$$\mathcal{D} = \bigcup_{i=0}^{3} \mathcal{D}_i(W)$$

and to prove our theorem it suffices to show that each of $\mathcal{D}_i$ is a closed subset of $\mathcal{D}$ and the restriction $\mathcal{E} \mid \mathcal{D}_i$ is continuous.

Observe that the homeomorphism

$$\mathcal{M} \longrightarrow \mathcal{M}, \quad (x, t, v) \mapsto (x, -t, -v)$$

sends $\mathcal{D}_0(W)$ to $\mathcal{D}_3(W)$ and the function $\mathcal{E}$ is invariant with respect to this homeomorphism. This settles the case of $\mathcal{D}_3(W)$.

As for $\mathcal{D}_2(W)$ observe that the same formula

$$(x, t, v) \mapsto (x, -t, -v)$$

defines a homeomorphism of $\mathcal{D}_2(W)$ onto $\mathcal{D}_0(W^{op})$. The case of $\mathcal{D}_2$ follows therefore by an application of Propositions 2.9 and 2.10 to the cobordism W^{op}.

The proof of Theorem 2.8 is now complete. □

Corollary 2.11. *Let W be a cobordism, and v_0 be a normal vector field. Let $U \subset W^{\circ}$ be an open subset. Let $x_0 \in W$ and assume that the curve $\gamma(x_0, \cdot; v_0)$ is defined on $[0, t_0]$ and $\gamma(x_0, t_0; v_0) \in U$. Then there is a neighbourhood $\mathcal{Z}$ of $(x_0, v_0) \in W \times \mathrm{Vect}^1_N(W)$ such that*

(1) *for every* $(x,v) \in \mathcal{Z}$ *the curve* $\gamma(x,\cdot;v)$ *is defined on* $[0,t_0]$ *and*
(2) $\gamma(x,t_0;v) \in U$.

Proof. Let us suppose that v_0 is a downward normal vector field (the case of upward vector fields is similar). Assuming that the property (1) does not hold, there is a sequence $z_n = (x_n, v_n) \in \mathcal{Z}$ and a sequence of real numbers $t_n < t_0$ such that $x_n \to x$, $v_n \to v$ and $\gamma(x_n, t_n, v_n) \in \partial_0 W$. Extracting a converging subsequence if necessary we can assume that $t_n \to t_0' \leqslant t_0$. Applying Theorem 2.8 we deduce that $\gamma(x, t_0'; v) \in \partial_0 W$, and this contradicts our assumptions. Moving to the point (2), observe that the map $(x,v) \mapsto \gamma(x,t_0;v)$ is continuous by Theorem 2.8 and our assertion follows. □

2.4. Moments of exit. Recall that for a downward normal vector field $v \in \mathrm{Vect}^1_\perp(W)$ and $x \in W$ we denote by $\tau(x,v) \in \mathbf{R}$ the moment when the curve $\gamma(x,\cdot;v)$ intersects $\partial_0 W$, if there exists such a moment. Observe that $\tau(x,v) \geqslant 0$.

Definition 2.12. The set of all pairs $(x,v) \in W \times \mathrm{Vect}^1_\perp(W)$ for which $\tau(x,v)$ is defined will be denoted by $\mathcal{F}$.

For $(x,v) \in \mathcal{F}$ we denote by $E(x,v)$ the point of intersection of the trajectory $\gamma(x,t;v)$ with $\partial_0 W$, so that

$$E(x,v) = \gamma\big(x, \tau(x,v); v\big) \in \partial_0 W.$$

The function E will be called the *exit function.*

Theorem 2.13. *The set* $\mathcal{F}$ *is open in* $W \times \mathrm{Vect}^1_\perp(W)$ *and the functions* $\tau : \mathcal{F} \to \mathbf{R}$, $E : \mathcal{F} \to W$ *are continuous.*

Proof. Let us first prove that $\mathcal{F}$ is open. A point (x,v) is in $\mathcal{F}$ if and only if there is $t \geqslant 0$, such that $(x,t,v) \notin \mathcal{D}$ (see the definition of $\mathcal{D}$ just before Theorem 2.8). Theorem 2.8 asserts that $\mathcal{D}$ is closed, therefore $\mathcal{F}$ is open.

As for continuity of τ, let $(x_n, v_n) \in \mathcal{F}$ be a sequence, converging to $(x,v) \in \mathcal{F}$. It suffices to show that any converging subsequence $\tau_{n_k} = \tau(x_{n_k}, v_{n_k})$ converges necessarily to $\tau(x,v)$ (the case $t = \infty$ is not excluded).

A) If $\lim_{k\to\infty} \tau_{n_k} = \theta < \tau(x,v)$, then the sequence $(x_{n_k}, \tau_{n_k}, v_{n_k}) \in \mathcal{D}$ converges to $(x,\theta,v) \in \mathcal{D}$. Theorem 2.8 implies that $\gamma(x,\theta;v) \in \partial_0 W$, and this leads to a contradiction, since $\theta < \tau(x,v)$.

B) If $\lim_{k\to\infty} \tau_{n_k} = \theta > \tau(x,v)$, (the case $\theta = \infty$ is not excluded), then there is $\alpha > \tau(x,v)$ such that for any n sufficiently large the points (x_n, α, v_n) are in $\mathcal{D}$ and therefore the point (x,α,v) is in $\mathcal{D}$, which contradicts the condition $\gamma(x, \tau(x,v); v) \in \partial_0 W$.

Thus we have proved that τ is continuous, and therefore E is also continuous. □

Corollary 2.14. *Let v be a downward normal vector field. Let $\Lambda(v)$ be the set of all x, where $\tau(x,v)$ is defined. Then $\Lambda(v) \subset W$ is open and the functions*

$$x \mapsto \tau(x,v), \quad \Lambda(v) \to \mathbf{R},$$
$$x \mapsto E(x,v), \quad \Lambda(v) \to \partial_0 W$$

are continuous. □

Now we shall apply the results obtained above to prove an analog of Corollary 1.6 for vector fields on cobordisms. Let v be a downward normal vector field on W, let $a \in W$ and assume that

$$(a,v) \in \mathcal{F} \subset W \times \mathrm{Vect}^1_N(W),$$

so that the v-trajectory starting at a reaches $\partial_0 W$ and intersects it at the moment $\tau(a,v)$ at the point $E(a,v)$. Theorem 2.13 implies that every pair $(x,w) \in W \times \mathrm{Vect}^1_N(W)$ sufficiently close to (a,v) is in $\mathcal{F}$, so that the moment of exit $\tau(x,w)$ and the point of exit $E(x,w)$ are defined. Put

$$\Gamma(x,w) = \gamma(x,[0,\tau(x,w)];w).$$

Proposition 2.15. *Let $(a,v) \in \mathcal{F}$. Let U be an open neighbourhood of $\Gamma(a,v)$, and $R \subset \partial_0 W$ be an open neighbourhood of $E(a,v)$ in $\partial_0 W$. Then there is a neighbourhood $\mathcal{S}$ of (a,v) in $\mathcal{F}$ such that for every $(x,w) \in \mathcal{S}$ we have:*

$$\Gamma(x,w) \subset U, \quad E(x,w) \subset R.$$

Proof. Assume that in any neighbourhood of (a,v) there are pairs (x,w) with $\Gamma(x,w) \not\subset U$. Then there are sequences

$$x_n \to a, \quad v_n \to v$$

and for every n there is a moment t_n such that $\gamma(x_n,t_n;v_n) \notin U$. We can assume that all pairs (x_n,v_n) are in a neighbourhood $\mathcal{S}_0 \subset \mathcal{F}$ of (a,v) so small that for every n we have $\tau(x_n,v_n) \leqslant C$ for some $C > 0$. Thus the sequence t_n is bounded from above; choose any converging subsequence $t_{n_k} \to t_0$. By Theorem 2.8 we have

$$\gamma(x_{n_k},t_{n_k};v_{n_k}) \to \gamma(a,t_0;v).$$

The set $W \setminus U$ is closed, therefore $\gamma(a,t_0;v) \notin U$, and we obtain a contradiction.

The inclusion $E(x,w) \subset R$ for every (x,w) in a neighbourhood of (a,v) follows from the previous argument, if we choose $U = (W \setminus \partial_0 W) \cup R$. □

In our applications we shall work with C^∞ vector fields.

Proposition 2.16. *Let* $v \in \mathrm{Vect}^1_N(W)$ *be a* C^∞ *vector field. Then the maps*

$$\tau(\cdot, v) : \Lambda(v) \to \mathbf{R}, \quad E(\cdot, v) : \Lambda(v) \to \partial_0 W$$

are of class C^∞.

Proof. We have already proved in Subsection 2.2 that the function $x \mapsto \tau(x, v)$ (and therefore the function $x \mapsto E(x, v)$) is of class C^∞ when restricted to a neighbourhood U of $\partial_0 W$. To deal with the general case, let $x \in W^\circ$ and assume that the v-trajectory starting at x reaches $\partial_0 W$. Then for some $t_0 \geqslant 0$ we have $\gamma(x, t_0; v) \in U_0 = U \setminus \partial_0 W$. Therefore for every y from some neighbourhood A of x we have: $\gamma(y, t_0; v) \in U_0$. The map $x \mapsto \gamma(x, t_0; v)$ is well defined and is of class C^∞ on A. Since $\tau(x, v) = \tau(\gamma(x, t_0; v)) - t_0$, the function $x \mapsto \tau(x, v)$ is also C^∞ on A, and the same is true for $E(\cdot, v)$. □

CHAPTER 2

Morse functions and their gradients

A critical point p of a C^∞ function f on a manifold is called *non-degenerate* if the bilinear form $f''(p) : T_p(M) \times T_p(M) \to \mathbf{R}$ is non-degenerate. The index of this form is called the *index of* p. A *Morse function* on a manifold is a C^∞ function, such that its critical points are all non-degenerate.

We begin with the classical *Morse lemma*, which says that every Morse function in a neighbourhood of its critical point of index k is diffeomorphic to the function $\mathscr{Q}_k + \text{const}$, where $\mathscr{Q}_k$ is a quadratic form of index k.

We prove then that the subset of all Morse functions on a closed manifold is open and dense in the set of all C^∞ functions on the manifold (Theorem 1.30; this result is deduced from a more general Theorem 1.25).

In the second section we introduce the gradients of Morse functions and forms. Recall that the gradient of a differentiable function $f : \mathbf{R}^m \to \mathbf{R}$ is the vector field

$$\operatorname{grad} f(x) = \Big(\frac{\partial f}{\partial x_1}(x), \ \ldots, \frac{\partial f}{\partial x_m}(x)\Big).$$

The notion of gradient generalizes immediately to smooth functions on Riemannian manifolds. For such a function $f : M \to \mathbf{R}$ the vector field $\operatorname{grad} f$ is defined by the formula

$$\langle \operatorname{grad} f(x), h \rangle = f'(x)(h)$$

(where $x \in M, h \in T_x M$ and $\langle \cdot, \cdot \rangle$ stands for the scalar product induced by the Riemannian metric on $T_x M$). This vector field will be called the *Riemannian gradient* of f. The function f is strictly increasing along any non-constant integral curve γ of $\operatorname{grad} f$, since

$$(f \circ \gamma)'(t) = f'(\gamma(t))(\gamma'(t)) = ||\operatorname{grad} f(\gamma(t))||^2.$$

Thus the properties of f and the flow generated by $\operatorname{grad} f$ are closely related to each other.

In general one can use functions increasing along each trajectory of a given vector field v to study the dynamics of the flow generated by v. This approach was deeply explored by A. M. Liapounov (see his thesis defended in 1892, and translated into French in [**88**]).

In Morse theory the notion of gradient descent was used already in the seminal article [**98**] of M.Morse. A very convenient class of gradient flows was introduced and extensively used by J.Milnor in his book [**92**]:

Definition 0.1 ([**92**], §3). Let M be a manifold, $f : M \to \mathbf{R}$ be a Morse function. A vector field v is called a *gradient-like vector field* for f, if

1) for every non-critical point x of f we have: $f'(x)(v(x)) > 0$,
2) for every critical point p of f there is a chart $\Psi : U \to V \subset \mathbf{R}^m$ for M, such that

$$\text{(F)} \quad f \circ \Psi^{-1}(x_1, \ldots, x_m) = f(p) - (x_1^2 + \cdots + x_k^2) + (x_{k+1}^2 + \cdots + x_m^2),$$

$$\text{(G)} \quad \Psi_*(v)(x_1, \ldots, x_m) = (-x_1, \ldots, -x_k, x_{k+1}, \ldots, x_m)$$

(where $k = \operatorname{ind} p$, $m = \dim M$).†

This definition requires that in some coordinate system in a neighbourhood of p both the function and the vector field have certain standard form. As for the function this requirement is not very restrictive: the Morse lemma says that the condition (F) can always be achieved. On the other hand the condition (G) is rather strong, in particular it implies the linearizability of v in a neighbourhood of p.

The notion of gradient-like vector field has many advantages from the point of view of differential topology. There is no need for auxiliary objects (like a Riemannian metric) to construct and study gradient-like vector fields. And the local structure of the vector field and its integral curves nearby critical points is much simpler than in the case of general Riemannian gradients. On the other hand the condition (G) above is obviously too rigid, and the class of gradient-like vector fields is in a sense smaller than one would like it to be.

We have suggested in [**120**] a new class of vector fields associated to each Morse function f. The definition of these vector fields (called *f-gradients*) is quite simple and uses only the condition 1) above and a certain non-degeneracy condition at every critical point. This class is wider than the class of Riemannian gradients, and the basic Morse theory is carried over to this class without many changes, once the local structure of the f-gradients around the critical points of f is established. In this book we work with f-gradients, however our exposition is built in such a way that the reader can stay if he wishes entirely in the domain of gradient-like vector fields (this could be a good idea for the first reading). All main results appear in three versions: for f-gradients, for Riemannian gradients and for gradient-like vector fields.

† It is not difficult to prove that every gradient-like vector field is a Riemannian gradient with respect to some Riemannian metric on M.

1. Morse functions and Morse forms

1.1. Functions on manifolds and their critical points. Let M be a manifold without boundary, and $f : M \to \mathbf{R}$ be a C^∞ function. Recall that the *derivative* or *differential* of f at p is a linear map $f'(p) : T_pM \to \mathbf{R}$, such that for every chart

$$\Psi : U \to V \subset \mathbf{R}^m, \quad p \in U, \quad \Psi(p) = a \tag{1}$$

we have

$$f'(p)(h) = g'(a)(\Psi'(p)h) \quad \text{for every} \quad h \in T_pM$$

where $g = f \circ \Psi^{-1}$, and g' is the usual differential of the C^∞ function of m real variables. A point $p \in M$ is called a *critical point* of f, if $f'(p) = 0$.

The *second derivative* $f''(p)$ of a function f at a critical point p is by definition a bilinear symmetric form on the tangent space T_pM, such that for any chart (1) we have

$$f''(p)(h,k) = g''(a)\big(\Psi'(p)h, \Psi'(p)k\big),$$

where g'' stands for the usual second differential of the C^∞ function $g = f \circ \Psi^{-1} : V \to \mathbf{R}$.

Exercise 1.1. Check that the above formula defines a bilinear form on T_pM which does not depend on a particular choice of the chart Ψ. What happens if p is not a critical point of f?

Recall that for a non-degenerate bilinear symmetric form B in a finite-dimensional vector space E its *index* is the maximal dimension of a vector subspace $L \subset E$ for which $B(h,h) \leqslant 0$ for every $h \in L$.

Equivalently, the index of B is the number of negative diagonal entries of the matrix $\mathcal{M}(B)$ of B with respect to any basis, where $\mathcal{M}(B)$ is diagonal.

Definition 1.2. A critical point p of a C^∞ function $f : M \to \mathbf{R}$ is called *non-degenerate* if the bilinear symmetric form $f''(p)$ is non-degenerate. In this case the index of $f''(p)$ is called the *index of* p and denoted $\operatorname{ind} p$ or $\operatorname{ind}_f p$.

A function $f : M \to \mathbf{R}$ is called a *Morse function* if all critical points of f are non-degenerate.

Exercise 1.3. Prove that non-degenerate critical points are isolated. Deduce that a Morse function on a closed manifold has only a finite number of critical points.

Definition 1.4. Let $f : M \to \mathbf{R}$ be a Morse function on a closed manifold. The set of critical points of f is denoted by $S(f)$, the set of critical points of f of index k is denoted by $S_k(f)$.

The number of critical points of f is denoted by $m(f)$, and the number of critical points of f of index k is denoted by $m_k(f)$. The numbers $m(f), m_k(f)$ are also called *Morse numbers of* f.

Example 1.5. The quadratic form $\mathscr{Q}_k : \mathbf{R}^m \to \mathbf{R}$ defined by

$$\mathscr{Q}_k(x_1, \dots, x_m) = -(x_1^2 + \dots + x_k^2) + (x_{k+1}^2 + \dots + x_m^2)$$

provides the simplest example. We have

$$\mathscr{Q}_k'(x)(h) = -2\sum_{i=1}^{k} x_i h_i + 2\sum_{j=k+1}^{m} x_j h_j.$$

The only critical point of this function is 0; the second derivative at this point is given by the next formula:

$$\mathscr{Q}_k''(0)(h, l) = -2\sum_{i=1}^{k} h_i l_i + 2\sum_{j=k+1}^{m} h_j l_j.$$

This bilinear form is non-degenerate and its index equals k.

The classical Morse lemma ([**98**], Lemma 4) asserts that the preceding example is essentially the only one.

Theorem 1.6 (Morse lemma). *Let $p \in M$ be a non-degenerate critical point of a C^∞ function $f : M \to \mathbf{R}$, and let $k = \operatorname{ind} p$. Then there is a chart $\Psi : U \to V$ defined in a neighbourhood U of p such that*

$$(\mathcal{MC}) \qquad \Psi(p) = 0, \quad \textit{and} \quad f \circ \Psi^{-1} = \mathscr{Q}_k + f(p).$$

Definition 1.7. A chart $\Psi : U \to V$ around p, which satisfies the condition $(\mathcal{MC})$ is called a *Morse chart* for f at p.

Proof of the theorem. The proof below is due to M. Morse; our exposition is close to [**61**] and [**77**]. It suffices to consider the case when the function f is defined in an open disc $B(0, R) \subset \mathbf{R}^m$, and $f(0) = 0$. Diagonalizing the bilinear form $f''(0)$ we can also assume that

$$f''(0)(h, k) = \mathscr{Q}_k''(0)(h, k) = -2\sum_{i=1}^{k} h_i k_i + 2\sum_{j=k+1}^{m} h_j k_j.$$

Write

$$f''(0)(h, k) = 2\langle Jh, k\rangle$$

where J is a linear symmetric map, and $\langle \cdot, \cdot \rangle$ stands for the standard scalar product in $\mathbf{R}^m$. The matrix of J is then diagonal with the entries

$$J_{ii} = -1 \text{ for } 1 \leqslant i \leqslant k \quad \text{and} \quad J_{ii} = 1 \text{ for } k+1 \leqslant i \leqslant m.$$

The Taylor formula with the integral rest† applied to the function f at 0 reads as follows (recall that $f(0)$ and $f'(0)$ vanish):

$$f(x) = \int_0^1 (1-t)f''(tx)(x,x)dt \quad \text{for every} \quad x \in B(0,R).$$

For $x \in B(0,R)$ define a symmetric linear operator $k(x) : \mathbf{R}^m \to \mathbf{R}^m$ by the following formula:

$$\langle k(x)\xi, \eta \rangle = \int_0^1 (1-t)f''(tx)(\xi,\eta)dt.$$

The map $x \mapsto k(x)$ is a C^∞ map from $B(0,R)$ to the vector space

$$\mathcal{L} = L(\mathbf{R}^m, \mathbf{R}^m)$$

of all the linear maps of $\mathbf{R}^m$ to itself, and $k(0) = J$. For every x the bilinear form $\langle k(x)\xi, \eta\rangle$ can be diagonalized in an appropriate basis, which depends on x. The next lemma shows that we can choose this basis to depend smoothly on x.

Lemma 1.8. *There is a C^∞ map $\mu : U \to \mathcal{L}$ defined in a neighbourhood $U \subset \mathbf{R}^m$ of 0, such that $\mu(0) = \mathrm{Id}$, and*

$$k(x) = \mu^*(x) \cdot J \cdot \mu(x)$$

for every x.

Proof. Let $S \subset \mathcal{L}$ be the vector subspace of all symmetric linear maps, and R be the space of all linear maps ξ, such that $J\xi$ is symmetric. The map $\xi \mapsto J\xi$ is an isomorphism of R onto S. Define a C^∞ map

$$B : R \to S, \quad \text{by} \quad B(\xi) = \xi^* J\xi.$$

We have:

$$B(\mathbf{1}) = J, \quad \text{and} \quad B'(\mathbf{1})(h) = h^*J + Jh = 2Jh$$

(where $\mathbf{1}$ denotes the identity map). Thus the map $B'(\mathbf{1}) : R \to S$ is an isomorphism, and therefore, B is a local diffeomorphism in a neighbourhood of $\mathbf{1}$. Set

$$\mu(x) = B^{-1} \circ k(x)$$

and the lemma is proved. □

Now we can deduce our theorem. Let $\Psi(x) = \mu(x)x$. The map Ψ is a local diffeomorphism in a neighbourhood of zero (it is easy to see that $\Psi'(0) = \mathrm{Id}$). We have

$$f(x) = \langle k(x)x, x\rangle = \langle J\mu(x)x, \mu(x)x\rangle = \langle J\Psi(x), \Psi(x)\rangle,$$

† See [**21**], Part 1, Th. 5.6.1.

therefore

$$f \circ \Psi^{-1} = \mathcal{Q}_k$$

and the proof of the theorem is finished. □

The next definition generalizes the notion of Morse function to the category of cobordisms.

Definition 1.9. Let W be a cobordism, and $f : W \to \mathbf{R}$ be a C^∞ function. We say that f is *a Morse function on* W if there are real numbers $a < b$ such that

(1) $f(W) = [a, b], \ f^{-1}(b) = \partial_1 W, \ f^{-1}(a) = \partial_0 W,$
(2) $f'(x) \neq 0$ whenever $x \in \partial W$.
(3) All the critical points of $f|(W \setminus \partial W)$ are non-degenerate.

A Morse function f on a cobordism W will be usually denoted as follows:

$$f : W \to [a, b].$$

Observe that for a Morse function f on a cobordism W every critical point is isolated, and since W is compact, the set of critical points is finite. The set of critical points of a Morse function f is denoted by $S(f)$, the set of critical points of index k by $S_k(f)$, and we put

$$m(f) = \text{card } S(f), \quad m_k(f) = \text{card } S_k(f).$$

1.2. Examples of Morse functions. In this subsection we discuss several classical examples of Morse functions. We omit the details, such as computation of indices of the critical points or the proof that the critical points are non-degenerate. It is a useful exercise for the reader to do these computations.

A Morse function $S^m \to \mathbf{R}$ *with two critical points*

Let $\langle \cdot, \cdot \rangle$ be the Euclidean scalar product in the space $\mathbf{R}^{m+1}$ and let S^m denote the unit sphere in $\mathbf{R}^{m+1}$ with respect to the corresponding norm, that is,

$$S^m = \{x \in \mathbf{R}^{m+1} \mid \langle x, x \rangle = 1\}..$$

S^m is a C^∞ submanifold of $\mathbf{R}^{m+1}$. The restriction to S^m of the linear map $x = (x_1, \dots, x_{m+1}) \mapsto x_{m+1}$ is a Morse function which has two critical points:

$$N = (0, \dots, 1), \quad \text{ind } N = m \quad \text{(the maximum)},$$
$$S = (0, \dots, -1), \quad \text{ind } S = 0 \quad \text{(the minimum)}.$$

Height functions

Generalizing the previous example consider any closed submanifold $M \subset \mathbf{R}^N$; the restriction to M of the linear map $(x_1, \dots, x_N) \mapsto x_N$ is called the *height function.* In many cases height functions have only non-degenerate critical points. Consider for example the standard embedding of the two-dimensional torus $\mathbf{T}^2 = S^1 \times S^1$ into $\mathbf{R}^3$:

$$E : \mathbf{T}^2 \hookrightarrow \mathbf{R}^3;$$

$$E(\phi, \theta) = \Big(r \sin\theta,\ (R + r\cos\theta)\cos\phi,\ (R + r\cos\theta)\sin\phi\Big),$$

where $0 < r < R$ and ϕ, θ are the angle variables. The image $E(\mathbf{T}^2)$ is depicted in Figure 2.

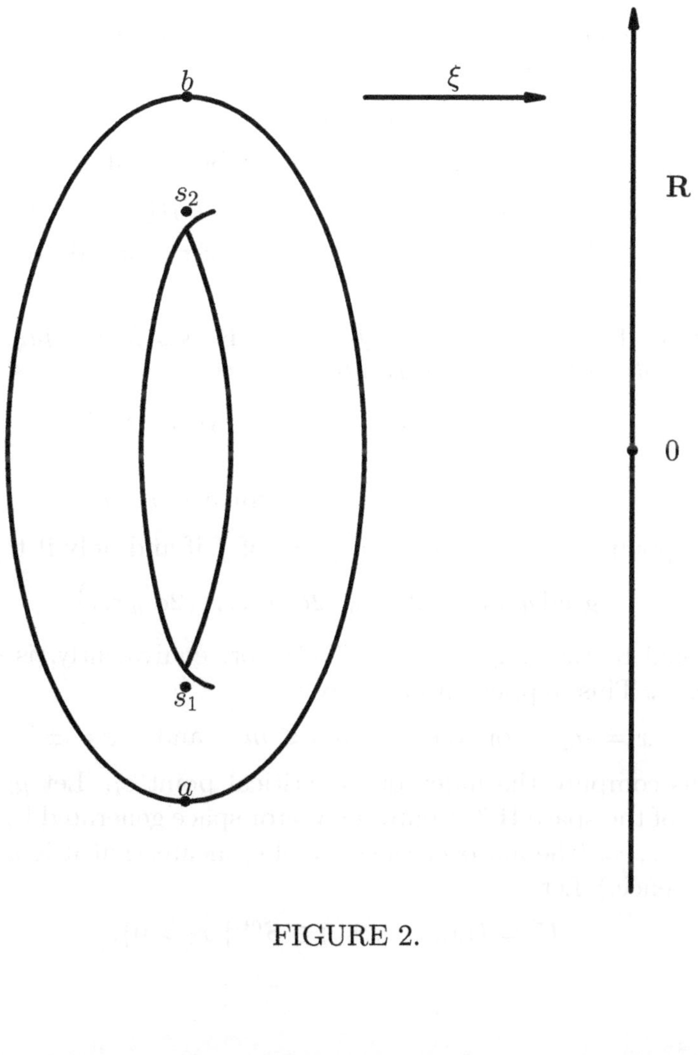

FIGURE 2.

The submanifold $E(\mathbf{T}^2)$ is obtained by rotation around the x-axis of the circle $S(A,r)$ in the plane (x,z), where $A=(0,0,R)$.

Exercise 1.10. Show that the height function $\zeta(x,y,z)=z$ has four non-degenerate critical points:

$$\begin{aligned}
b &= (0,0,R+r), & \text{ind } b &= 2 & &\text{(the maximum)},\\
s_2 &= (0,0,R-r), & \text{ind } s_2 &= 1 & &\text{(a saddle point)},\\
s_1 &= (0,0,-R+r), & \text{ind } s_1 &= 1 & &\text{(another saddle point)},\\
a &= (0,0,-R-r), & \text{ind } a &= 0 & &\text{(the minimum)}.
\end{aligned}$$

Quadratic forms on S^m

In this subsection we give more examples of Morse functions on the sphere

$$S^m = \{x \in \mathbf{R}^{m+1} \mid \langle x, x\rangle = 1\}.$$

Here it is convenient to use an alternative enumeration of the coordinates in $\mathbf{R}^{m+1}$, writing a generic point $x \in \mathbf{R}^{m+1}$ as

$$x = (x_0, x_1, \dots, x_m).$$

Let $\alpha_0 < a_1 < \cdots < \alpha_m$ be non-zero real numbers. Put

$$f = g \mid S^m, \qquad \text{where} \qquad g(x_0,\dots,x_m) = \alpha_0 x_0^2 + \cdots + \alpha_m x_m^2.$$

For $i=0,\dots,m$ let e_i denote the i-th vector of the standard orthonormal basis of $\mathbf{R}^{m+1}$.

Proposition 1.11. *The function $f : S^m \to \mathbf{R}$ is a Morse function, which has $2m$ non-degenerate critical points:*

$$S(f) = \{e_0, -e_0, e_1, -e_1, \ \dots \ , e_m, -e_m\},$$

and

$$\operatorname{ind} e_i = \operatorname{ind}(-e_i) = i \quad \textit{for every} \quad i.$$

Proof. A point $x \in S^m$ is a critical point of f if and only if the vector

$$\operatorname{grad} g(x) = \Big(2\alpha_0 x_0,\ 2\alpha_1 x_1, \dots, 2\alpha_m x_m\Big)$$

is orthogonal to the tangent space $T_x S^m$, or, equivalently, is collinear to the vector x. This happens if and only if

$$x = \varepsilon e_i \quad \text{for some} \quad 0 \leqslant i \leqslant m \quad \text{and} \quad \varepsilon = \pm 1.$$

Now let us compute the index of the critical point e_i. Let p_i denote the projection of the space $\mathbf{R}^{m+1}$ onto the vector space generated by the vectors $e_0, \dots, \widehat{e_i}, \dots, e_m$ (the hat over the symbol e_i means that it is *not* included in the sequence.) Let

$$U_i = \{(x_0, \dots, x_m) \in S^m \mid x_i > 0\}.$$

This set is an open neighbourhood of e_i in S^m. The restriction $\Psi_i = p_i \mid U_i$ is a local chart for S^m at e_i and

$$f \circ \Psi_i^{-1}(x_0, \dots, \widehat{x}_i, \dots, x_m) = \alpha_i + \sum_{j \neq i} (\alpha_j - \alpha_i) x_j^2.$$

Our assertion follows. □

A Morse function on $\mathbf{R}P^m$

The real projective space $\mathbf{R}P^m$ is a C^∞ manifold of dimension m, whose underlying topological space is the set of all straight lines passing through the origin in $\mathbf{R}^{m+1}$. The map $p : S^m \to \mathbf{R}P^m$ which associates to each point of S^m the straight line joining this point and the origin, is a regular covering with the structural group $\mathbf{Z}/2\mathbf{Z}$. The generator I of this group acts on S^m as follows:

$$I : S^m \to S^m, \quad I(x) = -x.$$

The Morse function $f : S^m \to \mathbf{R}$ from the previous example is invariant with respect to the involution, therefore it determines a Morse function $h : \mathbf{R}P^m \to \mathbf{R}$. The critical points of h are the images of the critical points of f under the projection p. Put

$$E_i = p(e_i) \quad \text{where} \quad 0 \leqslant i \leqslant m.$$

We have then

$$S(h) = \{E_0, \dots, E_m\}, \quad \text{where} \quad \operatorname{ind} E_i = i.$$

Here is an explicit formula for the function h in terms of the homogeneous coordinates in $\mathbf{R}P^m$:

$$h(x_0 : x_1 : \dots : x_m) = \frac{\sum_i \alpha_i |x_i|^2}{\sum_i |x_i|^2}.$$

Exercise 1.12. Let α_i be non-zero pairwise different real numbers. Show that the function $\phi : \mathbf{C}P^m \to \mathbf{R}$ defined by

$$\phi(z_0 : \dots : z_m) = \frac{\sum_i \alpha_i |z_i|^2}{\sum_i |z_i|^2}$$

is a Morse function. Find the critical points of ϕ and compute their indices.

The "pants" manifold

Consider the submanifold $W \subset \mathbf{R}^3$ depicted in Figure 3.

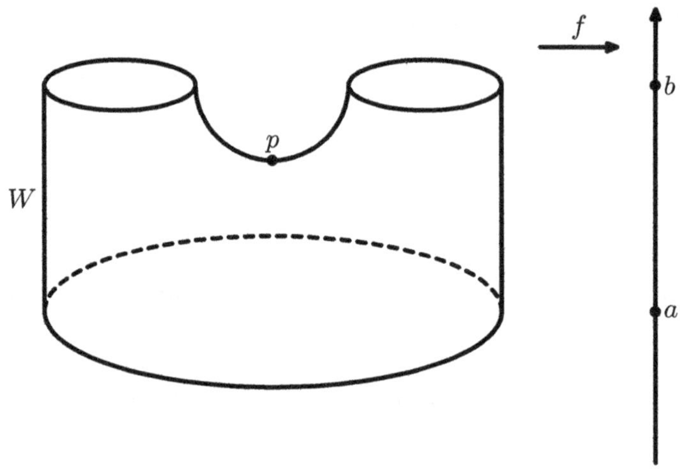

FIGURE 3.

This is a cobordism; $\partial_0 W$ is diffeomorphic to a circle, and $\partial_1 W$ is diffeomorphic to a disjoint union of two circles. The height function is a Morse function which has one critical point of index 1 (a saddle point). This function is among the most used in topology; we shall illustrate and test many of our constructions on this example. This Morse function is diffeomorphic to the restriction of the height function $\zeta : \mathbf{T}^2 \to \mathbf{R}$ on the torus to the cobordism

$$W' = \zeta^{-1}([-r/2 - R, 0]).$$

1.3. Circle-valued Morse functions. Let M be a manifold without boundary, and let $f : M \to S^1$ be a C^∞ function (the circle

$$S^1 = \{(x, y) \in \mathbf{R}^2 \mid x^2 + y^2 = 1\}$$

is a 1-dimensional submanifold of $\mathbf{R}^2$ and is endowed with the corresponding smooth structure).

For a point $x \in M$ choose a neighbourhood V of $f(x)$ in S^1 diffeomorphic to an open interval of $\mathbf{R}$, and let $U = f^{-1}(V)$. The map $f|U$ is then identified with a smooth map from U to $\mathbf{R}$. Thus all the local notions of the real-valued Morse theory, in particular the notions of *non-degenerate critical point*, and the *index of a critical point*, are carried over immediately to the framework of circle-valued functions.

Definition 1.13. A C^∞ map $f : M \to S^1$ is called a *Morse map*, if all its critical points are non-degenerate. For a Morse map $f : M \to S^1$ we denote by $S(f)$ the set of all critical points of f, and by $S_k(f)$ the set of all critical points of index k.

If M is compact, the set $S(f)$ is finite; in this case we denote by $m(f)$ the cardinality of $S(f)$ and by $m_k(f)$ the cardinality of $S_k(f)$.

The circle S^1 can be identified with the quotient $\mathbf{R}/\mathbf{Z}$, and circle-valued functions can be considered as *multi-valued real functions*, so that the value of the function at a point is defined only modulo the subgroup of integers. The natural way to work with the multi-valued functions (applied for example in the theory of Riemann surfaces for analytic functions) is to consider a covering space of the domain of definition such that the function becomes single-valued on the covering.

Recall the universal covering of the circle

$$\operatorname{Exp} : \mathbf{R} \to S^1, \ t \mapsto e^{2\pi i z}.$$

The structure group of this covering is the subgroup $\mathbf{Z} \subset \mathbf{R}$ acting on $\mathbf{R}$ by translations. It is convenient to use the multiplicative notation for the structure group and denote by t the generator corresponding to -1 in the additive notation. This generator will be called *the downward generator.*

Now let $f : M \to S^1$ be a circle-valued function. Let $\bar{M} \to M$ be the infinite cyclic covering induced from the universal covering $\operatorname{Exp} : \mathbf{R} \to S^1$ by the function f. By the definition of the induced covering we have a map $F : \bar{M} \to \mathbf{R}$, which makes the following diagram commutative:

$$\begin{array}{ccc} \bar{M} & \xrightarrow{F} & \mathbf{R} \\ \downarrow & & \downarrow \scriptstyle{\operatorname{Exp}} \\ M & \xrightarrow{f} & S^1 \end{array}$$

Here f is a Morse function if and only if F is a Morse function. The map F is *equivariant* with respect to the action of $\mathbf{Z}$ on $\bar{M}$ and on $\mathbf{R}$, that is $F(tx) = F(x) - 1$.

Although F is a real-valued Morse function, the standard Morse theory still can not be applied here, since the domain of definition of F is not compact (for example the set $S(F)$ is infinite if it is not empty). One of the possible ways to overcome this difficulty is to consider the restriction of F to the fundamental domain of $\bar{M}$ with respect to the action of $\mathbf{Z}$.

Definition 1.14. Let $a \in \mathbf{R}$ be a regular value of F. The set $W = F^{-1}([a-1,a])$ is a compact cobordism W with

$$\partial_1 W = F^{-1}(a),\ \partial_0 W = F^{-1}(a-1).$$

The set W is called the *fundamental cobordism.*

We have the Morse function

$$F \mid W : W \to [a-1,a].$$

The cobordism W can be described as follows. Let $\alpha = \operatorname{Exp}(a) \in S^1$; then α is a regular value of f, and $V = f^{-1}(\alpha)$ is a smooth submanifold of M. Cut the manifold M along V, and we obtain the cobordism W with both components of its boundary diffeomorphic to V.

Example 1.15. Let V be a compact manifold, $M = V \times S^1$, and $f : M \to S^1$ the projection onto the second factor. Then f is a Morse function without critical points, the covering $\bar{M}$ is identified with $V \times \mathbf{R}$ and the fundamental cobordism corresponding to a regular value a is identified with $V \times [a-1,a]$.

1.4. Morse forms. Now we are going to generalize the notion of Morse function still further. Observe that the definition of a real-valued Morse function can be reformulated entirely in terms of the differential of the function. This differential is a closed 1-form cohomologous to zero; if we omit this last restriction and allow non-exact closed 1-forms, then we are led to the Morse theory for closed 1-forms. In this subsection we give a brief introduction to this theory with the special emphasis on the case of circle-valued Morse function.

Let us begin with generalities about closed 1-forms. Recall that the de Rham cohomology group $H^1_{deRham}(M)$ is by definition the quotient of the space of all closed 1-forms by the subspace of the exact forms, that is the differentials of real-valued functions. There is the canonical isomorphism

$$H^1(M,\mathbf{R}) \approx H^1_{deRham}(M)$$

between the de Rham cohomology group and the singular cohomology group with real coefficients (see for example, [**161**], Ch. 2), and we shall identify these two groups. For a closed 1-form ω its image in the group $H^1(M,\mathbf{R})$ is called the *de Rham cohomology class of* ω and denoted $[\omega]$.

An important family of examples of closed 1-forms is provided by circle-valued smooth functions. The 1-form dx on $\mathbf{R}^1$ is invariant with respect to the action of $\mathbf{Z}$ on $\mathbf{R}$ and hence it defines a 1-form on $S^1 = \mathbf{R}/\mathbf{Z}$ which will be denoted by the same symbol dx. The de Rham cohomology class of dx is the image of the generator $\iota \in H^1(S^1, \mathbf{Z})$ with respect to the map $H^1(S^1, \mathbf{Z}) \hookrightarrow H^1(S^1, \mathbf{R})$. For a C^∞ function $f : M \to S^1$ set

$$df = f^*(dx).$$

Then df is a closed 1-form on M, which will be called *the differential of the circle-valued Morse map f*. Its de Rham cohomology class $[df]$ equals $f^*(\iota)$, therefore it is integral (i.e., it lies in the image of the injection of $H^1(M, \mathbf{Z})$ in $H^1(M, \mathbf{R})$).

Proposition 1.16. *The homotopy class of a C^∞ map $f : M \to S^1$ is determined by the de Rham cohomology class $[df]$ of its differential.*

Proof. Observe that the abelian group structure of S^1 allows us to define an abelian group structure on the set of all C^∞ maps $M \to S^1$ in such a way that for the sum $f + g$ of two C^∞ functions we have

$$d(f + g) = df + dg \in \Omega^1(M).$$

It suffices to prove that $[df] = 0$ implies that f is homotopic to a constant map. The condition $[df] = 0$ is equivalent to $df = dh$, with $h : M \to \mathbf{R}$ a real-valued C^∞ function. Therefore the de Rham cohomology class of df vanishes if and only if the map $f : M \to S^1$ lifts to a C^∞ function $\widetilde{f} : M \to \mathbf{R}$, which implies that f is homotopic to a constant map. □

Lemma 1.17. *Let $\omega \in \Omega^1(M)$ be a C^∞ closed 1-form. Then the two following conditions are equivalent.*

(1) *$\omega = df$ where $f : M \to S^1$ is a C^∞ function.*
(2) *$[\omega]$ is integral.*

Proof. Let us prove that any form ω within an integral class equals df for some $f : M \to S^1$. Assume for simplicity that M is connected, the general case is done similarly. Choose any point $p \in M$. For $x \in M$ and a C^∞ curve $\gamma : [a, b] \to M$ with $\gamma(a) = p$, $\gamma(b) = x$ set

$$F(p, \gamma) = \int_\gamma \omega.$$

For two paths γ_1, γ_2 the difference $F(p, \gamma_1) - F(p, \gamma_2)$ is an integer, therefore the complex number $e^{2\pi i F(p,\gamma)}$ does not depend on γ and determines a C^∞ map

$$f : M \to S^1, \ f(p) = e^{2\pi i F(p,\gamma)}$$

with $df = \omega$. □

Now let us turn to the case of closed 1-forms in an arbitrary de Rham cohomology class.

Definition 1.18. A C^∞ closed 1-form $\omega \in \Omega^1(M)$ on a manifold M is called a *Morse form* if for every $a \in M$ there is a neighbourhood U of a and a Morse function $f : U \to \mathbf{R}$ such that $\omega|U = df$.

For example, the differential of a C^∞ map $f : M \to S^1$ is a Morse form if and only if f is a Morse map. Observe that in this generalization of the Morse theory from real-valued Morse functions to closed 1-forms, the critical points of real-valued functions correspond to *zeros* of 1-forms.

Let ω be a closed 1-form, and let p be a zero of ω, that is, $\omega(p) = 0$. Pick a chart $\Psi : U \to V \subset \mathbf{R}^m$ for M at p, and let $a = \Psi(p)$. Let $\widetilde{\omega} = (\Psi^{-1})^*(\omega)$, then $\widetilde{\omega}$ is a 1-form on $V \subset \mathbf{R}^m$; write

$$\widetilde{\omega}(x) = \sum_i \widetilde{\omega}_i(x) dx_i, \qquad \text{and} \qquad J(\widetilde{\omega}, x) = \det \left| \frac{\partial \widetilde{\omega}_i}{\partial x_j}(x) \right|,$$

so that $\widetilde{\omega}_i(\cdot)$ and $J(\widetilde{\omega}, \cdot)$ are C^∞ functions $V \to \mathbf{R}$. The point p is called a *non-degenerate zero of* ω if

$$J(\widetilde{\omega}, a) \neq 0.$$

Exercise 1.19. Check that the above condition is invariant with respect to the choice of the local chart Ψ. Prove that a closed 1-form ω is a Morse form if and only if every zero of ω is non-degenerate.

For any non-degenerate zero p of a closed 1-form ω, *the index of* p is by definition the index of the non-degenerate bilinear form $f''(p)$ where $f : U \to \mathbf{R}$ is any C^∞ function in a neighbourhood of p with $df = \omega|U$.

Definition 1.20. Let ω be a Morse form. We denote the set of zeros of ω by $S(\omega)$, and its cardinality by $m(\omega)$. The subset of all zeros of index k is denoted $S_k(\omega)$ and its cardinality by $m_k(\omega)$.

Any Morse form locally in a neighbourhood of any of its zeros is diffeomorphic to the differential of the standard quadratic form. This is the contents of the next proposition, which follows immediately from the Morse lemma (Theorem 1.6 of the present Chapter).

Proposition 1.21. *Let ω be a Morse form on a manifold M, and let p be a zero of ω of index k. Then there is a chart $\Psi : U \to V$ for M at p such that*

$$\Psi(p) = 0 \quad \textit{and} \quad \left(\Psi^{-1}\right)^*(\omega) = -2\textstyle\sum_{i=1}^k x_i dx_i + 2\sum_{i=k+1}^m x_i dx_i. \qquad \square$$

Definition 1.22. Let ω be a Morse form on a manifold M, and let p be a zero of ω of index k. A chart $\Psi : U \to V$ for M at p satisfying the above condition is called a *Morse chart* for ω at p.

A generic closed 1-form is a Morse form. This is the contents of Theorem 1.25 below. The proof is due essentially to M. Morse ([**99**], Chapter VI, §8.1). It is based on the observation that, for a smooth function f defined on an open set in Euclidean space, it suffices to add a small linear form to f, in order to obtain a Morse function.

Let M be a manifold without boundary. The space $\Omega^1(M)$ will be endowed with the *weak* C^∞ topology. Recall that convergence of a sequence of 1-forms with respect to this topology means uniform convergence on every compact subset of M together with all partial derivatives.

Definition 1.23. Let M be a manifold and $\xi \in H^1(M, \mathbf{R})$. Denote by L_ξ the set of all closed 1-forms with de Rham cohomology class equal to ξ.

Observe that L_ξ is an affine subspace of $\Omega^1(M)$, and we endow L_ξ with the induced topology.

Definition 1.24. Let ω be a closed 1-form on a manifold M, and $X \subset M$. We say, that ω *is regular on* X, or X*-regular* if every zero $p \in X$ of ω is non-degenerate.

Theorem 1.25. *Let M be a manifold, and $\xi \in H^1(M, \mathbf{R})$. Let $X \subset M$ be a compact subset. The subset of all X-regular 1-forms is open and dense in L_ξ.*

Proof. We begin with the following lemma.

Lemma 1.26. *The set of X-regular closed 1-forms in L_ξ is open in L_ξ.*

Proof of the lemma. The contrary would mean that there is a sequence ω_n of 1-forms in L_ξ, converging to an X-regular form ω, and each ω_n has a degenerate zero $p_n \in X$. Since X is compact, we can assume that p_n converges to a point $p \subset X$. Choose a chart $\Psi : U \to V$ for M at p and a compact neighbourhood $Q \subset V$ of $a = \Psi(p)$. We obtain a sequence of smooth 1-forms $\lambda_n = (\Psi^{-1})^*(\omega_n)$ defined on V. These 1-forms converge on Q (together with all their partial derivatives) to the 1-form $\lambda = (\Psi^{-1})^*(\omega)$. The sequence $a_n = \Psi(p_n)$ converges to the point $a = \Psi(p)$, and therefore we have

$$J(\lambda_n, a_n) \to J(\lambda, a),$$

which is obviously contradictory, since $J(\lambda_n, a_n) = 0$, and $J(\lambda, a) \neq 0$. □

The density in L_ξ of X-regular forms follows from the next proposition.

Proposition 1.27. *Let ω be a closed 1-form on a manifold M, and $X \subset M$ be a compact subset of M.*

Then for every neighbourhood U of X, and every neighbourhood $\mathcal{V}$ of ω in the space $\Omega^1(M)$, there is a C^∞ function $h : M \to \mathbf{R}$ such that:

(1) supp $h \subset U$,
(2) $\xi = \omega - dh \in \mathcal{V}$,
(3) ξ *is* X*-regular.*

Proof. Let us first do a particular case when $M = \mathbf{R}^m$. In this case the 1-form ω can be identified with a C^∞ map

$$\mathbf{R}^m \xrightarrow{\omega} (\mathbf{R}^m)^* = L(\mathbf{R}^m, \mathbf{R})$$

from $\mathbf{R}^m$ to the space of linear forms on $\mathbf{R}^m$. By the Sard lemma the set of regular values of this map is dense in $(\mathbf{R}^m)^*$. Pick a regular value $\lambda \in (\mathbf{R}^m)^*$, then λ is a linear function $\mathbf{R}^m \to \mathbf{R}$. Pick a positive C^∞ function $g : \mathbf{R}^m \to \mathbf{R}$ such that $g(x) = 1$ for every x in some neighbourhood of X and supp $g \subset U$. Set

$$h(x) = g(x) \cdot \lambda(x).$$

We claim that the function h satisfies our requirements if λ is small enough. The first property holds by definition. The second property holds if λ is small enough. Finally, for every $x \in X$ we have $h'(x) = \lambda$, and the 1-form $\xi = \omega - dh$ is regular on X by our choice of λ.

Let us proceed to the case of arbitrary manifold M. The case when the compact X is contained in the domain of definition of some chart of M follows immediately from the previous argument.

The proposition in its full generality follows then from the next lemma. Let M be a manifold without boundary, and $X \subset M$ be a compact subset. We say that X is *nice* if the conclusion of the proposition holds for M, X and every closed 1-form ω on M.

Lemma 1.28. *If* X_1, X_2 *are nice, then* $X = X_1 \cup X_2$ *is nice.*

Proof. Let U be any neighbourhood of $X = X_1 \cup X_2$, and $\mathcal{V}$ be any neighbourhood of ω in $\Omega^1(M)$. Pick a C^∞ function $h_1 : M \to \mathbf{R}$ such that

(1) supp $h_1 \subset U$,
(2) the 1-form $\xi_1 = \omega - dh_1$ is in $\mathcal{V}$,
(3) $\xi_1 | X_1$ is regular.

Next pick a C^∞ function $h_2 : M \to \mathbf{R}$ such that

(1) supp $h_2 \subset U$,
(2) the 1-form $\xi_2 = \xi_1 - dh_2 = \omega - dh_1 - dh_2$ is in $\mathcal{V}$,
(3) $\xi_2 | X_2$ is regular.

By Lemma 1.26 the 1-form ξ_2 is still regular on X_1 if $dh_2 = \xi_2 - \xi_1$ is sufficiently small, and in this case the function $h_1 + h_2$ has the required properties. □

Corollary 1.29. *Let M be a closed manifold, and $\xi \in H^1(M, \mathbf{R})$. The set of Morse forms in L_ξ is open and dense in L_ξ.* □

Here are two corollaries of the theorem which refer to the cases of real-valued and circle-valued Morse functions.

Theorem 1.30. *Let M be a closed manifold. The set of all Morse functions $M \to \mathbf{R}$ is open and dense in the set of all C^∞ functions.*

Proof. Let $C^\infty(M)$ denote the space of all real-valued C^∞ functions, and $\mathcal{M}$ denote the subset of Morse functions. Let L denote the vector space of all C^∞ exact 1-forms. The map

$$d : C^\infty(M) \to L, \quad f \mapsto df$$

is continuous (both spaces are endowed with C^∞ topology). A function f is a Morse function if and only if df is a Morse form. Thus the set $\mathcal{M}$ is open in $C^\infty(M)$ by Lemma 1.26.

Now let us move to the proof of the density. It suffices to consider the case when M is connected. Let $f \in C^\infty(M)$ be any C^∞ function, $a \in M$ be any point. Denote by $F_a \subset C^\infty(M)$ the subspace of all functions g such that $g(a) = f(a)$. The restriction

$$d|F_a : F_a \to L$$

is a homeomorphism. The subspace of Morse forms is dense in L by Corollary 1.29, and this completes the proof. □

Theorem 1.31. *Let M be a closed manifold, and η a homotopy class of maps $M \to S^1$. In the set of all C^∞ functions in η the subset of Morse functions is open and dense.*

Proof. Let $\mathcal{F}_\eta$ denote the space of all C^∞ maps $M \to S^1$ within the homotopy class η, and $\mathcal{M}_\eta \subset \mathcal{F}_\eta$ denote the subspace of Morse maps. Let $\bar{\eta} \in H^1(M, \mathbf{Z}) \subset H^1(M, \mathbf{R})$ be the cohomology class corresponding to η via the canonical isomorphism

$$[M, S^1] \xrightarrow{\approx} H^1(M, \mathbf{Z}).$$

Recall that $L_{\bar{\eta}}$ denotes the subspace of $\Omega^1(M)$ consisting of closed 1-forms with de Rham cohomology class equal to $\bar{\eta}$. The map

$$\mathcal{D} : \mathcal{F}_\eta \to L_{\bar{\eta}}, \quad f \mapsto df$$

is continuous (both spaces are endowed with C^∞ topology). A function f is a Morse function if and only if df is a Morse form. Thus the set $\mathcal{M}_\eta$ is open in $\mathcal{F}_\eta$ by Lemma 1.26.

Now let us prove the density property. It suffices to consider the case when M is connected. Let $f : M \to S^1$ be any C^∞ function in the class η, and let $a \in M$ be any point. Let $\mathcal{F}_0$ be the subspace of $\mathcal{F}_\eta$ formed by all functions g with $g(a) = f(a)$. The restriction

$$\mathcal{D}|F_0 : F_0 \to L_{\bar{\eta}}$$

is a homeomorphism. Our assertion follows then from Corollary 1.29. □

2. Gradients of Morse functions and forms

2.1. Definition of f-gradients. Let W be a cobordism or a manifold without boundary, and $f : W \to \mathbf{R}$ a Morse function. Let v be a C^∞ vector field on W satisfying the condition

$$(2) \qquad f'(x)(v(x)) > 0 \text{ whenever } x \notin S(f).$$

The function $\phi(x) = f'(x)(v(x))$ vanishes on $S(f)$ and is strictly positive on $W \setminus S(f)$. Therefore every point $p \in S(f)$ is a point of local minimum of ϕ, and $\phi'(p) = 0$.

Definition 2.1. A vector field v is called an *f-gradient* if the condition (2) holds, and every $p \in S(f)$ is a non-degenerate minimum of the function $\phi(x) = f'(x)(v(x))$ (that is, the second derivative $\phi''(p)$ is a non-degenerate bilinear form on T_pW). The set of all f-gradients is denoted by $G(f)$.

We will now compute the second derivative $\phi''(p)$ in terms of the derivatives of f and v and give an alternative formulation of the non-degeneracy condition above.

Lemma 2.2. *Let v be an f-gradient, and $p \in S(f)$. Then $v(p) = 0$.*

Proof. Pick a Morse chart $\Psi : U \to V \subset \mathbf{R}^m$ for f at p, and let $w = \Psi_* v$. The function $f \circ \Psi^{-1} - f(p)$ equals the quadratic form

$$\mathcal{Q}_k(x,y) = -||x||^2 + ||y||^2; \quad x \in \mathbf{R}^k, y \in \mathbf{R}^{m-k}, \text{ where } k = \operatorname{ind} p.$$

We have $\mathcal{Q}'_k(x,y)(w(x,y)) > 0$ for every $(x,y) \neq (0,0)$, and we must prove $w(0,0) = 0$. Let $w(0,0) = (\xi, \eta)$, and write for $x \in \mathbf{R}^k$

$$w(x,0) = (\xi + \mathcal{O}_1(x), \eta + \mathcal{O}_2(x))$$

where $||\mathcal{O}_i(x)|| \leqslant C||x||$ nearby 0. We have

$$\mathcal{Q}'_k(x,0)(w(x,0)) = -\langle 2x, \xi + \mathcal{O}_1(x)\rangle \geqslant 0 \quad \text{for every} \quad x.$$

Set $x = t\xi$ with $t \to 0$ to see that it is possible only for $\xi = 0$. The proof that $\eta = 0$ is similar. □

Let $p \in S(f)$. Since $v(p) = 0$, there is a linear map $v'(p) : T_pW \to T_pW$, such that for every chart $\Psi : U \to V$ of W at p we have:

$$\Psi'(p)\big(v'(p)(h)\big) = \big(\Psi_*(v)\big)'(\Psi(p))(\Psi'(p)h)$$

(this map $v'(p)$ is called the *differential* of v at p).

Exercise 2.3. Use the above formula to *define* the linear map $v'(p)$. Apply the condition $v(p) = 0$ to show that the map $v'(p)$ defined in this way does not depend on the choice of the chart Ψ.

Lemma 2.4. *For $p \in S(f)$ we have*

$$\phi''(p)(h,k) = f''(p)(v'(p)h,k) + f''(p)(v'(p)k,h).$$

Proof. We can assume that W is an open subset of $\mathbf{R}^n$, and $p = 0 \in W$. It will be convenient to denote the canonical pairing between the linear forms and vectors in $\mathbf{R}^n$ by $\cdot$, so that $\phi(x) = f'(x) \cdot v(x)$. The Leibnitz rule gives:

$$\phi'(x)(h) = f''(x)(h) \cdot v(x) + f'(x) \cdot v'(x)(h).$$

Differentiate once more:

$$\begin{aligned}\phi''(x)(h,k) = f'''(x)(h,k) \cdot v(x) + f''(x)(h) \cdot v'(x)(k)\\ + f''(x)(k) \cdot v'(x)(h) + f'(x) \cdot v''(x)(h,k),\end{aligned}$$

where the first and the fourth term vanish for $x = p$, and we obtain the formula sought. □

The next proposition follows immediately.

Proposition 2.5. *Let $f : W \to \mathbf{R}$ be a Morse function (where W is a cobordism or a manifold without boundary). A vector field v on W is an f-gradient if and only if*

A) for every $x \notin S(f)$ we have $f'(x)(v(x)) > 0$ and

B) for every $p \in S(f)$ we have

$$f''(p)(v'(p)h,h) > 0 \text{ for every } h \neq 0.$$ □

2.2. Gradient-like vector fields and Riemannian gradients. As in the previous subsection let W be a cobordism or a manifold without boundary, and $f : W \to \mathbf{R}$ a Morse function. A C^∞ vector field v is called a *gradient-like vector field* for f if

1) for every non-critical point x of f we have: $f'(x)(v(x)) > 0$,
2) for every critical point p of f of index k there is a Morse chart $\Psi : U \to V \subset \mathbf{R}^m$ for f at p such that

(G) $$\Psi_*(v)(x_1, \dots, x_m) = (-x_1, \dots, -x_k, x_{k+1}, \dots, x_m).$$

Definition 2.6. A Morse chart for f satisfying (G) is called *a Morse chart for (f, v) at p.*

Here is a curious example.

Exercise 2.7. Let $\lambda_1, \dots, \lambda_m$ be non-zero positive real numbers; prove that the vector field

$$\mathscr{A}_k(x) = (-x_1, \dots, -x_k, x_{k+1}, \dots, x_m)$$

is a gradient-like vector field for the Morse function

$$f(x_1, \dots, x_m) = -\sum_{i=1}^{k} \lambda_i x_i^2 + \sum_{i=k+1}^{m} \lambda_i x_i^2; \quad f : \mathbf{R}^m \to \mathbf{R}.$$

The corresponding Morse chart is given by

$$\Psi(x_1, \dots, x_m) = (\sqrt{\lambda_1} \cdot x_1, \dots, \sqrt{\lambda_m} \cdot x_m).$$

Lemma 2.8. *A gradient-like vector field for f is an f-gradient.*

Proof. Let p be a critical point of f, and Ψ be a Morse chart for (f, v) at p. Let $\phi(y) = f'(y)(v(y))$. Then

$$(\phi \circ \Psi^{-1})(x) = 2||x||^2$$

and this function has a non-degenerate minimum at 0. □

Let us proceed to another important family of examples: Riemannian gradients. Recall that the Riemannian gradient of f with respect to a Riemannian metric $\langle \cdot, \cdot \rangle$ on W is defined by the following formula:

$$\langle \operatorname{grad} f(x), h \rangle = f'(x)(h)$$

(where x is any point of W, and $h \in T_x W$).

Proposition 2.9. *Every Riemannian gradient is an f-gradient.*

Proof. The condition A) from Proposition 2.5 is obviously satisfied. The condition B) will be deduced from the next lemma.

Lemma 2.10. *Let p be a critical point of f, let v be the Riemannian gradient of f with respect to a Riemannian metric on W. Then*

(3) $$\langle v'(p)h, k \rangle = f''(p)(h, k),$$

where $h, k \in T_p W$.

Proof. It suffices to consider the case when W is an open subset of $\mathbf{R}^m$ and $p = 0 \in W$. The Riemannian metric is then identified with a bilinear positive definite form $\mathcal{G}(x)$ depending on $x \in W$, and the map $x \mapsto \mathcal{G}(x)$ is of class C^∞. By the definition of Riemannian gradient we have

$$\mathcal{G}(x)(v(x), k) = f'(x)(k) \quad \text{for every} \quad x \in W, \quad k \in \mathbf{R}^m.$$

Differentiating the above equality we obtain the following:

$$\mathcal{G}'(x)(h)(v(x), k) + \mathcal{G}(x)(v'(x)h, k) = f''(x)(h, k).$$

Now just substitute $x = 0$ and the lemma is proven. □

To complete the proof of our proposition observe first that the formula (3) implies that the linear map $v'(p)$ is an isomorphism. Next substitute $k = v'(p)h$ in the formula (3), to obtain

$$f''(p)(v'(p)h, h) = \langle v'(p)h, v'(p)h \rangle,$$

and the expression in the right-hand side of the previous formula can vanish only for $h = 0$. □

Every gradient-like vector field is a Riemannian gradient, as the following proposition shows.†

Proposition 2.11. *Let W be a cobordism or a manifold without boundary. Let $f : W \to \mathbf{R}$ be a Morse function, and v a gradient-like vector field for f. Then v is the Riemannian gradient for f with respect to some Riemannian metric on W.*

Proof. The proof is obtained by constructing the required Riemannian metric locally, and then gluing the results together with the help of the partition of unity. We shall do the proof in several steps. Let us call the triple (W, f, v) *regular* if the conclusion of the proposition holds for (W, f, v).

A. Let us consider first the simplest possible case when $W = U \times]a, b[$, where U is a manifold without boundary, and

$$f(x, t) = t, \quad v(x, t) = (0, 1).$$

This triple (W, f, v) will be called *trivial*. It is obviously a regular triple: for every Riemannian metric $\mathcal{G}$ on U the product of $\mathcal{G}$ with the Euclidean metric on $]a, b[$ satisfies the requirement.

B. Let (W, f, v) be any triple where W is a manifold without boundary or a cobordism. Assume that for each point $a \in W$ there is a neighbourhood $U(a)$ of a such that the triple $(U(a), f|U(a), v|U(a))$ is regular. Then (W, f, v) is regular. Indeed, let us extract from the covering $\{U(a)\}_{a \in W}$ a locally finite covering $\{U_i\}$ of W. Let $\{h_i\}_{i \in I}$ be the corresponding partition of unity. Let $\mathcal{G}_i$ be a Riemannian metric in U_i such that $v|U_i$ is the

† For a quicker proof of this proposition see Exercise 2.22 of Chapter 4 (page 131).

Riemannian gradient of $f|U_i$ with respect to $\mathcal{G}_i$. Then the sum $\sum_i h_i\mathcal{G}_i$ is the required metric on W.

C. Now we can do the case of arbitrary triples (W, f, v) without critical points. Observe that if a triple (W, f, v) is regular, and h is a C^∞ everywhere positive function on W, then the triple $(W, f, h \cdot v)$ is regular. (Indeed if v is the Riemannian gradient for f with respect to $\mathcal{G}$, then $h \cdot v$ is the Riemannian gradient for f with respect to $h^{-1} \cdot \mathcal{G}$.) Therefore it suffices to consider only the gradient-like vector fields u with $f'(x)(u(x)) = 1$ for every $x \in W$.

In view of the point B above the proof in this case will be completed if we show that for any point $a \in W$ there is a neighbourhood $U(a)$ such that the triple $(U(a), f|U(a), v|U(a))$ is diffeomorphic to a trivial triple. To this end, let $c = f(a)$ and let V be the level surface $f^{-1}(c)$. There is $\epsilon > 0$ and a neighbourhood U of a in V, such that the map

$$\Psi : U \times]-\epsilon, \epsilon[\to W; \quad (x, t) \mapsto \gamma(x, t; v)$$

is a diffeomorphism of $U \times]-\epsilon, \epsilon[$ onto its image. Denote this image by $U(a)$; then Ψ^{-1} gives a diffeomorphism of the triple $(U(a), f|U(a), v|U(a))$ to a trivial one.

D. Now let us move to the case of Morse functions with a non-empty set of critical points. For every critical point a of such function f there is a neighbourhood $U(a)$ of a such that the triple $(U(a), f|U(a), v|U(a))$ is regular. (Indeed, let $U(a)$ be the domain of definition of a Morse chart Ψ for f satisfying the condition (G) from page 34. Then $v|U(a)$ is the Riemannian gradient of $f|U(a)$ with respect to the Riemannian metric $\Psi_*^{-1}(2E)$, where E is the Euclidean metric in $U(a)$.) In view of point B this proves already our proposition for the case of manifolds without boundary.

E. It remains to prove our proposition in the case when W is a cobordism. In view of points B and D it suffices to prove that there is a neighbourhood U of $\partial_0 W$ and a metric $\mathcal{G}$ on U such that $v|U$ is the Riemannian gradient of $f|U$ with respect to $\mathcal{G}$. Similarly to point C above we can assume that in a neighbourhood V of ∂W we have $f'(x)(v(x)) = 1$. Let α be sufficiently small so that the map

$$\Psi : \partial_0 W \times [0, \alpha[\to W; \Psi(x, t) = \gamma(x, t; v)$$

is a diffeomorphism of $\partial_0 W \times [0, \alpha[$ onto its image. Then the Ψ-image of the product metric on $\partial_0 W \times [0, \alpha]$ will do. $\square$

2.3. Topological properties of the space of all f-gradients. Let W be a cobordism, and $f : W \to [a, b]$ a Morse function. We denote by $\text{Vect}^\infty(W)$ the vector space of all C^∞ vector fields on W, endowed with C^∞ topology, that is, the topology of uniform convergence with all partial derivatives. This topology is determined by a countable family of semi-norms and the resulting vector space is metrizable and complete.

Definition 2.12. We denote by $\mathrm{Vect}^\infty(W, S(f))$ the subspace of all the vector fields vanishing on $S(f)$.

Then $\mathrm{Vect}^\infty(W, S(f)) \subset \mathrm{Vect}^\infty(W)$ is a closed linear subspace of finite codimension $m(f) = \mathrm{card}\, S(f)$. The set $G(f)$ of all f-gradients is a subset of $\mathrm{Vect}^\infty(W, S(f))$.

Proposition 2.13. *The set $G(f)$ is an open convex subset of* $\mathrm{Vect}^\infty(W, S(f))$.

Proof. Convexity is easy to prove. To prove that $G(f)$ is open, we have to show that given an f-gradient v, any vector field $u \in \mathrm{Vect}^\infty(W, S(f))$ sufficiently close to v satisfies the conditions A) and B) of Proposition 2.5. As for condition B) this is obvious. Let us proceed to the condition A). The cobordism W is compact, therefore it suffices to prove that for every $a \in W$ the following property holds:

> ($\mathcal{P}$) There is a neighbourhood $U(a)$ of a and a neighbourhood $\mathcal{R}$ of v in $\mathrm{Vect}^\infty(W, S(f))$ such that for every $u \in \mathcal{R}$ and every $x \in U(a) \setminus S(f)$ we have:
> $$f'(x)(u(x)) > 0.$$

Let us first check this property for any point $a \notin S(f)$. Let $U(a)$ be a neighbourhood of a such that $\overline{U(a)} \cap S(f) = \varnothing$. There is a number $C > 0$ such that

$$f'(x)(v(x)) > C \quad \text{in} \quad \overline{U(a)}.$$

This inequality holds obviously if we replace v by any vector field sufficiently close to v in C^∞ topology, and thus we have verified the property ($\mathcal{P}$) for a.

Now let us proceed to the case $a \in S(f)$. The property ($\mathcal{P}$) being local, it suffices to consider the case of functions defined in an open subset of a Euclidean space. For a function $g : U \to E$ defined in an open set $U \subset \mathbf{R}^n$, with values in a normed vector space E, put

$$||g||_U = \sup_{x \in U} ||g(x)||.$$

Lemma 2.14. *Let $\phi : U \to \mathbf{R}$ be a C^∞ function defined in a neighbourhood U of 0 in $\mathbf{R}^m$. Assume that $\phi(0) = 0$ and that ϕ has a non-degenerate minimum in 0. Then there is a compact neighbourhood $V \subset U$ of 0 and a number $\delta > 0$ such that every C^∞ function $\psi : U \to \mathbf{R}$ satisfying*

$$\psi(0) = 0, \ \psi'(0) = 0, \ ||\phi'' - \psi''||_U < \delta, \ ||\phi''' - \psi'''||_U < \delta,$$

satisfies also $\psi(x) > 0$ if $x \in V$ and $x \neq 0$.

Proof. For any function ψ with $\psi(0) = 0, \psi'(0) = 0$ the Taylor formula reads as follows:

$$\psi(x) = \frac{1}{2}\psi''(0)(x,x) + \int_0^1 \frac{(1-t)^2}{2}\psi'''(tx)(x,x,x)dt$$

(see [**21**], Part 1, Th. 5.6.1). The function ϕ having a non-degenerate minimum in 0, there exists a compact neighbourhood V of 0 and a constant $\alpha > 0$ such that $\phi(x) \geqslant \alpha||x||^2$ for any $x \in V$. The Taylor formula above imply for the difference $\psi(x) - \phi(x)$ the following estimate:

$$\psi(x) - \phi(x) \leqslant \frac{1}{2}||\phi''(0) - \psi''(0)|| \cdot ||x||^2 + \frac{1}{2}\sup_{x\in U}||\psi'''(x) - \phi'''(x)|| \cdot ||x||^3$$

and our result follows with any δ such that $\frac{1}{2}\delta(1 + \text{diam}V) \leqslant \alpha/2$. □

At this moment we have already three types of vector fields, associated with a given Morse function f: Riemannian gradients, gradient-like vector fields, and f-gradients. We are going to add one more notion to this collection.

Definition 2.15. A vector field v is called *a weak gradient for f* or *a weak f-gradient* if

$$f'(x)(v(x)) > 0 \quad \text{for every} \quad x \notin S(f).$$

Thus the weak f-gradients form the widest class of vector fields along whose trajectories the function f increases.†

Lemma 2.16. *Let v be a weak f-gradient, and $p \in S(f)$. Then $v(p) = 0$.*

Proof. The proof of Lemma 2.2 is valid here without any changes. □

Let us denote the space of all gradient-like vector fields for f by $GL(f)$, the space of all Riemannian gradients for f by $GR(f)$, and the space of all weak gradients for f by $GW(f)$. We have the inclusions

$$GL(f) \subset GR(f) \subset G(f) \subset GW(f) \subset \text{Vect}^\infty(W, S(f)),$$

and in the rest of this section we study their topological properties.

Proposition 2.17. *The set $G(f)$ is dense in $GW(f)$.*

Proof. Let v be any weak gradient for f, and v_0 be any f-gradient. We shall now prove that for any $\epsilon > 0$ the vector field $w = v + \epsilon v_0$ is an f-gradient. Indeed, condition A) from Proposition 2.5 holds by obvious reasons. Proceeding to condition B) let us introduce the following notation:

$$\phi(x) = f'(x)(v(x)), \; \psi(x) = f'(x)(v_0(x)), \; \theta(x) = f'(x)(w(x)).$$

† The notion of weak gradient was introduced by D. Schütz in his paper [**141**].

For every critical point p of f the function $\phi(x)$ has a local minimum at p, therefore $\phi''(p)(h,h) \geqslant 0$ for every $h \in T_pW$. The non-degeneracy condition for v implies that $\psi''(p)(h,h) > 0$ for every $h \in T_pW$. Therefore

$$\theta''(p)(h,h) = \phi''(p)(h,h) + \epsilon\psi''(p)(h,h) > 0 \quad \text{for every} \quad h \in T_pW.$$

It remains to note that the vector field w converges to v as $\epsilon \to 0$. □

Now let us turn to the comparison of the spaces $GR(f), GL(f)$ and $G(f)$. It is clear that the main difference between the definitions of f-gradients, Riemannian gradients and gradient-like vector fields is in the conditions imposed on the behaviour of the vector field nearby the critical points. In particular, if $f : W \to \mathbf{R}$ has no critical points at all, then $GR(f) = GL(f) = G(f) = GW(f)$. A simple necessary condition for an f-gradient to be Riemannian gradient, respectively a gradient-like vector field, is given in the next lemma.

Lemma 2.18. 1. *Let w be a gradient-like vector field for a Morse function $f : W \to \mathbf{R}$, and $p \in S(f)$. Then the linear map $w'(p) : T_pW \to T_pW$ is diagonalizable over $\mathbf{R}$, and its eigenvalues are equal to ± 1.*
2. *Let v be a Riemannian gradient for f, and $p \in S(f)$. Then the linear map $v'(p) : T_pW \to T_pW$ is diagonalizable over $\mathbf{R}$.*

Proof. The first point is obvious from the definition. As for the second point, recall from Lemma 2.10 that the linear map $v'(p) : T_pW \to T_pW$ is symmetric with respect to the scalar product induced on T_pW by the Riemannian metric, therefore $v'(p)$ is diagonalizable. □

Example 2.19. Consider a Morse function

$$f : \mathbf{R}^2 \to \mathbf{R}, \quad f(x,y) = x^2 + y^2,$$

and three f-gradients:

$$u(x,y) = (x - y, x + y), \quad w(x,y) = (2x, y), \quad v(x,y) = (x,y);$$

the first is an f-gradient, the second is a Riemannian gradient for f (with respect to the Riemannian metric $dx^2 + 2dy^2$), and the third is a gradient-like vector field for f. Figure 4 shows the behaviour of their integral curves.

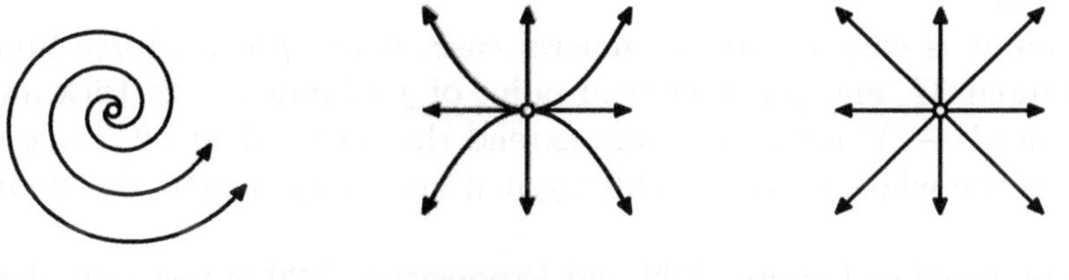

FIGURE 4.

Proposition 2.20. *If $f : W \to \mathbf{R}$ has at least one critical point, the subset $GL(f)$ is not dense in $GR(f)$. If f has at least one critical point p with*

$$\operatorname{ind} p \geqslant 2 \quad \text{or} \quad \operatorname{ind} p \leqslant \dim W - 2,$$

then $GR(f)$ is not dense in $G(f)$.

Proof. The following lemma says that the spectrum of a matrix is a continuous function of the matrix, the proof is easy and will be omitted.

Lemma 2.21. *Let A_n be a sequence of complex $m \times m$-matrices, converging to a matrix A. Let U be a neighbourhood of the spectrum $\sigma(A)$. Then $\sigma(A_n) \subset U$ for all n, except maybe a finite number of integers n.* □

Now the first part of the proposition follows easily, since for every gradient-like vector field v and $\lambda > 0$ the vector field $w = \lambda v$ is a Riemannian gradient, and for every critical point p the eigenvalues of $w'(p)$ have absolute value λ. The second part is more delicate; it follows from the next lemma.

Lemma 2.22. *Let $f : W \to \mathbf{R}$ be a Morse function on a cobordism W, and $p \in S(f)$. Assume that $\operatorname{ind} p \geqslant 2$ or $\operatorname{ind} p \leqslant \dim W - 2$. Then there is an f-gradient v such that $v'(p)$ has at least one imaginary eigenvalue.*

Proof. We shall do the proof in the case $\operatorname{ind} p \geqslant 2$, the case $\operatorname{ind} p \leqslant \dim W - 2$ is similar. Consider first the case when $M = \mathbf{R}^m$, and the Morse function is the standard quadratic form

$$\mathcal{Q}_k(x) = -\sum_{i=1}^{k} x_i^2 + \sum_{i=k+1}^{m} x_i^2.$$

Recall the $\mathcal{Q}_k$-gradient

$$\mathcal{A}_k(x) = (-x_1, \ldots, -x_k, x_{k+1}, \ldots, x_m)$$

in $\mathbf{R}^m$. Put

$$w(x) = (-x_1 + x_2, -x_2 - x_1, -x_3, \ldots, -x_k, x_{k+1}, \ldots, x_m).$$

Then w is also a $\mathcal{Q}_k$-gradient and the linear map $w'(0)$ has imaginary eigenvalues.

Now it is easy to do the general case, when f is a Morse function on a manifold M, and p is a critical point of f of index $\geqslant 2$. Pick any Morse chart $\Psi : U \to V$ for f at p, and extend the vector field $(\Psi^{-1})_*(w)$ defined in U to the whole of W, modifying it if necessary nearby the boundary of U.

The proof of Lemma 2.22 and Proposition 2.20 is now complete. □

Exercise 2.23. Prove that if W is a cobordism of dimension 1, and $f : W \to [a, b]$ is a Morse function, then $GR(f) = G(f)$.

Remark 2.24. Proposition 2.20 implies that $GR(f)$ is not dense in $G(f)$ for every Morse function f on a cobordism W of dimension $\geqslant 3$. For $\dim W = 2$ it remains to consider the case when every critical point of f has index 1; this is done in Exercise 1.26 of Chapter 3 (page 81).

2.4. Gradients of Morse forms. The proofs in the previous subsection are all based on the computations in local charts. Since locally any Morse form is the differential of a Morse function, the generalizations to the case of Morse forms are straightforward. In this subsection we gathered the corresponding definitions and theorems. Let M be a closed manifold without boundary, and ω a Morse form on M.

Definition 2.25. A C^∞ vector field v on M is an *ω-gradient* if

A) for every $x \notin S(\omega)$ we have $\omega(x)(v(x)) > 0$ and
B) for every $p \in S(\omega)$ the real-valued function $\phi(x) = \omega(x)(v(x))$ has a non-degenerate local minimum at p.

The space of all ω-gradients will be denoted $G(\omega)$.

Definition 2.26. Let $\langle \cdot, \cdot \rangle$ be a Riemannian metric on M. The vector field $\operatorname{grad} \omega$ on M, defined by the formula

$$\omega(x)(h) = \langle \operatorname{grad} \omega(x), h \rangle \quad \text{for} \quad x \in M, \ h \in T_xM,$$

is called the *Riemannian gradient* for ω (with respect to the Riemannian metric $\langle \cdot, \cdot \rangle$). The space of all Riemannian ω-gradients will be denoted $GR(\omega)$.

A Riemannian gradient for ω is an ω-gradient (similarly to Proposition 2.9).

Definition 2.27. A C^∞ vector field v is called a *gradient-like vector field for* ω, if

1) $\omega(x)(v(x)) > 0$ whenever $x \notin S(\omega)$, and
2) for every $p \in S(f)$ there is a chart $\Psi : U \to V \subset \mathbf{R}^m$ for M at p, such that $\Psi(p) = 0$ and

$$(\Psi^{-1})^*(\omega)(x_1, \ldots, x_m) = -2 \sum_{i=1}^{k} x_i dx_i + 2 \sum_{i=k+1}^{m} x_i dx_i,$$

$$\Psi_*(v)(x_1, \ldots, x_m) = (-x_1, \ldots, -x_k, x_{k+1}, \ldots, x_m)$$

(where $k = \operatorname{ind} p$).

A chart Ψ satisfying the above condition will be called *a Morse chart for* (ω, v).

The space of all gradient-like vector fields for ω will be denoted $GL(\omega)$.

Similarly to Proposition 2.11, any gradient-like vector field for ω is a Riemannian gradient for ω with respect to some Riemannian metric on M.

Definition 2.28. A vector field v on M is called a *weak gradient for* ω if

$$\omega(x)(v(x)) > 0 \text{ whenever } x \notin S(\omega).$$

The space of all weak gradients for ω will be denoted $GW(\omega)$.

Definition 2.29. Denote by $\mathrm{Vect}^\infty(M, S(\omega))$ the space of all C^∞ vector fields on M vanishing on $S(\omega)$.

Then $\mathrm{Vect}^\infty(M, S(\omega))$ is a closed vector subspace of codimension $m(\omega)$ in the space $\mathrm{Vect}^\infty(M)$ of all C^∞ vector fields on M. We have the inclusions

$$GL(\omega) \subset GR(\omega) \subset G(\omega) \subset GW(\omega) \subset \mathrm{Vect}^\infty(M, S(\omega)).$$

The proofs of the next propositions are completely similar to the proofs of Propositions 2.13, 2.17 and 2.20.

Proposition 2.30. *The set $G(\omega)$ is a convex open subset of $\mathrm{Vect}^\infty(M, S(\omega))$. The set $G(\omega)$ is dense in $GW(\omega)$.* □

Proposition 2.31. *If ω has at least one zero, then the set $GL(\omega)$ is not dense in $GR(\omega)$. If ω has at least one zero p with*

$$\mathrm{ind}\, p \geqslant 2 \quad \text{or} \quad \mathrm{ind}\, p \leqslant \dim M - 2$$

then the set $GR(\omega)$ is not dense in $G(\omega)$. □

2.5. Gradients of circle-valued Morse functions. The differentials of circle-valued Morse functions constitute the most important class of Morse forms. We shall be dealing mostly with this particular case, and for the convenience of future references we gathered in this subsection the corresponding definitions and results.

For a circle-valued Morse function $f : M \to S^1$ on a closed manifold M, its differential df is a closed 1-form; its value at a point $x \in M$ will be also denoted by $f'(x) \in T^*_x(M)$.

Definition 2.32. A C^∞ vector field v on M is an *f-gradient* if

A) for every $x \notin S(f)$ we have $f'(x)(v(x)) > 0$ and
B) for every $p \in S(f)$ the real-valued function $\phi(x) = f'(x)(v(x))$ has a non-degenerate local minimum at p.

The space of all f-gradients will be denoted $G(f)$.

Definition 2.33. Let $\langle\cdot,\cdot\rangle$ be a Riemannian metric on M. The vector field $\operatorname{grad} f$ on M, defined by the formula

$$f'(x)(h) = \langle \operatorname{grad} f(x), h\rangle \quad \text{for} \quad x \in M,\ h \in T_xM$$

is called the *Riemannian gradient* for f (with respect to the metric $\langle\cdot,\cdot\rangle$). The space of all Riemannian f-gradients will be denoted $GR(f)$.

Definition 2.34. A C^∞ vector field v is called a *gradient-like vector field for* f, if

1) $f'(x)(v(x)) > 0$ whenever $x \notin S(f)$, and
2) for every $p \in S(f)$ there is a chart $\Psi : U \to V \subset \mathbf{R}^m$ for M at p, such that $\Psi(p) = 0$ and

$$d(f \circ \Psi^{-1})(x_1, \dots, x_m) = -2\sum_{i=1}^{k} x_i dx_i + 2\sum_{i=k+1}^{m} x_i dx_i,$$

$$\Psi_*(v)(x_1, \dots, x_m) = (-x_1, \dots, -x_k, x_{k+1}, \dots, x_m)$$

(where $k = \operatorname{ind} p$).

A chart Ψ satisfying the above condition will be called *a Morse chart for* (f, v).

The space of all gradient-like vector fields for f will be denoted $GL(f)$.

Any gradient-like vector field for f is a Riemannian gradient for f with respect to some Riemannian metric on M.

Definition 2.35. A vector field v on M is called a *weak gradient for* f if

$$f'(x)(v(x)) > 0 \text{ whenever } x \notin S(f).$$

The space of all weak gradients for ω will be denoted $GW(f)$.

Definition 2.36. Denote by $\operatorname{Vect}^\infty(M, S(f))$ the space of all C^∞ vector fields on M vanishing on $S(f)$.

Then $\operatorname{Vect}^\infty(M, S(f))$ is a closed vector subspace of codimension $m(f)$ in the space $\operatorname{Vect}^\infty(M)$ of all C^∞ vector fields on M. We have

$$GL(f) \subset GR(f) \subset G(f) \subset GW(f) \subset \operatorname{Vect}^\infty(M, S(f)).$$

The next propositions are particular cases of Propositions 2.30, respectively 2.31.

Proposition 2.37. *The set* $G(f)$ *is a convex open subset of the space* $\operatorname{Vect}^\infty(M, S(f))$. *The set* $G(f)$ *is dense in* $GW(f)$. □

Proposition 2.38. *If f has at least one critical point, then the set $GL(f)$ is not dense in $GR(f)$. If f has at least one zero p with*

$$\operatorname{ind} p \geqslant 2 \quad or \quad \operatorname{ind} p \leqslant \dim M - 2,$$

then the set $GR(f)$ is not dense in $G(f)$. □

3. Morse functions on cobordisms

Theorem 1.25 (page 47) says that Morse functions form an open and dense subset of the space of all C^∞ functions on a closed manifold M. The aim of this section is to generalize this result to the case of cobordisms.

Definition 3.1. A C^∞ function $\phi : W \to [a, b]$ on a cobordism W is called *regular* if $\phi^{-1}(a) = \partial_0 W$, $\phi^{-1}(b) = \partial_1 W$. The space of all regular functions endowed with the C^∞ topology is denoted by $\mathcal{R}(W)$.

In particular any Morse function on W is regular. We shall prove that the set of Morse functions $\mathcal{M}(W)$ is an open and dense subset of $\mathcal{R}(W)$. The argument is in a sense more delicate than the one we used for Theorem 1.25 in view of the additional restriction that the functions in question are constant on each component of the boundary. As the first step let us show that regular functions exist:

Lemma 3.2. *Let $a < b$. There are regular functions $\phi : W \to [a, b]$.*

Proof. Pick any downward normal field v on W. Define a C^∞ function f_0 in a neighbourhood of $\partial_0 W$ by the following formula:

$$f_0(x) = a + \tau(x, v)$$

(recall that $\tau(x, v)$ stands for the moment of exit of the v-trajectory starting at x). Define a C^∞ function f_1 in a neighbourhood of $\partial_1 W$ by the following formula:

$$f_1(x) = b - \tau(x, -v).$$

Pick any smooth function $\psi : W \setminus \partial W \to]a, b[$. Glue the functions ψ, f_0, f_1 together with the help of a usual argument using partitions of unity and the proof is complete. □

Theorem 3.3. *Let W be a cobordism. The set $\mathcal{M}(W)$ is open and dense in $\mathcal{R}(W)$ with respect to C^∞ topology.*

Proof. The proof that $\mathcal{M}(W)$ is open repeats the proof of Lemma 1.26 without any changes. Now let us prove that $\mathcal{M}(W)$ is dense in $\mathcal{R}(W)$. Let $f : W \to [a, b]$ be any regular function on W. Pick any upward normal vector field v on W. We begin by perturbing the function f in such a way that the resulting function has no critical points in a neighbourhood of ∂W:

Lemma 3.4. *In any neighbourhood $U \subset \mathcal{R}(W)$ of f there is a regular function g such that*

$$g'(x)(v(x)) > 0 \quad \textit{for every} \quad x \in \partial W. \tag{4}$$

Proof. Let $\epsilon > 0$. The function $k(x) = f(x) + \epsilon\tau(x, v)$ (defined in some neighbourhood of $\partial_0 W$) has the property $k'(x)(v(x)) > 0$ for every $x \in \partial_0 W$. (Indeed, for every v-trajectory γ starting at a point $x \in \partial_0 W$ the function $f(\gamma(t))$ has a local minimum at the point $t = 0$, since f is regular, therefore the derivative of the function $k(\gamma(t))$ at zero is strictly positive.) Put

$$g_0(x) = \epsilon h(x)\tau(x, v) + f(x)$$

where h is a C^∞ function which is equal to 1 in a neighbourhood of $\partial_0 W$ and has its support in the domain of definition of $\tau(x, v)$. The function g_0 satisfies the condition

$$g_0'(x)(v(x)) > 0$$

for x in a neighbourhood of $\partial_0 W$ and is arbitrarily close to f if ϵ is sufficiently small. Applying a similar argument to the upper component of the boundary, we obtain a function g which was sought. □

It suffices therefore to prove the next lemma.

Lemma 3.5. *Let g be a regular function satisfying (4). Let $\mathcal{V} \subset \mathcal{R}(W)$ be a neighbourhood of g. Then there is a Morse function $h : W \to [a, b]$ in $\mathcal{V}$.*

Proof. Put

$$W_\delta = g^{-1}([a + \delta, b - \delta]).$$

Let $\delta > 0$ be so small that $g'(x)(v(x)) > d > 0$ in $W \setminus W_\delta$. Pick a smooth function $\alpha : W \to [0, 1]$ such that

$$\alpha(x) = 1 \quad \text{if} \quad x \in W_\delta, \qquad \text{and} \qquad \operatorname{supp} \alpha \subset W \setminus W_{\delta/2}.$$

Proposition 1.27 of the present Chapter implies that there is a function $\widetilde{g} : W^\circ \to \mathbf{R}$ such that:

(1) The restriction $\widetilde{g}|W_\delta$ has only non-degenerate critical points.
(2) The function

$$h(x) = g(x) + \alpha(x) \cdot (\widetilde{g}(x) - g(x))$$

does not have critical points in $W_{\delta/2} \setminus W_\delta$.
(3) $h \in \mathcal{V}$.

The function h satisfies the requirements of our lemma. The proof of Theorem 3.3 is now complete. □

We finish this subsection with a result to be used in the sequel.

Definition 3.6. Let W be a cobordism. A C^∞ function g defined in a neighbourhood U of $\partial_0 W$ will be called a *bottom germ* if

(i) g has no critical points, and

(ii) for some $a \in \mathbf{R}$ we have: $g(U) \subset [a, \infty[$, and $g^{-1}(a) = \partial_0 W$.

We shall say that two germs are *equivalent* if they coincide in a neighbourhood of $\partial_0 W$. Similarly one defines the notion of a *top germ*.

Any Morse function on W generates (by restriction) a bottom germ and a top germ.

Proposition 3.7. *Assume that the cobordism W is endowed with an upward normal vector field v such that all its trajectories reach the upper boundary.*

Let f_0 be any bottom germ, and f_1 any top germ. Assume that $f_0(\partial_0 W) < f_1(\partial_1 W)$.

Then there is a Morse function ϕ on W such that v is a ϕ-gradient and the germs generated by ϕ are equivalent to f_0, f_1.

Proof. Multiplying if necessary the vector field v by a nowhere vanishing positive function we can assume that:

(1) $f_0'(x)(v(x)) = 1$ for every x in a neighbourhood of $\partial_0 W$.
(2) $f_1'(x)(v(x)) = 1$ for every x in a neighbourhood of $\partial_1 W$.
(3) For every $x \in \partial_0 W$ we have $\tau(x, v) \geqslant 1$.

Assuming these three conditions we shall construct a diffeomorphism of the product cobordism $\partial_0 W \times [0,1]$ onto W, which sends each segment $x \times [0,1]$ of the product cobordism to the trajectory of v starting at x (suitably reparameterized). Here is the formula:

$$\Phi : \partial_0 W \times [0,1] \to W, \quad \Phi(x,t) = \gamma\Big(x, \chi(t, \tau(x,v)); v\Big)$$

where $\chi : [0,1] \times [1, \infty[\to \mathbf{R}_+$ is a C^∞ function having the following property:

$(\mathcal{P})$: For every given y the function $t \mapsto \chi(t,y)$ is a diffeomorphism of $[0,1]$ onto $[0,y]$ such that

$$\frac{\partial \chi}{\partial t} = 1 \text{ in some neighbourhood of } \{0,1\}.$$

See the graph of the function $t \mapsto \chi(t,y)$ in Figure 5.

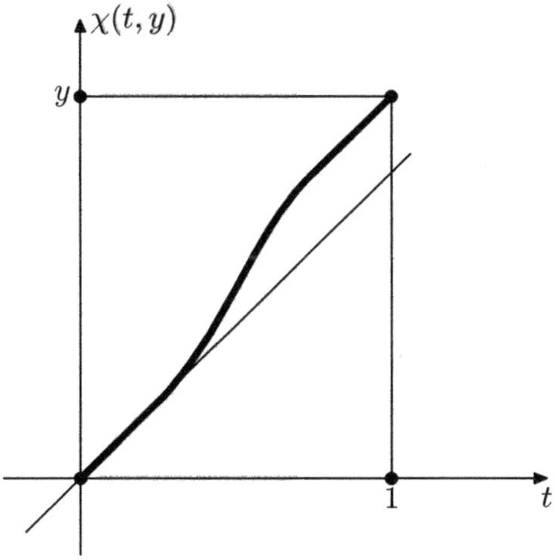

FIGURE 5.

The existence of such a function is almost obvious from the picture. We postpone the precise proof until Lemma 3.8, and now observe that the property $(\mathcal{P})$ implies that the map Φ is indeed a diffeomorphism of $\partial_0 W \times [0,1]$ onto W. We have:

$$f_0 \circ \Phi(x,t) = t + a, \quad \text{for} \quad t \quad \text{in a neighbourhood of} \quad 0;$$
$$f_1 \circ \Phi(x,t) = b + t - 1, \quad \text{for} \quad t \quad \text{in a neighbourhood of} \quad 1.$$

Pick a C^∞ diffeomorphism $\theta : [0,1] \to [a,b]$ such that $\theta'(t) = 1$ for any t in a neighbourhood of $\{0,1\}$. The composition of the maps

$$W \xrightarrow{\Phi^{-1}} \partial_0 W \times [0,1] \to [0,1] \xrightarrow{\theta} [a,b]$$

has the required properties. To complete the proof it remains therefore to prove the following lemma.

Lemma 3.8. *There exists a function χ satisfying the property $(\mathcal{P})$ above.*

Proof. Pick a C^∞ function $h : \mathbf{R}_+ \to [0,1]$ such that

$$\begin{aligned} h(t) &= 0 && \text{for} \quad t \leqslant 1/3, \\ h(t) &= 1 && \text{for} \quad 2/3 \leqslant t \leqslant 1, \\ h'(t) &> 0 && \text{for} \quad 1/3 < t < 2/3. \end{aligned}$$

Put

$$\chi(t, y) = t + (y-1)h(t);$$

it is obvious that $(\mathcal{P})$ holds for this function. □

CHAPTER 3

Gradient flows of real-valued Morse functions

Let $f : W \to [a, b]$ be a Morse function on a cobordism W and v an f-gradient. The dynamical system generated by v is called a *gradient flow*.[†] This chapter is about basic geometric constructions related to gradient flows, like stable manifolds, gradient descent maps, and others. We will see that each critical point of f is a hyperbolic fixed point of the flow generated by v, so the local theory of hyperbolic dynamical systems can be applied. In the first section we give a brief account of the basic results of this theory with a special emphasis on the applications to gradient flows. We discuss the Hadamard-Perron theorem, the Lyapunov stability theory and the Grobman-Hartman theorem. In the second section we study the global stable manifolds of the critical points of Morse functions, and the third section is devoted to the notion of gradient descent, one of the main tools of the Morse theory.

1. Local properties of gradient flows

A linear map of a finite-dimensional real vector space to itself is called *elementary* if it does not have eigenvalues with zero real part. It is called *negative elementary*, respectively *positive elementary*, if the real part of every eigenvalue is strictly negative, respectively strictly positive. Observe that A is elementary if and only if the linear map $\exp(A)$ is *hyperbolic*, that is, the spectrum of $\exp(A)$ does not contain complex numbers of absolute value equal to 1. A zero p of a vector field v on a manifold M is called *elementary* if the differential $v'(p)$ is an elementary linear map of T_pM to itself.

The local structure of a flow generated by a vector field in a neighbourhood of an elementary zero is a classical and thoroughly studied subject. The first substantial progress in this domain was made by A. M. Lyapunov in his doctoral thesis "The General Problem of the Stability of Motion" **[88]** (1892). He developed efficient methods for investigation of stability of equilibrium states of dynamical systems (*Lyapunov's direct method*). He

[†] It is worth noting that for $\partial W \neq \varnothing$ the gradient flow is only a *partial flow*, that is, the integral curves of v may not be defined on the whole of $\mathbf{R}$.

proved in particular that any negative elementary zero of a vector field is an asymptotically stable equilibrium state. The Lyapunov direct method uses the construction of a function which is increasing along each non-constant trajectory of the flow; these ideas are among the predecessors of the Morse theory.

The next step in the development of the local theory of hyperbolic dynamical systems was made by J. Hadamard. In his paper [**52**] (1902) he considered the dynamical system determined by a diffeomorphism $A : \mathbf{R}^2 \to \mathbf{R}^2$ to itself with a hyperbolic fixed point in the origin. Assuming that the eigenvalues λ, μ of $A'(0)$ are real, and satisfy $\lambda < 1 < \mu$, he constructed two A-invariant submanifolds, S (*the stable invariant manifold*) and U (*the unstable invariant manifold*), such that S and U both contain the origin, $A|U$ is expanding, and $A|S$ is contracting. In 1928 O. Perron, using a different method, constructed stable and unstable manifolds for elementary zeros of vector fields in Euclidean space of arbitrary dimension. The Hadamard-Perron theorem on the existence of stable manifolds is now a basic working tool in the theory of hyperbolic dynamical systems. Our reference for this theorem is the book [**1**] of R. Abraham and J. Robbin. A new elegant proof of the Hadamard-Perron theorem was given in 1970 by M. C. Irwin in his article [**66**]. We recommend to the reader the systematic and comprehensive exposition of this proof in the book [**67**] (see also Chapter 2 of the book [**122**]).

The Hadamard-Perron theorem suggests that in a neighbourhood of an elementary zero the dynamical system generated by the flow is equivalent to the product of its stable and unstable part. This is true, but only in the topological category: the map which gives the equivalence is in general a non-differentiable homeomorphism. Here is an equivalent statement: the flow generated by a vector field in a neighbourhood of an elementary zero is topologically equivalent to its linear part. This fundamental result is due to D. M. Grobman [**50**], [**51**] and is often referred to as the Grobman-Hartman theorem. We discuss it in Subsection 1.3.

1.1. The Hadamard-Perron theorem. Let M be a manifold without boundary. Let v be a C^∞ vector field on M, and p a zero of v.

Definition 1.1. 1. The set of all points $x \in M$ such that the v-trajectory starting at x is defined on $\mathbf{R}_+$ and $\lim_{t\to\infty} \gamma(x,t;v) = p$ is called the *stable set of v at p* and denoted $W^{st}(v,p)$.
2. The set of all points $x \in M$ such that the $(-v)$-trajectory starting at x is defined on $\mathbf{R}_+$ and $\lim_{t\to\infty} \gamma(x,t;-v) = p$, is called the *unstable set of v at p* and denoted $W^{un}(v,p)$.

Stable sets are often called *stable manifolds* (similarly for the unstable sets). This terminology is justified by the fact that often these sets are submanifolds of the ambient manifold, as it is the case for example in the situation of the Hadamard-Perron theorem.To state this theorem we need some recollections from linear algebra. Let $L : E \to E$ be an elementary linear map. It is easy to check that there is a unique direct sum decomposition $E = E^- \oplus E^+$ into L-invariant vector subspaces, such that $L|E^- : E^- \to E^-$ is negative elementary and $L|E^+ : E^+ \to E^+$ is positive elementary. The space E^- (respectively E^+) is called the *negative subspace* (respectively the *positive subspace*) with respect to to L. The dimension of E^- will be also called the *index* of L.

Definition 1.2. Let v be a C^∞ vector field on a manifold M, and let p be an elementary zero of v. The negative and positive subspaces of T_pM with respect to $v'(p)$ will be denoted $T_p^-(M,v)$, respectively $T_p^+(M,v)$. The index of $v'(p)$ will be called the *index of* p.

Here is the statement of the Hadamard-Perron theorem.

Theorem 1.3. *Let v be a C^∞ vector field on a manifold M, and p an elementary zero of v. Then there is a neighbourhood U of p, such that*

(1) *$W^{st}(v|U,p)$ and $W^{un}(v|U,p)$ are submanifolds of U.*

(2) $W^{st}(v|U,p) \cap W^{un}(v|U,p) = \{p\}$.

(3) $T_pW^{st}(v|U,p) = T_p^-(M,v), \quad T_pW^{un}(v|U,p) = T_p^+(M,v)$.

For the proof we refer to [**1**], Theorem 27.1 of Chapter 6. This theorem is formulated in [**1**] for the case of vector fields of class C^r with $r < \infty$; the C^∞ case follows easily. The point (3) above is not included in the statement of the Theorem 27.1 of [**1**]; it follows from the Lemma 27.2 of the book, in which the stable manifolds are described in local coordinates.

Example 1.4. Let v be a *linear vector field* on a finite-dimensional real vector space E, that is, $v(x) = Lx$ where $L : E \to E$ is a linear map. If L is an elementary linear map, then 0 is obviously an elementary zero of v and the space E itself is a neighbourhood satisfying the conditions (1) – (3) above. We have

$$W^{un}(v,0) = E^+, \quad W^{st}(v,0) = E^-.$$

In general however the neighbourhood U from Theorem 1.3 above is not equal to M. The next example is borrowed from the book [**1**, figure 27-2]; here the stable set of the point P coincides with the unstable set and is homeomorphic to the Bernoulli lemniscate, therefore is not a topological manifold:

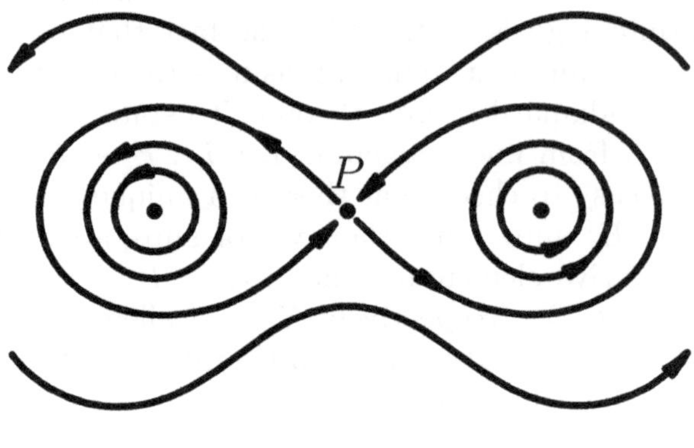

FIGURE 6.

Let us call a neighbourhood U of p *regular* if the sets $W^{st}(v|U,p)$, $W^{un}(v|U,p)$ satisfy the conditions (1)–(3) of Theorem 1.3. It follows from this theorem that for every elementary zero p of v there are arbitrarily small regular neighbourhoods. Moreover, one can prove that any open neighbourhood of p which is a subset of a regular neighbourhood is regular itself (see Exercise 1.11). For a given v the manifold $W^{st}(v|U,p)$ certainly depends on U, but it is easy to prove that its *germ* at the point p does not. This is the subject of the next corollary.

Corollary 1.5. *Let U_1, U_2 be regular neighbourhoods of an elementary zero p of a vector field v. Put*

$$N_1 = W^{st}(v|U_1,p), \ N_2 = W^{st}(v|U_2,p).$$

Then there is a submanifold $N \subset M$ of dimension $\operatorname{ind} p$ *such that $p \in N \subset N_1 \cap N_2$, and N is an open subset of N_1 and of N_2.*

Proof. Pick any regular neighbourhood $U \subset U_1 \cap U_2$ of p. The stable set $N = W^{st}(v|U,p)$ is then a manifold of dimension $k = \operatorname{ind} p$, and we have obviously $N \subset N_1, N \subset N_2$, therefore N is an open subset of both N_1 and N_2. □

The manifold $W^{st}(v|U,p)$ (where U is a regular neighbourhood of p) will be called the *local stable manifold for v at p* to distinguish it from the *global* stable set $W^{st}(v,p)$. We shall sometimes denote the local stable manifold by $W^{st}_{loc}(v,p)$, if the value of U is clear from the context or not important.

Similar properties hold for the unstable manifolds. The local unstable manifold for v at p will be denoted by $W^{un}_{loc}(v,p)$.

1.2. Lyapunov discs. Let p be an elementary zero of a vector field v. In this section we shall study the restriction of the flow to the stable and unstable manifolds of v at p. The equilibrium point p of the restriction of v to the stable manifold is indeed *stable*: as we shall see later on in this section there are arbitrarily small compact v-invariant neighbourhoods of p in $W^{st}_{loc}(p)$. The main aim of this subsection is the construction and investigation of special types of such invariant neighbourhoods: *the Lyapunov discs.*

We begin with preliminaries about vector fields in $\mathbf{R}^k$. Let w be a C^∞ vector field on the closed disc $D = D^k(0,R)$, which *points inward* D, that is

$$(1) \qquad \forall a \in S^{k-1}(0,R) = \partial D^k \ : \ \langle a, w(a)\rangle < 0$$

(where $\langle \cdot, \cdot \rangle$ denotes the Euclidean scalar product in $\mathbf{R}^k$). The next lemma is an easy exercise in the theory of differential equations in Euclidean space; the proof will be omitted (see Figure 7).

FIGURE 7.

Lemma 1.6. *Let w be a vector field satisfying (1). Then*

1. *For every $x \in D$ the w-trajectory $\gamma(x, \cdot; w)$ is defined on the whole of $\mathbf{R}_+$.*
2. *For every $x \in D$ we have:*

$$\gamma(x,t;w) \in D \setminus \partial D \quad \textit{for} \quad t > 0.$$

3. *$D \setminus \partial D$ is w-invariant.*
4. *Any w-trajectory can intersect ∂D only once.* □

Now let M be a manifold, and v a C^∞ vector field on M. Let $\phi : D = D^k(0,R) \hookrightarrow M$ be a C^∞ embedding of a closed disc to M, let $\mathcal{D}$ denote the image of this embedding. We say that *v is tangent to $\mathcal{D}$* if for every $x \in D^k(0,R)$ we have: $v(\phi(x)) \in T_{\phi(x)}(\mathcal{D})$. If v is tangent to $\mathcal{D}$, then $v|\mathcal{D}$ is the ϕ-image of a C^∞ vector field w on D, which is determined by the following formula:

$$\phi'(y)(w(y)) = v(\phi(y)) \quad \text{for} \quad y \in D.$$

The vector field w will be denoted by $\phi_*^{-1}(v)$. We shall say that *v points inward $\mathcal{D}$* if v is tangent to $\mathcal{D}$ and $\phi_*^{-1}(v)$ has the property (1). The following lemma is an immediate consequence of Lemma 1.6.

Lemma 1.7. *Let v be a tangent vector field to $\mathcal{D}$, which points inward $\mathcal{D}$. Then*

(1) *$\mathcal{D}$ and $\mathcal{D} \setminus \partial\mathcal{D}$ are v-invariant.*
(2) *Any v-trajectory can intersect $\partial\mathcal{D}$ only once.* □

Now we can introduce *Lyapunov charts* and *Lyapunov discs.*

Definition 1.8. Let v be a C^∞ vector field on a manifold M and p be an elementary zero of v. *A stable Lyapunov chart* for v at p is a C^∞ embedding $j : D \hookrightarrow M$ of a k-dimensional closed disc $D = D^k(0,R)$ to M (where $k = \operatorname{ind} p$) such that:

(1) $j(0) = p$.
(2) For some regular neighbourhood U of p we have $\mathcal{D} = j(D) \subset W^{st}(v|U, p)$.
(3) v points inward $\mathcal{D}$.

The image $\mathcal{D}$ of a stable Lyapunov chart is called *a stable Lyapunov disc.* A stable Lyapunov chart for $(-v)$ at p is called an *unstable Lyapunov chart for v at p.* The image of an unstable Lyapunov chart is called *an unstable Lyapunov disc.*

In Figure 8 we have depicted several trajectories and a Lyapunov disc (shaded) for a stable manifold of a zero of a vector field.

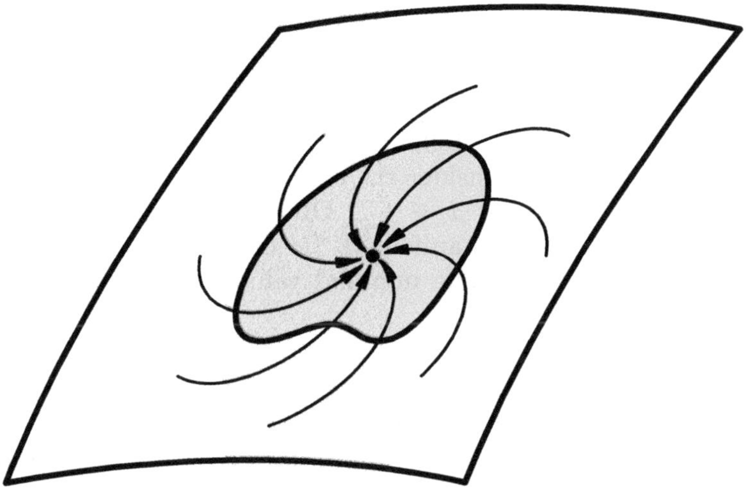

FIGURE 8.

Proposition 1.9. *Let v be a C^∞ vector field on a manifold M and p an elementary zero of v. For any neighbourhood A of p in M there is a stable Lyapunov disc $\mathcal{D}$ for v at p such that $\mathcal{D} \subset A$.*

Proof. Choose any chart $\Psi : U \longrightarrow V$ of the manifold $W^{st}_{loc}(v, p)$ at p, with $\Psi(p) = 0$. The vector field $w = \Psi_*(v|U)$ is of class C^∞ in V, and the eigenvalues of $w'(0)$ have negative real part. According to a theorem due to Lyapunov (see for example **[6]**, §22 and §23) there is a positive definite quadratic form ϕ on $\mathbf{R}^k$ (where $k = \operatorname{ind} p$) which is a Lyapunov function for w in a neighbourhood of 0, that is,

$$\phi'(x)(w(x)) < 0$$

for every $x \neq 0$ in some neighbourhood V_0 of 0. For any $\lambda > 0$ the set $B = \phi^{-1}([0, \lambda])$ is a k-dimensional ellipsoid, and there is a diffeomorphism

$$g : D = D^k(0, R) \xrightarrow{\approx} B,$$

such that $g(0) = 0$. Choose λ sufficiently small so that $B \subset V_0$; the map $\Psi^{-1} \circ g$ is then a stable Lyapunov chart for v at p, and $\mathcal{D} = \Psi^{-1}(g(D)) \subset R$ is the Lyapunov disc sought. □

Lemma 1.10. *Let v be a C^∞ vector field on a manifold M and p an elementary zero of v. Let $\mathcal{D}$ be a stable Lyapunov disc for v at p. Then*

(1) *$\mathcal{D}$ and $\mathcal{D} \setminus \partial\mathcal{D}$ are v-invariant.*
(2) *Any v-trajectory intersects $\partial\mathcal{D}$ at most once.*
(3) *For any $x \in W^{st}(v,p) \setminus \mathcal{D}$ there exists a unique $t > 0$ such that $\gamma(x,t;v) \in \partial\mathcal{D}$.*

Proof. The first two points follow immediately from Lemma 1.7. For the proof of (3) it suffices to consider the case when $\mathcal{D}$ is a subset of a regular neighbourhood U of p and $x \in U$. Observe that the set $\mathcal{D} \setminus \partial\mathcal{D}$ is a neighbourhood of p in the manifold $W^{st}(v|U,p)$. Therefore there is $\theta > 0$ such that $\gamma(x,\theta;v) \in \mathcal{D}$. The minimal value of such θ is the number t sought. □

Exercise 1.11. Let p be an elementary zero of a vector field v on a manifold M. Let U be a regular neighbourhood of p (for the terminology see page 70). Prove that any neighbourhood $V \subset U$ of p is also regular.

Recall ([**6**], §22 and §23) that a zero p of a vector field v on a manifold M is called *stable* (or *Lyapunov stable*) if for every neighbourhood V of p there is a neighbourhood $V' \subset V$ such that every v-trajectory starting at a point of $x \in V'$ stays at V. The zero p is called *asymptotically stable* if it is stable, and there is a neighbourhood W of p such that for every $x \in W$ we have $\gamma(x,t;v) \to p$ as $t \to \infty$. The following corollary is easy to prove using the properties of Lyapunov discs.

Corollary 1.12. *Let p be an elementary zero of a vector field v on a manifold M. Then the point p is an asymptotically stable zero of the restriction of v to any local stable manifold $W^{st}_{loc}(v,p)$.* □

Now we shall use Lyapunov charts to deduce a fundamental property of stable sets.

Proposition 1.13. *Let M be a closed manifold. Let v be a C^∞ vector field, and p an elementary zero of v of index k. The set $W^{st}(v,p)$ is the image of a C^∞ immersion of an open k-dimensional disc to M.*

Proof. Pick a stable Lyapunov disc $\mathcal{D}_0$ for v at p. Put

$$\mathcal{B}_0 = \mathcal{D}_0 \setminus \partial\mathcal{D}_0.$$

For $\lambda > 0$ put

$$\mathcal{D}_\lambda = \Phi(-v,\lambda)(\mathcal{D}_0) = \{\gamma(x,\lambda;-v) \mid x \in \mathcal{D}_0\}.$$

(Recall that $\Phi(-v,\cdot)$ denotes the one-parameter group of diffeomorphisms generated by the vector field $-v$.) Then $\mathcal{D}_\lambda$ is an embedded k-dimensional closed disc, v is tangent to $\mathcal{D}_\lambda$ and points inward $\mathcal{D}_\lambda$. The set $W_\lambda = \mathcal{D}_\lambda \setminus \mathcal{B}_0$ is a k-dimensional cobordism with

$$\partial_1 W_\lambda = \partial\mathcal{D}_0, \ \partial_0 W_\lambda = \partial\mathcal{D}_\lambda.$$

For $x \in W^{st}(v,p) \setminus \mathcal{B}_0$ denote by $t(x)$ the moment when $\gamma(x,\cdot;v)$ reaches $\partial\mathcal{D}_0$, and by $A(x)$ the point of intersection of this curve with $\partial\mathcal{D}_0$. The restrictions

$$A|W_\lambda : W_\lambda \to \partial\mathcal{D}_0, \ t|W_\lambda : W_\lambda \to \mathbf{R},$$

are C^∞ maps (since $\tau(x)$ is the exit moment and $A(x)$ is the exit point for the v-trajectory $\gamma(x,\cdot;v)$ on the cobordism W_λ).

Let $\lambda, \mu > 0$ and let $f : [0,\lambda] \to [0,\mu]$ be any C^∞ map, which equals identity in a neighbourhood of 0. Define a map

$$G_f : \mathcal{D}_\lambda \to \mathcal{D}_\mu$$

by the following formula:

$$\begin{aligned} G_f(x) &= x && \text{for} \quad x \in \mathcal{D}_0, \\ G_f(x) &= \gamma\big(A(x), f(t(x)); -v\big) && \text{for} \quad x \in \mathcal{D}_\lambda \setminus \mathcal{D}_0. \end{aligned}$$

It is clear that G_f is a C^∞ map.

Lemma 1.14. *If $f : [0,\lambda] \to [0,\mu]$ is a diffeomorphism, then G_f is a diffeomorphism of $\mathcal{D}_\lambda$ onto $\mathcal{D}_\mu$.*

Proof. The map $G_{f^{-1}}$ is the inverse for G_f. □

Now we can construct an immersion of the open k-dimensional disc $\mathcal{B}_1 = \mathcal{D}_1 \setminus \partial\mathcal{D}_1$ onto $W^{st}(v,p)$. Choose any C^∞ diffeomorphism $h : [0,1[\to \mathbf{R}_+$, such that $h(y) = y$ for $y \in \mathbf{R}$ in a neighbourhood of 0. Define a C^∞ map $G : \mathcal{B}_1 \to M$ by the following formula:

$$\begin{aligned} G(x) &= x && \text{for} \quad x \in \mathcal{B}_0, \\ G(x) &= \gamma\big(A(x), h(t(x)); -v\big) && \text{for} \quad x \in \mathcal{B}_1 \setminus \mathcal{D}_0. \end{aligned}$$

The restriction of G to any disc $\mathcal{D}_\lambda$ equals G_f where $f = h|[0,\lambda]$; therefore $G(\mathcal{D}_\lambda) = \mathcal{D}_{h(\lambda)}$ and

$$G(\mathcal{B}_1) = \bigcup_{0<\lambda<1} \mathcal{D}_{h(\lambda)} = W^{st}(v,p).$$

The map G is an injective immersion as it follows from Lemma 1.14, and the proof of our proposition is now complete. □

Intuitively the map G constructed above can be described as follows. For every flow line Γ of $(-v)$ belonging to $W^{st}(v,p)$, let Γ_1 be the intersection of Γ with $\mathcal{D}_1$. The map G "stretches" Γ_1 so as to cover the whole curve Γ. This construction will be used several times in the sequel, and we reserve for it the term *the stretching construction.*

The subject of the next proposition is the continuous dependence of Lyapunov charts on the vector field. Let M be a closed manifold and p an elementary zero of a C^∞ vector field v on M. Let $\mathcal{V}$ denote the vector space of all C^∞ vector fields on M endowed with the weak C^∞ topology. Let $\mathcal{V}_p$ denote the subspace of all vector fields on M vanishing in p. Denote by $C^1(D, M)$ the space of C^1 maps of the disc D to M endowed with the usual C^1 topology.

Proposition 1.15. *Let $j : D \hookrightarrow M$ be a stable Lyapunov chart for v at p. There exists a neighbourhood $\mathcal{U}$ of v in $\mathcal{V}_p$ and for each $u \in \mathcal{U}$ there is a stable Lyapunov chart $j_u : D \hookrightarrow M$, such that $j_v = j$ and the map*

$$u \mapsto j_u; \quad \mathcal{U} \to C^1(D, M)$$

is continuous.

For the proof we refer to the book [**1**]. Proposition 1.15 follows from Lemma 27.5 of this book. A similar proposition holds for unstable Lyapunov charts.

1.3. The Grobman-Hartman theorem. Let v be a C^1 vector field on a manifold M and $a \in M$. We denote by $I(a, v)$ the interval of definition of the maximal integral curve $\gamma(a, \cdot; v)$.

Definition 1.16. Let v_1 be a C^1 vector field on a manifold M_1, and v_2 a C^1 vector field on a manifold M_2. A homeomorphism $\Psi : M_1 \to M_2$ is called a *flow equivalence* between v_1 and v_2 if for every point $a \in M_1$ we have:

(1) $I(a, v_1) = I(\Psi(a), v_2)$.
(2) For every $t \in I(a, v_1)$ we have:

$$\gamma(\Psi(a), t; v_2)) = \Psi(\gamma(a, t; v_1)).$$

Here is one of the versions of the Grobman-Hartman theorem

Theorem 1.17. [[**67**], Theorem 5.25] *Let E be a finite-dimensional real vector space, and $L : E \to E$ be an elementary linear map. For every C^1 vector field $\eta : E \to E$ with sufficiently small C^1 norm there is a flow equivalence between L and $L + \eta$.*

It is important to note that in general the flow equivalence in Theorem 1.17 is not differentiable.

The next proposition says that the flow equivalence class of the flow generated by an elementary linear map L depends only on the index of L. This is a well-known fact from the theory of linear differential equations (see [**6**], or [**67**]); we outline a proof for completeness.

Proposition 1.18. *Let E be an m-dimensional real vector space. Let $L : E \to E$ be an elementary linear map of index k. Then there is a flow equivalence between L and the elementary linear map $\mathscr{A}_k : \mathbf{R}^m \to \mathbf{R}^m$ given by the formula*

$$\mathscr{A}_k(x_1, \dots, x_k, x_{k+1}, \dots, x_m) = (-x_1, \dots, -x_k, x_{k+1}, \dots, x_m).$$

Proof. Let $E = E^- \oplus E^+$ be the L-invariant decomposition where E^- and E^+ are the negative, respectively positive, subspaces of E with respect to L. We have linear vector fields $L^- = L|E^-, L^+ = L|E^+$ on E^-, respectively on E^+. There is a flow equivalence between L^- and $\mathscr{A}_k^-$, and also between L^+ and $\mathscr{A}_k^+$ (this is one of the versions of the Lyapunov stability theorem, see for example, [**67**], Theorem 4.40, or [**6**], §22). The direct product of these equivalences is a flow equivalence between L and $\mathscr{A}_k$. □

Now we shall use Theorem 1.17 together with Proposition 1.18 to prove that every vector field in a neighbourhood of its elementary zero of index k is flow equivalent to the vector field $\mathscr{A}_k$. Let v be a C^1 vector field on an m-dimensional manifold M, let p be an elementary zero of v of index k, and U be an open neighbourhood of p. Put

$$\mathcal{B}(R) = B^k(0, R) \times B^{m-k}(0, R) \subset \mathbf{R}^k \times \mathbf{R}^{m-k}.$$

Definition 1.19. A homeomorphism

$$\Phi : \mathcal{B}(R) \to U$$

will be called *a box* for v at p, if Φ is a flow equivalence between the flow $\mathscr{A}_k|\mathcal{B}(R)$ and the flow $v|U$.

Theorem 1.20. *Let v be a C^1 vector field on a manifold M, let p be an elementary zero of v. For every neighbourhood V of p there is a box $\Phi : \mathcal{B}(R) \to M$ such that $\Phi(\mathcal{B}(R)) \subset V$.*

Proof. The assertion is clearly local therefore it suffices to prove our theorem for the case when $M = \mathbf{R}^m$ and $p = 0$ is the unique zero of the vector field v. Let $L = v'(0)$, then L is an elementary linear map. The difference $v - v'(0)$ is C^1 small in a small neighbourhood of 0. The idea is to apply Theorem 1.17 and deduce that v and L are flow equivalent in a neighbourhood of 0. Unfortunately Theorem 1.17 applies only to vector fields defined on the whole of $\mathbf{R}^m$. So we shall modify the vector field v outside a small neighbourhood of 0 so that the resulting vector field will be C^1 close to L everywhere in $\mathbf{R}^m$. Pick a C^∞ function

$$h : \mathbf{R}^m \to [0,1] \quad \text{with} \quad h(x) = 0 \quad \text{for} \quad ||x|| \leqslant 1,$$
$$\text{and} \quad h(x) = 1 \quad \text{for} \quad ||x|| \geqslant 2,$$

(where $||\cdot||$ is the usual Euclidean norm in $\mathbf{R}^m$). For $\lambda > 0$ define a vector field w_λ on $\mathbf{R}^m$ by the following formula:

$$w_\lambda(x) = (1 - h(x/\lambda)) \cdot v(x) + h(x/\lambda) \cdot Lx.$$

The vector fields w_λ and v are equal in the disc $D(0,\lambda)$ and outside the disc $D(0,2\lambda)$ we have $w_\lambda(x) = Lx$.

Lemma 1.21. *The C^1 norm of the difference $w_\lambda - L$ converges to zero as $\lambda \to 0$.*

Proof. We have:

$$w_\lambda(x) - Lx = \big(1 - h(x/\lambda)\big)\big(v(x) - Lx\big).$$

The Taylor formula says

$$v(x) - Lx = R(x) \qquad \text{with} \qquad ||R(x)|| \leqslant C||x||^2$$

(recall that $v'(0) = L$ by definition). Thus the C^0-norm of $w_\lambda - L$ is less than $4C\lambda^2$. As for the differential of $w_\lambda - L$, we have:

$$(w_\lambda - L)'(x)(\xi) = -(1/\lambda) \cdot h'(x/\lambda)(\xi) \cdot (v(x) - Lx) \\ + \big(1 - h(x/\lambda)\big) \cdot \big(v'(x)(\xi) - L\xi\big).$$

It is easy to see that the C^0-norm of each of both terms of the right-hand side of the above equality converge to zero when $\lambda \to 0$. □

Applying Theorem 1.17 and Proposition 1.18 we deduce that for λ sufficiently small there is a flow equivalence between $\mathscr{A}_k$ and w_λ, that is, there a homeomorphism

$$\Phi : \mathbf{R}^m \xrightarrow{\approx} \mathbf{R}^m$$

which sends every integral curve of $\mathscr{A}_k$ to an integral curve of w_λ. Restricting Φ to $\mathcal{B}(R)$ with R sufficiently small gives the equivalence sought. □

1.4. Gradient-like vector fields. In this and the following subsection we discuss the Hadamard-Perron and Grobman-Hartman theorems in the particular situation of gradient flows of a Morse function nearby a critical point.

Let us begin with the simplest case of gradient-like vector fields. By the very definition any such field is locally diffeomorphic to the linear vector field $\mathscr{A}_k$. This leads to a short proof of the Hadamard-Perron theorem for this particular case. Moreover the stable and unstable manifolds for a gradient-like vector field are immediately reconstructed from the Morse charts and the corresponding local data.

Let us first consider the vector field $\mathscr{A}_k$. Writing a generic point in $\mathbf{R}^m = \mathbf{R}^k \oplus \mathbf{R}^{m-k}$ as a pair (x, y) with $x \in \mathbf{R}^k$, $y \in \mathbf{R}^{m-k}$, we have

$$\mathscr{A}_k(x, y) = (-x, y).$$

Proposition 1.22. *Let $U = B^k(0, R) \times B^{m-k}(0, R)$. Then*

(1) *$B^k(0, R) \times 0$ is the stable manifold for $\mathscr{A}_k \mid U$.*
(2) *$0 \times B^{m-k}(0, R)$ is the unstable manifold for $\mathscr{A}_k \mid U$.*

Proof. Here is an explicit formula for the integral curves of the vector field $v = \mathscr{A}_k$: for $z_0 = (x_0, y_0)$ we have

$$\gamma(z_0, t; v) = (x_0 e^{-t}, y_0 e^t).$$

Thus if $y_0 = 0$, then $\gamma(z_0, t; v) \in U$ for every $t \geqslant 0$, and the curve $\gamma(z_0, t; v)$ converges to 0 as $t \to \infty$. Therefore

$$W^{st}(v|U, 0) \supset B^k(0, R) \times 0.$$

To prove the inverse inclusion, observe that if $y_0 \neq 0$, then the curve $\gamma(z_0, \cdot; v)$ quits U at the moment $t_0 = \ln R - \ln ||y_0||$. Therefore $\gamma(z_0, \cdot; v|U)$ is not defined on the whole of $\mathbf{R}_+$ and $z_0 \notin W^{st}(v|U, 0)$. A similar argument proves part (2) of the proposition. □

Now let W be a manifold without boundary or a cobordism, f a Morse function on W, $p \in W$ a critical point of f of index k, and v a gradient-like vector field for f. Let

$$\Psi : V \to \mathcal{B}(R) = B^k(0, R) \times B^{m-k}(0, R) \subset \mathbf{R}^m$$

be a Morse chart for (f, v) at p (that is, both f and v have the standard form in the local coordinate system defined by Ψ). It is immediate from the preceding proposition, that

$$W^{st}(v|U, p) = \Psi^{-1}\big(B^k(0, R) \times 0\big), \quad W^{un}(v|U, p) = \Psi^{-1}\big(0 \times B^{m-k}(0, R)\big),$$

and therefore Ψ^{-1} is a box for v at p. This box is not only a homeomorphism, but a C^∞ diffeomorphism.

1.5. f-gradients. Now let us proceed to the case of general f-gradients. Let W be a manifold without boundary or a cobordism, $f : W \to \mathbf{R}$ a Morse function, and v an f-gradient.

Proposition 1.23. *Every zero of v is elementary.*

Proof. Let $p \in S(f)$. By Proposition 2.5 of Chapter 2 (page 51) we have

$$f''(p)\big(v'(p)h, h\big) > 0 \quad \text{for every} \quad h \in T_pM.$$

Now our assertion follows immediately from the next lemma.

Lemma 1.24. *Let $A : E \times E \to \mathbf{R}$ be a symmetric bilinear form on a finite-dimensional real vector space E. Let $L : E \to E$ be a linear map, such that*

$$A(Lh, h) > 0 \quad \textit{for every} \quad h \neq 0.$$

Then

(1) *L is elementary.*
(2) *The restriction of A to the negative L-invariant subspace E^- is a negative definite bilinear form.*
(3) *The restriction of A to to the positive L-invariant subspace E^+ is a positive definite bilinear form.*

Proof. To prove the point (1), assume that the complexification $L^{\mathbf{C}} : E^{\mathbf{C}} \to E^{\mathbf{C}}$ has an eigenvector $h \in E^{\mathbf{C}}, h \neq 0$ with a purely imaginary eigenvalue $i\lambda, \lambda \in \mathbf{R}$. Let

$$\alpha = h + \bar{h}, \ \beta = i(h - \bar{h}),$$

so that α, β are in the real vector space E. Then

$$L\alpha = \lambda\beta, \ L\beta = -\lambda\alpha, \quad \text{and} \quad A(L\alpha, \alpha) = -A(L\beta, \beta).$$

Since $A(Lh, h) > 0$ for every $h \neq 0$ we deduce that $\alpha = \beta = 0$, and this leads to a contradiction.

Proceed now to the point (2). For $h \in E^-, h \neq 0$ consider the C^∞ curve $\gamma(t) = e^{tL}h$, lying entirely in E^-. Observe that $A(\gamma(t), \gamma(t))$ is a strictly increasing function of t (since the derivative of this function equals $2A(Le^{tL}h, e^{tL}h) > 0$). For every $t > 0$ the linear map $e^{tL}|E^-$ is hyperbolic, and every eigenvalue λ of this map satisfies $|\lambda| < 1$. Therefore

$$\gamma(t) = e^{tL}h \to 0, \quad \text{and} \quad A(\gamma(t), \gamma(t)) \to 0, \quad \text{as} \quad t \to \infty.$$

Thus $0 > A(\gamma(0), \gamma(0)) = A(h, h)$. The proof of the third point is completely similar. □

Thus Theorem 1.3 applies, and we obtain in particular the local stable and unstable manifolds for the critical points of f. In the case of general f-gradients the geometry of these manifolds is not as simple as in the case of gradient-like vector fields, where by definition the pair (f, v) is locally C^∞ diffeomorphic to the standard pair $(\mathcal{Q}_k, \mathcal{A}_k)$. Recall however that the pair (f, v) is locally *topologically equivalent* to the standard one by Theorem 1.20; this result will be very useful in the sequel.

Exercise 1.25. Let W be a Riemannian cobordism, $f : W \to [a, b]$ a Morse function, and v the Riemannian gradient of f. Let $p \in S(f)$. Prove that the subspaces $T_p^+(W)$, $T_p^-(W)$ are orthogonal with respect to the bilinear form $f''(p)$, that is, $f''(p)(h, k) = 0$ whenever $h \in T_p^+(W), k \in T_p^-(W)$.

Construct an example of an f-gradient for which $T_p^+(W)$, $T_p^-(W)$ are not orthogonal.

Exercise 1.26. Using the preceding exercise, show that if f is a Morse function on a cobordism of dimension 2, which has a critical point of index 1, then $GR(f)$ is not dense in $G(f)$.

2. Descending discs

Having investigated the local structure of gradient flows, we can now proceed to global properties. In this section we study the stable sets of critical points of Morse functions on cobordisms; these sets are also called *descending discs.*

2.1. Lockers. An important working tool for investigation of descending discs is the notion of a *locker.*

Definition 2.1. Let $f : W \to [a, b]$ be a Morse function on a cobordism W and v an f-gradient. Let p be a critical point of f. A compact neighbourhood K of p is called a *locker for* (f, v) *at* p, if the following property holds:

> If $x \in \operatorname{Int} K$ and for some $t \geqslant 0$ we have $\gamma(x, t; v) \notin K$, then there is $t_0 \in [0, t]$ such that $\gamma(x, t_0; v) \in K$ and $f(\gamma(x, t_0; v)) > f(p)$.

In other words K is a locker, if a v-trajectory can quit K only passing through a point of K where the value of f is greater than $f(p)$.

It is easy to construct lockers for the pair $(\mathcal{Q}_k, \mathcal{A}_k)$, where $\mathcal{Q}_k$ is the standard quadratic form on $\mathbf{R}^m = \mathbf{R}^k \times \mathbf{R}^{m-k}$:

$$\mathcal{Q}_k(x, y) = -||x||^2 + ||y||^2, \quad \text{where} \quad x \in \mathbf{R}^k, y \in \mathbf{R}^{m-k}$$

and

$$\mathcal{A}_k(x, y) = (-x, y).$$

Namely, for any $r > 0$ the set

$$Q(r) = D^k(0, r) \times D^{m-k}(0, r)$$

is a locker for $(\mathcal{Q}_k, \mathcal{A}_k)$ at $(0, 0)$. The proof follows immediately from the explicit formula describing the integral curves of $\mathcal{A}_k$:

$$(x(t), y(t)) = (e^{-t}x(0), e^t y(0)).$$

The integral curves of the flow are depicted in Figure 9.

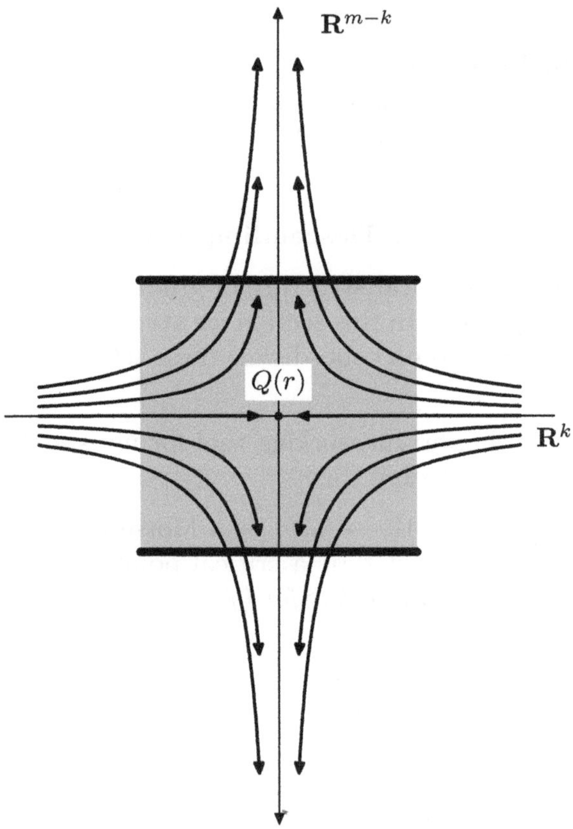

FIGURE 9.

The same geometrically obvious idea leads to the proof of the next proposition.

Proposition 2.2. *Let f be a Morse function on a cobordism W and v an f-gradient. For every critical point p of f and every open neighbourhood V of p there is a locker $K \subset V$ for (f, v) at p.*

Proof. Let

$$\Phi : \mathcal{B}(R) = B^k(0, R) \times B^{m-k}(0, R) \to U$$

be a box for (f, v) at p (see Definition 1.19). Here $k = \operatorname{ind} p$. For $r, \rho \in \mathbf{R}_+$ put

$$Q(\rho, r) = \{(x, y) \in \mathbf{R}^k \times \mathbf{R}^{m-k} \mid ||x|| \leqslant \rho, ||y|| \leqslant r\}.$$

The set-theoretical boundary of $Q(\rho, r)$

$$Fr(Q(\rho, r)) = Q(\rho, r) \setminus \text{Int } Q(\rho, r)$$

is the disjoint union of two sets

$$Q_-(\rho, r) = \{(x, y) \mid ||x|| = \rho, ||y|| < r\},$$
$$Q_+(\rho, r) = \{(x, y) \mid ||x|| \leqslant \rho, ||y|| = r\}.$$

An argument similar to the above shows that an $\mathscr{A}_k$-trajectory starting at a point of $Q(\rho, r)$ can quit this set only through a point of $Q_+(\rho, r)$. For $r, \rho < R$ put

$$\mathcal{F} = \Phi(Q(\rho, r)), \quad \mathcal{F}_+ = \Phi(Q_+(\rho, r)), \quad \mathcal{F}_- = \Phi(Q_-(\rho, r)).$$

Since the vector fields $\mathscr{A}_k|\mathcal{B}(R)$ and $v|U$ are flow equivalent, we deduce that a v-trajectory starting at a point of $\mathcal{F}$ can leave the set $\mathcal{F}$ only through a point of $\mathcal{F}_+$.

Therefore the subset $\mathcal{F}$ is a locker for (f, v) at p if $f(x) > f(p)$ for every point $x \in \mathcal{F}_+$.

Lemma 2.3. *Let $0 < r < R$. For any $\rho > 0$ sufficiently small we have:*

$$f(x) > f(p) \quad \textit{for every} \quad x \in \mathcal{F}_+ = \Phi(Q_+(\rho, r)).$$

Proof. Let $g = f \circ \Phi$, then g is a continuous function on $\mathcal{B}(R)$. We have:

$$\Phi(0 \times D^{m-k}(0, r)) \subset W^{un}(v, p),$$

therefore there is $\epsilon > 0$ such that

$$g(x) > g(p) + \epsilon \quad \text{for every} \quad x \in 0 \times \partial D^{m-k}(0, r).$$

The distance from every point of $Q_+(\rho, r)$ to $0 \times \partial D^{m-k}(0, r)$ is not more than ρ, therefore for any $\rho > 0$ sufficiently small we have:

$$g(x) > g(p) + \epsilon \quad \text{for every} \quad x \in Q_+(\rho, r),$$

and the proof is over. □

It follows that for a given $r \in]0, R[$ the set $\mathcal{F} = \Phi(Q(\rho, r))$ is a locker for (f, v) at p, for any ρ sufficiently small. □

Here is the first application of lockers.

Proposition 2.4. *Let $f : W \to [a, b]$ be a Morse function on a cobordism W and v an f-gradient. Then every v-trajectory either reaches $\partial_1 W$ or converges to a critical point of f.*

Proof. Assume that some v-trajectory $\gamma(t)$ does not reach $\partial_1 W$. Then it is defined on the whole of $\mathbf{R}_+$. This implies that the set $S(f)$ is not empty. Indeed, if f does not have critical points, then the function $f'(x)(v(x))$ has no zeros and is bounded on W from below by some strictly positive constant C. Therefore

$$(f \circ \gamma)'(t) = f'(\gamma(t))\big(v(\gamma(t))\big) \geqslant C > 0 \quad \text{for every } t,$$

and $f(\gamma(t)) \xrightarrow[t\to\infty]{} \infty$ which is impossible.

A similar argument proves that for every neighbourhood U of $S(f)$ there are arbitrarily large values T with $\gamma(T) \in U$. Since $S(f)$ is finite, there is one critical point $p \in S(f)$ and a sequence $t_n \in \mathbf{R}$ with

$$\gamma(t_n) \to p, \quad t_n \to \infty.$$

We can assume that γ is not constant. Then $f(\gamma(t))$ is a strictly increasing function of t, and the above condition implies that $f(\gamma(t)) < f(p)$ for every t. Let K be any locker for (f, v) at p. For some n the point $\gamma(t_n)$ is in $\operatorname{Int} K$, and therefore $\gamma(t) \in K$ for every $t \geqslant t_n$ (otherwise $f(\gamma(t)) > f(p)$ for some $t \in \mathbf{R}$ by the definition of locker). Proposition 2.2 implies that the locker K may be chosen arbitrarily small, therefore $\gamma(t) \to p$, and the curve γ is in the stable manifold of p. □

2.2. Descending discs.

Definition 2.5. Let $f : W \to [a, b]$ be a Morse function on a cobordism W and v an f-gradient. Let p be a critical point of f. The stable set of v with respect to p will be denoted $D(p, v)$ and called also the *descending disc* of p (with respect to v).† The unstable set of v with respect to p will be denoted $D(p, -v)$ and called the *ascending disc* of p (with respect to v). Thus

$$D(p, v) = \{x \in W \mid \gamma(x, t; v) \xrightarrow[t\to\infty]{} p\},$$
$$D(p, -v) = \{x \in W \mid \gamma(x, t; -v) \xrightarrow[t\to\infty]{} p\}.$$

Our next aim is to prove that the descending disc is a manifold with boundary. We shall do the proof in several steps. Recall that we denote by W° the manifold $W \setminus \partial W$.

Proposition 2.6. *The set $D(p, v) \cap W^\circ$ is a submanifold of W° of dimension $k = \operatorname{ind} p$.*

Proof. We begin with a lemma.

† We call the set $D(p, v)$ the *descending disc* since it is situated *below* the critical level surface $f^{-1}(f(p))$.

Lemma 2.7. *There is a neighbourhood V of p, such that $D(p,v) \cap V$ is a submanifold of V of dimension k.*

Proof. Let U be an open neighbourhood of p such that the stable set $S = W^{st}(v|U, p)$ is a submanifold of U. Let $K \subset U$ be a locker for (f, v) at p. We are going to prove that

$$S \cap \text{Int } K = D(p,v) \cap \text{Int } K, \tag{2}$$

which implies our assertion (with $V = \text{Int } K$). The inclusion

$$S \cap \text{Int } K \subset D(p,v) \cap \text{Int } K$$

is obvious, so it suffices to establish the inverse inclusion. Let $x \in D(p,v) \cap \text{Int } K$. The trajectory $\gamma(x, \cdot; v)$ cannot leave K (otherwise at some moment t_0 we would have $f(\gamma(x, t_0; v)) > f(p)$). Therefore $\gamma(x, \cdot; v)$ stays in K forever and converges to p, which means $x \in S \cap \text{Int } K$. □

Now let $x \in D(p,v) \cap W^\circ$, $x \neq p$. Let T be a sufficiently large number so that $\gamma(x, T; v) \in V$. The map $y \mapsto \gamma(y, T; v)$ is a diffeomorphism of a neighbourhood Q of x onto a neighbourhood $Q' \subset V$ of $x' = \gamma(x, T; v)$ which sends $Q \cap D(p,v)$ onto $Q' \cap D(p,v)$. Thus we have a diffeomorphism of pairs

$$(Q, Q \cap D(p,v)) \approx (Q', Q' \cap D(p,v)),$$

and since $Q' \cap D(p,v)$ is a submanifold of Q' by the lemma above, our proposition is proved. □

Proposition 2.8. *Let $c = f(p)$. For any $\epsilon > 0$ sufficiently small, the set*

$$\Delta_\epsilon = D(p,v) \cap f^{-1}([c - \epsilon, c])$$

is a manifold with boundary. It is diffeomorphic to the closed k-dimensional disc D^k.

Proof. Denote the restriction $f|D(p,v) \cap W^\circ$ by g. For every $x \in D(p,v)$ the function f is strictly increasing along the trajectory $\gamma(x, t; v)$ which converges to p as $t \to \infty$. Therefore the function g has a maximum at p, and this maximum is non-degenerate by Lemma 1.24. It follows from the Morse lemma that Δ_ϵ is a manifold with boundary diffeomorphic to D^k. □

Corollary 2.9. *The topological space $D(p,v)$ is contractible.*

Proof. We will prove that the inclusion $\Delta_\epsilon \hookrightarrow D(p,v)$ is a homotopy equivalence. For every $x \in D(p,v) \setminus \Delta_\epsilon$ let $\tau(x)$ be the moment when the curve $\gamma(x, \cdot; v)$ reaches the set Δ_ϵ. The function τ is continuous (see Theorem 2.13 of Chapter 1, page 30). Define now a strong deformation retraction

$$H : D(p,v) \times [0,1] \to \Delta_\epsilon$$

of $D(p, v)$ onto Δ_ϵ by the following formula (where $\theta \in [0, 1]$):

$$H(x, \theta) = \gamma(x, \theta \cdot \tau(x); v) \quad \text{for} \quad x \notin \Delta_\epsilon,$$
$$H(x, \theta) = x \quad \text{for} \quad x \in \Delta_\epsilon. \qquad \square$$

Now we are going to investigate the set $D(p, v)$ nearby the lower boundary $\partial_0 W = f^{-1}(a)$ of W.

Proposition 2.10. *Let $D_0 = D(p, v) \cap \partial_0 W$. Let $W' = f^{-1}([a, c])$ where the interval $[a, c]$ is regular.*[†] *There is a diffeomorphism*

$$G : \partial_0 W \times [a, c] \to W'$$

such that

(1) $G(D_0 \times [a, c]) = W' \cap D(p, v)$.
(2) *$f \circ G$ is equal to the projection map $\partial_0 W \times [a, c] \to [a, c]$.*

Proof. Let us first do the proof in the case when

$$f'(x)(v(x)) = 1 \quad \text{for every} \quad x \in W'.$$

In this assumption the map

$$G(x, t) = \gamma(x, t - a; v), \quad G : \partial_0 W \times [a, c] \to W'$$

is the diffeomorphism sought. The general case will be reduced to this particular one.

Definition 2.11. Two vector fields u, w on a cobordism W are called *equivalent*, if $w(x) = h(x)u(x)$, where $h : W \to \mathbf{R}$ is a C^∞ function such that $h(x) > 0$ for every $x \in W$, and $h(x) = 1$ in a neighbourhood of the set of zeros of u.

If u is an f-gradient for a Morse function f, and w is a vector field equivalent to u, then w is also an f-gradient, and the descending discs $D(p, u)$ and $D(p, w)$ are equal for every p.

Returning to the proof of our proposition, it suffices to find for any f-gradient v an equivalent vector field w such that $f'(x)(w(x)) = 1$ for every $x \in W'$. Let $w = h \cdot v$ where $h : W \to \mathbf{R}$ is a C^∞ strictly positive function with

$$h(x) = \Big(f'(x)(v(x))\Big)^{-1} \quad \text{for} \quad x \quad \text{in a neighbourhood of} \quad W'$$

and $h(x) = 1$ for x in a neighbourhood of $S(f)$. Then the vector field w is equivalent to v and we have $f'(x)(w(x)) = 1$ for x in a neighbourhood of W'. The proof is now complete. $\square$

[†] Recall that an interval is called *regular* if it contains no critical values of f.

Corollary 2.12. *Let $f : W \to [a,b]$ be a Morse function on a cobordism W. For a regular value λ of f the set $D(p,v) \cap f^{-1}(\lambda)$ is a submanifold of $f^{-1}(\lambda)$.*

Proof. If $\lambda > a$, then the submanifold $D(p,v) \cap W^\circ$ of W° is transverse to $f^{-1}(\lambda)$, and it suffices to apply Proposition 2.6. Now let $\lambda = a$, and pick some $c > a$ such that the interval $[a,c]$ is regular. Proposition 2.10 implies that for every $\lambda \in]a,c[$ there is a diffeomorphism of pairs

$$\big(f^{-1}(\lambda), f^{-1}(\lambda) \cap D(p,v)\big) \approx \big(\partial_0 W \times \lambda, D_0 \times \lambda\big),$$
$$\text{where} \quad D_0 = \partial_0 W \cap D(p,v)$$

and our assertion follows. □

Proposition 2.13. *Let $f : W \to [a,b]$ be a Morse function on a cobordism W and v an f-gradient. The set $D(p,v)$ is a submanifold with boundary of W, and $\partial D(p,v) = D(p,v) \cap \partial_0 W$.*[†]

Proof. Proposition 2.6 implies that the set $D(p,v) \cap W^\circ$ is a submanifold of W°. Thus it remains to explore the local structure of $D(p,v)$ nearby $\partial_0 W$. Put $D_0 = D(p,v) \cap \partial_0 W$. Pick any $c \in]a,b[$ such that the interval $[a,c]$ is regular, and let $U = f^{-1}\big([a,c[\big)$. The space U is a manifold with boundary, $\partial U = \partial_0 W$. Recall the diffeomorphism G from Proposition 2.10, and observe that the restriction

$$G \,\Big|\, \partial_0 W \times [a,c[\;:\; \partial_0 W \times [a,c[\xrightarrow{\approx} U$$

is a diffeomorphism which sends $D_0 \times [a,c[$ onto $D(p,v) \cap U$. Since $D_0 \times [a,c[$ is a submanifold with boundary of $\partial_0 W \times [a,c[$, we deduce that $D(p,v) \cap U$ is a submanifold with boundary of U. □

Now we shall describe the diffeomorphism type of the space $D(p,v)$ in two special cases.

Definition 2.14. Let p be a critical point of a Morse function $f : W \to \mathbf{R}$.

1) We shall say that p is of type (α) if for every $x \in D(p,v) \setminus \{p\}$ the trajectory $\gamma(x, \cdot; -v)$ reaches $\partial_0 W$.
2) We shall say that p is of type (ω) if for every $x \in D(p,v)$ the trajectory $\gamma(x, \cdot; -v)$ does not reach $\partial_0 W$.

For example, if $\partial_0 W = \varnothing$, every critical point p is of type (ω).

Exercise 2.15. Let $f : W \to [a,b]$ be a Morse function having only one critical value. Show that every critical point of f is of type (α).

[†] The set $D(p,v) \cap \partial_0 W$ is often called *the sole* of the descending disc $D(p,v)$.

Proposition 2.17 below implies that in the case (ω) the descending disc is diffeomorphic to $\mathbf{R}^k$ where $k = \operatorname{ind} p$. Let us begin with an example.

Example 2.16. Consider the height function $f : S^m \to \mathbf{R}$ on the sphere $S^m \subset \mathbf{R}^{m+1}$ (see page 38). This function has two critical points: the maximum N and the minimum S. Every v-trajectory starting at a point different from S converges to N, and therefore

$$D(N, v) = S^m \setminus \{S\} \approx \mathbf{R}^m.$$

Let us have a closer look at the geometry of this example for $m = 2$. Let v be the Riemannian gradient of the height function (we endow the sphere S^2 with the Riemannian metric induced from the Euclidean metric in $\mathbf{R}^3$). At any point $x \in S^2$ the vector $v(x)$ is the orthogonal projection to $T_x S^2$ of the vector $(0, 0, 1)$. The flow lines of $(-v)$ are the big half-circles of our sphere. The family of these flow lines is depicted in Figure 10 (the arrows show the direction of the field $(-v)$). The set $D(N, v) \setminus N$ is the union of a family of flow lines joining N and S; this family is parameterized by the boundary of the shaded disc. Using the *stretching construction* (page 76) it is not difficult to construct a diffeomorphism of the interior of the shaded disc onto $D(N, v) \setminus N$: stretch each radius of the shaded disc along the flow line to which it belongs so as to cover the whole of the flow line. This geometric method is the essence of the proof of the following proposition.

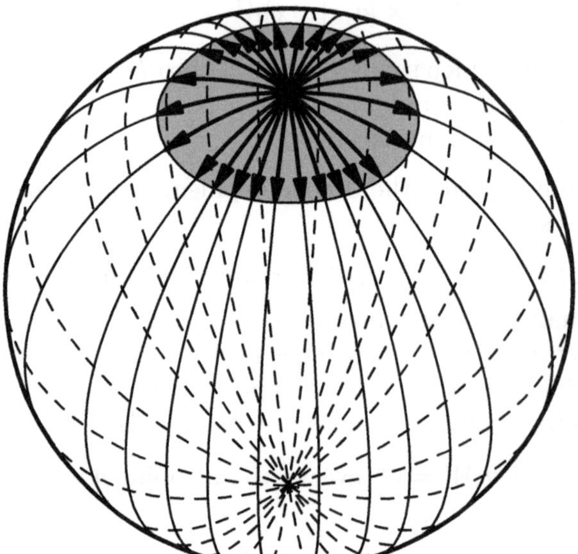

FIGURE 10.

Proposition 2.17. *If* $\operatorname{ind} p = k$ *and* p *is of type* (ω), *then the manifold* $D(p,v)$ *is diffeomorphic to an open* k*-dimensional disc* $B^k(0,1)$ *via a diffeomorphism which sends* p *to* $0 \in B^k(0,1)$.

Proof. Let $c = f(p)$. For every $\epsilon > 0$ sufficiently small the set

$$\Delta = f_0^{-1}([c-\epsilon, c])$$

is a manifold with boundary diffeomorphic to the closed disc $D^k(0,1)$ (Proposition 2.8). The vector field $v|\partial\Delta$ points inward Δ, therefore Δ is a stable Lyapunov disc for v. An argument similar to the proof of Proposition 1.13 (the stretching construction) applies here and we obtain an injective immersion $j : \mathcal{B} \to W^\circ$ of an open k-dimensional disc $\mathcal{B}$ to W° such that $j(\mathcal{B}) = D(p,v)$. Since $D(p,v)$ is a C^∞ submanifold of W° (by Proposition 2.13), the map $j : \mathcal{B} \to D(p,v)$ is a diffeomorphism. □

An open disc in $\mathbf{R}^k$ is foliated by its radii, and the set $D(p,v)$ is foliated by the images of the flow lines. It is natural to ask whether the diffeomorphism constructed above preserves these structures. For $\xi \in \mathbf{R}^k$, $||\xi|| = r$ the open interval

$$]0,\xi[= \{t\xi \mid 0 < t < 1\}$$

will be called the *radius of* $B^k(0,r)$ *corresponding to* ξ.

Definition 2.18. A map $G : B^k(0,r) \to D(p,v)$ is called *ray-preserving* if $G(0) = p$ and for every radius ρ of $B^k(0,r)$ the image $G(\rho)$ is in a flow line of v.

In general the diffeomorphism G constructed in the proof of Proposition 2.17 is not ray-preserving.

Exercise 2.19. Show that there is no ray-preserving diffeomorphism for the case of the vector field u from the example 2.19 of Chapter 2 (page 57).

Proposition 2.20. *Let* v *be a gradient-like vector field for a Morse function* f, *and* $p \in S_k(f)$ *be a critical point of* f *of type* (ω). *Then there is a ray-preserving diffeomorphism* $\psi : B^k(0,1) \to D(p,v)$.

Proof. The Morse function $f_0 = f|D(p,v)$ has only one non-degenerate critical point p which is a maximum. The vector field $v_0 = v|D(p,v)$ is a gradient-like vector field for f_0. Pick any Morse chart

$$\Psi : U \to B^k(0,R), \quad \Psi(p) = 0,$$

for (f_0, v_0) at p. The inverse diffeomorphism Ψ^{-1} is then a ray-preserving map. For any $r < R$ the restriction of Ψ^{-1} to a closed disc $D^k(0,r)$ is a stable Lyapunov disc for $v|D(p,v)$ at p. Applying to this disc the stretching construction we obtain a ray-preserving diffeomorphism sought. □

Let us proceed to the case of critical points of type (α).

Proposition 2.21. *Let $f : W \to [a, b]$ be a Morse function on a cobordism W and v an f-gradient. Let p be a critical point of f of type (α) of index k. Then $D(p, v)$ is diffeomorphic to the closed k-dimensional disc D^k.*

Proof. The proof is again an application of the stretching construction. Let $c = f(p)$ and

$$\mathcal{D}_0 = f_0^{-1}\big([c - \epsilon, c]\big),$$

so that $\mathcal{D}_0$ is a Lyapunov disc for v. Its boundary $\partial\mathcal{D}_0$ is an embedded sphere in $f^{-1}(c - \epsilon)$. Every $(-v)$-trajectory starting at $\partial\mathcal{D}_0$ reaches $\partial_0 W$, so that $\partial\mathcal{D}_0$ is in the domain of definition of the function $x \mapsto \tau(x, -v)$. It is easy to prove that $\tau(x, -v) > 1$ for every $x \in \partial\mathcal{D}_0$ if ϵ is chosen sufficiently small. As in the proof of Proposition 1.13 put

$$\mathcal{D} = \Phi(-v, 1)(\mathcal{D}_0) = \{\gamma(x, 1; -v) \mid x \in \mathcal{D}_0\}.$$

We will construct a diffeomorphism of $\mathcal{D}$ onto $D(p, v)$. For $x \in \mathcal{D} \setminus \mathcal{D}_0$ denote by $t(x)$ the moment when $\gamma(x, \cdot; v)$ reaches $\partial\mathcal{D}_0$, and by $A(x)$ the point of intersection of this curve with $\partial\mathcal{D}_0$. Denote by $T(x)$ the moment of exit $\tau(A(x), -v)$ of the $(-v)$-trajectory starting at $A(x)$. Then

$$t : \mathcal{D} \setminus \mathcal{D}_0 \to]0, 1], \quad T : \mathcal{D} \setminus \mathcal{D}_0 \to]1, \infty[, \quad A : \mathcal{D} \setminus \mathcal{D}_0 \to \partial\mathcal{D}_0$$

are C^∞ maps. Recall from the page 64 the C^∞ function

$$\chi : [0, 1] \times]1, \infty[\to \mathbf{R}_+,$$

it has the property that for each $y \in]1, \infty[$ the map $t \mapsto \chi(t, y)$ is a diffeomorphism of $[0, 1]$ onto $[0, y]$ which is equal to the identity map in a neighbourhood of 0.

Define now a map $G : \mathcal{D} \to D(p, v)$ by the following formula:

$$G(x) = x \qquad \text{for} \quad x \in \mathcal{D}_0,$$
$$G(x) = \gamma\Big(A(x), \chi\big(t(x), T(x)\big); -v\Big) \qquad \text{for} \quad x \in \mathcal{D} \setminus \mathcal{D}_0.$$

It is clear that G is a C^∞ map of $\mathcal{D}$ onto $D(p, v)$. For each $(-v)$-trajectory γ the map G stretches the part of γ contained in $\mathcal{D}_0$ so as to cover the whole of the curve γ. Using the same technique as in the proof of Proposition 1.13 it is easy to construct a C^∞ inverse for G; therefore G is the diffeomorphism sought. □

Corollary 2.22. *Let $f : W \to [a, b]$ be a Morse function on a cobordism W and v an f-gradient. A critical point p of f is of type (α) if and only if $D(p, v)$ is compact.*

Proof. If p is of type (α), then $D(p, v)$ is compact as it follows from the preceding proposition. To prove the inverse implication, assume that p

is not of type (α). Then there is a critical point $q \in S(f)$ and a $(-v)$-trajectory $\gamma : \mathbf{R} \to W$ such that

$$\lim_{t\to\infty} \gamma(t) = q, \qquad \lim_{t\to-\infty} \gamma(t) = p.$$

The point q is then in the set $\overline{D(p,v)} \setminus D(p,v)$ and therefore $D(p,v)$ is not compact. □

In the rest of this subsection we discuss orientations of descending discs. Every descending disc $D(p,v)$ is contractible, therefore an orientable manifold. To introduce an orientation in $D(p,v)$ it suffices to orient the tangent space to $D(p,v)$ at the point p. An orientation of $D(p,v)$ induces an orientation of $\partial D(p,v)$. If we choose an orientation of $T_pD(p,v)$, the tangent space to the ascending disc $T_pD(p,-v)$ acquires a *coorientation* (that means that the quotient $T_pW/T_pD(p,-v)$ is oriented). The disc $D(p,-v)$ being a contractible submanifold of W, the normal fiber space to $D(p,-v)$ acquires an orientation, and the manifold $D(p,-v)$ becomes a *cooriented* submanifold. In view of Proposition 2.10 the submanifold $D(p,-v) \cap \partial_1 W \subset \partial_1 W$ inherits this coorientation.

2.3. Stratification of manifolds by descending discs. Let M be a closed manifold, $f : M \to \mathbf{R}$ a Morse function, v an f-gradient. Every non-constant v-trajectory converges to a critical point of f (see Proposition 2.4), therefore the descending discs form a partition of M:

$$M = \bigsqcup_{p\in S(f)} D(p,v),$$

where each submanifold $D(p,v)$ is diffeomorphic to the open disc $B^k(0,R) \subset \mathbf{R}^k$ of dimension $k = \operatorname{ind} p$.

Proposition 2.23. *Let λ be a regular value of f. The set*

$$D_\lambda = \bigcup_{f(p)<\lambda} D(p,v)$$

is compact.

Proof. Consider the cobordism

$$W = f^{-1}(]-\infty,\lambda]), \quad \partial_1 W = f^{-1}(\lambda), \quad \partial_0 W = \varnothing.$$

Let U be the set of all points $x \in W$, such that the v-trajectory $\gamma(x,\cdot;v)$ reaches $\partial_1 W$. Corollary 2.14 of Chapter 1 (page 31) implies that U is open, and since $D_\lambda = W \setminus U$, the set D_λ is compact. □

The proposition above implies that for every $p \in S(f)$ the set

$$\mathscr{B}(D(p,v)) = \overline{D(p,v)} \setminus D(p,v)$$

is contained in the union of the descending discs $D(q,v)$ with $f(q) \leqslant f(p)$, $q \neq p$. A stronger property is proved in the next proposition.

Proposition 2.24. *Let* $p \in S(f)$. *Then*

$$\mathscr{B}(D(p,v)) \subset \bigcup_{f(q)<f(p)} D(q,v).$$

Proof. In view of Proposition 2.23 it suffices to prove that for every critical point $q \neq p$ with $f(q) \geqslant f(p)$ we have

$$\mathscr{B}(D(p,v)) \cap D(q,v) = \varnothing.$$

Assume that we have a sequence of points $x_n \in D(p,v)$ converging to some $x \in D(q,v)$. Pick a locker K for (f,v) at q. Let T be a sufficiently large real number so that $\gamma(x,T;v) \in \operatorname{Int} K$. Then we have also $\gamma(x_n,T;v) \in \operatorname{Int} K$ if n is sufficiently large. The trajectory $\gamma(x_n,\cdot;v)$ converges to p, therefore it leaves necessarily the set K and must pass through a point $y \in K$ with $f(y) > f(q)$. We deduce that $f(p) > f(q)$. □

Thus the manifold M is the result of the procedure of successive "attachments" of discs $D(p,v)$, where each disc $D(p,v)$ is attached to the union of the discs $D(q,v)$ with $f(q) < f(p)$. We will show later that for a generic f-gradient v this partition is a cellular decomposition of M.

2.4. Examples.

Quadratic forms on S^m

Recall that the function

$$f = g \mid S^m, \qquad \text{where} \qquad g(x_0,\dots,x_m) = \alpha_0 x_0^2 + \dots + \alpha_m x_m^2$$

and α_i are pairwise distinct non-zero real numbers, is a Morse function on the sphere S^m (see page 40). Endow S^m with the Riemannian metric induced from the Euclidean metric in $\mathbf{R}^{m+1}$. Let v be the Riemannian gradient of g with respect to this metric. We are going to describe the descending and ascending discs for the critical points of g with respect to this gradient. To simplify the notation we assume that $\alpha_0 < \alpha_1 < \dots < \alpha_m$. The critical points of f were found in Proposition 1.11 of Chapter 2 (page 40): for every i with $0 \leqslant i \leqslant m$ the set $S_i(f)$ consists of two points:

$$N_i = (0,\dots,\underset{i}{1},\dots,0), \quad S_i = (0,\dots,\underset{i}{-1},\dots,0).$$

Lemma 2.25. *We have*

$$v(x_0,\dots,x_m) = \Big(\mu_0(x)x_0,\dots,\mu_m(x)x_m\Big)$$

where

$$\mu_i(x) = 2\alpha_i - 2\sum_{j=0}^{m} \alpha_j x_j^2.$$

Proof. The vector $\mathrm{grad} f(x)$ of f at $x \in S^m$ is the orthogonal projection to $T_x S^m$ of the Riemannian gradient of g. We have

$$\mathrm{grad} g(x) = \big(2\alpha_0 x_0, \dots, 2\alpha_m x_m\big),$$

and our formula follows by a simple computation. $\square$

For $0 \leqslant i \leqslant m$ put

$$L_i = \{(x_0, \dots, x_m) \mid x_{i+1} = x_{i+2} = \dots = x_m = 0\},$$

and let Σ_i be the unit sphere in the vector subspace L_i, so that

$$\Sigma_i = S^m \cap L_i, \quad \dim \Sigma_i = i.$$

Consider the upper and lower half-spheres of Σ_i:

$$\Sigma_i^+ = \{(x_0, \dots, x_i, \dots, 0) \in \Sigma_i \mid x_i > 0\},$$
$$\Sigma_i^- = \{(x_0, \dots, x_i, \dots, 0) \in \Sigma_i \mid x_i < 0\}.$$

We have

$$\Sigma_i = \Sigma_i^+ \cup \Sigma_i^- \cup \Sigma_{i-1}, \quad N_i \in \Sigma_i^+, \; S_i \in \Sigma_i^-.$$

Proposition 2.26.

$$D(N_i, v) = \Sigma_i^+, \quad D(S_i, v) = \Sigma_i^-.$$

Proof. Let us prove the first assertion, the second one is similar. Observe that each sphere Σ_i is $\pm v$-invariant (this is immediate from Lemma 2.25). Since

$$\Sigma_i = \Sigma_i^+ \cup \Sigma_i^- \cup \Sigma_{i-1},$$

it follows that the subsets Σ_i^+, Σ_i^- also are $\pm v$-invariant. Let

$$x = (x_0, \dots, x_i, 0, \dots, 0) \in \Sigma_i^+, \quad x \neq N_i.$$

The v-trajectory $\gamma(t)$ starting at x stays in Σ_i^+, therefore it converges to a critical point in $\overline{\Sigma_i^+} = \Sigma_i^+ \cup \Sigma_{i-1}$. We are going to show that it converges to N_i.

The i-th coordinate $\gamma_i(t)$ of the trajectory $\gamma(t)$ is always positive. Moreover, the differential equation $\gamma'(t) = v(\gamma(t))$ implies

$$\gamma_i'(t) = \mu_i(\gamma(t))\gamma_i(t).$$

Since $x \neq N_i$, we have

$$\gamma_i(t) \neq 1 \quad \text{for all} \quad t.$$

Recall that $\sum_{j=0}^{i}\gamma_j(t)^2 = 1$, therefore

$$\mu_i(\gamma(t)) = 2\sum_{j=0}^{i-1}(\alpha_i - \alpha_j)\ \left(\gamma_j(t)\right)^2 > 0 \quad \text{for all} \quad t.$$

Thus the function $\gamma_i(t)$ is strictly increasing and the curve $\gamma_i(t)$ cannot converge to a critical point in Σ_{i-1}, therefore $\lim_{t\to\infty}\gamma(t) = N_i$.

Thus the stable manifold $D(N_i, v)$ contains Σ_i^+. Both $D(N_i, v)$ and Σ_i^+ have the same dimension and contain the point N_i. Therefore for some neighbourhood U of N_i we have

$$D(N_i, v) \cap U \subset \Sigma_i^+.$$

Since

$$D(N_i, v) = \bigcup_{t>0} \Phi(-v, t)\Big(D(N_i, v) \cap U\Big)$$

and Σ_i^+ is $(-v)$-invariant, we obtain $D(N_i, v) \subset \Sigma_i^+$ and our proposition follows. □

Thus the stratification of the manifold S^m by the descending discs of f is a cellular decomposition of S^m (one of the most usual ones). It is not a minimal cellular decomposition, but it has the advantage of being invariant with respect to the involution $x \mapsto -x$, and therefore it induces a cellular decomposition of $\mathbf{R}P^m$ which has exactly one cell in each dimension.

Let us give the description of the ascending discs of $-v$. Let

$$\widehat{L}_j = \{(x_0, \dots, x_m) \mid x_0 = \dots = x_{j-1} = 0\}.$$

Put $\widehat{\Sigma}_j = S^m \cap \widehat{L}_j$, so that $\widehat{\Sigma}_j$ is the unit sphere in the vector space $\widehat{L}_j$, and $\dim \widehat{\Sigma}_j = m - j$. Put

$$\widehat{\Sigma}_j^+ = \{(0, \dots, 0, x_j, \dots, x_m) \in \Sigma_j \mid x_j > 0\},$$

$$\widehat{\Sigma}_j^- = \{(0, \dots, 0, x_j, \dots, x_m) \in \Sigma_j \mid x_j < 0\}.$$

Thus

$$\widehat{\Sigma}_j = \widehat{\Sigma}_j^+ \cup \widehat{\Sigma}_j^- \cup \widehat{\Sigma}_{j+1}.$$

Proposition 2.27.

$$D(N_j, -v) = \widehat{\Sigma}_j^+, \quad D(S_j, -v) = \widehat{\Sigma}_j^-.$$

Proof. Apply Proposition 2.26 to the function $-f$ and its gradient $(-v)$. □

Figure 11 illustrates our results for the case $m = 2$. The fat curves with arrows depict the integral curves of $(-v)$.

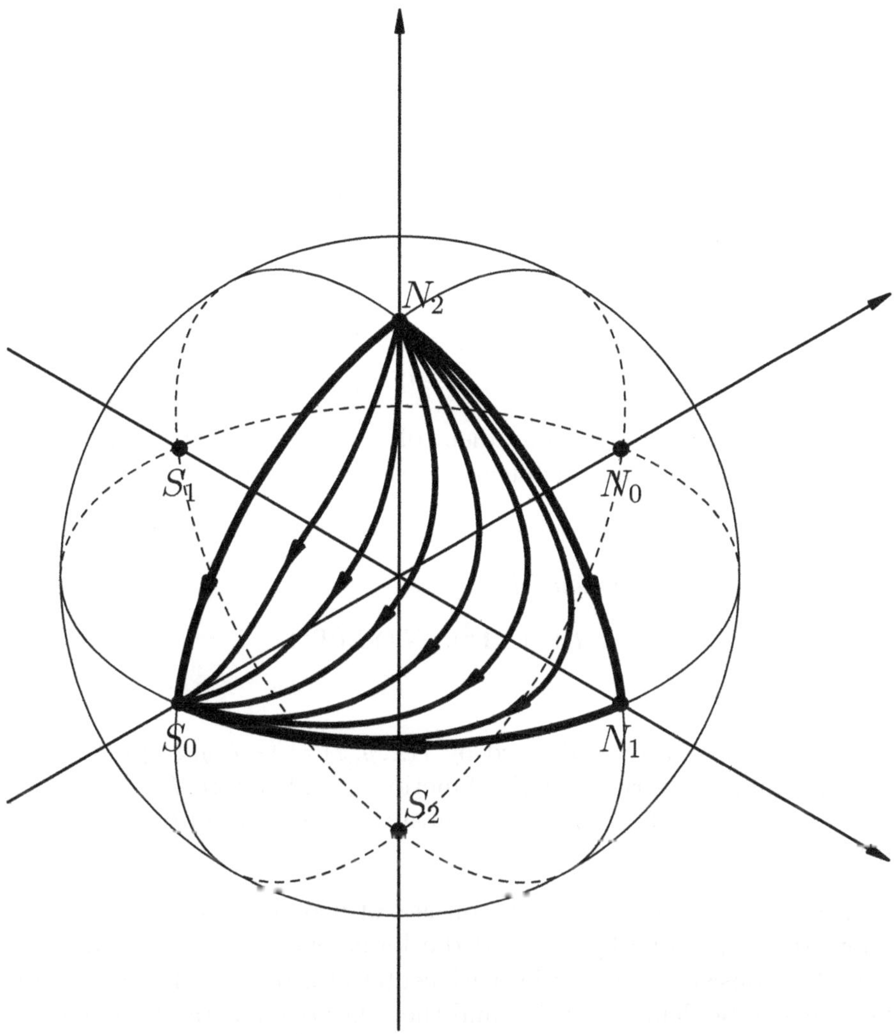

FIGURE 11.

The height function on the torus

In our next example the stratification of the manifold by the descending discs is *not* a cellular decomposition. We consider the two-dimensional torus $\mathbf{T}^2 = S^1 \times S^1$. Recall the embedding

$$E : \mathbf{T}^2 \hookrightarrow \mathbf{R}^3, \quad (\phi, \theta) \mapsto \Big(r\sin\theta,\ (R+r\cos\theta)\cos\phi,\ (R+r\cos\theta)\sin\phi\Big),$$

where ϕ, θ are the angle coordinates on two copies of S^1, and the height function

$$\zeta : E(\mathbf{T}^2) \to \mathbf{R}, \quad \zeta(x, y, z) = z.$$

This function has four critical points (see page 40):

$$\begin{array}{lll} b = (0,0,R+r), & \text{ind } b = 2 & \text{(the maximum)}, \\ s_2 = (0,0,R-r), & \text{ind } s_2 = 1 & \text{(a saddle point)}, \\ s_1 = (0,0,-R+r), & \text{ind } s_1 = 1 & \text{(another saddle point)}, \\ a = (0,0,-R-r), & \text{ind } a = 0 & \text{(the minimum)}. \end{array}$$

Let v be the Riemannian gradient of ζ with respect to the Riemannian metric induced from the Euclidean metric on $\mathbf{R}^3$. Let

$$L = \{(x,y,z) \mid x = 0\}, \quad N = \{(x,y,z) \mid y = 0\}.$$

Lemma 2.28. *The submanifolds*

$$L \cap E(\mathbf{T}^2), \ N \cap E(\mathbf{T}^2)$$

are $\pm v$*-invariant.*

Proof. The involution I defined by $I(x,y,z) = (-x,y,z)$ preserves both the Riemannian metric, and the function ζ. Therefore $I_*(v) = v$, hence the intersection $E(\mathbf{T}^2) \cap L$ is $(\pm v)$-invariant. The proof of the invariance of $N \cap E(\mathbf{T}^2)$ is similar. □

The intersection $E(\mathbf{T}^2) \cap N$ consists of two circles of radius r. Let us denote the upper one by C_2, and the lower one by C_1. The intersection $E(\mathbf{T}^2) \cap L$ consists of two circles with center at zero: one of them has radius $R-r$, it will be denoted by Z_1, and the other one has radius $R+r$, it will be denoted by Z_2. It is not difficult to deduce the next proposition.

Proposition 2.29.

$$\begin{aligned} D(s_2, v) &= Z_1 \setminus \{s_2\}, \\ D(s_1, v) &= C_1 \setminus \{s_1\}, \\ D(b, v) &= E(\mathbf{T}^2) \setminus \Big(C_1 \cup Z_1\Big). \end{aligned}$$

□

Observe that the partition of $E(\mathbf{T}^2)$ into the union of the descending discs is not a cellular decomposition; indeed, we have

$$s_1 \in \overline{D(s_2, v)}$$

that is, the closure of the 1-dimensional disc $D(s_2, v)$ intersects another one-dimensional disc $D(s_1, v)$ in its interior. We shall see later that the Morse function ζ can be modified so that the partition of $E(\mathbf{T}^2)$ into the union of the descending discs will be a cellular decomposition.

2.5. Thickenings of descending discs. In this subsection W is a Riemannian cobordism, $f : W \to [a, b]$ is a Morse function on W, and v is an f-gradient. Let $a \in W^\circ$ and $\delta > 0$ be sufficiently small so that the exponential map $\exp_a : T_a W \to W$ is well defined and for some $\epsilon > \delta$ the restriction of $\exp_a$ to the closed Euclidean disc $D(0, \epsilon)$ is a diffeomorphism onto its image. We shall denote the closed Riemannian disc with center at a by $D_\delta(a)$ and the open Riemannian disc with center at a by $B_\delta(a)$. Each time we use this notation we shall assume that δ is sufficiently small so that the above restriction holds.

Definition 2.30. For $p \in S(f)$ set

$$B_\delta(p, v) = \{x \in W \mid \exists t \geqslant 0 : \gamma(x, t; v) \in B_\delta(p)\},$$
$$D_\delta(p, v) = \{x \in W \mid \exists t \geqslant 0 : \gamma(x, t; v) \in D_\delta(p)\}.$$

Set

$$D_\delta(v) = \bigcup_{p \in S(f)} D_\delta(p, v), \quad B_\delta(v) = \bigcup_{p \in S(f)} B_\delta(p, v),$$

$$D(v) = \bigcup_{p \in S(f)} D(p, v).$$

Lemma 2.31. *$B_\delta(p, v)$ is open.*

Proof. The assertion follows immediately from the continuous dependence of the integral curves of vector fields on their initial values. □

The set $D(p, v)$ is not closed in general: if there is an integral curve of v joining two critical points p and q with $f(q) < f(p)$, then clearly $q \in \overline{D(p, v)} \setminus D(p, v)$. Similarly, the set $D_\delta(p, v)$ is not closed in general. But the union $D(v)$ of all the sets $D(p, v)$ is closed, as well as the set $D_\delta(v)$:

Lemma 2.32. (1) *$D(v)$ is compact.*
(2) *$D_\delta(v)$ is compact.*
(3) *$D_\delta(v) = \overline{B_\delta(v)}$.*

Proof. 1) The complement $W \setminus D(v)$ consists of all points $x \in W$, such that $\gamma(x, \cdot; v)$ reaches $\partial_1 W$. This set is open by Corollary 2.14 of Chapter 1 (page 31). A similar argument proves the point (2). As for the third point, it follows from the obvious inclusions $B_\delta(v) \subset D_\delta(v) \subset \overline{B_\delta(v)}$. □

Intuitively it is quite clear that the δ-thickenings $D_\delta(v)$ of the set $D(v)$ converge to $D(v)$ as $\delta \to 0$. To make this assertion precise, let us first recall the corresponding notions (see [**28**], Ch.3, Sect. 6)

Definition 2.33. Let K be a subset of a topological space X. A system of open neighbourhoods $\{U_\sigma\}_{\sigma \in S}$ of K is called a *fundamental system of neighbourhoods of* K, if for every open set $B \supset K$ there is $\sigma \in S$, such that $U_\sigma \subset B$.

Proposition 2.34. *Let $f : W \to [a, b]$ be a Morse function on a cobordism W and v an f-gradient.*

(1) *For every $\theta > 0$ we have:*

$$D_\theta(v) = \bigcap_{\delta > \theta} B_\delta(v).$$

(2) $D(v) = \bigcap_{\delta > 0} B_\delta(v)$.
(3) $\{B_\delta(v)\}_{\delta > \theta}$ *is a fundamental system of neighbourhoods of* $D_\theta(v)$.
(4) $\{B_\delta(v)\}_{\delta > 0}$ *is a fundamental system of neighbourhoods of* $D(v)$.

Proof. (1) A point $x \in W$ *does not* belong to $D_\theta(v)$ if and only if the trajectory $\gamma(x, \cdot; v)$ reaches $\partial_1 W$ without intersecting the discs $D_\theta(p), p \in S(f)$. Then it does not intersect the discs $D_\delta(p)$ for some $\delta > \theta$. A similar argument proves (2).

(3) Denote $B_\delta(v)$ by R_δ, and $D_\theta(v)$ by K. Then
(A) $\overline{R_\delta} \subset R_{\delta'}$, if $\delta < \delta'$.
(B) $\bigcap_{\delta > \theta} R_\delta = K$.
(C) K and W are compact.

These three properties imply that $\{R_\delta\}_{\delta > \theta}$ form a fundamental system of neighbourhoods of K. Indeed, let $U \supset K$ be a neighbourhood, then the sets $\{W \setminus \overline{R_\delta}\}_{\delta > \theta}$ form an open covering of the compact $W \setminus U$. Thus for some δ_0 we have: $W \setminus \overline{R_{\delta_0}} \supset W \setminus U$, and $\overline{R_{\delta_0}} \subset U$.

A similar argument proves (4). □

Exercise 2.35. Let W be a Riemannian cobordism. Let $f : W \to [a, b]$ be a Morse function and v be an f-gradient. Let $p \in S(f)$.

(1) Prove that

$$\bigcap_{\delta > 0} D_\delta(p, v) = D(p, v).$$

(2) Construct an example where the sets $D_\delta(p,v)$ do not form a fundamental system of neighbourhoods of $D(p,v)$.

3. The gradient descent

Let W be a cobordism, $f : W \to [a,b]$ a Morse function on W, and v an f-gradient. Let $U \subset W$ be the set of all x, such that the $(-v)$-trajectory starting at x reaches $\partial_0 W$. It follows from Proposition 2.4 (page 83) that

$$U = W \setminus D(-v),$$

where $D(-v)$ is the union of all the ascending discs of v. For every $x \in U$ the moment of intersection of $\gamma(x,\cdot;-v)$ with $\partial_0 W$ is denoted by $\tau(x,-v)$, and the point of intersection is denoted by $E(x,-v)$. Proposition 2.16 of Chapter 1 (page 32) implies that the functions $\tau(\cdot,-v), E(\cdot,-v)$ are of class C^∞ on U. The restriction of the map $E(\cdot,-v)$ to the set

$$U_1 = \partial_1 W \cap U = \partial_1 W \setminus D(-v)$$

will be denoted by $(-v)^{\leadsto}$ and called the *transport map* associated to $(-v)$, so that

$$(-v)^{\leadsto}(x) = E(x,-v) = \gamma\big(x,\tau(x,-v);-v\big).$$

The transport map is a diffeomorphism

$$(-v)^{\leadsto} : U_1 \xrightarrow{\approx} U_0$$

of U_1 onto the open subset U_0 of $\partial_0 W$:

$$U_0 = \partial_0 W \cap U = \partial_0 W \setminus D(v).$$

The inverse diffeomorphism is the transport map corresponding to v:

$$\overset{\leadsto}{v} : U_0 \xrightarrow{\approx} U_1.$$

These transport maps prove very useful, although they are not in general everywhere defined in $\partial_1 W$, and, respectively, $\partial_0 W$. In many cases the domain of definition of $(-v)^{\leadsto}$ is dense in $\partial_1 W$. This is for example the case when the function f has no local minima. Indeed, in this case the indices of all critical points of f are not less than 1, and the subset $D(-v) \cap \partial_1 W$ is a finite union of submanifolds of $\partial_1 W$ of codimension not less than 1. For a subset $A \subset \partial_1 W$ we write (by a certain abuse of notation):

$$(-v)^{\leadsto}(A) = (-v)^{\leadsto}(A \cap U_1).$$

Note that if $L \subset \partial_1 W$ is a submanifold of $\partial_1 W$, then $(-v)^{\leadsto}(L)$ is a submanifold of $\partial_0 W$, but compactness of L does not imply compactness of $(-v)^{\leadsto}(L)$.

If $\alpha < \beta$ are regular values of f one can consider the cobordism

$$W' = f^{-1}([\alpha, \beta])$$

and apply the construction above to the Morse function $f|W'$ and its gradient $v|W'$. The corresponding transport map will be denoted

$$(-v)^{\rightsquigarrow}_{[\beta,\alpha]} : \partial_1 W' \setminus D(-v) \xrightarrow{\approx} \partial_0 W' \setminus D(v),$$

which is a diffeomorphism of an open subset

$$\partial_1 W' \setminus D(-v) \subset \partial_1 W' = f^{-1}(\beta),$$

onto an open subset

$$\partial_0 W' \setminus D(v) \subset \partial_0 W' = f^{-1}(\alpha).$$

3.1. Tracks and C^0-stability. Sometimes it is useful to keep track of the whole $(-v)$-trajectory of a point $x \in W$, not only of its point of exit.

Definition 3.1. Let $f : W \to [a, b]$ be a Morse function on a cobordism W and v an f-gradient. Let $X \subset W$. The set

$$T(X, -v) = \{\gamma(x, t; -v) \mid x \in X, t \geqslant 0\}$$

is called the *track of X with respect to $(-v)$*.

We have:

$$T(\overline{A}, -v) \subset \overline{T(A, -v)}.$$

For a compact set X its track $T(X, -v)$ is not necessarily compact; for example for one-point set $X = \{x\}$, with $x \in D(-v)$ the track $T(X, -v)$ is homeomorphic to a half-open interval.

Lemma 3.2. *Let $f : W \to [a, b]$ be a Morse function on a cobordism W and v an f-gradient.*

(1) *If X is compact, then $T(X, -v) \cup D(v)$ is compact.*
(2) *If X is compact, and every $(-v)$-trajectory starting from a point of X reaches $\partial_0 W$, then $T(X, -v)$ is compact.*
(3) *For any X we have*

$$\overline{T(X, -v) \cup D(v)} = T(\overline{X}, -v) \cup D(v).$$

(4) *For any X and $\delta > 0$ we have*

$$\overline{T(X, -v) \cup B_\delta(v)} = T(\overline{X}, -v) \cup D_\delta(v).$$

Proof.

(1) A point $y \in W$ *does not* belong to $T(X, -v) \cup D(v)$, if and only if the trajectory $\gamma(x, \cdot; -v)$ reaches $\partial_1 W$ without intersecting X. Therefore the complement of $T(X, -v) \cup D(v)$ in W is open (by Proposition 2.15 of Chapter 1, page 31).

(2) The moment of exit function $x \mapsto \tau(x, -v)$ is continuous in x. Therefore the set

$$\{(x,t) \in X \times \mathbf{R} \mid x \in X, t \in [0, \tau(x,-v)]\}$$

is compact. The track $T(X, -v)$ is the image of this set via the continuous map $(x,t) \mapsto \gamma(x,t;-v)$, therefore the track is compact.

(3) Since $\overline{X}$ is compact, $T(\overline{X}, -v) \cup D(v)$ is also compact, and therefore

$$\overline{T(X,-v) \cup D(v)} \subset T(\overline{X}, -v) \cup D(v).$$

The inverse inclusion is also obvious.

(4) Similar to (3). □

Proposition 3.3. *Let $X \subset W \setminus \partial W$ be a compact set such that*

(A) *Every trajectory starting at a point of X reaches $\partial_0 W$;*

(B) *A $(-v)$-trajectory can intersect the set X at most once, that is, for every $a \in X$ we have*

$$\gamma(a,t;-v) \notin X \quad \text{if} \quad t > 0.$$

Then there is a homeomorphism

$$F : X \times [0,1] \xrightarrow{\approx} T(X, -v)$$

such that $F(x \times 1) = x$ for every $x \in X$ and $F(X \times 0) \subset \partial_0 W$.

Proof. Define F by the following formula:

$$F(x,t) = \gamma\big(x, (1-t) \cdot \tau(x,-v); -v\big)$$

(recall that $\tau(x,-v)$ stands for the moment of exit for the point x, that is, $\gamma(x, \tau(x,-v); -v) \in \partial_0 W$). Condition (A) implies that this map is well defined, continuous and surjective; condition (B) implies that the map is injective. By the previous proposition the set $T(X,-v)$ is compact, and therefore the map F is a homeomorphism. □

Remark 3.4. There is another homeomorphism from $X \times [0,1]$ to $T(X,-v)$ which suits better for some applications. Define a map

$$G : X \times [0,1] \to T(X,-v); \quad G(x,t) = \gamma\big(x, t \cdot \tau(x,-v); -v\big),$$

so that

$$G(x,0) = x, \quad G(x,1) \in \partial_0 W \quad \text{for every} \quad x.$$

It is easy to deduce from the above proposition that G is a homeomorphism.

Now we shall prove that the track $T(X,-v)$ is in a sense stable with respect to small perturbations of the vector field v.

Proposition 3.5. *Let $X \subset W$ be a compact subset, and assume that every $(-v)$-trajectory starting at a point of X reaches $\partial_0 W$. Let U be an open subset of W, and R be an open subset of $\partial_0 W$, such that*

$$T(X, -v) \subset U, \quad T(X, -v) \cap \partial_0 W \subset R.$$

Then there is $\epsilon > 0$ such that for every f-gradient w with $||v - w|| < \epsilon$[†] *we have:*

1) *Every $(-w)$-trajectory starting at a point of X reaches $\partial_0 W$.*
2) $T(X, -w) \subset U, \quad T(X, -w) \cap \partial_0 W \subset R.$
3) *Assume that $X \subset \partial_1 W$. Then the maps*

$$(-v)^{\rightsquigarrow}|X, \ (-w)^{\rightsquigarrow}|X : X \to R$$

are homotopic.

Proof. 1) For every $y \in X$ there is a neighbourhood $U = U(y)$ of y and $\epsilon > 0$ such that for every f-gradient w with $||v - w|| < \epsilon$ and every $z \in U$ the $(-w)$-trajectory starting at z reaches $\partial_0 W$ (by Theorem 2.13 of Chapter 1, page 30). Choose a finite family of subsets $U(y)$ covering X and the proof is over.

2) The argument is similar to the proof of Proposition 2.15 of Chapter 1 (page 31). Assume that arbitrarily close to v (in C^0-topology) there are vector fields w with $T(X, -w) \not\subset U$. Choose a sequence v_n of such vector fields converging to v. Let

$$\gamma(x_n, t_n; -v_n) \notin U \quad \text{where} \quad x_n \in X, t_n \geqslant 0.$$

Since X is compact, we can assume (extracting a subsequence if necessary) that $x_n \to x_0 \in X$. Observe that the sequence t_n is bounded from above. (Indeed, the condition $t_n \to \infty$ would imply, by Theorem 2.8 of Chapter 1 (page 27), that $\gamma(x, \cdot; -v)$ is defined on the whole of $\mathbf{R}_+$.) Therefore, extracting a subsequence if necessary, we can assume that t_n converges to a number t_0, and therefore $\gamma(x_n, t_n; -v_n) \to \gamma(x_0, t_0; -v)$. Since $W \setminus U$ is compact, we have $\gamma(x_0, t_0; -v) \in W \setminus U$ and this leads to a contradiction. We have proved therefore that for each vector field w sufficiently close to v in C^0-topology, we have $T(X, -w) \subset U$.

The inclusion $T(X, -w) \cap \partial_0 W \subset R$ for every w sufficiently close to v follows from the previous argument, if we choose $U = (W \setminus \partial_0 W) \cup R$.

3) Choose $\epsilon > 0$ so small that the points 1) and 2) hold. Let w be an f-gradient with $||w - v|| < \epsilon$. For $t \in [0, 1]$ put

$$v_t = (1 - t)v + tw,$$

[†] Recall that $||\cdot||$ stands for the C^0-norm.

so that v_t is an f-gradient. The map

$$H : X \times [0,1] \to \partial_1 W, \quad H(x,t) = (-v_t)^{\leadsto}(x)$$

is continuous as it follows from Theorem 2.13 of Chapter 1 (page 30) and

$$H(x,0) = (-v)^{\leadsto}, \quad H(x,1) = (-w)^{\leadsto},$$

therefore H is a homotopy from $(-v)^{\leadsto}$ to $(-w)^{\leadsto}$. □

The next proposition is a generalization of the previous one; the proof is similar and will be left to the reader.

Proposition 3.6. *Let X, X' be compact subsets of $\partial_1 W$ with $X' \subset X$. Assume that every $(-v)$-trajectory starting at a point of X reaches $\partial_0 W$ and let $Q \subset \partial_0 W$ be an open subset of $\partial_0 W$ with*

$$(-v)^{\leadsto}(X') \subset Q.$$

Then there is $\epsilon > 0$ such that for every f-gradient w with $||v - w|| < \epsilon$ we have:

1) *Every $(-w)$-trajectory starting at a point of X reaches $\partial_0 W$.*
2) *$(-w)^{\leadsto}(X') \subset Q$.*
3) *The homomorphisms in the relative homology, induced by the maps of pairs*

$$(-v)^{\leadsto}, (-w)^{\leadsto} : (X, X') \to (\partial_0 W, Q)$$

are equal. □

Proposition 3.7. *Let $X \subset W$ be a compact subset. Let U be an open subset of W, and R be an open subset of $\partial_0 W$. Assume that*

$$T(X, -v) \cup D(v) \subset U, \quad \big(T(X, -v) \cup D(v)\big) \cap \partial_0 W \subset R.$$

Then there is $\epsilon > 0$ such that for every f-gradient w with $||v - w|| < \epsilon$ we have:

$$T(X, -w) \cup D(w) \subset U, \quad \big(T(X, -w) \cup D(w)\big) \cap \partial_0 W \subset R.$$

Proof. Let $K = W \setminus U$. It is easy to see that the condition

$$T(X, -w) \cup D(w) \subset U \tag{3}$$

holds if and only if every w-trajectory starting at a point of K reaches $\partial_1 W$ and $T(K, w) \cap X = \varnothing$. Applying the preceding proposition to the vector field $(-v)$ and the compact K and the open neighbourhood $W \setminus X$ of $T(K, v)$ we deduce that the condition (3) holds for every w sufficiently C^0-close to v.

The inclusion

$$\big(T(X,-w)\cup D(w)\big)\cap \partial_0 W \subset R$$

for every w sufficiently close to v follows from the previous argument, if we choose $U = (W\setminus \partial_0 W)\cup R$. □

The preceding proposition implies in particular that the descending discs of an f-gradient are stable with respect to small perturbations of the gradient. In the next proposition we show that the same holds for the δ-thickenings of the descending discs.

Proposition 3.8. *Let $f : W \to [a,b]$ be a Morse function on a cobordism W and v an f-gradient. Let $\delta > 0$. Let U be an open neighbourhood of $D_\delta(v)$. Let R be an open neighbourhood of $D_\delta(v)\cap \partial_0 W$ in $\partial_0 W$. Then there is $\epsilon > 0$ such that for every f-gradient w with $||w - v|| < \epsilon$ we have:*

$$D_\delta(w)\subset U,$$
$$D_\delta(w)\cap \partial_0 W \subset R.$$

Proof. Let $K = W\setminus U$, then K is a compact subset of W. Every v-trajectory starting at a point of K reaches $\partial_1 W$, and

$$T(K,v)\subset W\setminus \mathcal{D}_\delta, \quad \text{where} \quad \mathcal{D}_\delta = \bigcup_{p\in S(f)} D_\delta(p).$$

Applying Proposition 3.5 we deduce that there is $\epsilon > 0$ such that

$$T(K,w)\subset W\setminus \mathcal{D}_\delta \quad \text{for every} \quad ||w-v|| < \epsilon.$$

Therefore for every f-gradient w with $||w - v|| < \epsilon$ we have: $U \supset D_\delta(w)$. The second inclusion is proved similarly. □

Here is still another C^0-stability property.

Proposition 3.9. *Let $K \subset W$ be a compact subset. Let $\delta > 0$. Assume that $K\subset B_\delta(v)$. Then there is $\epsilon > 0$ such that for every f-gradient w with $||w - v|| < \epsilon$ we have:*

$$K\subset B_\delta(w).$$

Proof. Let

$$\mathcal{B}_\delta = \bigcup_{p\in S(f)} B_\delta(p).$$

For any $x\in K$ there is $t = t(x)\geqslant 0$ such that $\gamma(x,t;v)\in \mathcal{B}_\delta$. By Corollary 2.11 of Chapter 1 (page 29) there is a neighbourhood $U(x)$ of x and $\epsilon = \epsilon(x) > 0$ such that for every $y\in U(x)$ and every f-gradient w with $||w - v|| < \epsilon(x)$ we have $\gamma(y, t(x); w)\in \mathcal{B}_\delta$. The subsets $\{U(x)\}_{x\in K}$ form an open covering of K. Choose a finite subcovering $\{U(x_i)\}_{1\leqslant i\leqslant n}$; then the number $\epsilon = \min_i \epsilon(x_i)$ satisfies the requirements of the proposition. □

Corollary 3.10. *Let $0 < \eta < \delta$. There is $\epsilon > 0$ such that for every f-gradient w with $||w - v|| < \epsilon$ we have:*

$$D_\eta(v) \subset B_\delta(w). \qquad \square$$

3.2. A homotopy derived from the gradient descent. Let $f : W \to [a, b]$ be a Morse function on a cobordism W and v an f-gradient. If $\partial W = \varnothing$, then every maximal integral curve of $(-v)$ is defined on the whole of $\mathbf{R}$ and we obtain a one-parameter group $\Phi(-v, t)$ of diffeomorphisms of W, defined by the formula

$$\Phi(-v, t)(x) = \gamma(x, t; -v) \quad \text{for} \quad x \in W, \ t \in \mathbf{R}.$$

The aim of the present section is to generalize this construction to the case when $\partial W \neq \varnothing$.

Recall that for $x \in W$ we denote by $\tau(x, -v)$ the moment of the intersection of the curve $\gamma(x, \cdot; -v)$ with $\partial_0 W$, if this curve reaches $\partial_0 W$. Extend the function $\tau(\cdot, -v)$ to the whole of W, setting $\tau(x, -v) = \infty$ if the trajectory never reaches $\partial_0 W$. Define a map $L : W \times \mathbf{R}^+ \longrightarrow W$ as follows:

$$\begin{aligned} L(x, t) &= \gamma(x, t; -v) && \text{if } t \leqslant \tau(x, -v), \\ L(x, t) &= \gamma(x, \tau(x, -v); -v) && \text{if } t \geqslant \tau(x, -v). \end{aligned}$$

In other words, the curve

$$t \mapsto L(x, t), \quad \mathbf{R}_+ \to W$$

follows the $(-v)$-trajectory until the moment when this trajectory reaches $\partial_0 W$; at this moment the curve stops and does not move any more. The map L depends obviously on v and if we want to stress this dependence we add the symbol v to the notation and write $L(x, t; -v)$ instead of $L(x, t)$. The properties of the map L are listed in the next proposition (the proof is obvious).

Proposition 3.11. 1. *$L : W \times \mathbf{R}^+ \to W$ is a continuous map.*

2. *$L(x, t) = x$ for $x \in \partial_0 W$.*
3. *For any $t, t' \in \mathbf{R}_+$ and any $x \in W$ we have*

$$L\big(L(x, t), t'\big) = L(x, t + t').$$

4. *If $\partial_0 W = \varnothing$, then $L(x, t) = \Phi(-v, t)(x)$.* $\square$

This proposition shows that the map

$$L : W \times \mathbf{R}_+ \to W$$

is a one-parameter semigroup of continuous maps, which is a natural generalization of the one-parameter group of diffeomorphisms associated to $(-v)$ in the case $\partial W = \varnothing$.

The next proposition states one of the basic properties of the gradient descent. Roughly, it says that moving downwards along the integral curves of the gradient will shrink the whole manifold to a small neighbourhood of the union of descending discs if we wait sufficiently long.

Proposition 3.12. *Let $f : W \to [a,b]$ be a Morse function on a cobordism W and v an f-gradient. Let U be a neighbourhood of $D(v)$. There is $T > 0$ such that for $t \geqslant T$ and every $x \in W$ we have: $L(x,t) \in U \cup \partial_0 W$.*

Proof. Let $\delta > 0$ be so small that $B_\delta(v) \subset U$ (such δ exists since the sets $B_\delta(v)$ form a fundamental system of neighbourhoods of $D(v)$, see Proposition 2.34, page 98.

Lemma 3.13. *Put $K = W \setminus B_\delta(v)$. For every $x \in K$ there is $t \geqslant 0$ such that $L(x,t) \in B_\delta(v) \cup \partial_0 W$.*

Proof. The contrary would mean that for some $x \in K$ the $(-v)$-trajectory starting at x does not reach $\partial_0 W$ and does not converge to a critical point of f; this is impossible. □

Lemma 3.14. *For every $x \in K$ there is a neighbourhood $U(x)$ of x and a number $t = t(x) \geqslant 0$ such that $L(y,t) \in B_\delta(v) \cup \partial_0 W$ for every $y \in U(x)$.*

Proof. Let $x \in K$ and pick $t \geqslant 0$ such that $L(x,t) \in B_\delta(v) \cup \partial_0 W$. Let us begin with the case $L(x,t) \in \partial_0 W$. The curve $\gamma(x,\cdot;-v)$ reaches $\partial_0 W$ at some moment $t_0 \leqslant t$.Therefore there is a neighbourhood $U(x)$ of x such that for every $y \in U(x)$ the curve $\gamma(y,\cdot;-v)$ reaches $\partial_0 W$ at the moment $t(y) \leqslant t_0 + 1$. Thus the conclusion of our lemma holds with $t = t_0 + 1$.

Consider now the other case: $\gamma(x,t;-v) \in B_\delta(v) \setminus \partial_0 W$. Since the set $B_\delta(v) \setminus \partial_0 W$ is open, there is an open neighbourhood $U(x)$ of x such that for every $y \in U(x)$ we have $\gamma(y,t;-v) \in B_\delta(v) \setminus \partial_0 W$ and the conclusion of the lemma holds in this case as well. □

Now we can complete the proof of the proposition. The neighbourhoods $U(x)$ constructed in Lemma 3.14 form an open covering of the compact K. Choose a finite covering $U(x_i)$ and put $T = \max_i t(x_i)$. For every $x \in K$ the $(-v)$-trajectory starting at x intersects $B_\delta(v) \cup \partial_0 W$ at least once at some moment $\leqslant T$. Since $B_\delta(v) \cup \partial_0 W$ is $(-v)$-invariant, we have

$$L(x,T) \in B_\delta(v) \cup \partial_0 W \quad \text{for every} \quad x \in K.$$

□

3.3. A deformation retraction derived from the gradient descent. The proposition below is very important for the Morse theory. It says that the homotopy type of the manifold $f^{-1}([a, x])$ does not change when the value of x varies without passing through a critical point. The proof is one more variation on the gradient descent theme.

Proposition 3.15. *Let $f : W \to [a, b]$ be a Morse function on a cobordism W and v an f-gradient. Let c, d be real numbers such that $a \leqslant c \leqslant d \leqslant b$ and the interval $[c, d]$ is regular. Then the cobordism $W_0 = f^{-1}\big([a, c]\big)$ is a deformation retract of the cobordism $W_1 = f^{-1}\big([a, d]\big)$ and also of the open set $U = f^{-1}(\ [a, d[\ )$.*

Proof. Let $W' = f^{-1}([c, d])$. The Morse function

$$f|W' : W' \to [c, d]$$

has no critical points and the domain of definition of the functions $\tau(\cdot, -v|W')$ and $E(\cdot, -v|W')$ is the whole of W'. The deformation retraction

$$H : W_1 \times [0, 1] \to W_1$$

is defined as follows:

$$H(x, t) = \gamma\Big(x, t \cdot \tau(x, -v|W'); -v|W'\Big) \qquad \text{for} \quad x \in W',$$

$$H(x, t) = x \qquad \text{for} \quad x \in W_0.$$

The same formula defines a deformation retraction of U onto W_0. □

Exercise 3.16. Prove that the cobordism W_0 from the previous proposition is *diffeomorphic* to W_1.

Exercises to Part 1

(1) Let W be a cobordism, f a Morse function on W, and v a weak f-gradient, such that every zero of v is elementary.
  (a) Show that the main results of Section 2 of Chapter 3 generalize to this setting. Prove that the descending manifold of any critical point is a submanifold of W with boundary.[†]
  (b) Denote by $g(f)$ the set of all weak gradients, such that every zero of v is elementary. Construct an example when $g(f)$ is not an open subset of $\mathcal{V}(f)$.
  Hint: Consider the function
  $$f : \mathbf{R}^2 \to \mathbf{R}, \quad f(x, y) = xy$$
  and a weak f-gradient
  $$v(x, y) = (-x + y^3, y + x^3).$$
  Consider the vector field
  $$v_\alpha(x, y) = (-x + \alpha y + y^3, y + x^3)$$
  with α a small negative number.

(2) The aim of this exercise is to prove a generalization of Proposition 2.4 (page 83). Let $f : W \to [a, b]$ be a C^∞ function on a cobordism such that $f^{-1}(b) = \partial_1 W, f^{-1}(a) = \partial_0 W$. Assume that the set $S(f)$ of critical points of f is finite and $S(f) \cap \partial W = \varnothing$. (We do not assume here that f is a Morse function.) Let v be any C^∞ vector field on W, such that $f'(x)(v(x)) > 0$ for every $x \notin S(f)$. Prove that for every v-trajectory $\gamma(x, \cdot; v)$ one of the following possibilities holds:
  (a) $\lim_{t\to\infty} \gamma(x, t; v) \to p$ for some $p \in S(f)$.
  (b) The trajectory $\gamma(x, \cdot; v)$ reaches the upper boundary $\partial_1 W$.

† The Morse theory for vector fields satisfying the condition above was developed by G. Minervini [**97**].

Part 2

Transversality, handles, Morse complexes

The descending discs of a gradient of a Morse function on a closed manifold form a stratification of the manifold, which is the Morse-theoretic counterpart of CW decompositions. In this part we discuss these stratifications and related classical issues, such as the Kupka-Smale theorem, Rearrangement Lemma, handles, and Morse complexes.

CHAPTER 4

The Kupka-Smale transversality theory for gradient flows

Let M be a closed manifold, $f : M \to \mathbf{R}$ a Morse function and v an f-gradient. The manifold M is stratified by the stable manifolds of critical points, and each stable manifold $D(p,v)$ is diffeomorphic to an open disc of dimension $\operatorname{ind} p$. This stratification is a cellular decomposition provided that the exterior boundary $\mathscr{B}(D(p,v)) = \overline{D(p,v)} \setminus D(p,v)$ of each disc $D(p,v)$ is contained in the union of all discs of dimensions strictly less than $\dim D(p,v) = \operatorname{ind} p$, that is:

$$\mathscr{B}(D(p,v)) \subset \bigcup_{\operatorname{ind} q < \operatorname{ind} p} D(q,v).$$

It turns out that this condition can be reformulated entirely in terms of the flow lines of the vector field v. Let us first introduce two basic notions.

Definition 0.1. Let γ be a flow line of v, and $p, q \in S(f)$. We say that γ is a *v-link from p to q* (or *γ joins p with q*) if

$$\lim_{t\to-\infty} \gamma(x,t;v) = p, \qquad \lim_{t\to+\infty} \gamma(x,t;v) = q.$$

Definition 0.2. We say that v satisfies the *almost transversality condition*, (or *v is almost transverse*) if there are no v-links from p to q for $\operatorname{ind} p \geqslant \operatorname{ind} q, p \neq q$.[†]

We prove in this chapter that v is almost transverse if and only if the corresponding stratification by the descending discs is a cellular decomposition (Theorem 3.51). We show that the subset of almost transverse gradients is dense in the set $G(f)$ of all f-gradients with respect to C^∞ topology, and open in $G(f)$ with respect to C^0 topology (Subsection 3.6). These results are related to one of the classical topics of differential topology: the Kupka-Smale Transversality Theory, which was developed in the original papers [**152**] by S. Smale and [**80**] by I. Kupka. Recall that a vector field on a manifold is called *transverse* if it has only hyperbolic zeros, and

[†] The almost transverse gradients were introduced in the author's paper [**111**], where they were called *almost good* gradients.

all stable and unstable manifolds of zeros of the vector field are transverse to each other. The Kupka-Smale Transversality Theory implies that the transversality condition is generic; in precise terms, the set of all transverse C^∞ vector fields is residual in the set of all C^∞ vector fields (with respect to C^∞ topology).

We shall be working only with vector fields which are gradients of Morse forms, and we need a version of the Kupka-Smale theory adapted for our purposes. The basic result here is Theorem 2.5, which says that for a given Morse form ω the transversality condition is generic in the set of ω-gradients. In precise terms, the set of all transverse ω-gradients is residual in the set of all ω-gradients. The main ideas of the proof (Sections 1 and 2) are the same as in the classical papers of I. Kupka and S. Smale. Our presentation is close to the paper of M. Peixoto [**123**].

We shall need one more classical result of differential topology (also related to the transversality theory). This is the *Rearrangement Lemma* (Theorem 3.17 of Section 3), which implies in particular that for every almost transverse gradient v of a Morse function f, there is a Morse function g such that v is a g-gradient and

$$g(p) < g(q) \quad \text{whenever} \quad \operatorname{ind} p < \operatorname{ind} q.$$

This theorem appeared first in S.Smale's work on the h-cobordism conjecture in the context of handle decompositions of manifolds (see for example [**149**], Theorem 7.1, and [**150**], Theorem B). Our exposition follows §4 of J. Milnor's book [**92**], with minor modifications.

1. Perturbing the Lyapunov discs

Let ω be a closed 1-form on a manifold M without boundary, and u be an ω-gradient. Let $p, q \in S(\omega)$. Let Δ be a Lyapunov stable disc for u at p and ∇ a Lyapunov unstable disc for u at q, such that $\Delta \cap \nabla = \varnothing$. For a vector field v and $T > 0$ let us denote

$$\vec{v}_T(\nabla) = \bigcup_{\tau \in [0,T[} \Phi(v,\tau)(\nabla).$$

The set $\vec{v}_T(\nabla)$ is therefore the trajectory of the set ∇ with respect to the action of diffeomorphisms $\Phi(v,\tau)$ with $\tau \in [0, T[$. If supp $(u - v)$ does not intersect ∇, then ∇ is also an unstable Lyapunov disc for v at q and

$$\vec{v}_T(\nabla) = \Phi(v,T)(\nabla \setminus \partial\nabla).$$

Thus the manifold $\vec{v}_T(\nabla)$ is a part of the unstable manifold for v at q.

Theorem 1.1. *For any $T > 0$, any neighbourhood U of $\partial\Delta$ and any neighbourhood $\mathcal{B}$ of u in $G(\omega)$ there is a vector field $v \in \mathcal{B}$ such that*

(1) $$\operatorname{supp}(v-u) \subset U \setminus (\Delta \cup \nabla),$$

(2) $$\Delta \pitchfork \vec{v}_T(\nabla).$$ †

The proof of this theorem occupies the present section.

1.1. Admissible vector fields. Let Σ denote the boundary of the Lyapunov disc Δ. Pick a C^∞ embedding

$$\psi : \Sigma \times B^{m-k}(0,1) \to M, \qquad \text{with} \quad \psi(\sigma, 0) = \sigma \quad \text{for every} \quad \sigma \in \Sigma$$

transverse to the vector field u (where $k = \operatorname{ind} p$, and $m = \dim M$). Its image is an $(m-1)$-dimensional submanifold of M. Let $0 < \rho < 1$; put

$$\mathcal{F}_\rho = \psi\big(\Sigma \times B^{m-k}(0,\rho)\big),$$

then $\mathcal{F}_\rho$ is also an $(m-1)$-dimensional submanifold transverse to u. It will be called the *fence.* Diminishing ρ if necessary we can assume that $\mathcal{F}_\rho \cap \Delta = \Sigma$. These objects are depicted in Figure 12. The Lyapunov disc Δ is shaded, several flow lines are depicted by rays. The fence $\mathcal{F}_\rho$ is represented by the boundary surface of the cylinder. The thick curve on the cylinder depicts the subset

$$\mathcal{Y} = \vec{v}_T(\nabla) \cap \mathcal{F}_\rho.$$

Lemma 1.2. *Let v be a vector field, transverse to $\mathcal{F}_\rho$. Assume that* supp $(u-v)$ *does not intersect ∇.*

Then $\mathcal{Y}$ is a submanifold of $\mathcal{F}_\rho$, and transversality of $\vec{v}_T(\nabla)$ to Δ is equivalent to transversality of $\mathcal{Y}$ to Σ inside $\mathcal{F}_\rho$.

Proof. Since supp $(u-v)$ does not intersect ∇ the vector field v is tangent to $\vec{v}_T(\nabla)$, hence the transversality of v to $\mathcal{F}_\rho$ implies the transversality of $\vec{v}_T(\nabla)$ and $\mathcal{F}_\rho$, and $\mathcal{Y}$ is therefore a submanifold of $\mathcal{F}_\rho$.

If $\mathcal{Y}$ and Σ are transverse as submanifolds of $\mathcal{F}_\rho$, then at every point $x \in \mathcal{Y} \cap \Sigma$ we have

$$T_x\mathcal{Y} + T_x\Sigma = T_x\mathcal{F}_\rho,$$

and since

$$T_x\vec{v}_T(\nabla) = T_x\mathcal{Y} \oplus \{v(x)\}, \quad T_x\Delta = T_x\Sigma \oplus \{v(x)\}$$

† Recall that $\pitchfork$ means *"is transverse to"*.

the manifolds $v_T(\nabla)$ and Δ are transverse in any point $x \in \Sigma$. Applying the flow diffeomorphism it is easy to deduce that they are transverse also at any point of their intersection.

The proof of the inverse implication

$$\vec{v}_T(\nabla) \pitchfork \Delta \Rightarrow \mathcal{Y} \pitchfork_{\mathcal{F}_\rho} \Sigma$$

is similar. □

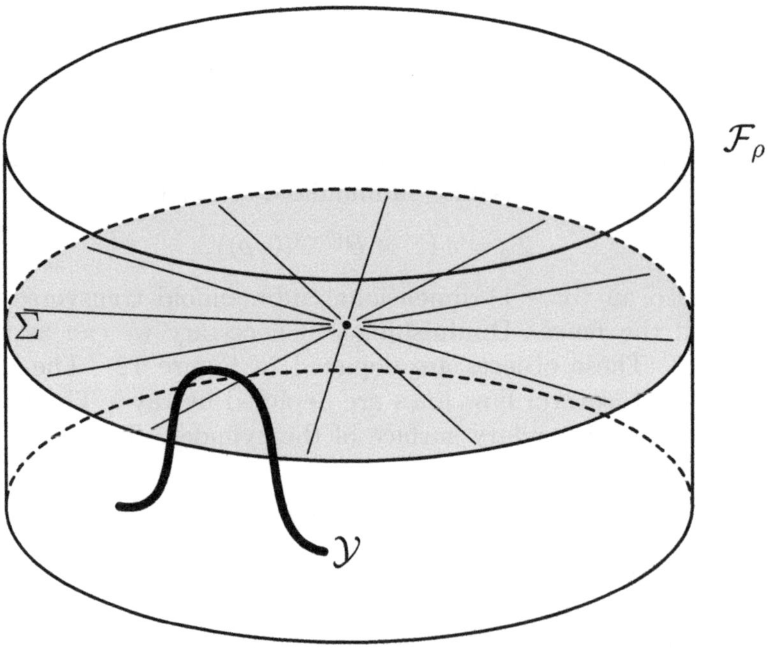

FIGURE 12.

Define a C^∞ map

$$H : \mathcal{F}_\rho \times]-\epsilon, \epsilon[\to M, \quad H(x,\tau) = \gamma(x,\tau;u).$$

We shall assume that the positive numbers ϵ, ρ are chosen sufficiently small so that the map H is a C^∞ embedding onto an open subset of M, and that the image of H does not intersect ∇. Note that by the definition

$$(H^{-1})_*(u) = (0,1).$$

Definition 1.3. Put

$$\mathcal{C}_\rho = H\Big(\mathcal{F}_\rho \times [-\epsilon/2, 0]\Big).$$

A vector field v on M will be called *admissible* if

(1) supp $(u - v) \subset \operatorname{Int} \mathcal{C}_\rho$;
(2) The second component of $(H^{-1})_*(v)$ is equal to 1.

An admissible vector field is in a sense a perturbation of the initial vector field u, and the integral curves of an admissible vector field v can be viewed as perturbations of the integral curves of u. Namely, let γ be a maximal integral curve of v. The set

$$\gamma^{-1}(\mathcal{C}_\rho) = \{t \in \mathbf{R} \mid \gamma(t) \in \mathcal{C}_\rho\}$$

is a finite or countable union of closed intervals $I_\alpha, \alpha \in \mathcal{J}$ (where $\mathcal{J}$ is a finite or countable set) each of length $\epsilon/2$. We have:

$$\gamma(t) \in \operatorname{supp}(u - v) \Rightarrow t \in \operatorname{Int} I_\alpha \quad \text{for some} \quad \alpha.$$

The intervals I_α will be called *deviation intervals* for γ, since outside these intervals the curve γ is also an integral curve of u. An integral curve of v can enter $\mathcal{C}_\rho$ only through a point of

$$\mathcal{H}_\rho = H(\mathcal{F}_\rho \times (-\epsilon/2)),$$

and it must quit $\mathcal{C}_\rho$ through a point of $\mathcal{F}_\rho$. Figure 13 is an illustration.

Figure 13 depicts the case $\dim M = 2$. In the lightly shaded rectangle the field u has coordinates $(-x, y)$, so that its integral curves are usual hyperbolas. The set $\mathcal{C}_\rho$ is dark shaded. In this domain the integral curves of an admissible vector field v deviate from the integral curves of u. One integral curve of an admissible vector field v is shown in the picture by a thick line.

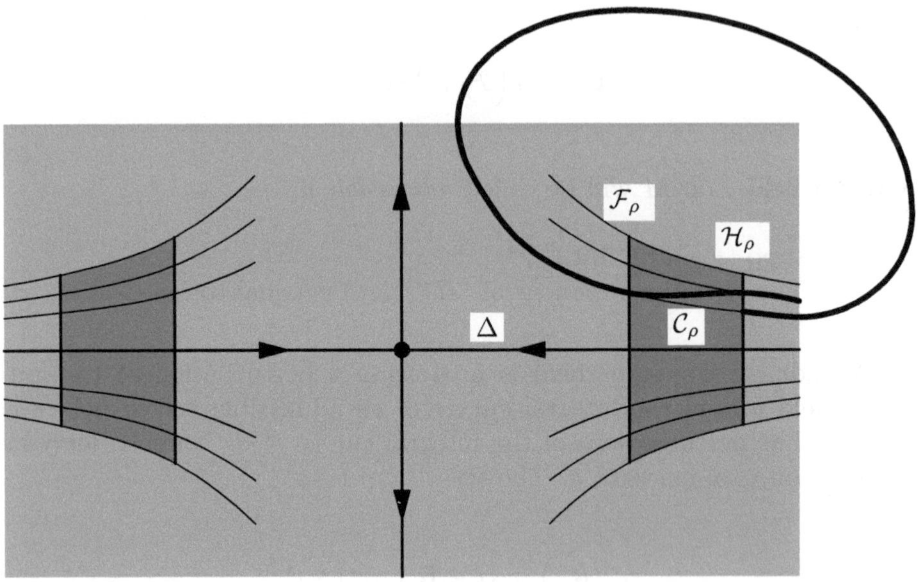

FIGURE 13.

1.2. Time of the first return for admissible fields.

Definition 1.4. Let v be a C^∞ vector field on M and $K \subset M$ be a compact subset. A point $x \in K$ is called an *exit point* if $\gamma(x,t;v) \notin K$ for all t in some interval $]0,\epsilon[$. The set of exit points will be denoted $E(K,v)$. For $x \in E(K,v)$ the moment of the first return of $\gamma(x,\cdot;v)$ to K will be denoted by

$$Ret(x,v) = \min\{t > 0 \mid \gamma(x,t;v) \in K\}.$$

If $\gamma(x,t;v) \notin K$ for every $t > 0$ we put $Ret(x,v) = \infty$. If $E(K,v) \neq 0$ then the number

$$Ret(K,v) = \inf_{x \in E(K)} Ret(x,v)$$

is called *the time of the first return to K with respect to v.*

Thus a v-trajectory which has quit K at some moment t_0 will stay outside of K for the values of parameter $t \in]t_0, Ret(K,v) + t_0[$.

The next lemma says that if the fence $\mathcal{F}_\rho$ is very low, the time of the first return to the set

$$\mathcal{C}_\rho = H\Big(\mathcal{F}_\rho \times [-\epsilon/2, 0]\Big)$$

for any admissible vector field is very large.

Lemma 1.5. *Let $T > 0$. There is $\rho = \rho(T)$ such that for every admissible vector field v the time of the first return to $\mathcal{C}_\rho$ with respect to v is not less than T.*

Proof. Let us first outline the main idea of the proof. For every admissible vector field v the set Δ is a Lyapunov disc for v, therefore, the v-trajectory starting at any point of $\Sigma = \partial\Delta$ converges to p and never returns to Σ, so that $Ret(\Sigma, v) = \infty$. If the fence $\mathcal{F}_\rho$ is very low, it is in a small neighbourhood of Σ hence the time of the first return to $\mathcal{F}_\rho$ must be very large.

Now let us proceed to precise arguments. Pick some $\rho > 0$. It is clear that the set of exit points $E(\mathcal{C}_\rho, v)$ is equal to $\mathcal{F}_\rho$. Moreover for any $\epsilon > 0$ sufficiently small and any admissible vector field v we have

$$\gamma(x,t;u) = \gamma(x,t;v) \notin \mathcal{C}_\rho \quad \text{for every} \quad t \in]0,\epsilon[, \quad \text{and every} \quad x \in \mathcal{F}_\rho.$$

For every $x \in \Sigma$ the u-trajectory starting at x converges to p and never returns to Σ. In particular

$$\gamma(x,t;u) \notin \overline{\mathcal{C}_\rho} \quad \text{for every} \quad t \in [\epsilon, T] \quad \text{and} \quad x \in \Sigma.$$

Both Σ and $\overline{\mathcal{C}_\rho}$ are compact, therefore there is a neighbourhood A of Σ such that

$$\gamma(x,t;u) \notin \overline{\mathcal{C}_\rho} \quad \text{for every} \quad t \in [\epsilon, T],\ x \in A.$$

Let ρ' be sufficiently small, so that $\mathcal{F}_{\rho'} \subset A$. Then for every $x \in \mathcal{F}_{\rho'}$ the u-trajectory starting at x is also a v-trajectory on the interval $[\epsilon, T]$. Therefore

$$\gamma(x,t;u) = \gamma(x,t;v) \quad \text{for} \quad \tau \in [\epsilon, T]$$

and

$$\gamma(x,t;v) \notin \overline{\mathcal{C}_\rho} \quad \text{for every} \quad t \in [\epsilon, T] \quad \text{and} \quad x \in \mathcal{F}_{\rho'}.$$

According to our choice of $\epsilon > 0$ we have also $\gamma(x,t;v) \notin \mathcal{C}_\rho$ for every $t \in]0,\epsilon[$ and every $x \in \mathcal{F}_{\rho'}$. Put $\rho(T) = \rho'$ and the proof of the lemma is complete. □

Let us now return to the proof of Theorem 1.1. Choose ρ and ϵ sufficiently small so that

$$H\Big(\mathcal{F}_\rho \times]-\epsilon, \epsilon[\Big) \subset U.$$

Then every admissible vector field satisfies the condition (1) of the theorem. Our aim is therefore to construct an admissible vector field v sufficiently close to u such that

$$\vec{v}_T(\nabla) \pitchfork \Delta.$$

By Lemma 1.2 this condition is equivalent to the following one:

$$\Sigma \pitchfork \left(\vec{v}_T(\nabla) \cap \mathcal{F}_\rho\right) \quad \text{in} \quad \mathcal{F}_\rho.$$

Let us associate to each admissible vector field v a diffeomorphism

$$\mathcal{T}_v : \mathcal{H}_\rho \to \mathcal{F}_\rho; \quad \mathcal{T}_v(x) = \gamma(x, \epsilon/2; v).^\dagger$$

It is clear that

$$\vec{v}_T(\nabla) \cap \mathcal{F}_\rho \subset \mathcal{T}_v\left(\vec{v}_T(\nabla) \cap \mathcal{H}_\rho\right).$$

Lemma 1.6. *Let v be an admissible vector field and assume that the time of the first return to $\mathcal{C}_\rho$ with respect to v is more than T. Then*

$$\vec{u}_T(\nabla) \cap \mathcal{H}_\rho = \vec{v}_T(\nabla) \cap \mathcal{H}_\rho.$$

Proof. We will prove the inclusion

$$\vec{u}_T(\nabla) \cap \mathcal{H}_\rho \supset \vec{v}_T(\nabla) \cap \mathcal{H}_\rho,$$

the proof of the inverse inclusion is similar. Let

$$x = \gamma(z, t_0; v) \in \mathcal{H}_\rho, \quad \text{where} \quad 0 \leqslant t_0 < T, \quad \text{and} \quad z \in \nabla.$$

We wish to prove that

$$x = \gamma(z, t_0; u).$$

Observe that t_0 is the left extremity of an interval of deviation for the integral curve $\gamma(z, \cdot; v)$. If $x \neq \gamma(z, t_0; u)$, then there is a point $t_1 \in [0, t_0[$ belonging to another interval of deviation of $\gamma(z, \cdot; v)$. We have $\gamma(z, t_1; v) \in \mathcal{C}_\rho$, and therefore the time of the first return for $\mathcal{C}_\rho$ is less than t_0 which contradicts the assumptions of the lemma. □

Let

$$N = \vec{u}_T(\nabla) \cap \mathcal{H}_\rho.$$

To prove Theorem 1.1 it suffices therefore to construct an admissible vector field $v \in \mathcal{B}$ such that

$$\mathcal{T}_v(N) \pitchfork \Sigma \quad \text{in} \quad \mathcal{F}_\rho.$$

This is all about vector fields and submanifolds in $\mathcal{C}_\rho$, and we will do the construction using a suitable parametrization of this domain. Recall that by definition the fence $\mathcal{F}_\rho$ is the image of an embedding

$$\psi : \Sigma \times B \to \mathcal{F}_\rho, \quad \text{where} \quad B = B^{m-k}(0, \rho).$$

---

† Recall that by definition $\mathcal{H}_\rho = H^{-1}(\mathcal{F}_\rho \times (-\epsilon/2))$.

Consider a diffeomorphism

$$\Gamma : \Sigma \times B \times \,]-\epsilon, \epsilon[\; \xrightarrow{\approx} V,$$
$$\Gamma(\sigma, b, t) = \gamma\big(\psi(\sigma, b), t; u\big)$$

(here V is some open neighbourhood of Σ). Then

$$\Gamma\Big(\Sigma \times B \times 0\Big) = \mathcal{F}_\rho, \quad \Gamma\Big(\Sigma \times B \times (-\epsilon/2)\Big) = \mathcal{H}_\rho,$$

and we have

$$\Gamma_*^{-1}(u) = (0, 0, 1).$$

For an admissible vector field v let us denote by $\widetilde{v}$ the image $\Gamma_*^{-1}(v)$, then

(A) $\quad$ supp $(\widetilde{v} - \widetilde{u}) \subset \Sigma \times B \times]-\epsilon/2, 0[$,

(B) $\quad$ the third coordinate of $\widetilde{v}$ equals 1.

We have a diffeomorphism

$$z \mapsto \gamma(z, \epsilon/2; \widetilde{v}); \quad B \times (-\epsilon/2) \to B \times 0,$$

which will be denoted by $\mathscr{T}_{\widetilde{v}}$. Let $\widetilde{N} = \Gamma^{-1}(N)$. Our task now translates as follows: given any neighbourhood $\mathcal{W}$ of $\widetilde{u} = (0, 0, 1)$ in the space of C^∞ vector fields on $\Sigma \times B \times]-\epsilon, \epsilon[$ construct a vector field $w \in \mathcal{W}$ satisfying (A) and (B) above and such that

$$\mathscr{T}_w(\widetilde{N}) \pitchfork \Sigma \quad \text{in} \quad \Sigma \times B \times 0.$$

This is achieved by a standard transversality argument based on the Sard lemma. The reader familiar with transversality theory will easily complete the argument. For the reader less familiar with differential topology we have included Subsection 1.3 where we discuss in detail the necessary notions. The property which we need follows from Proposition 1.9. □

1.3. Time-dependent vector fields and isotopies. In this subsection we review the connection between time-dependent vector fields and isotopies; the main aim is Proposition 1.9, which is used in the proof of Theorem 1.1.

Let M be a manifold without boundary, $I \subset \mathbf{R}$ a non-empty interval (open or closed). The vector space $\mathrm{Vect}^\infty(M \times I)$ of all C^∞ vector fields on $M \times I$ can be considered as the space of *time-dependent vector fields on* M. The vector field with the coordinates $(0, 1)$ will be denoted by v_0.

Definition 1.7. Let M be a manifold, $I \subset \mathbf{R}$ an interval, and $I' = [\alpha, \beta]$ a closed interval contained in the interior of I. Let $K \subset M$ be a compact subset. We denote by $\mathrm{Vect}^\infty(M \times I; K, I')$ the subset of $\mathrm{Vect}^\infty(M \times I)$ formed by all vector fields v such that:

(1) supp $(v - v_0)$ is a subset of $K \times I'$,

(2) the second coordinate of v is equal to 1.

The space $\mathrm{Vect}^\infty(M \times I)$ will be endowed with the weak C^∞ topology. Recall that a sequence $v_n \in \mathrm{Vect}^\infty(M \times I)$ converges to v in this topology if and only if v_n converges to v together with all partial derivatives on every compact $K \subset M \times I$. The subspace $\mathrm{Vect}^\infty(M \times I; K, I')$ will be endowed with the induced topology.

Let $v \in \mathrm{Vect}^\infty(M \times I; K, I')$. For a point $x \in M$, and the values $t_1, t_2 \in I$ such that

$$I' \subset [t_1, t_2] \subset I$$

consider the v-trajectory starting at $(x, t_1) \in M \times t_1$. The value of this trajectory at $t = t_2 - t_1$ is a point (y, t_2) and we will denote the point $y \in M$ by $\mathscr{T}_v(x)$. (It is clear that this point does not depend on the particular choice of t_1, t_2). The map

$$x \mapsto \mathscr{T}_v(x)$$

is then a diffeomorphism of M to itself, which will be called the *shift diffeomorphism* associated to v. For any vector field $v \in \mathrm{Vect}^\infty(M \times I; K, I')$ the diffeomorphism $\mathscr{T}_v$ is isotopic to the identity via the isotopy $\mathscr{T}_{v(t)}$ where $v(t) = (1-t)v + tv_0$.

Let us consider an example. In the next lemma we construct a vector field in the Euclidean space such that the corresponding shift diffeomorphism sends the origin to a given vector h. We show that this vector field can be chosen small if h is small.

Lemma 1.8. *Put $B = B^l(0, R) \subset \mathbf{R}^l$, let $h \in B$. Let I, I' be intervals of $\mathbf{R}$ such that $I' \subset \mathrm{Int}\, I$. Let $0 < r < R$.*

There are continuous maps

$$\sigma, \theta : B^l(0, r/2) \to \mathrm{Vect}^\infty\Big(B \times I; D^l(0, r), I'\Big)$$

such that $\sigma(0) = \theta(0) = v_0$ and for every $h \in B^l(0, r/2)$ we have

$$\mathscr{T}_{\sigma(h)}(0) = h, \quad \mathscr{T}_{\theta(h)}(h) = 0.$$

Proof. We shall construct the map σ, the construction of θ is similar. Pick a C^∞ function

$$\mu : \mathbf{R} \to \mathbf{R}_+ \quad \text{with} \quad \mathrm{supp}\, \mu \subset I', \quad \int \mu(t)dt = 1.$$

For $h \in B^l(0, r/2)$ define a vector field $\lambda(h)$ on $\mathbf{R}^l \times \mathbf{R}$ by the following formula:

$$\lambda(h)(x, t) = (\mu(t) \cdot h, 1).$$

Then the projection to $\mathbf{R}^l$ of the maximal integral curve of $\lambda(h)$ passing through $(h,a) \in \mathbf{R}^l \times \mathbf{R}$ is the segment in $\mathbf{R}^l$, joining 0 and h. The corresponding shift diffeomorphism sends 0 to h. Pick any C^∞ function

$$\chi : \mathbf{R}_+ \to \mathbf{R}_+, \quad \text{with} \quad \chi(u) = 1 \quad \text{for} \quad u \in [0, r/2]$$
$$\text{and} \quad \chi(u) = 0 \quad \text{for} \quad u \in [r, \infty[.$$

The vector field $\sigma(h)$ defined in the next formula has the required properties.

$$\sigma(h)(x,t) = \Big(\chi(||x||) \cdot \mu(t) \cdot h, 1\Big).$$

□

Now let us proceed to the results from transversality theory cited in the previous subsection. Let Σ be a compact manifold, and put

$$I =]-\epsilon, \epsilon[, \quad I' = [-\epsilon/2, 0].$$

As before we denote $B^l(0,R)$ by B. Put $v_0 = (0,0,1)$. Let $K = D^l(0,r)$ where $0 < r < R$. To each vector field

$$w \in \mathrm{Vect}^\infty(\Sigma \times B \times I; \Sigma \times K, I')$$

we have associated a diffeomorphism

$$\mathscr{T}_w : \Sigma \times B \to \Sigma \times B.$$

Proposition 1.9. *Let*

$$N \subset \Sigma \times B$$

be a submanifold. Let $\mathcal{V}$ be a neighbourhood of $v_0 = (0,0,1)$ in the space $\mathrm{Vect}^\infty\Big(\Sigma \times B \times I; \Sigma \times K, I'\Big)$. Then there is a vector field $w \in \mathcal{V}$ such that

$$\mathscr{T}_w(N) \pitchfork \Sigma \times 0 \quad \text{in} \quad \Sigma \times B.$$

Proof. For $h \in B^l(0, r/2)$ let

$$\theta(h) \in \mathrm{Vect}^\infty(\ B \times I; K, I'\ )$$

be the vector field constructed in the previous lemma. Define a vector field $w(h)$ by the following formula:

$$w(h) = (0, \theta(h)) \in \mathrm{Vect}^\infty\Big(\Sigma \times B \times I; \Sigma \times K, I'\Big).$$

We have then

$$\mathscr{T}_{w(h)}(\Sigma \times h) = \Sigma \times 0,$$

therefore $\mathscr{T}_{w(h)}(N)$ is transverse to $\Sigma \times 0$ if and only if N is transverse to $\Sigma \times h$. Observe that $N \pitchfork \Sigma \times h$ if and only if h is a regular value of a restriction to N of the projection $\Sigma \times B \to B$. According to the Sard

Lemma there are arbitrarily small vectors h satisfying this condition. Thus we can choose a vector field $w = w(h)$ arbitrarily close to v_0. □

2. The transverse gradients are generic

Let ω be a Morse form on a closed manifold M, and v be an ω-gradient. For every zero p of ω the stable set $W^{st}(v,p)$ and the unstable set $W^{un}(v,p)$ are images of open discs in Euclidean spaces with respect to C^∞ immersions.

Definition 2.1. We say that v *satisfies the transversality condition* (or v is *transverse*) if

$$W^{st}(v,p) \pitchfork W^{un}(v,q) \text{ for every } p,q \in S(\omega).$$

The set of all ω-gradients, satisfying the transversality condition, is denoted by $G_T(\omega)$.

The main aim of the present section is to prove that the transversality property is generic. We begin with the introductory Subsection 2.1 which reviews some general topology.

2.1. Residual subsets. Recall that a subset of a topological space X is called *residual* in X, if it contains the intersection of a countable family of open and dense subsets of X. The classical Baire theorem ([**72**], Ch. 6, Th. 34) says that if X is a complete metric space, then any residual subset of X is dense in X. We shall need an assertion that is slightly more general than the Baire theorem cited above:

Proposition 2.2. *If $Y \subset X$ is an open subset of a complete metric space, and $A \subset Y$ is a residual subset of Y, then A is dense in Y.*

Proof. Let $B = B(a,r) \subset Y$ be any open ball. We are going to show that $A \cap B \neq \varnothing$. We know that

$$A \supset \bigcap_{n \geqslant 1} A_n$$

where each of $A_n \subset Y$ is open and dense in Y. Since A_1 is open and dense in Y there is $a_1 \in A_1 \cap B$. Let $r_1 > 0$ be so small that

$$D(a_1,r_1) \subset A_1 \cap B \quad \text{and} \quad r_1 \leqslant r/2.$$

Since A_2 is open and dense in Y there exists $a_2 \in A_2 \cap B(a_1,r_1)$. Pick r_2 so small that

$$D(a_2,r_2) \subset B(a_1,r_1) \cap A_2 \quad \text{and} \quad r_2 \leqslant r_1/2.$$

Continuing this procedure, we obtain a sequence of closed balls $D(a_i,r_i)$ such that

$$D(a_i,r_i) \subset B(a_{i-1},r_{i-1}) \cap A_i, \quad \text{and} \quad r_i < r/2^i.$$

In particular we have:

$$D(a_i, r_i) \subset \bigcap_{j=1}^{i} A_j \quad \text{for every} \quad i$$

and $r_i \to 0$ as $i \to \infty$. Since X is a complete metric space, we have

$$\bigcap_{i=1}^{\infty} D(a_i, r_i) = \{b\},$$

with $b \in D(a_1, r_1) \subset B$. Then $b \in B \cap \left(\bigcap_i A_i\right)$ and our proposition is proved. □

Now we are going to prove that the property of being a residual subset of a topological space is in a sense *local.*

Definition 2.3. A subset $A \subset X$ of a topological space X is called *locally residual* if for every point $x \in X$ there is a neighbourhood U of x such that $U \cap A$ is residual in U.

Recall that a *base* of a topological space X is a family $\mathcal{B} = \{U_i\}_{i \in I}$ of open subsets, such that any open subset of X is the union of some members of $\mathcal{B}$. The Lindelöf theorem ([**72**], Ch. 1, Th. 15) says that if a topological space X has a countable base, then every open covering of X contains a countable subcovering.

Lemma 2.4 ([**122**], Lemma 3.3)**.** *Let X be a topological space with a countable base. Any locally residual subset $Y \subset X$ is residual in X.*

Proof. For each point $a \in X$ let $U(a)$ be an open neighbourhood of a such that $Y \cap U(a)$ is residual in $U(a)$. Choose a sequence $\{a_i\}_{i \in \mathbf{N}}$ of points in X such that the sets $U_i = U(a_i)$ cover the whole of X. For each U_i choose open and dense subsets $W_{ij} \subset U_i$ such that $\cap_j W_{ij} \subset Y \cap U_i$. The set

$$V_{ij} = W_{ij} \cup (X \setminus \overline{U_i})$$

is open and dense in X, so it suffices to show that Y contains $\cap_{i,j} V_{ij}$. It is clear that

$$\bigcap_j V_{ij} \subset A_i = (Y \cap U_i) \cup (X \setminus \overline{U_i}),$$

and the intersection of the sets A_i is a subset of Y (indeed, if $z \notin Y$ is in this intersection, then for every i the point z is in $X \setminus \overline{U_i}$ which is impossible). □

2.2. The genericity theorem. Let ω be a Morse form on a closed manifold M. Recall that $G(\omega)$ denotes the set of all ω-gradients, and $G_T(\omega)$ denotes the set of all transverse ω-gradients.

Theorem 2.5. *The set $G_T(\omega)$ is residual in $G(\omega)$.*

Before proceeding to the proof of the theorem let us mention an important corollary.

Corollary 2.6. *The set $G_T(\omega)$ is dense in $G(\omega)$.*

Proof. Endowed with the C^∞ topology the space $\mathrm{Vect}^\infty(M)$ is a complete metric space (see [**142**], §1, or [**53**], page 68). The same is true for its subspace $\mathrm{Vect}^\infty(M, S(\omega))$ formed by all the vector fields vanishing in $S(\omega)$. Recall that $G(\omega)$ is an open subset of $\mathrm{Vect}^\infty(M, S(\omega))$ (Proposition 2.30 of Chapter 2 (page 60)). Therefore Theorem 2.5 together with Proposition 2.2 imply that the set $G_T(\omega)$ is dense in $G(\omega)$. $\square$

Proof of Theorem 2.5. It suffices to show that $G_T(\omega)$ is locally residual (by Lemma 2.4). For a given gradient v_0 we will now define a neighbourhood $\mathcal{V}$ of v_0, together with a countable family of open and dense subsets $\mathcal{T}_n(p,q) \subset \mathcal{V}$, such that the intersection of this family is in $G_T(\omega)$.

We have proved in Proposition 1.15 of Chapter 3 (page 76) that there is a neighbourhood $\mathcal{V}$ of v_0 in $G(\omega)$, and for every $u \in \mathcal{V}$ and every $p \in S(\omega)$ there are stable and unstable Lyapunov charts

$$j_{st}(u,p) : D^k \hookrightarrow M, \quad j_{un}(u,p) : D^{m-k} \hookrightarrow M$$
$$j_{st}(u,p)(0) = p = j_{un}(u,p)(0)$$

(where $k = \operatorname{ind} p$ and $m = \dim M$) depending continuously on u in C^1 topology. Denote the corresponding Lyapunov discs by

$$\Delta_p(u) = j_{st}(u,p)(D^k), \quad \nabla_p(u) = j_{un}(u,p)(D^{m-k}).$$

We can assume that the Lyapunov discs corresponding to different critical points are disjoint.

Definition 2.7. Let $n \in \mathbf{N}$ and $p, q \in S(\omega)$. The set $\mathcal{T}_n(p,q)$ consists of all the ω-gradients $u \in \mathcal{V}$ such that

$$\Phi(-u,n)\big(\Delta_p(u)\big) \pitchfork \Phi(u,n)\big(\nabla_q(u)\big).$$

(Recall that $\Phi(u,t)(x) = \gamma(x,t;u)$.)

We have then

$$\bigcap_{n\in\mathbf{N}, p,q \in S(f)} \mathcal{T}_n(p,q) = G_T(\omega) \cap \mathcal{V}.$$

Theorem 2.5 follows from the next assertion.

Theorem 2.8. *The set $\mathcal{T}_n(p,q)$ is an open and dense subset of $\mathcal{V}$.*

Proof. Let us first prove that $\mathcal{T}_n(p,q)$ is open. The embeddings

$$\Phi(-u,n)\circ j_{st}(u,p) : D^k \hookrightarrow M, \quad \text{where} \quad k=\operatorname{ind} p,$$

$$\Phi(u,n)\circ j_{un}(u,q) : D^{m-l} \hookrightarrow M, \quad \text{where} \quad l=\operatorname{ind} q$$

depend continuously on u in C^1 topology. The transversality property is open with respect to C^1 topology (see [**61**], Ch. 3, §2), and therefore the set $\mathcal{T}_n(p,q)$ is an open subset of $\mathcal{V}$.

The proof of the density property is more complicated. It relies in an essential way on the material of the previous section.

Lemma 2.9. *Let $u\in\mathcal{V}$. Let $t>0$. There is a neighbourhood $\mathcal{V}'\subset\mathcal{V}$ of u such that for any $v\in\mathcal{V}'$ satisfying*

$$\text{supp}\,(v-u)\cap(\Delta_p(u)\cup\nabla_q(u))=\varnothing \tag{3}$$

the condition

$$\Phi(-v,t)\big(\Delta_p(v)\big) \pitchfork \Phi(v,t)\big(\nabla_q(v)\big) \tag{T}$$

follows from the condition

$$\Delta_p(u) \pitchfork \Phi(v,2t+1)\big(\nabla_q(u)\big). \tag{T'}$$

Proof. For any v satisfying (3) $\Delta_p(u)$ is a stable Lyapunov disc for v. It may not coincide with the disc $\Delta_p(v)$ of our family, but we will show now that there is a neighbourhood $\mathcal{V}'\subset\mathcal{V}$ of u such that for any $v\in\mathcal{V}'$ satisfying (3) we have

$$\Delta_p(v)\subset\Phi(-v,1/2)(\Delta_p(u)), \quad \nabla_q(v)\subset\Phi(v,1/2)(\nabla_q(u)). \tag{4}$$

The embedded disc $K=\Phi(-v,1/2)(\Delta_p(u))$ is a stable Lyapunov disc for v at p. For any point $x\in\Delta_p(v)$ the v-trajectory starting at x converges to p. Thus if $x\notin K$ this trajectory intersects the set $\partial K=\Phi(-v,1/2)(\partial\Delta_p(u))$. Observe however that for every v sufficiently close to u we have:

$$\partial K\cap\Delta_p(v)=\Phi(-v,1/2)(\partial\Delta_p(u))\cap\Delta_p(v)=\varnothing.$$

(Indeed, this equality holds for $v=u$. Further, the diffeomorphism $\Phi(-v,$ $1/2)$ and the embedded disc $\Delta_p(v)$ depend continuously on v.) This contradiction shows that

$$\Delta_p(v)\subset K=\Phi(-v,1/2)(\Delta_p(u)).$$

The second inclusion of (4) is similar.

Observe now that for $v\in\mathcal{V}'$ the condition (T') implies the condition

$$\Delta_p(v) \pitchfork \Phi(v,2t)\big(\nabla_q(v)\big),$$

which is obviously equivalent to (T). Our lemma follows. □

Now we can prove that $\mathcal{T}_n(p,q)$ is dense in $\mathcal{V}$. Let $u \in \mathcal{V}$ and let $\mathcal{B} \subset \mathcal{V}$ be any neighbourhood of u. Let $\mathcal{V}' \subset \mathcal{B}$ be a neighbourhood of u in $\mathcal{V}$ such that the conclusion of Lemma 2.9 holds with $t = n$. Applying Theorem 1.1 to the Lyapunov discs $\nabla_q(u)$ and $\Delta_p(u)$ we obtain an ω-gradient $v \in \mathcal{B}$ such that

$$\Delta_p(u) \pitchfork \Phi(v, 2n+1)\big(\nabla_q(u)\big).$$

By Lemma 2.9 we conclude that

$$\Phi(-v,n)\big(\Delta_p(v)\big) \pitchfork \Phi(v,n)\big(\nabla_q(v)\big).$$

Therefore

$$v \in \mathcal{B} \cap \mathcal{T}_n(p,q)$$

and we have proved that the transversality property is dense.

The proof of Theorem 2.5 is now complete. □

The cases of real-valued and circle-valued functions are of special importance for us:

Corollary 2.10. *Let M be a closed manifold and f a real-valued or circle-valued Morse function on M. The subset $G_T(f)$ of transverse f-gradients is dense (with respect to the C^∞ topology) in the space $G(f)$ of all f-gradients.* □

2.3. The case of gradient-like vector fields. Let ω be a Morse form on a closed manifold M, and v a gradient-like vector field for ω. Theorem 2.5 implies that by a small perturbation of v we can obtain an ω-gradient $\widetilde{v}$ satisfying transversality condition. The theorem does not guarantee that $\widetilde{v}$ is a gradient-like vector field for ω, but a closer look at the proof shows that we can achieve it. This is the aim of the present section. We shall first develop the corresponding terminology.

Definition 2.11. A *Morse atlas* for ω is a collection of Morse charts

$$\mathcal{A} = \Big\{\Psi_p : U_p \to B^k(0,R_p) \times B^{m-k}(0,R_p)\Big\}_{p \in S(\omega)}$$

for ω:

$$(\Psi_p^{-1})^*(\omega) = -2\sum_{i=1}^{k} x_i dx_i + 2\sum_{i=k+1}^{m} x_i dx_i, \qquad \Psi_p(p) = 0,$$

one chart for each critical point p of ω, such that the subsets $\{\overline{U_p}\}_{p \in S(\omega)}$ are pairwise disjoint.

Definition 2.12. Let $\mathcal{A}$ be a Morse atlas for ω. An ω-gradient v is called *adjusted to* $\mathcal{A}$ if for every chart Ψ_p of the atlas $\mathcal{A}$ we have

$$(\Psi_p)_*(v)(x_1,\dots,x_m) = (-x_1,\dots,-x_k,x_{k+1},\dots,x_m) \quad \text{where} \quad k = \operatorname{ind} p.$$

In other words, v is adjusted to $\mathcal{A}$ if every chart Ψ_p of $\mathcal{A}$ is a Morse chart for (ω, v) at p.

Every gradient-like vector field is adjusted to some Morse atlas. The set of all gradient-like vector fields for ω, adjusted to a given Morse atlas $\mathcal{A}$ will be denoted $G(\omega,\mathcal{A})$.

Theorem 2.13. *Let ω be a Morse form on a closed manifold M. Let $\mathcal{A}$ be a Morse atlas for ω. The set of all $v \in G(\omega,\mathcal{A})$ satisfying the transversality condition is residual in $G(\omega,\mathcal{A})$.*

Proof. The argument here is similar to the proof of Theorem 2.5 for ω-gradients, and even somewhat simpler, since for every $p \in S(\omega)$ we can choose the same Lyapunov disc for every vector field in $G(\omega,\mathcal{A})$. Let $p \in S(\omega)$, $\operatorname{ind} p = k$. Let

$$\Psi_p : U_p \to B^k(0,R) \times B^{m-k}(0,R)$$

be the corresponding chart from $\mathcal{A}$. Let $\mu : B^k(0,R) \to M$ denote the restriction of Ψ_p^{-1} to the subset $B^k(0,R)\times 0$, which will be identified with $B^k(0,R)$.

Lemma 2.14. *The map μ can be extended to a stable Lyapunov chart $\bar\mu : D^k(0,R) \to M$ for v at p.*

Proof. For $\xi \in \mathbf{R}^k$, $\xi \neq 0$ denote by $\widehat{\xi}$ the vector of length $R/2$ collinear to ξ, that is,

$$\widehat{\xi} = \frac{1}{2}R\frac{\xi}{||\xi||}.$$

Let u_0 denote the vector field $u_0(x) = -x$. For every $\xi \neq 0$ the point ξ and $\widehat{\xi}$ are on the same integral curve of u_0, and

$$\xi = \gamma(\widehat{\xi}, \tau(\xi); u_0) \quad \text{with} \quad \tau(\xi) = \log\Big(\frac{R}{2||\xi||}\Big).$$

Since μ sends the vector field u_0 to $-v$, the map

$$\xi \mapsto \gamma(\mu(\widehat{\xi}\,), \tau(\xi); v)$$

is the extension sought. □

The resulting Lyapunov disc $\bar\mu(D^k(0,R))$ will be denoted Δ_p. Similarly, the embedding $\Psi_p^{-1}|B^{m-k}(0,R)$ can be extended to an unstable Lyapunov chart $D^{m-k}(0,R) \to M$ for v at p, and the corresponding Lyapunov disc will be denoted by ∇_p. For $p, q \in S(\omega)$ and a natural number n let

$$\mathcal{T}_n(p,q) = \{v \in G(\omega, \mathcal{A}) \mid \Delta_p \pitchfork \Phi(v,n)(\nabla_q)\}.$$

It suffices to prove that this set is open and dense in $G(\omega, \mathcal{A})$ for every p, q, n. The proof that $\mathcal{T}_n(p,q)$ is open is similar to the proof of Theorem 2.8.

To prove its density let $v \in G(\omega, \mathcal{A})$, and let $\mathcal{U}$ be any neighbourhood of v in $G(\omega, \mathcal{A})$. Let $\epsilon > 0$, and put

$$\widetilde{\Delta}_p = \Phi(v, -\epsilon)(\Delta_p) = \{\ \gamma(x, -\epsilon; v) \mid x \in \Delta_p\ \}.$$

Then $\widetilde{\Delta}_p$ is also a stable Lyapunov disc for ω at p. For ϵ sufficiently small $\widetilde{\Delta}_p$ does not intersect the images of other charts of the atlas $\mathcal{A}$. In particular, $\widetilde{\Delta}_p \cap \nabla_q = \varnothing$. Apply Theorem 1.1 to obtain a vector field $u \in \mathcal{U}$ satisfying

$$\widetilde{\Delta}_p \pitchfork \Phi(u,n)(\nabla_q).$$

We can assume that supp $(v-u)$ does not intersect the images of the charts of the atlas $\mathcal{A}$, so that u is also a gradient-like vector field for ω adjusted to $\mathcal{A}$. The density of the subset $\mathcal{T}_n(p,q)$ in $G(\omega, \mathcal{A})$ is proved. □

Corollary 2.15. *The set $G_T(\omega, \mathcal{A})$ is a dense subset of $G(\omega, \mathcal{A})$.*

Proof. Pick any $v_0 \in G(\omega, \mathcal{A})$. Let

$$\mathcal{V}(\mathcal{A}) = \{v \in \mathrm{Vect}^\infty(M) \mid v|U_p = v_0|U_p \text{ for every } p \in S(\omega)\}.$$

Then $\mathcal{V}(\mathcal{A})$ is a closed subset of $\mathrm{Vect}^\infty(M)$ therefore a complete metric space, and it is easy to see that $G(\omega, \mathcal{A})$ is an open subset of this space. Applying Proposition 2.2 we complete the proof. □

Corollary 2.16. *The subset $GL_T(\omega) \subset GL(\omega)$ of transverse gradient-like vector fields for ω is dense in $GL(\omega)$.* □

Let us single out a corollary concerning the important cases of real-valued and circle-valued Morse functions.

Corollary 2.17. *Let M be a closed manifold and f be a real-valued or circle-valued Morse function on M. The subset $GL_T(f)$ of transverse gradient-like vector fields for f is dense (with respect to the C^∞ topology) in the space $GL(f)$ of all gradient-like vector fields for f.* □

2.4. The case of Riemannian gradients. Let M be a closed manifold and ω a Morse form on M. In this subsection we present two versions of the Kupka-Smale theory for Riemannian gradients of ω. The first version (Proposition 2.23) says that the subset $GR_T(\omega)$ of transverse Riemannian ω-gradients is residual in the set $GR(\omega)$ of all Riemannian ω-gradients. This statement is not entirely satisfactory, since we do not know if the topological space $GR(\omega)$ has the Baire property, and we can not deduce from this proposition the density of $GR_T(\omega)$ in $GR(\omega)$.

We can improve this result working with the set $\mathcal{R}(M)$ of all Riemannian metrics. For a Riemannian metric $R \in \mathcal{R}(M)$ we denote by $\mathrm{grad}_R(\omega)$ the corresponding Riemannian gradient of ω. Put

$$\mathcal{R}_T(M,\omega) = \{R \in \mathcal{R}(M) \mid \mathrm{grad}_R(\omega) \in G_T(\omega)\}.$$

We will prove that the subset $\mathcal{R}_T(M,\omega)$ is residual in $\mathcal{R}(M)$ (Proposition 2.24) and the density of $GR_T(\omega)$ in $GR(\omega)$ follows.

The set $\mathcal{R}(M)$ is an open subset of the space of C^∞ sections of the vector bundle $\mathscr{S} : S^2T^*M \to M$ (the second symmetric degree of the cotangent bundle) and will be endowed with the C^∞ topology. Let us denote by

$$\mathcal{G} : \mathcal{R}(M) \to GR(\omega); \quad \mathcal{G}(R) = \mathrm{grad}_R(\omega)$$

the map which associates to each Riemannian metric its ω-gradient; this map is continuous with respect to C^∞ topology on both spaces, and also with respect to C^0 topology on the both spaces. The map $\mathcal{G}$ is surjective, and can be thought of as a fibration.

Exercise 2.18. Show that for every $v \in GR(\omega)$ the fiber $\mathcal{G}^{-1}(v)$ is a convex subset of $\mathcal{R}(M)$ which is closed both in C^∞ and in C^0 topology.

We begin with a lemma which allows us to construct local sections for $\mathcal{G}$. Let R be a Riemannian metric, and $v = \mathrm{grad}_R(\omega)$ be the corresponding Riemannian gradient. Let U be any neighbourhood of $S(\omega)$, and $K = M \setminus U$. Let

$$G_{K,R} = \{u \in G(\omega) \mid \mathrm{supp}\ (u - v) \subset K\}.$$

This subset is closed in $G(\omega)$. The 1-form ω determines for each $x \in M$ a linear form $T_xM \to \mathbf{R}$, which will be denoted by ω_x. For a point $x \in M \setminus S(\omega)$ put

$$L_x = \mathrm{Ker}\ \Big(\omega_x : T_xM \to \mathbf{R}\Big) \subset T_xM,$$

so that L_x is a vector subspace of T_xM of codimension 1.

Lemma 2.19. *Let $u \in G_{K,R}$. There is a unique Riemannian metric $\mathcal{F}(u)$ satisfying the following conditions:*

(1) *u is the Riemannian gradient of ω with respect to $\mathcal{F}(u)$.*
(2) *The metrics $\mathcal{F}(u)$ and R are equal in U.*

(3) *For every $x \in K$ the restrictions to L_x of the scalar products induced by $\mathcal{F}(u)$ and R are equal.*

Proof. Let us have a closer look at the requirements imposed on $\mathcal{F}(u)$ by the conditions (1)–(3). For any $x \in K$ the tangent space T_xM is the direct sum of L_x and the one-dimensional subspace generated by $u(x)$. These two spaces must be $\mathcal{F}(u)$-orthogonal, since u is the Riemannian gradient of ω with respect to $\mathcal{F}(u)$. The scalar product on L_x induced by $\mathcal{F}(u)$ equals the one induced by $R = \mathcal{F}(v)$ and the square of the $\mathcal{F}(u)$-length of $u(x)$ equals $\omega_x(u(x))$.

It is now obvious that there exists a unique scalar product on T_xM satisfying these requirements, and it is not difficult to obtain an explicit formula for this product. Let us denote by $\langle \cdot, \cdot \rangle$ the scalar product on T_xM corresponding to R and $\{\cdot, \cdot\}$ the scalar product on T_xM corresponding to $\mathcal{F}(u)$. Put

$$\mu(x) = \omega_x(u(x)),$$

then for $x \in K$ we have

$$(5)\quad \{h, k\} = \langle h, k\rangle - \frac{\omega_x(h)}{\mu(x)}\langle u(x), k\rangle - \frac{\omega_x(k)}{\mu(x)}\langle u(x), h\rangle + \frac{\omega_x(k)\omega_x(h)}{(\mu(x))^2}\langle u(x), u(x)\rangle + \frac{\omega_x(k)\omega_x(h)}{\mu(x)}.$$

Indeed, it is clear that the right-hand side of the formula (5) determines a bilinear symmetric positive definite form on T_xM which equals $\langle \cdot, \cdot \rangle$ on L_x, and which equals $\omega_x(u(x))$ when evaluated on the pair $(u(x), u(x))$. It is also obvious that $\{u(x), k\} = 0$ whenever $k \in L_x$.

Since $\mu(x), u(x)$ and ω_x are C^∞ functions of x, the above formula implies that the scalar product $\{\cdot, \cdot\}$ on the space T_xM determines a Riemannian metric on M of class C^∞. □

Proposition 2.20. *The map*

$$u \mapsto \mathcal{F}(u); \quad G_{K,R} \to \mathcal{R}(M)$$

satisfies

$$\mathcal{G} \circ \mathcal{F} = \mathrm{id} \ : G_{K,R} \to G_{K,R}.$$

The map $\mathcal{F}$ is continuous with respect to C^∞ topology on $G_{K,R}$ and on $\mathcal{R}(M)$, and also continuous with respect to C^0 topology on these spaces.

Proof. It follows immediately from the formula (5) since the function $\mu \in C^\infty(M)$ depends continuously on the vector field u, in C^∞ topology, as well as in C^0 topology. □

Let us state a corollary of the previous result in view of the further applications.

Corollary 2.21. *Let ω be a Morse form on a closed manifold M. Let v be a Riemannian gradient of ω with respect to a Riemannian metric R. Let $K \subset M \setminus S(\omega)$ be compact and $\mathcal{U}$ be a neighbourhood of R in the space $\mathcal{R}(M)$ with respect to the C^0 topology. Then there is $\delta > 0$ such that every ω-gradient u with $||u - v|| < \delta$ and* supp $(u - v) \subset K$ *is the Riemannian gradient of ω with respect to a Riemannian metric $R' \in \mathcal{U}$.* □

Exercise 2.22. Derive from Corollary 2.21 a quick proof of Proposition 2.11 of Chapter 2 (page 53).

Now we will apply these techniques to Kupka-Smale transversality theory. The next proposition is not difficult to prove following the lines of the proof of Theorem 2.5 and using the previous corollary.

Proposition 2.23. *Let ω be a Morse form on a closed manifold M. Then the set $GR_T(\omega)$ is residual in $GR(\omega)$.* □

Let us proceed to the second version.

Proposition 2.24. *Let ω be a Morse form on a closed manifold M. The subset $\mathcal{R}_T(M, \omega) \subset \mathcal{R}(M)$ is residual in $\mathcal{R}(M)$.*

Proof. We shall prove that $\mathcal{R}_T(M, \omega)$ is locally residual in $\mathcal{R}(M)$. Let v_0 be a Riemannian gradient of ω with respect to a Riemannian metric R_0. Recall from Subsection 2.2 the neighbourhood $\mathcal{V} \subset G(\omega)$ of v_0 and the subsets $\mathcal{T}_n(p, q) \subset \mathcal{V}$ (page 124). The set

$$\mathcal{R}_n(p, q) = \mathcal{G}^{-1}\big(\mathcal{T}_n(p, q) \cap GR(\omega)\big)$$

is an open neighbourhood of R_0 in $\mathcal{R}(M)$. We are going to prove that $\mathcal{R}_n(p, q)$ is dense in

$$\widetilde{\mathcal{V}} = \mathcal{G}^{-1}(\mathcal{V} \cap GR(\omega)).$$

Let $Q \in \widetilde{\mathcal{V}}$ be a Riemannian metric on M, and let $u = \mathcal{G}(Q)$ be the corresponding ω-gradient. Since $\mathcal{T}_n(p, q)$ is dense in $\mathcal{V}$ (see Subsection 2.2) there exists a sequence of ω-gradients $u_k \in \mathcal{T}_n(p, q)$ converging to u in C^∞ topology. Checking through the proof of density of $\mathcal{T}_n(p, q)$ it is not difficult to show that we can choose the sequence u_k in such a way that for some neighbourhood A of the set $S(\omega)$ we have

$$\text{supp } (u_k - u) \cap \bar{A} = \varnothing.$$

Using Proposition 2.20 we obtain a sequence

$$\mathcal{F}(u_k) \in \mathcal{R}_n(p, q)$$

of Riemannian metrics converging to Q; the proof is over. □

Corollary 2.25. *Let ω be a Morse form on a closed manifold M. The subset $\mathcal{R}_T(M,\omega) \subset \mathcal{R}(M)$ is dense in $\mathcal{R}(M)$. The subset $GR_T(\omega)$ is dense in $GR(\omega)$.*

Proof. The set $\mathcal{R}(M)$ is an open subset of the complete metric space of C^∞ sections of the vector bundle $S^2T^*M \to M$, and $\mathcal{R}_T(M,\omega)$ is a residual subset of $\mathcal{R}(M)$. By Proposition 2.2 $\mathcal{R}_T(M,\omega)$ is dense in $\mathcal{R}(M)$. The second assertion follows immediately. □

The most important for us are the cases of real-valued and circle-valued Morse functions:

Corollary 2.26. *Let M be a closed manifold and f be a real-valued or circle-valued Morse function on M. The subset $GR_T(f)$ of all transverse Riemannian gradients for f is dense (with respect to the C^∞ topology) in the space $GR(f)$ of all Riemannian gradients for f.* □

3. Almost transverse gradients and Rearrangement Lemma

Let $f : W \to [a,b]$ be a Morse function on a cobordism W and v an f-gradient. The function f induces a partial order $\underset{f}{\prec}$ on the set of its critical points:

$$p \underset{f}{\prec} q \Leftrightarrow f(p) \leqslant f(q). \tag{6}$$

This order is an important invariant of the Morse function f and its gradient v. The Rearrangement Lemma 3.38 which is the first main result of the present section implies that if v is almost transverse, then there is a Morse function g, such that v is a g-gradient and

$$p \underset{g}{\prec} q \Leftrightarrow \operatorname{ind} p \leqslant \operatorname{ind} q.$$

Then we prove that the set of almost transverse f-gradients is open and dense in the set of all f-gradients with respect to C^0 topology. In the end of the section we study *Morse stratifications*, that is, cellular decompositions of manifolds arising from Morse functions.

3.1. Separated Morse functions.

Definition 3.1. A Morse function $f : W \to [a,b]$ is called *separated*, if for any $p, q \in S(f)$, $p \neq q$ we have $f(p) \neq f(q)$.

In this subsection we discuss a simplest example of the *rearrangement techniques*: given a Morse function f and an f-gradient v we construct a separated Morse function g having the same gradient v.

Definition 3.2. Let $f : W \to [a, b]$ be a Morse function on a cobordism W and v an f-gradient. A Morse function $\phi : W \to [\alpha, \beta]$ is called *adjusted to* (f, v), if:

1) $S(\phi) = S(f)$,
2) the function $f - \phi$ is constant in a neighbourhood of $\partial_0 W$, in a neighbourhood of $\partial_1 W$, and in a neighbourhood of each point of $S(f)$,
3) v is also a ϕ-gradient.

Proposition 3.3. *Let $f : W \to [a, b]$ be a Morse function on a cobordism W and v an f-gradient. Then there is a separated Morse function $g : W \to [a, b]$ adjusted to (f, v).*

Proof. The proposition is an immediate consequence of the following lemma:

Lemma 3.4. *Let $p \in S(f)$, let U be a neighbourhood of p. Then for every $\lambda \in \mathbf{R}$ sufficiently small there is a Morse function ϕ such that*

(A) *supp $(\phi - f) \subset U$,*
(B) *$\phi(x) = f(x) + \lambda$ for x in a neighbourhood of p,*
(C) *v is a ϕ-gradient.*

Proof. We can assume that p is the only critical point of f in $\overline{U}$. Pick a C^∞ function h such that supp $h(x) \subset U$ and $h(x) = 1$ for x in some neighbourhood U_0 of p. Let $\phi(x) = f(x) + \lambda h(x)$, then the conditions (A) and (B) hold. Now let us prove that ϕ is a Morse function if λ is sufficiently small. Since

$$\text{supp } (f' - \phi') = \text{supp } (h') \subset U \setminus U_0$$

it suffices to show that ϕ has no critical points in $U \setminus U_0$ if λ is sufficiently small. Observe that there is $C > 0$ such that

$$f'(x)(v(x)) > C \quad \text{for} \quad x \in \overline{U} \setminus U_0.$$

Therefore we have

$$\phi'(x)(v(x)) > C/2 \quad \text{for} \quad x \in \overline{U} \setminus U_0,$$

if

$$|\lambda| \cdot \Big(\sup_{x \in W} \big|h'(x)(v(x))\big| \Big) < \frac{C}{2}.$$

For such values of λ the function $\phi'(x)$ have no zeros in $\overline{U}\setminus U_0$, and the vector field v is also a ϕ-gradient, as required in the point (C). This completes the proof of Lemma 3.4 and of Proposition 3.3. $\square$

3.2. Links between critical points. Let $f : W \to [a,b]$ be a Morse function on a cobordism W and v an f-gradient. Let $p, q \in S(f)$. Recall that a v-link from p to q is an integral curve γ of v such that

$$\lim_{t\to-\infty} \gamma(x,t;v) = p, \qquad \lim_{t\to+\infty} \gamma(x,t;v) = q.$$

Definition 3.5. We say that two critical points $p, q \in S(f)$ are *v-linked* (or simply *linked* when no confusion is possible) if there is a v-link from p to q or from q to p.

We shall use the following terminology:

$$D_\delta(p, \pm v) = D_\delta(p, v) \cup D_\delta(p, -v),$$
$$D(p, \pm v) = D(p, v) \cup D(p, -v).$$

Observe that two critical points p, q are not linked if and only if

$$D(p, \pm v) \cap D(q, \pm v) = \varnothing.$$

Proposition 3.6. *Let $p \in S(f)$. Assume that p is not linked to any other critical point of f. Then there is $\delta > 0$ such that*

$$D_\delta(p, \pm v) \cap D_\delta(q, \pm v) = \varnothing \quad \textit{for every} \quad q \in S(f),\ q \neq p.$$

Proof. In view of Proposition 3.3 we can assume that f is separated, and in particular p is the only critical point on the critical level $c = f(p)$. Let $\epsilon > 0$ be so small, that c is the unique critical value of f in $[c-\epsilon, c+\epsilon]$. Put

$$W_0 = f^{-1}\big([a, c-\epsilon]\big), \quad v_0 = v|W_0,$$
$$W_1 = f^{-1}\big([c-\epsilon, c+\epsilon]\big), \quad v_1 = v|W_1,$$
$$V = f^{-1}(c-\epsilon).$$

We shall prove that for $\delta > 0$ sufficiently small we have

$$(C) \qquad D_\delta(p, \pm v) \cap D_\delta(q, \pm v) = \varnothing \quad \text{for every} \quad q \in W_0,$$

the case of the critical points q with $f(q) > c$ is similar. It is not difficult to prove that for δ sufficiently small the condition (C) is equivalent to the following one:

$$(C') \qquad D_\delta(p, v) \cap D_\delta(q, -v) = \varnothing \quad \text{for every} \quad q \in W_0.$$

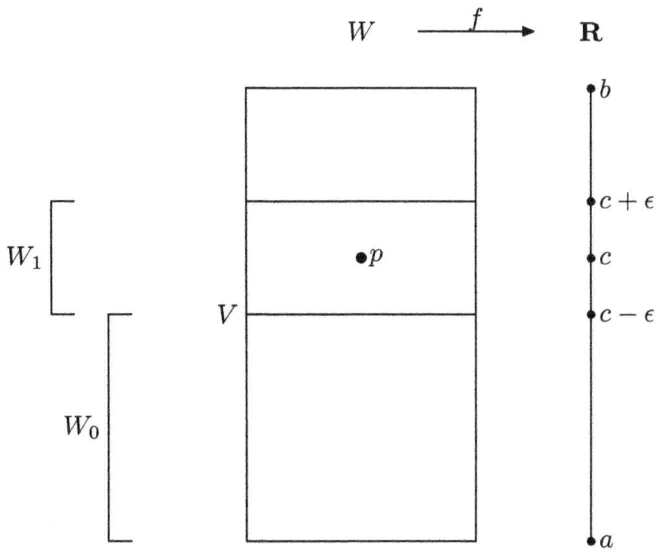

FIGURE 14.

To prove (C') observe that since the point p is not linked to other critical points, the descending disc of p does not intersect the ascending discs of the critical points in W_0:

$$\big(D(p, v_1) \cap V\big) \cap \big(D(-v_0) \cap V\big) = \varnothing.$$

Proposition 2.34 of Chapter 3 (page 98) implies that the sets $D_\nu(p, v_1)$ form a fundamental system of neighbourhoods of $D(p, v_1)$, therefore for every $\nu > 0$ sufficiently small

$$\big(D_\nu(p, v_1) \cap V\big) \cap \big(D(-v_0) \cap V\big) = \varnothing.$$

Fix some $\nu > 0$ so that this equality holds. One more application of the same proposition (this time to the cobordism W_0) gives that

$$(D_\nu(p, v_1) \cap V) \cap (D_\delta(-v_0) \cap V) = \varnothing,$$

for every $\delta > 0$ sufficiently small; therefore

$$D_\delta(p, v) \cap D_\delta(q, -v) = \varnothing$$

for every $q \in W_0 \cap S(f)$ and $\delta > 0$ sufficiently small. □

Lemma 3.7. *Let $f : W \to [a, b]$ be a Morse function on a cobordism W and v an f-gradient. Let $p \in S(f)$ and $\theta > 0$. Assume that*

$$D_\theta(p, \pm v) \cap D(q, \pm v) = \varnothing \quad \textit{for every} \quad q \in S(f),\ q \neq p.$$

Then $D_\theta(p,v)$ *and* $D_\theta(p,-v)$ *are compact sets.*

Proof. The proposition above implies that there is $\delta < \theta$ such that

$$D_\delta(p, \pm v) \cap D_\delta(q, \pm v) = \varnothing \quad \text{for every} \quad q \in S(f).$$

Let us first prove that for every $\alpha < \delta$ the sets $D_\alpha(p,v)$ and $D_\alpha(p,-v)$ are compact. Put

$$A = D_\alpha(p,v), \quad B = \bigcup_{q \neq p} D_\alpha(q,v).$$

The sets A, B have the following properties:

(1) $A \cup B$ is compact.

(2) There are open sets A', B' with

$$A \subset A', \; B \subset B', \; A' \cap B' = \varnothing$$

(for example, we can choose $A' = B_\delta(p,v), \; B' = \cup_{q \neq p} B_\delta(q,v)$). It is an easy exercise from elementary topology to check that these two properties imply that both A and B are compact. Thus $D_\alpha(p,v)$ is compact for any $\alpha < \delta$.

Now let us return to $D_\theta(p,v)$. The set $Q = D_\theta(p) \setminus B_\alpha(p,v)$ is compact and every $(-v)$-trajectory starting at a point of Q reaches $\partial_0 W$. Therefore the track

$$T(Q,-v) = \{\gamma(x,t;-v) \mid x \in Q, \; t \geqslant 0\}$$

is compact, and the set

$$D_\theta(p,v) = T(Q,-v) \cup D_\alpha(p,v)$$

is also compact. □

Corollary 3.8. *Let* $f : W \to [a,b]$ *be a Morse function on a cobordism* W *and* v *an* f*-gradient. Assume that the critical points of* f *are not* v*-linked to each other. Then there is* $\delta > 0$ *such that for every* $\theta < \delta$ *the sets* $D_\theta(p,v)$ *(where* $p \in S(f)$*) are pairwise disjoint compact subsets of* W. □

We shall often work with functions satisfying the conditions of the preceding corollary.

Definition 3.9. Let $f : W \to [a,b]$ be a Morse function on a cobordism W and v an f-gradient. If the critical points of f are not v-linked to each other, then the pair (f,v) is called *elementary*, and W is called an *elementary cobordism*.†

† A similar terminology is used in [**92**], although the term *elementary* is reserved there for a particular case of cobordisms endowed with a Morse function with a unique critical point.

Examples 3.10. (1) If a Morse function $f : W \to [a, b]$ has only one critical value, then the pair (f, v) is elementary for any f-gradient v.

(2) If all critical points of f have the same index, and v is an almost transverse f-gradient, then the pair (f, v) is elementary.

Any cobordism W is a union of elementary ones. Indeed, pick a Morse function $f : W \to [a, b]$. Choose a sequence of regular values $a = a_0 < a_1 < \cdots < a_{N+1} = b$ such that in each interval $[a_i, a_{i+1}]$ the function f has only one critical value. Then each cobordism $W_s = f^{-1}([a_s, a_{s+1}])$ is elementary, and W is the union of the cobordisms W_s.

3.3. The Rearrangement Lemma. Let $f : W \to [a, b]$ be a Morse function on a cobordism W and v an f-gradient. Let p be a critical point of f. Proposition 3.11 below says that if

$$D_\delta(p, \pm v) \cap D(q, \pm v) = \varnothing \quad \text{for every} \quad q \neq p, \tag{7}$$

then it is possible to modify the function inside $D_\delta(p, \pm v)$ so as to increase or diminish its values on $D_\delta(p)$, and v still will be a gradient of the resulting function. This proposition is the crucial step for the proof of the Rearrangement Lemma, Theorem 3.17. Before we give the statement let us explain the basic idea with a simple example.

Consider the "pants" submanifold $W \subset \mathbf{R}^3$ as shown in Figure 15, and let $f : W \to \mathbf{R}$ be the height function, that is, the restriction to W of the projection $(x, y, z) \mapsto z$.

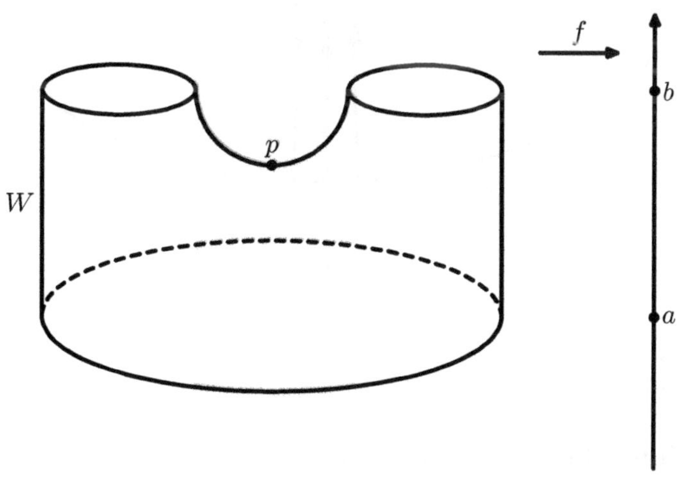

FIGURE 15.

Figure 16 depicts the manifold W viewed *from the top*. The manifold itself is shaded, and the set $D_\delta(p, \pm v)$ is dark shaded. The two integral curves that constitute $D(p, v) \setminus \{p\}$ are shown by the two vertical segments with arrows. (Observe that the condition (7) is satisfied obviously, since p is the only critical point of f.) Both the function f and the new function g are strictly increasing along these two curves. But the speed of growth of g will be different from that of f so that g will have the prescribed value at p. The speed of growth will be changed also in the ascending disc $D(p, -v)$ (shown by the two horizontal segments on the picture) and in the subset $D_\delta(p, \pm v)$ in such a way that outside the dark shaded domain the function g equals the initial function f.

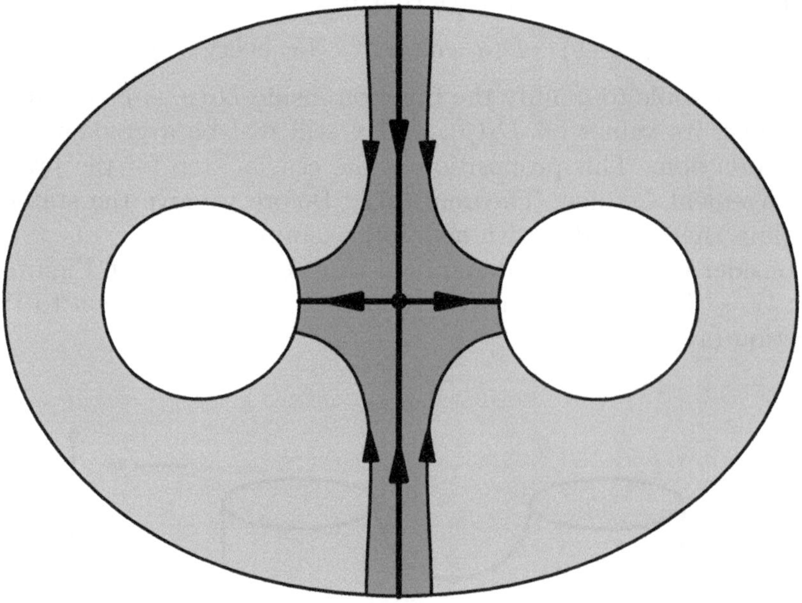

FIGURE 16.

Proposition 3.11. *Let $f : W \to [a, b]$ be a Morse function on a cobordism W and v an f-gradient. Let $p \in S(f)$ and $\delta > 0$. Assume that*

$$D_\delta(p, \pm v) \cap D(q, \pm v) = \varnothing \quad \textit{for every} \quad q \neq p. \tag{8}$$

Let $0 < \nu < \delta$. *Let* $[\mu, \varkappa] \subset [a, b]$ *be an interval and let* $c \in]\mu, \varkappa[$. *Then there is a Morse function* $g : W \to [a, b]$, *adjusted to* (f, v) *and such that*

(A) $f(x) = g(x)$ *for* $x \notin D_\delta(p, \pm v)$,
(B) $g(p) = c$ *and* $g(D_\nu(p)) \subset]\mu, \kappa[$.

Proof. It is sufficient to do the proof on the assumption $a < \mu < \kappa < b$. We will define the function g in three open subsets of W, whose union is the whole of W.

The set $W \setminus D_\delta(p, \pm v)$.

Here our task is easy: put

(9) $$g(x) = f(x) \quad \text{for every} \quad x \in W \setminus D_\delta(p, \pm v),$$

as required in (A). Observe that $W \setminus D_\delta(p, \pm v)$ is open by Lemma 3.7.

The set $B_\nu(p, \pm v) = B_\nu(p, v) \cup B_\nu(p, -v)$.

Choose any interval $[\alpha, \beta] \subset]a, b[$ so that $f(D_\delta(p)) \subset]\alpha, \beta[$. Pick a C^∞ diffeomorphism $\lambda : [a, b] \to [a, b]$ with the following properties:

$\lambda(t) = t$ for t in a neighbourhood of a and b;

$\lambda([\alpha, \beta]) \subset]\mu, \kappa[$;

$\lambda(f(p)) = c$ and $\lambda'(t) = 1$ for t in a neighbourhood of $f(p)$.

Set

(10) $$g(x) = \lambda(f(x)) \quad \text{for} \quad x \in B_\nu(p, \pm v).$$

The condition (B) above is now fulfilled.

The rest.

Now it remains to define the function g on the rest of W, that is , in

$$X = D_\delta(p, \pm v) \setminus B_\nu(p, \pm v).$$

For every $y \in X$ the maximal integral curve $\gamma(y, \cdot; v)$ is defined on a compact interval $[a_y, b_y]$ of $\mathbf{R}$ and the function $f \circ \gamma : [a_y, b_y] \to \mathbf{R}$ is strictly increasing. We shall define the function g along each curve $\gamma(y, \cdot; v)$ in such a way that $g \circ \gamma$ will be strictly increasing, and that the definition will be compatible with the formulas (9) and (10) above. The set $D_\nu(p,v)$ is compact by Lemma 3.7, and the set $B_\delta(p,v)$ is its open neighbourhood. Pick a C^∞ function $h : \partial_0 W \to [0,1]$ such that

$$h(x) = 1 \quad \text{for} \quad x \in D_\nu(p,v) \cap \partial_0 W,$$
$$h(x) = 0 \quad \text{for} \quad x \notin B_\delta(p,v) \cap \partial_0 W.$$

Extend the function h to the whole of W as follows:

$$\begin{aligned} h(x) &= 1 && \text{for} \quad x \in D(p, \pm v), \\ h(x) &= 0 && \text{for} \quad x \in D(q,-v) \quad \text{with} \quad q \neq p, \\ h(x) &= h(E(x,-v)) && \text{for} \quad x \notin D(-v). \end{aligned}$$

The last condition implies that for every integral curve γ of v the restriction of h to γ is constant and equal to the value of h at the point of intersection of γ with $\partial_0 W$.

Lemma 3.12. *The function h is of class C^∞.*

Proof. The restriction of h to $W \setminus D(-v)$ is of class C^∞, since the function $x \mapsto E(x,-v)$ is C^∞. Thus it remains to show that h is of class C^∞ in a neighbourhood of the set $D(-v)$. Observe that

$$h(x) = 1 \quad \text{for} \quad x \in D_\nu(p, \pm v),$$

therefore h is of class C^∞ on the open set $B_\nu(p,\pm v)$, which is a neighbourhood of $D(p,-v)$. Further,

$$h(x) = 0 \quad \text{for} \quad x \notin D_\delta(p,\pm v),$$

since for these points x we have:

$$E(x,-v) \subset \partial_0 W \setminus B_\delta(p,v) \quad \text{or} \quad x \in D(q,-v) \quad \text{for} \quad q \neq p.$$

Therefore h is of class C^∞ also on the open set $W \setminus D_\delta(p,v)$ which is a neighbourhood of $D(q,\pm v)$ for every $q \neq p$. □

Now we can give a formula which defines the function g on the whole of W. Set

$$g(x) = h(x) \cdot \lambda(f(x)) + (1 - h(x)) \cdot f(x)$$

(the function λ is defined on page 139). It is easy to check that the new definition is compatible with the formulas (9) and (10). It remains to prove that v is a g-gradient. Let $\gamma : [\alpha,\beta] \to W$ be any non-constant integral curve of v. Since h is constant on γ, and both functions $f(x)$ and $\lambda(f(x))$ are strictly increasing along γ, the function g is also strictly increasing along γ, and therefore

$$g'(x)(v(x)) > 0 \quad \text{for every} \quad x \notin S(f).$$

Thus the vector field v satisfies the condition (A) from Proposition 2.5 of Chapter 2 (page 51). The condition (B) follows obviously from the fact that the function $f - g$ is constant in a neighbourhood of each critical point. Since $f = g$ nearby ∂W and in a neighbourhood of $S(f)$, the function g is adjusted to (f, v). The proof of our proposition is complete. $\square$

Let us deduce some corollaries to be used in the sequel.

Corollary 3.13. *Let $f : W \to [a, b]$ be a Morse function on a cobordism W and v an f-gradient. Let p be a critical point of f, which is not linked to any other critical point. Then there is a Morse function $\phi : W \to [a, b]$, adjusted to (f, v) such that $\phi(p) > \phi(q)$ for every $q \in S(\phi)$, $q \neq p$.*

Proof. Apply first Proposition 3.6, then Proposition 3.11. $\square$

Corollary 3.14. *Let $f : W \to [a, b]$ be a Morse function on a cobordism W and v an f-gradient. Let p be a critical point of f, which is not linked to any other critical point q with $f(q) > f(p)$. Then there is a Morse function $\phi : W \to [a, b]$, adjusted to (f, v) such that $\phi(p) > \phi(q)$ for every $q \in S(\phi)$, $q \neq p$.*

Proof. Consider the cobordism $W_0 = f^{-1}\big([f(p) - \epsilon, b]\big)$ where $\epsilon > 0$ is so small that $f(p)$ is the smallest critical value of $f|W_0$. Applying to W_0 the preceding corollary we obtain a Morse function $\psi : W_0 \to [f(p) - \epsilon, b]$ adjusted to $(f|W_0, v|W_0)$ and such that $\psi(q) > \psi(p)$ for every $q \in S(\psi)$, $q \neq p$. Now here is the definition of the function ϕ sought:

$$\phi(x) = f(x) \quad \text{for} \quad x \notin W_0,$$
$$\phi(x) = \psi(x) \quad \text{for} \quad x \in W_0. \qquad \square$$

Recall that a critical point p of f is *of type* (α) if every non-constant $(-v)$-trajectory in $D(p, v)$ reaches $\partial_0 W$.

Corollary 3.15. *Let $f : W \to [a, b]$ be a Morse function on a cobordism W and v an f-gradient. Let p be a critical point of f of type (α). Then there is a Morse function $\phi : W \to [a, b]$ adjusted to (f, v) and a regular value λ of ϕ, such that p is the only critical point of ϕ in $\phi^{-1}([a, \lambda])$.*

Proof. Let $\mu > f(p)$ be any regular value of f such that the interval $]f(p), \mu]$ is regular. Let $W_0 = f^{-1}([a, \mu])$. Then the point p is not linked to other critical points of $f|W_0$. By Proposition 3.6 there is $\delta > 0$ such that

$$D_\delta(p, \pm v|W_0) \cap D_\delta(q, \pm v|W_0) = \varnothing \quad \text{for every} \quad q \in S(f|W_0).$$

Apply Proposition 3.11 to the function $f|W_0 : W_0 \to [a, \mu]$ to obtain a Morse function $\psi : W_0 \to [a, \mu]$ adjusted to $(f|W_0, v|W_0)$ and a regular value λ of ψ such that p is the only critical point of ψ in $\psi^{-1}([a,\lambda])$. Now here is the definition of the function ϕ sought:

$$\phi(x) = f(x) \quad \text{for} \quad x \notin W_0,$$
$$\phi(x) = \psi(x) \quad \text{for} \quad x \in W_0. \qquad \square$$

Corollary 3.16. *Let $f : W \to [a,b]$ be a Morse function on a cobordism W and v an f-gradient. Let p be a critical point of f of type (α). Then the family $\{B_\delta(p,v)\}_{\delta>0}$ is a fundamental system of neighbourhoods of $D(p,v)$.*

Proof. Apply the preceding corollary. □

The main aim of the present subsection is the following theorem.

Theorem 3.17 (Rearrangement Lemma). *Let $f : W \to [a,b]$ be a Morse function on a cobordism W and v an f-gradient. Assume that (f,v) is elementary. Let $S(f) = \{p_1,\dots,p_k\}$, and let $\alpha_1,\dots,\alpha_k$ be arbitrary real numbers in $]a,b[$. Then there is a Morse function $g : W \to [a,b]$, adjusted to (f,v), such that for every i we have $g(p_i) = \alpha_i$.*

Proof. The proof is an easy induction argument based on Proposition 3.11. □

Corollary 3.18. *Assume that (f,v) is elementary. Let $p \in S(f)$. Then the family $\{B_\delta(p,v)\}_{\delta>0}$ is a fundamental system of neighbourhoods of $D(p,v)$.* □

3.4. Broken flow lines and an ordering in $S(f)$. Let $f : W \to [a,b]$ be a Morse function on a cobordism W and v an f-gradient.

Definition 3.19. Let $p,q \in S(f)$. We say that there exists a *broken flow line* of v joining q and p, and we write

$$q \lhd p$$

if there is a sequence of critical points

$$q = q_0,\ q_1, \dots,\ q_k = p,$$

and for every i there is a v-link from q_i to q_{i+1}.

The relation $\lhd$ is a partial order on the set $S(f)$; observe that we have

$$q \lhd p \Rightarrow f(q) < f(p).$$

Definition 3.20. A subset $A \subset S(f)$ is called an *initial segment of $S(f)$* if

$$\Big(p \in A \ \&\ q \lhd p\Big) \Rightarrow \Big(q \in A\Big).$$

Example 3.21. If v is almost transverse, then the set of all critical points of f of indices less than a given integer is an initial segment of $S(f)$.

Definition 3.22. For a set $A \subset S(f)$ denote

$$D(A,v) = \bigcup_{p\in A} D(p,v), \quad B_\delta(A,v) = \bigcup_{p\in A} B_\delta(p,v),$$
$$D_\delta(A,v) = \bigcup_{p\in A} D_\delta(p,v).$$

The main aim of the present subsection is to generalize the material of Subsection 2.5 of Chapter 3 (page 97) and to deduce some basic topological properties of the sets $D(A,v)$, where A is an initial segment of $S(f)$. Let us begin with an auxiliary lemma.

Proposition 3.23. *Let $f : W \to [a,b]$ be a Morse function on a cobordism W and v an f-gradient. Let $A \subset S(f)$. Then the following two conditions are equivalent:*

(1) *There is a Morse function ϕ adjusted to (f,v) and a regular value λ of ϕ such that*
$$A = S(\phi) \cap \phi^{-1}([a,\lambda]) = S(f) \cap \phi^{-1}([a,\lambda]).$$
(2) *A is an initial segment of $S(f)$ with respect to v.*

Proof. The implication (1)⇒(2) is obvious. The inverse implication is proved by induction on the number of elements in $B = S(f) \setminus A$. When B is a one-point set, the result follows by an application of Corollary 3.14. Assume now that our proposition holds whenever card $B < k$ and let us prove it for the case when card $B = k$. We can assume that f is separated, that is,

$$f(p) \neq f(q) \quad \text{if} \quad p,q \in S(f),\ p \neq q.$$

Let $p_0 \in B$ be the point of B with the minimal value of f, that is

$$f(p_0) < f(p) \quad \text{for every} \quad p \in B,\ p \neq p_0.$$

Put

$$A' = A \cup \{p_0\}, \quad B' = B \setminus \{p_0\}.$$

Then we have for every $p \in A'$:

$$q \lhd p \Rightarrow q \in A'$$

so that A' is an initial segment of $S(f)$ and therefore we can apply the induction assumption to the sets A', B' and obtain a Morse function $\phi : W \to [a,b]$, adjusted to (f,v) and having a regular value λ such that

$$A' \subset \phi^{-1}([a,\lambda]), \quad B' \subset \phi^{-1}([\lambda,b]).$$

Now apply Corollary 3.14 to the cobordism $\phi^{-1}([a,\lambda])$ and the point p_0 and the proof is over. □

Proposition 3.24. *Let $f : W \to [a,b]$ be a Morse function on a cobordism W and v an f-gradient. Let A be an initial segment of $S(f)$. Then*

(1) *$D(A,v)$ is compact.*

(2) *$D(A,v) = \bigcap_{\theta>0} B_\theta(A,v)$.*

(3) *The family*
$$\{B_\theta(A,v)\}_{\theta>0}$$
is a fundamental system of neighbourhoods of $D(A,v)$.

Proof. By Proposition 3.23 there is a function $\phi : W \to [a,b]$ adjusted to (f,v) and a regular value λ of ϕ, such that $p \in S(f)$ is in A if and only if $\phi(p) < \lambda$. Let W_0 denote the cobordism $\phi^{-1}([a,\lambda])$. Then
$$D(A,v) = D(A,v|W_0),$$
and this set is compact by Lemma 2.32 of Chapter 3 (page 97). Similarly, the other items follow from Proposition 2.34 of Chapter 3 (page 98), applied to the cobordism W_0. □

Now let us proceed to the thickenings of the descending discs.

Proposition 3.25. *Let $f : W \to [a,b]$ be a Morse function on a cobordism W and v an f-gradient. Let A be an initial segment of $S(f)$. Then for every $\delta > 0$ sufficiently small we have:*

(1) *$D_\delta(A,v)$ is compact.*

(2) *$D_\delta(A,v) = \bigcap_{\theta>\delta} B_\theta(A,v)$.*

(3) *The family*
$$\{B_\theta(A,v)\}_{\theta>\delta}$$
is a fundamental system of neighbourhoods of $D_\delta(A,v)$.

Proof. Similarly to Proposition 3.24, pick a function $\phi : W \to [a,b]$ adjusted to (f,v) and a regular value λ of ϕ such that $p \in S(f)$ is in A if and only if $\phi(p) < \lambda$. Let W_0 denote the cobordism $\phi^{-1}([a,\lambda])$. Let $\delta > 0$ be sufficiently small so that for every critical point $p \in A$ we have
$$D_\delta(p) \subset W_0 \setminus \partial W_0 = \phi^{-1}(]a,\lambda[).$$
Then
$$D_\delta(A,v) = D_\delta(A,v|W_0)$$
and this set is compact by Lemma 2.32 of Chapter 3 (page 97). The second point follows by an application of Proposition 2.34 of Chapter 3 (page 98). □

3.5. Transverse gradients. Let $f : W \to [a, b]$ be a Morse function on a cobordism W and v an f-gradient.

Definition 3.26. We say that v satisfies the *transversality condition,* if

$$D(p, v) \pitchfork D(q, -v) \quad \text{for every} \quad p, q \in S(f).$$

A gradient satisfying transversality condition will also be called *transverse.* The set of all transverse f-gradients is denoted $G_T(f)$.

The aim of this subsection is to prove an analog of the Kupka-Smale theorem for gradients of Morse functions on cobordisms. We shall reduce the case of cobordisms to the case of closed manifolds (considered in Section 2), embedding the cobordism in question into its double.

Theorem 3.27. *The set $G_T(f)$ is residual in $G(f)$.*

Proof. Recall the construction of *the double* of W (see for example [**100**], Ch. 1, §1). Consider the quotient space M of the space $W \times \{0, 1\}$ with respect to the following equivalence relation:

$$x \times 0 \sim x \times 1 \quad \text{if} \quad x \in \partial W.$$

Endowed with the quotient topology, the space M is obviously a compact m-dimensional topological manifold (where $m = \dim W$). Let us introduce on M a structure of a C^∞ manifold. Pick any f-gradient v_0, such that $f'(x)(v_0(x)) = 1$ for x in a neighbourhood of ∂W. Let $\epsilon > 0$ be sufficiently small, so that both $\partial_1 W$ and $\partial_0 W$ have collars of height ϵ with respect to v_0. For any C^∞ chart $\phi : U \to V \subset \mathbf{R}^{m-1}$ of the manifold $\partial_1 W$ define a continuous map $\widehat{\phi} : V \times]-\epsilon, \epsilon[\to M$ as follows:

$$\widehat{\phi}(x, t) = \gamma\big(\phi^{-1}(x), t; v_0\big) \times 0 \qquad \text{for} \quad t \leqslant 0;$$
$$\widehat{\phi}(x, t) = \gamma\big(\phi^{-1}(x), t; -v_0\big) \times 1 \qquad \text{for} \quad t \geqslant 0.$$

It is clear that $\widehat{\phi}$ is a homeomorphism on its image, and the inverse map $\widehat{\phi}^{-1}$ is therefore a chart of M defined on a neighbourhood of U. Similarly for any C^∞ chart $\psi : U' \to V' \subset \mathbf{R}^{m-1}$ of the manifold $\partial_0 W$ we construct a chart $\widehat{\psi}^{-1}$ of M defined on a neighbourhood of U'.

The collection of the maps $\widehat{\phi}^{-1}, \widehat{\psi}^{-1}$ together with the C^∞ charts of $(W \setminus \partial W) \times 0$ and $(W \setminus \partial W) \times 1$ is a C^∞ atlas of the manifold M. The image of $W \times 0$ in M is a ∂-submanifold of M, we shall identify it with the cobordism W itself. The image of $W \times 1$ in M will be denoted by $\widetilde{W}$; it is also a ∂-submanifold of M. The map

$$I : M \to M; \quad x \times 0 \mapsto x \times 1, \quad x \times 1 \mapsto x \times 0$$

is a C^∞ involution of M, such that

$$I(W) = \widetilde{W}, \; I(\widetilde{W}) = W.$$

Now we shall extend the function f to M. To simplify the notation we shall assume that $f(W) = [0, 1/2]$. Define a Morse map $F : M \to S^1 = \mathbf{R}/\mathbf{Z}$ as follows:

$$F(x) = f(x) \quad \text{and} \quad F(I(x)) = 1 - f(x) \quad \text{for} \quad x \in W.$$

Extend the vector field v_0 to the whole of M setting

$$v_0(I(x)) = -I_*(v_0(x)) \quad \text{for} \quad x \in W;$$

we obtain then an F-gradient, satisfying the condition $v_0 = -I_*(v_0)$.

Let a be a real number, $0 < a < 1/4$, such that the intervals $[0, a]$, $[1/2 - a, 1/2]$ are regular, and put $W' = f^{-1}([a, 1/2 - a])$.

Definition 3.28. Let v be an f-gradient. An F-gradient $\widetilde{v}$ is called a *quasi-extension* of v, if $\widetilde{v} \mid W' = v$.

Lemma 3.29. (1) *Every f-gradient v has a quasi-extension.*

(2) *If $\widetilde{v}$ is a quasi-extension of $v \in G(f)$, then for every neighbourhood A of $\widetilde{v}$ in $Vect^\infty(M)$ there is a neighbourhood B of v in $G(f)$ such that every $u \in B$ has a quasi-extension in A.*

Proof. Let $h : W \to [0, 1]$ be a C^∞ function which vanishes in a neighbourhood of ∂W and equals 1 in a neighbourhood of W'. Let v be an f-gradient, The vector field w defined in the following formula is a quasi-extension for v.

$$w(x) = h(x) \cdot v(x) + (1 - h(x)) \cdot v_0(x).$$

Now let us proceed to the second point. If the assertion were not true, then there would exist a sequence of f-gradients v_n converging to v in C^∞ topology, and such that v_n do not have quasi-extensions in A. Consider the vector field

$$w_n(x) = h(x) \cdot (v_n(x) - v(x)) + \widetilde{v}(x).$$

This vector field is a quasi-extension of v_n, and it converges to $\widetilde{v}$ as $n \to \infty$. We have obtained a contradiction with our assumption. □

Now we can proceed to the proof of our theorem. It suffices to prove that the set $G_T(f)$ is locally residual in $G(f)$.

Lemma 3.30. *Let $u \in G(f)$. There is a neighbourhood $\mathcal{U} \in G(f)$ of u and for every $v \in \mathcal{U}$ and every $p \in S(f)$ there are Lyapunov charts*

$$j_{st}(v, p) : D \to \text{Int } W', \quad j_{un}(v, p) : D' \to \text{Int } W'$$

for v at p depending continuously on v in C^1-topology.

Proof. Apply Proposition 1.15 of Chapter 3 (page 76) to any quasi-extension $\widetilde{u}$ of u. □

The stable Lyapunov discs obtained in the preceding lemma will be denoted by $\Delta_p(v)$, and the unstable ones by $\nabla_p(v)$. Let $T_n(p,q)$ denote the set of all f-gradients $v \in \mathcal{U}$ having a quasi-extension $\widetilde{v}$ such that

$$\Phi(\widetilde{v}, -n)\Big(\Delta_p(v)\Big) \pitchfork \Phi(\widetilde{v}, n)\Big(\nabla_q(v)\Big) \tag{11}$$

Lemma 3.31. *We have*

$$\bigcap_{n,p,q} T_n(p,q) \subset G_T(f).$$

Proof. Let v be a vector field belonging to the intersection of all sets $T_n(p,q)$. For each n choose a quasi-extension $\widetilde{v}_n$ of v such that

$$\Phi(\widetilde{v}_n, -n)\Big(\Delta_p(v)\Big) \pitchfork \Phi(\widetilde{v}_n, n)\Big(\nabla_q(v)\Big). \tag{12}$$

Observe that for every p, q the submanifolds $D(p,v)$ and $D(q,-v)$ are transverse if and only if their intersections with W' are transverse. These intersections can be expressed in terms of the flow generated by $\widetilde{v}_n$ as follows:

$$W' \cap D(p,v) \subset \bigcup_n \Big(\Phi(\widetilde{v}_n, -n)(\Delta_p(v))\Big),$$

$$W' \cap D(q,-v) \subset \bigcup_n \Big(\Phi(\widetilde{v}_n, n)(\nabla_q(v))\Big),$$

and in view of (12) the proof of the lemma is complete. □

Now let us prove that $T_n(p,q)$ is open. Let $v \in G_T(f)$ and let $\widetilde{v}$ be a quasi-extension of v such that (11) holds. Apply Theorem 2.8 (page 125) and point (2) of Lemma 3.29 above and the assertion follows.

The proof of density is also quite simple. Let v be any f-gradient. Let $\widetilde{v}$ be any quasi-extension of v. Let w_k be a sequence of transverse F-gradients converging to $\widetilde{v}$. Then the restrictions $w_k \mid W$ are in $T_n(p,q)$ and converge to v.

Thus the set $T_n(p,q)$ is open and dense in $G(f)$, and the proof of our theorem is finished. □

Corollary 3.32. *Let $f : W \to [a,b]$ be a Morse function on a cobordism W. The subset $G_T(f)$ is dense in $G(f)$.*

Proof. The proof repeats the arguments of Corollary 2.6. □

Exercise 3.33. Consider the case of gradient-like vector fields and Riemannian gradients.

3.6. Almost transverse gradients. Let $f : W \to [a, b]$ be a Morse function on a cobordism W and v an f-gradient. Recall that v is called *almost transverse* if for any $p, q \in S(f)$ we have

$$\big(\operatorname{ind} p \leqslant \operatorname{ind} q \ \&\ p \neq q\big) \Rightarrow \big(D(p, v) \cap D(q, -v) = \varnothing\big).$$

Proposition 3.34. *An f-gradient v is almost transverse if and only if for any $p, q \in S(f)$ we have*

$$\big(\operatorname{ind} p \leqslant \operatorname{ind} q\big) \Rightarrow \big(D(p, v) \pitchfork D(q, -v)\big).$$

Proof. Our proposition follows immediately from the next lemma.

Lemma 3.35. *Let $p, q \in S(f)$, $\operatorname{ind} p \leqslant \operatorname{ind} q$, $p \neq q$. Then*

$$D(p, v) \pitchfork D(q, -v)$$

if and only if

$$D(p, v) \cap D(q, -v) = \varnothing.$$

Proof. If $D(p, v)$ and $D(q, -v)$ have empty intersection, they are transverse. If the intersection of these manifolds is not empty, then necessarily $f(p) > f(q)$. In this case pick any regular value λ of f with $f(q) < \lambda < f(p)$. The intersections

$$A = D(p, v) \cap f^{-1}(\lambda), \quad B = D(q, -v) \cap f^{-1}(\lambda)$$

are submanifolds of $V = f^{-1}(\lambda)$, and $D(p, v)$ is transverse to $D(q, -v)$ if and only if A is transverse to B as submanifolds of V. Observe that

$$\dim A + \dim B = (\operatorname{ind} p - 1) + (\dim W - \operatorname{ind} q - 1) < \dim V,$$

therefore the condition $A \pitchfork B$ is equivalent to $A \cap B = \varnothing$, which is in its turn equivalent to $D(p, v) \cap D(q, -v) = \varnothing$. □

Corollary 3.36. *A transverse f-gradient is almost transverse.* □

The set of all almost transverse f-gradients is denoted $G_A(f)$. Since $G_T(f) \subset G_A(f)$, and $G_T(f)$ is dense in $G(f)$ with respect to C^∞ topology, the set $G_A(f)$ is also dense in $G(f)$ with respect to C^∞ topology. More surprising is the fact that the set $G_A(f)$ is *open in $G(f)$ with respect to C^0 topology.* The proof which is based on the theory of ordered Morse functions, will be given in the end of this subsection.

Definition 3.37. *An ordered Morse function* is a Morse function $\phi : W \to [a, b]$ on a cobordism W together with a sequence $a_0, \dots, a_{m+1}$ of regular values (where $m = \dim W$), such that $a = a_0 < a_1 < \cdots < a_{m+1} = b$ and

$$S_i(f) \subset \phi^{-1}(\,]a_i, a_{i+1}[\,) \tag{13}$$

for every i. The sequence $a_0, \dots, a_{m+1}$ is called the *ordering sequence* for ϕ.

Note that we consider the ordering sequence as a part of the data of the ordered Morse function.

Proposition 3.38. *Let $f : W \to [a, b]$ be a Morse function on a cobordism W and v an f-gradient. Then the two following properties are equivalent:*

(i) *v is almost transverse.*
(ii) *There is an ordered Morse function $\phi : W \to [a, b]$, such that ϕ is adjusted to (f, v), and for every k the value of ϕ on all its critical points of index k is the same.*

Proof. (ii) $\Rightarrow$ (i). Let ϕ be a Morse function satisfying (ii). Then every two critical points x, y of f such that $\operatorname{ind} x \leqslant \operatorname{ind} y$ satisfy also the condition $f(x) \leqslant f(y)$ and there can exist no v-link from y to x.

(i) $\Rightarrow$ (ii). The proof will be obtained by successive applications of the Rearrangement Lemma. Apply Proposition 3.3 to construct a separated Morse function g adjusted to (f, v). Two critical points p, q of g will be called *neighbours* if there is no critical value of g between $g(p)$ and $g(q)$. An *inversion* is a pair (p, q) of neighbours such that $g(p) < g(q)$ and $\operatorname{ind} p > \operatorname{ind} q$. For an inversion (p, q) let

$$W' = g^{-1}(\ [g(p) - \epsilon, g(q) + \epsilon]\ )$$

where $\epsilon > 0$ is sufficiently small, so that $g(p), g(q)$ are the only critical values of $g|W'$. The pair $(g|W', v|W')$ is elementary, since v satisfies the almost transversality condition. Therefore, applying Theorem 3.17 to the cobordism W' we can diminish the number of inversions by one. Applying this procedure several times we end with a Morse function h, adjusted to (f, v) and having no inversions. Thus the function h is ordered, and applying again the Rearrangement Lemma, Theorem 3.17, we put all the critical points of a given index i on the same level surface of f. □

It is sometimes useful to rearrange the critical points in a different way.

Proposition 3.39. *Let $f : W \to [a, b]$ be a Morse function on a cobordism W and v an f-gradient. Then the two following properties are equivalent:*

(i) *v is almost transverse.*
(ii) *For any linear ordering $\prec$ of the set of all critical points such that*

$$\operatorname{ind} p < \operatorname{ind} q \Rightarrow p \prec q,$$

there is a Morse function ϕ, adjusted to (f, v) and such that $\phi(p) < \phi(q)$ whenever $p \prec q$ and $p \neq q$.

Proof. The proof is similar to 3.38 and will be omitted. □

Proposition 3.40. *Let $f : W \to [a,b]$ be a Morse function, and v be an f-gradient. Let $\phi : W \to [\alpha, \beta]$ be a Morse function, adjusted to (f,v). Then there is $\delta > 0$, such that every f-gradient w with $||w - v|| < \delta$ is also a ϕ-gradient*† *so that ϕ is adjusted to (f,w).*

Proof. Let us first show that for every f-gradient w sufficiently C^0-close to v we have $\phi'(x)(w(x)) > 0$ for every $x \in W \setminus S(f)$. Let U be an open neighbourhood of $S(f)$ such that $f|U = \phi|U$. Pick an $\epsilon > 0$ such that

$$\phi'(x)(v(x)) > \epsilon \quad \text{for every} \quad x \in W \setminus U$$

(such ϵ exists, since $W \setminus U$ is compact and the function $\phi'(x)(v(x))$ is strictly positive on this set). It is clear now that for $||w - v||$ sufficiently small we have

$$\phi'(x)(w(x)) > \epsilon/2 \quad \text{for every} \quad x \in W \setminus U.$$

More precisely, this inequality holds if

$$||w - v|| < \frac{\epsilon}{2 \cdot \sup_{x \in W \setminus U} ||\phi'(x)||}\ .$$

Thus we have verified that the condition A) from Proposition 2.5 of Chapter 2 (page 51) holds for ϕ and w if w is sufficiently close to v in C^0 topology. It remains to observe that, since $f - \phi$ is constant in a neighbourhood of each critical point, the condition B) from the same proposition holds for w and ϕ. The proof is over. □

Now we can prove that the almost transversality condition is C^0-open.

Theorem 3.41. *Let $f : W \to [a,b]$ be a Morse function on a cobordism W. The set $G_A(f)$ is dense in $G(f)$ with respect to C^∞ topology, and open in $G(f)$ with respect to C^0 topology.*

Proof. The density property follows immediately from Theorem 3.27. Now let us prove that $G_A(f)$ is C^0-open. Let $v \in G_A(f)$. Pick an ordered Morse function $\phi : W \to [a,b]$, adjusted to (f,v), and such that for a given k all the critical points of ϕ of index k lie on the same level surface of ϕ. Proposition 3.40 implies that every f-gradient sufficiently C^0-close to v, is also a ϕ-gradient, therefore (by Proposition 3.38) it satisfies the almost transversality condition. □

Now we proceed to topological properties of descending discs of almost transverse gradients.

---

† Recall that $||\cdot||$ stands for the C^0-norm.

Definition 3.42. Let $f : W \to [a, b]$ be a Morse function on a cobordism W and v an f-gradient. Set

$$D(\text{ind}{\leqslant}k\ ; v) = \bigcup_{\text{ind}\, p \leqslant k} D(p, v), \quad D_\delta(\text{ind}{\leqslant}k\ ; v) = \bigcup_{\text{ind}\, p \leqslant k} D_\delta(p, v).$$

We shall also use a similar notation like $B_\delta(\text{ind}{\leqslant}k\ ; v)$ or $D(\text{ind}{=}k\ ; v)$ which is clear without special definition.

Proposition 3.43. *Let $f : W \to [a, b]$ be a Morse function on a cobordism W and v an almost transverse f-gradient. Then for every k*

(1) *The set $D(\text{ind}{\leqslant}k\ ; v)$ is compact.*
(2) *We have*

$$\bigcap_{\theta > 0} B_\theta(\text{ind}{\leqslant}k\ ; v) = D(\text{ind}{\leqslant}k\ ; v).$$

(3) *The family*

$$\{B_\theta(\text{ind}{\leqslant}k\ ; v)\}_{\theta > 0}$$

is a fundamental system of neighbourhoods of $D(\text{ind}{\leqslant}k\ ; v)$.

Proof. Let A_k be the set of all critical points with indices $\leqslant k$. Then A_k is an initial segment of $S(f)$, and

$$D(\text{ind}{\leqslant}k\ ; v) = D(A_k, v), \quad B_\delta(\text{ind}{\leqslant}k\ ; v) = B_\delta(A_k, v).$$

Apply Proposition 3.24 (page 144) and the proof is finished. □

Corollary 3.44. *Assume that v satisfies the almost transversality condition. Let $p \in S_k(f)$. Then the set $D(p, v) \cup D(\text{ind}{\leqslant}k{-}1\ ; v)$ is compact.*

Proof. Similar to Proposition 3.43. □

Corollary 3.45. *Assume that v satisfies the almost transversality condition. Let $p \in S_k(f)$. Let U be an open subset of W, such that*

$$D(\text{ind}{\leqslant}k{-}1\ , v) \cup \partial_0 W \subset U.$$

Then the set $D(p, v) \setminus U$ is compact.

Proof. We have

$$D(p, v) \setminus U = \Big(D(p, v) \cup D(\text{ind}{\leqslant}k{-}1\ , v)\Big) \setminus U.$$

The set in the right-hand side of the preceding formula is compact by Corollary 3.44. □

Now let us proceed to the thickenings of the descending discs.

Proposition 3.46. *Assume that v satisfies the almost transversality condition. Then for every k and every $\delta > 0$ sufficiently small:*

(1) *The set* $D_\delta(\text{ind}{\leqslant}k\ ; v)$ *is compact.*
(2) *We have*

$$\bigcap_{\theta>\delta} B_\theta(\text{ind}{\leqslant}k\ ; v) = D_\delta(\text{ind}{\leqslant}k\ ; v).$$

(3) *The family*

$$\{B_\theta(\text{ind}{\leqslant}k\ ; v)\}_{\theta>\delta}$$

is a fundamental system of neighbourhoods of $D_\delta(\text{ind}{\leqslant}k\ ; v)$.

Proof. Similar to Proposition 3.43. □

3.7. Morse stratifications. In this subsection we study stratifications of manifolds formed by descending discs of gradients of Morse functions. Let us first recall some basic notions of the theory of cellular decompositions. Let X be a Hausdorff topological space. *A k-cell of X* is a subspace of X homeomorphic to the k-dimensional open disc B^k. The number k is called *the dimension of the cell. A characteristic map* ϕ for a k-cell $E \subset X$ is a continuous map $\phi : D^k \to X$ of the closed k-dimensional disc to X, such that the restriction $\phi | B^k$ is a homeomorphism of B^k onto E. (Not every cell has a characteristic map.) A disjoint family $\{E_\alpha\}_{\alpha\in I}$ of subspaces of some topological space X will be called *topologically disjoint* if every set E_α is an open and closed subset of the topological space $\cup_\alpha E_\alpha$.

Definition 3.47. *A finite cellular decomposition* of a topological space X is a finite sequence of disjoint cells of X, such that

(A) the union of all cells is X,
(B) for each k the family of all k-dimensional cells is topologically disjoint,
(C) for each k the union of the cells of dimension $\leqslant k$ is a closed subspace of X.

The second condition is included to avoid configurations like the one depicted in Figure 17, where the 1-cells are not topologically disjoint.

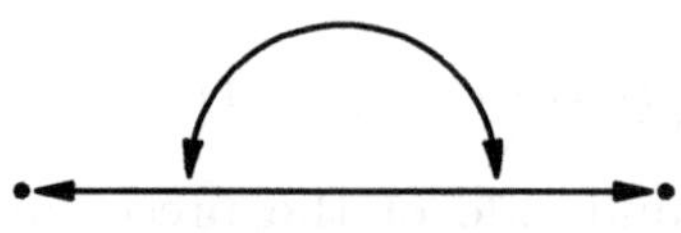

FIGURE 17.

Definition 3.48. Let X be a topological space, and $Y \subset X$ a subset. The set
$$\mathscr{B}(Y) = \overline{Y} \setminus Y$$
is called the *exterior boundary* of Y.

Proposition 3.49. *Let X be a topological space endowed with a finite cellular decomposition, and let E be one of cells. Then the exterior boundary*
$$\mathscr{B}(E) = \overline{E} \setminus E$$
of E is contained in the union of all cells of dimensions strictly smaller than $\dim E$.

Proof. Let $k = \dim E$ and let $\{E_1, \dots, E_n\}$ be the set of all k-dimensional cells, with $E_1 = E$. Condition (C) implies that $\overline{E}$ is contained in the union of all cells of dimension $\leqslant k$, thus it can not intersect the cells of dimension $> k$. It remains to prove that

$$\overline{E_i} \cap E_j = \varnothing \quad \text{for} \quad j \neq i. \tag{14}$$

The sets E_i are open and closed in $Z = E_1 \cup \dots \cup E_n$, therefore there exist open subsets $W_i \subset X$ such that $W_i \cap Z = E_i$. Assume that (14) does not hold. Then the closed subset $X \setminus W_j$ of X contains E_i, but does not contain $\overline{E_i}$, this leads to a contradiction. □

Definition 3.50. A finite cellular decomposition is called a *finite CW complex* if every cell has a characteristic map.†

Now let us return to the Morse theory.

Theorem 3.51. *Let $f : M \to \mathbf{R}$ be a Morse function on a closed manifold M, and v be an f-gradient. The two following properties are equivalent:*

(1) *v is almost transverse.*
(2) *The family*
$$\{D(p, v)\}_{p \in S(f)}$$
is a cellular decomposition of M.

Proof. (2) ⇒ (1). Assume that the condition (2) holds, but v is not almost transverse. Then there is a v-link from p to q with $p, q \in S(f)$ and $\operatorname{ind} p \geqslant \operatorname{ind} q$. In this case the point p is in the exterior boundary of the descending disc $D(q, v)$, that is,
$$p \in \mathscr{B}\big(D(q, v)\big) = \overline{D(q, v)} \setminus D(q, v)$$

† We shall work only with finite cellular decompositions and CW complexes and our definitions are adapted to this class of topological spaces. See **[90]**, **[153]**, **[29]** for the definitions in the general case.

and this is impossible by Proposition 3.49.

(1) $\Rightarrow$ (2). As we know from Proposition 2.17 of Chapter 3 (page 89) each set $D(p,v)$ is a k-cell, where $k = \operatorname{ind} p$. The condition (C) of Definition 3.47 follows from Proposition 3.43. To check the condition (B) it suffices to show that for every $p \in S_k(f)$ the subspace $D(p,v)$ is closed in $D(\text{ind}{=}k\ ; v)$. Let $a_n \in D(p,v)$ be a sequence of points converging to $a \in D(\text{ind}{=}k\ ; v)$.

Choose an ordered Morse function ϕ adjusted to (f,v), and let α, β be regular values of ϕ, such that all the critical points of ϕ of index k are in the cobordism $W = \phi^{-1}([\alpha,\beta])$. Choose $T > 0$ sufficiently large, so that $\gamma(a,T;v) \in W \setminus \partial W$. The sequence $\gamma(a_n,T;v) \in D(p,v) \cap W$ converges to $\gamma(a,T;v)$. Every $(-v)$-trajectory in $D(p,v)$ reaches $\partial_0 W$, therefore $D(p,v) \cap W$ is compact. Thus

$$\gamma(a,T;v) = \lim_{n\to\infty} \gamma(a_n,T;v) \in D(p,v) \cap W,$$

and $a \in D(p,v)$. □

Definition 3.52. When the f-gradient v satisfies the almost transversality condition, the cellular decomposition

$$M = \bigcup_{p \in S(f)} D(p,v)$$

will be called the *Morse stratification* of M corresponding to f and v. The cells of this decomposition will be called *Morse cells* of M. The set $D(\text{ind}{\leqslant}k\ ; v)$ will be called the *k-th skeleton* of the stratification.

It would be nice to have a theorem asserting that every Morse stratification is a finite CW complex. To prove this theorem one should construct for every critical point p of a given Morse function $f : M \to \mathbf{R}$ a homeomorphism $B^{\operatorname{ind} p} \xrightarrow{\approx} D(p,v)$, extendable to a continuous map $D^{\operatorname{ind} p} \longrightarrow M$ of the closed disc to our manifold. In many particular cases it is easy to construct such homeomorphisms. Recall for instance the height function on the two-dimensional sphere from Example 2.16 of Chapter 3 (page 88) (the required extension sends the whole boundary of the disc to the south pole of the sphere). It turns out however that the proof of a general result is a difficult task. (See the works [**85**], [**55**], [**56**], [**83**], [**97**] for the state of the art.†) The problem is that naturally arising homeomorphisms like the ray-preserving homeomorphisms for the case of gradient-like vector fields, are in general not extendable to continuous maps of the closed disc, see the examples below.

† The idea of a cellular decomposition associated with a Morse function appeared first in the paper [**157**] of R. Thom.

Fortunately, many results of the Morse theory do not depend on the construction of such CW complexes. The Morse stratification is a geometric structure naturally associated to a Morse function and its gradient, and this structure itself is sufficient for many purposes. We shall see that many results and ideas from the theory of CW complexes have their natural analogs in the framework of Morse stratifications. For example the Morse complex (see Chapter 6) is an analog of the cellular chain complex of a CW complex. Let us mention also the notion of the *dual Morse stratification* (see Subsection 1.6 of Chapter 6, page 205). This is indeed a remarkable construction: the Morse-theoretical proof of the Poincaré duality isomorphism is simpler than the classical one which is based on the triangulation theory.

Example 3.53. Consider a non-standard embedding of the 2-dimensional sphere into $\mathbf{R}^3$ as shown in Figure 18. Let M denote the image of this embedding.

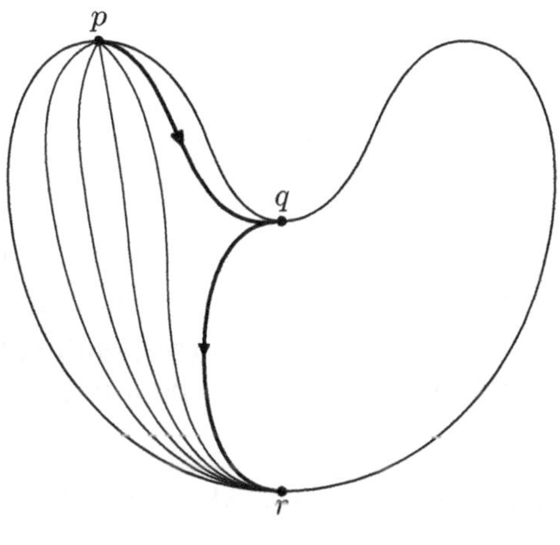

FIGURE 18.

The height function $f : M \to \mathbf{R}$ is a Morse function with four critical points: two local maxima, one local minimum and one saddle point. On the figure above we have depicted several flow lines of a gradient-like vector field v for f (the arrows show the direction of $(-v)$). All the $(-v)$-trajectories except one converge to r, and the exceptional one converges to q. There is a natural ray-preserving homeomorphism H of a 2-dimensional open disc onto $D(p, v)$ (see Proposition 2.20 of Chapter 3, page 89). In Figure 19 we have drawn the rays of the 2-dimensional disc, which are carried by H to the curves depicted on the previous figure. The ray L is sent to the flow line joining p to q, and the other rays are sent to the flow lines converging to r.

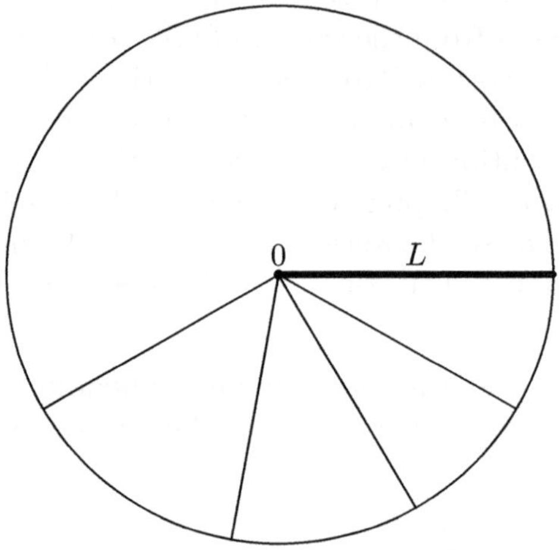

FIGURE 19.

If H were extendable to a continuous map of the closed disc D^2 to M, then this map would send the ends of all the rays to the ends of the corresponding flow lines. For the ray L the end of the corresponding flow line is the point q. For any other ray the end of the corresponding flow line is the point r. Thus such an extension does not exist.

Exercise 3.54. Study the case of the Morse function

$$f : S^m \to \mathbf{R}, \quad f(x_0, \dots, x_m) = \sum_{i=0}^{m} \alpha_i x_i^2$$

and its Riemannian gradient (see Subsection 2.4 of Chapter 3, page 92, and Figure 11 on page 95).

Exercise 3.55. Recall the embedding of the 2-torus into $\mathbf{R}^3$:

$$E : \mathbf{T}^2 \hookrightarrow \mathbf{R}^3, \quad (\phi, \theta) \mapsto \Big(r \sin\theta,\ (R + r\cos\theta)\cos\phi,\ (R + r\cos\theta)\sin\phi\Big).$$

The height function

$$\zeta : \mathbf{T}^2 \to \mathbf{R}, \quad \zeta(\phi, \theta) = (R + r\cos\theta)\sin\phi$$

corresponding to this embedding is a Morse function (see page 39), and we have seen (page 97) that the partition of $E(\mathbf{T}^2)$ into the union of the descending discs is not a cellular decomposition.

1) Check that the Riemannian gradient corresponding to this embedding does not satisfy the transversality condition.
2) Let A_α be the rotation of the space $\mathbf{R}^3$ by the angle α around the y-axis. Show that for every $\alpha \in [0, \pi/2[$ the height function ζ_α corresponding to the embedding $A_\alpha \circ E$ is a Morse function, which has four critical points of the same indices as the function $\zeta = \zeta_0$, namely, the maximum $M(\alpha)$ the minimum $m(\alpha)$ and two saddle points $s_1(\alpha), s_2(\alpha)$.
3) Show that there is α such that $\zeta_\alpha(s_1(\alpha)) = \zeta_\alpha(s_2(\alpha))$. For this value of α any gradient of ζ_α is necessarily almost transverse, and the Morse stratification of $\mathbf{T}^2$ is therefore a cellular decomposition, having two cells of dimension 1, one cell of dimension 2 and one zero-dimensional cell.

3.8. Morse-Smale filtrations. Let $f : W \to [a, b]$ be a Morse function on a cobordism W and v an almost transverse f-gradient. In this section we associate to v a certain class of filtrations of the cobordism W, called *Morse-Smale filtrations.* The k-th term of a Morse-Smale filtration is a ∂-manifold of dimension $\dim W$ which can be considered as a thickening of the k-th skeleton $D(\text{ind}{\leqslant}k\ ; v)$ of the Morse stratification. Technically these thickenings are very useful, in particular they allow us to extract topological information from the Morse stratification without organizing the Morse cells into a CW complex.

Definition 3.56. Let $\phi : W \to [a, b]$ be an ordered Morse function on a cobordism W, let $m = \dim W$. Let $\{a_i\}_{0 \leqslant i \leqslant m}$ be the ordering sequence for ϕ. For $0 \leqslant k \leqslant m$ put

$$W^{(k)} = \phi^{-1}([a_0, a_{k+1}]).$$

The filtration

$$\partial_0 W = W^{(-1)} \subset W^{(0)} \subset \cdots \subset W^{(m)} = W$$

is called *the Morse-Smale filtration induced by* ϕ. The filtration $(W^{(k)}, \partial_0 W)$ of the pair $(W, \partial_0 W)$ will also be called Morse-Smale filtration.

Since the value of ϕ on every critical point of index $\leqslant k$ is less than a_{k+1}, we have $D(\text{ind}{\leqslant}k\ ; v) \subset W^{(k)}$. The set $W^{(k)}$ is a compact ∂-manifold and its topological properties are in a sense simpler than those of $D(\text{ind}{\leqslant}k\ ; v)$. For many purposes the set $W^{(k)}$ is a good substitute for $D(\text{ind}{\leqslant}k\ ; v)$; the first proposition of this kind is Theorem 3.58 below which asserts that we can always choose a Morse-Smale filtration so that its successive terms are as close as we like to the skeletons of the Morse stratification.

Definition 3.57. Let $f : W \to [a,b]$ be a Morse function on a cobordism W and v an almost transverse f-gradient. A filtration $W^{(k)}$ of the cobordism W is called a *Morse-Smale filtration adjusted to* (f,v) if there is an ordered Morse function ϕ on W, adjusted to (f,v), and such that $W^{(k)}$ is the Morse-Smale filtration induced by ϕ.

Theorem 3.58. *Let $f : W \to [a,b]$ be a Morse function on a cobordism W and v an almost transverse f-gradient. Let $\{U_k\}_{0\leqslant k\leqslant m-1}$ be a family of open subsets of W, such that*

$$D(\text{ind}_{\leqslant k}\,; v) \cup \partial_0 W \subset U_k \quad \textit{for every} \quad k \leqslant m-1,$$

where $m = \dim W$. Then there is a Morse-Smale filtration $W^{(k)}$ adjusted to (f,v) such that for every $k \leqslant \dim W - 1$ we have $W^{(k)} \subset U_k$.

Proof. Start with any ordered Morse function $\psi : W \to [a,b]$ adjusted to (f,v), and let $W_0^{(k)}$ be the corresponding Morse-Smale filtration. The terms $W^{(k)}$ of the new filtration will be obtained by pushing $W_0^{(k)}$ downwards along the flow lines of v.

We begin by studying the gradient descent deformation in a somewhat more general context than necessary for the present theorem, in view of further applications. Let $h : W \to \mathbf{R}_+$ be a C^∞ function, such that $\operatorname{supp} h \subset W \setminus \big(\partial W \cup S(f)\big)$. Let $\tau \in \mathbf{R}$; define a diffeomorphism $H_\tau : W \to W$ by the following formula:

$$H_\tau = \Phi(w,\tau), \qquad \text{where} \qquad w(x) = h(x)v(x).$$

Thus $H_\tau(x) = x$ if $x \notin \operatorname{supp} h$ (in particular this is the case for x in a neighbourhood of $\partial W \cup S(f)$). In the domain where $h(x) > 0$ the points $H_\tau(x)$ and x are on the same v-trajectory. Put

$$\psi_\tau = \psi \circ H_\tau.$$

The function ψ_τ is obviously an ordered Morse function on W with the same critical points and the same ordering sequence as ψ.

Proposition 3.59. *The Morse function ψ_τ is adjusted to (f,v).*

Proof. By definition the functions ψ_τ and ψ are equal nearby $S(f)$ and nearby ∂W, therefore we have only to show that

$$\psi'_\tau(y)(v(y)) > 0 \quad \text{for every} \quad y \notin S(f).$$

Since

$$\psi'_\tau(y)(v(y)) = \psi'(H_\tau(y))\Big(H'_\tau(y)\big(v(y)\big)\Big),$$

the inequality above follows from the next lemma.

Lemma 3.60. *Define a vector field* $\widetilde{v}$ *on* W *by the following formula:*

$$\widetilde{v} = (H_\tau)_*(v).$$

Then for every $x \in W \setminus S(f)$ *we have:*

$$\widetilde{v}(x) = \lambda(x)v(x) \quad \text{with} \quad \lambda(x) > 0. \tag{15}$$

Proof. Put

$$y = \gamma(x, -\tau; w), \qquad \text{so that} \qquad x = H_\tau(y).$$

Denote by Z the set of zeros of h. Let us first consider the case when $x \notin Z$, that is, $h(x) > 0$. Then $h(y) > 0$, and we have:

$$\widetilde{v}(x) = H'_\tau(y)(v(y)) = \frac{1}{h(y)}\Big(H'_\tau(y)(w(y))\Big).$$

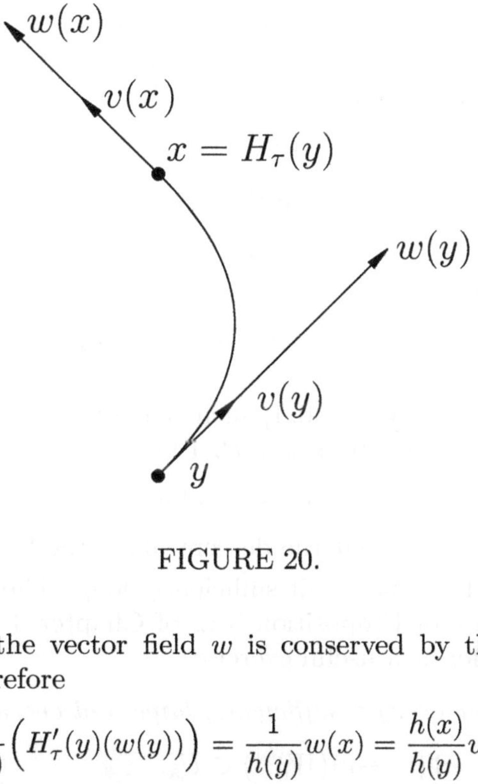

FIGURE 20.

Observe that the vector field w is conserved by the diffeomorphism $H_\tau = \Phi(w, \tau)$, therefore

$$\frac{1}{h(y)}\Big(H'_\tau(y)(w(y))\Big) = \frac{1}{h(y)} w(x) = \frac{h(x)}{h(y)} v(x), \tag{16}$$

and (15) is proved for $h(x) > 0$.

Now let us consider the case when $x \in \operatorname{Int} Z$, that is, the function h vanishes in some neighbourhood U of x. In this case we have $H_\tau(y) = y$ for every $y \in U$, and therefore $\widetilde{v}(y) = v(y)$ for $y \in U$; thus the equality (15) is proved also in this case.

It remains to consider the case when $x \in Z \setminus \mathrm{Int}\, Z$, so that in particular, $h(x) = 0$ and $x \notin S(f)$. In this case $y = x$ and there is a sequence of points

$$x_n \to x \quad \text{with} \quad x_n \notin S(f) \quad \text{and} \quad h(x_n) > 0.$$

We have:

$$\widetilde{v}(x_n) = \lambda(x_n)v(x_n), \quad \text{where} \quad \lambda(x_n) > 0 \quad \text{and}$$
$$\widetilde{v}(x_n) \to \widetilde{v}(x) = H'_\tau(y)(v(y)) \neq 0, \;\; v(x_n) \to v(x) \neq 0.$$

It is not difficult to deduce from this property that the sequence $\lambda(x_n)$ converges to a number $\lambda = \lambda(x) > 0$. We have therefore $\widetilde{v}(x) = \lambda(x)v(x)$, and our lemma is proved also in this case. □

The proof of Proposition 3.59 is now complete. □

Now let us return to the proof of the theorem. Diminishing U_k if necessary, we can assume that for every k the set U_k is a subset of the k-th term $W_0^{(k)}$ of the Morse-Smale filtration corresponding to ψ. Pick a collar C_0 for $\partial_0 W$ and a collar C_1 for $\partial_1 W$ with respect to v such that

$$C_0 \subset U_k \quad \text{and} \quad W_0^{(k)} \subset W \setminus C_1 \quad \text{for every} \quad k.$$

Let $\delta > 0$ be sufficiently small, so that

$$B_\delta(\mathrm{ind}{\leqslant}k\;; v) \subset U_k \quad \text{for every} \quad k.$$

Put

$$R = C_0 \cup C_1 \cup \bigcup_{p \in S(f)} B_\delta(p).$$

Pick a C^∞ function $h : W \to [0,1]$ such that $h(x) = 0$ in a neighbourhood of $\partial W \cup S(f)$ and $h(x) > 0$ for $x \notin R$. Put

$$w(x) = h(x)v(x).$$

The next lemma asserts that the descent along the flow lines of $(-w)$ will push $W_0^{(k)}$ to U_k if we shall wait sufficiently long. The proof is completely similar to the proof of Proposition 3.12 of Chapter 3 (page 106) and will be left to the reader as a useful exercise.

Lemma 3.61. *For every τ sufficiently large and every k we have:*

$$\Phi(w, -\tau)\big(W_0^{(k)}\big) \subset U_k.$$

As we have already shown the function $\psi_\tau = \psi \circ \Phi(w, \tau)$ is an ordered Morse function adjusted to (f, v). The k-th term $W^{(k)}$ of the filtration corresponding to ψ_τ equals

$$W^{(k)} = \Phi(w, -\tau)(W_0^{(k)})$$

and by the previous lemma this set is in U_k if τ is sufficiently large. This completes the proof of Theorem 3.58. □

CHAPTER 5

Handles

Let $f : W \to [a,b]$ be a Morse function on a cobordism W and v an f-gradient. Let p be a critical point of f of type (α), that is, each non-constant $(-v)$-trajectory in $D(p,v)$ reaches $\partial_0 W$. Then $D(p,v)$ is diffeomorphic to the k-dimensional closed disc D^k with $k = \operatorname{ind} p$. A *handle* for p is a special compact neighbourhood $\mathcal{H}$ of $D(p,v)$ homeomorphic to $D^k \times D^{m-k}$ where $m = \dim W$. Thus handles form one more species in the family of thickenings of descending discs. Compared to the thickenings already considered, handles have the advantage of a simple topological structure, which moreover depends only on the index of the critical point in question. The construction of handles is explained in the first section of this chapter.

In the second section we study Morse functions which have only critical points of type (α). In this case $H_*(W, \partial_0 W)$ is a free abelian group generated by the homology classes of the descending discs modulo their boundaries. We show that the basis formed by these classes is stable with respect to C^0-small perturbations of the gradient.

1. Construction of handles

Definition 1.1. Let $f : W \to [a,b]$ be a Morse function on a cobordism W and v an f-gradient. Let $p \in S(f)$, and $k = \operatorname{ind} p$. A *handle* for p is a compact $(-v)$-invariant neighbourhood $\mathcal{H}$ of $D(p,v)$ together with a homeomorphism $\Phi : D^k \times D^{m-k} \to \mathcal{H}$ of the product of two closed discs

$$D^k = D^k(0, r_1), \quad D^{m-k} = D^{m-k}(0, r_2)$$

onto $\mathcal{H}$, such that

(1) $\Phi(D^k \times 0) = D(p,v)$;
(2) $\Phi(0 \times 0) = p$;
(3) $\Phi(S^{k-1} \times D^{m-k}) = \mathcal{H} \cap \partial_0 W$.

Observe that if a critical point p has a handle, then $D(p,v)$ is compact, and therefore p is necessarily of type (α), see Corollary 2.22 of Chapter 3 (page 90).

Figure 21 is an illustration. The cobordism in question is the "pants" submanifold of $\mathbf{R}^3$, and the Morse function is the height function. The critical point is shown by a black circle and the handle is shaded.

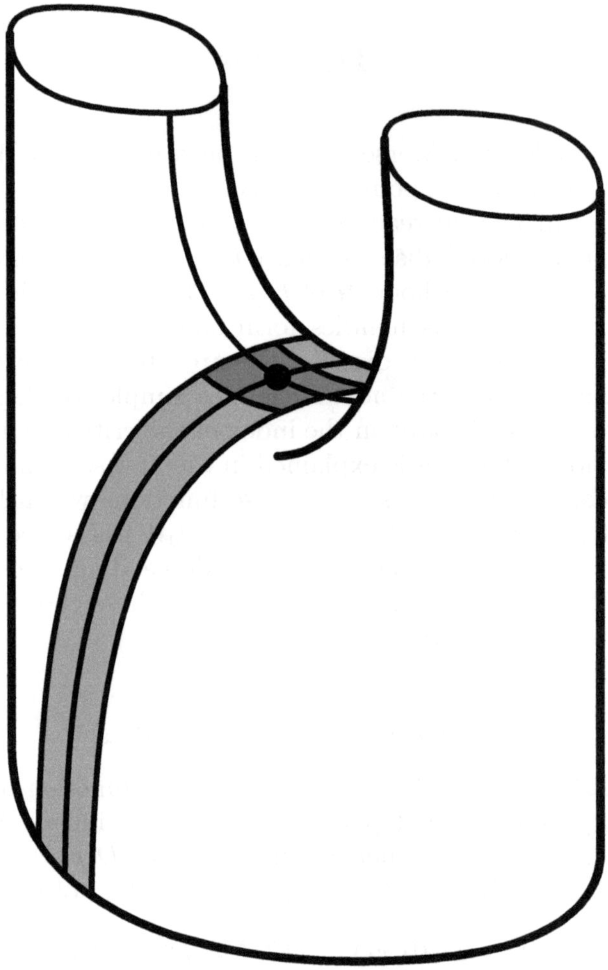

FIGURE 21.

Theorem 1.2. *Let $f : W \to [a, b]$ be a Morse function on a cobordism W and v an f-gradient. Let p be a critical point of f of type (α). Let U be a neighbourhood of $D(p, v)$. Then there exists a handle $\mathcal{H} \subset U$ for p.*

One can guess from Figure 21 the basic idea of the proof. We choose a small compact neighbourhood of p (shown by the dark grey color on the picture) and the handle is defined as the track of this neighbourhood with respect to $(-v)$.

We first prove the theorem for the particular case of gradient-like vector fields (the next subsection). The proof in the general case (Subsection 1.2) is based on similar ideas but is slightly more complicated.

Note the condition of $(-v)$-invariance of $\mathcal{H}$ in the theorem. This condition is technically quite important; omitting it leads to an assertion which can be easily proved using elementary differential topology. Indeed, the submanifold $D(p,v) \subset W$ is a neat submanifold with boundary of the manifold W. Moreover it is easy to show that the normal bundle to $D(p,v)$ in W is trivial. Thus the tubular neighbourhood of $D(p,v)$ in W is homeomorphic to the product $D(p,v) \times D^{m-k}$ (see Section 6 of Chapter 4 of [**61**] for details).

1.1. Proof of the existence theorem: the case of gradient-like vector fields. Applying Corollary 3.15 of Chapter 4 (page 141) it is easy to reduce our assertion to the particular case when p is the unique critical point of f, and we shall assume this during the proof. Let

$$\Psi : V \to \mathcal{B}(R) = B^k(0,R) \times B^{m-k}(0,R)$$

be a Morse chart for (f,v) at p, such that $V \subset U$. Fix any $r \in]0,R[$, and put

$$\begin{aligned} Q &= D^k(0,r) \times D^{m-k}(0,r) \subset \mathbf{R}^k \times \mathbf{R}^{m-k}, \\ Q_- &= S^{k-1}(0,r) \times D^{m-k}(0,r) \subset \mathbf{R}^k \times \mathbf{R}^{m-k}, \\ \mathcal{F} &= \Psi^{-1}(Q), \quad \mathcal{F}_- = \Psi^{-1}(Q_-). \end{aligned}$$

We shall prove that the track $T(\mathcal{F},-v)$ is a handle.[†] A $(-v)$-trajectory can quit the set $\mathcal{F}$ only by passing through a point of $\mathcal{F}_-$ (see Section 2 of Chapter 3, page 81 and Figure 22)

[†] An exercise for the reader: locate the subsets $\mathcal{F}$, $\mathcal{F}_-$, $T(\mathcal{F},-v)$ in Figure 21.

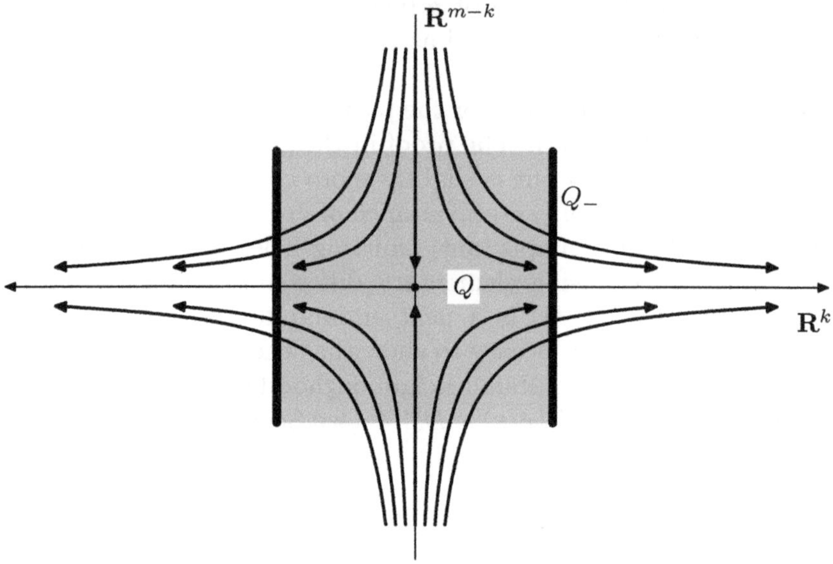

FIGURE 22.

and therefore

$$T(\mathcal{F}, -v) = \mathcal{F} \cup T(\mathcal{F}_-, -v).$$

Our first aim is to study the set $T(\mathcal{F}_-, -v)$.

Lemma 1.3. (1) *Any $(-v)$-trajectory starting at a point of $\mathcal{F}_-$ reaches $\partial_0 W$.*

(2) *For any $x \in \mathcal{F}_-$ and $t > 0$ we have: $\gamma(x, t; -v) \notin \mathcal{F}$.*

Proof. For every $x \in \mathcal{F}_-$ we have $f(x) \leqslant f(p)$, therefore any $(-v)$-trajectory starting at a point of $\mathcal{F}_-$ cannot converge to p and reaches $\partial_0 W$; the first point is proved. As for the second one, let $x \in \mathcal{F}_-$. Assume that there is $\tau > 0$ such that $y = \gamma(x, \tau; -v) \in \mathcal{F}$. Pick the minimal τ with this property, so that $\gamma(x,]0, \tau[; -v)$ does not intersect $\mathcal{F}$. Then the point y is in the boundary of $\mathcal{F}$, and since $f(y) < f(x) \leqslant f(p)$, the point y is in $\mathcal{F}_-$ and $f(y) < f(p)$. However for such points y the vector $-v(y)$ points *outside* $\mathcal{F}$ (see Figure 22), and therefore y can not be the point through which the curve $\gamma(x, \cdot; -v)$ enters the domain $\mathcal{F}$. □

The map

$$G : \mathcal{F}_- \times [0, 1] \to T(\mathcal{F}_-, -v), \quad G(x, t) = \gamma(x, t \cdot \tau(x, -v); -v)$$

is a homeomorphism (see Remark 3.4 of Chapter 3, page 101). The set $\mathcal{F}_- \times [0,1]$ is homeomorphic to $Q_- \times [0,1]$ via the homeomorphism $B = (\Psi^{-1}|Q_-) \times \mathrm{id}$. The composition of these two homeomorphisms will be denoted

$$C = G \circ B : Q_- \times [0,1] \to T(\mathcal{F}_-, -v).$$

Let

$$K = Q \sqcup \Big(Q_- \times [0,1]\Big)$$

and consider the quotient space $K/\sim$ where $\sim$ is the following equivalence relation: $x \sim (x,0)$ if $x \in Q_-$. The space $K/\sim$ is homeomorphic to Q as illustrated in Figure 23. The space Q is depicted on the left and the space $Q_- \times [0,1]$ is on the right (here the dimensions of the two discs D^k, D^{m-k} are equal to 1). The subset Q_- is the lateral surface of the cylinder Q which is glued to the interior lateral surface $Q_- \times 0$ on the right.

FIGURE 23.

The two maps

$$\Psi^{-1}|Q : Q \to \mathcal{F}, \quad C : Q_- \times [0,1] \to T(\mathcal{F}_-, -v)$$

coincide obviously on the subset $Q_- = Q_- \times 0$, therefore they can be glued together to a homeomorphism Φ as shown on the next diagram

$$\begin{array}{ccc} K/\sim & \xrightarrow{\quad\Phi\quad} & T(\mathcal{F}, -v) \\ \Big\| & & \Big\| \\ Q \cup \big(Q_- \times [0,1]\big) & \xrightarrow{\Psi^{-1}|Q \,\cup\, C} & \mathcal{F} \cup T(\mathcal{F}_-, -v) \end{array}$$

The set $T(\mathcal{F}, -v)$ together with the homeomorphism Φ is the handle sought. $\square$

1.2. Proof of the existence theorem: the general case. As in the previous subsection, we shall assume that p is the unique critical point of f. Let

$$\phi : \mathcal{B}(R) = B^k(0,R) \times B^{m-k}(0,R) \to V$$

be a box for (f,v) at p such that $V \subset U$ (see Definition 1.19 of Chapter 3 (page 77)). Fix any $r \in]0,R[$, and put

$$Q(\rho) = D^k(0,r) \times D^{m-k}(0,\rho) \subset \mathbf{R}^k \times \mathbf{R}^{m-k},$$
$$Q_-(\rho) = S^{k-1}(0,r) \times D^{m-k}(0,\rho) \subset \mathbf{R}^k \times \mathbf{R}^{m-k},$$
$$\mathcal{F}(\rho) = \phi(Q(\rho)), \quad \mathcal{F}_-(\rho) = \phi(Q_-(\rho))$$

(where $k = \operatorname{ind} p$, $m = \dim W$ and ρ is a number in $[0,R[$). We shall prove that for every $\rho > 0$ sufficiently small the set $T(\mathcal{F}(\rho), -v)$ is a handle.

It is easy to see that a $(-v)$-trajectory can quit $\mathcal{F}(\rho)$ only by passing through a point of $\mathcal{F}_-(\rho)$. Therefore

$$T\big(\mathcal{F}(\rho), -v\big) = \mathcal{F}(\rho) \cup T\big(\mathcal{F}_-(\rho), -v\big).$$

Lemma 1.4. *For any ρ sufficiently small we have:*

(1) *Any $(-v)$-trajectory starting at a point of $\mathcal{F}_-(\rho)$ reaches $\partial_0 W$.*
(2) *For any $x \in \mathcal{F}_-(\rho)$ and $t > 0$ we have: $\gamma(x,t;-v) \notin \mathcal{F}(\rho)$.*

Proof. (1) Any non-constant $(-v)$-trajectory in $D(p,v)$ reaches the boundary. In particular this is the case for the trajectories starting at any point of $\mathcal{F}_-(0)$. The set $\mathcal{F}_-(0)$ is compact, and therefore there is a neighbourhood R of $\mathcal{F}_-(0)$ such that every $(-v)$-trajectory starting at a point of R reaches $\partial_0 W$. For $\rho > 0$ sufficiently small we have $\mathcal{F}_-(\rho) \subset R$, and the first point of our lemma is proved.

(2) Now let us move to the second point. The contrary would imply that there are sequences

$$\rho_n \searrow 0, \quad x_n \in \mathcal{F}_-(\rho_n), \quad y_n \in \mathcal{F}(\rho_n), \quad t_n > 0,$$
$$\text{such that} \quad \gamma(x_n, t_n; -v) = y_n.$$

Observe that there is $\epsilon > 0$ such that the map

$$(x,t) \mapsto \gamma(x,t,-v); \quad \mathcal{F}_-(\rho) \times]-\epsilon, \epsilon[\to W$$

is injective for every $\rho < R$. Therefore $t_n \geqslant \epsilon$. Further, choose any $C > 0$ such that every $(-v)$-trajectory starting at a point of the boundary $\partial\mathcal{F}(0) = \mathcal{F}_-(0)$ of the topological disc $\mathcal{F}(0)$ reaches $\partial_0 W$ at the moment $t \leqslant C$. Then every t_n is less than $2C$ if ρ is sufficiently small. Extracting subsequences if necessary, we can assume that

$$x_n \to a, \quad y_n \to b, \quad t_n \to \tau \geqslant \epsilon.$$

Then

$$a \in \mathcal{F}_-(0), \quad b = \gamma(a, \tau; -v) \in \mathcal{F}(0), \quad \tau > 0.$$

However it follows from the definition of boxes that a $(-v)$-trajectory can intersect the subset $\mathcal{F}_-(0) = \partial\mathcal{F}(0)$ only once. This leads to a contradiction. □

The end of the proof is completely similar to the case of gradient-like vector fields. The map

$$G : \mathcal{F}_-(\rho) \times [0,1] \to T(\mathcal{F}_-(\rho), -v), \quad G(x,t) = \gamma(x, t \cdot \tau(x, -v); -v)$$

is a homeomorphism (see Remark 3.4 of Chapter 3, page 101). Let B denote the homeomorphism

$$\big(\phi | Q_-(\rho)\big) \times \mathrm{id} : Q_-(\rho) \times [0,1] \to \mathcal{F}_-(\rho) \times [0,1].$$

Let

$$K(\rho) = Q(\rho) \sqcup \Big(Q_-(\rho) \times [0,1]\Big)$$

and consider the quotient space $K(\rho)/\sim$ with respect to the following equivalence relation: $x \sim (x, 0)$ if $x \in Q_-(\rho)$. The space $K(\rho)/\sim$ is then homeomorphic to $Q(\rho)$ itself. The maps

$$\phi | Q(\rho) : Q(\rho) \to \mathcal{F}(\rho), \quad B : Q_-(\rho) \times [0,1] \to \mathcal{F}_-(\rho) \times [0,1]$$

can be glued together to a homeomorphism

$$\Phi : K(\rho)/\sim \ \xrightarrow{\approx} T(\mathcal{F}_-(\rho), -v).$$

The set $T(\mathcal{F}_-(\rho), -v)$ together with the homeomorphism Φ is the handle sought. □

1.3. Homotopy type of elementary cobordisms. Let $f : W \to [a, b]$ be a Morse function on a cobordism W, and assume that (f, v) is elementary, that is, the critical points of f are not linked to each other. By Corollary 3.8 of Chapter 4 (page 136) there is $\delta > 0$ such that the sets $B_\delta(p, v)$ are pairwise disjoint. For each $p \in S(f)$ pick a handle $\mathcal{H}_p$ for p. We can assume that $\mathcal{H}_p \subset B_\delta(p, v)$ for every $p \in S(f)$, so that the handles are pairwise disjoint. Set

$$X = \partial_0 W \cup \bigcup_{p \in S(f)} D(p, v), \quad Y = \partial_0 W \cup \bigcup_{p \in S(f)} \mathcal{H}_p;$$

we have $\partial_0 W \subset X \subset Y \subset W$. The homeomorphism type of X is easy to describe: X is obtained from $\partial_0 W$ by one of the most common constructions in algebraic topology.

Lemma 1.5. *X is homeomorphic to a space obtained from $\partial_0 W$ by attaching $m(f)$ cells, one cell of dimension* $\operatorname{ind} p$ *for each critical point of index p.*

Proof. Obvious. □

Next we shall show that W is homotopy equivalent to X.

Proposition 1.6. *X is a strong deformation retract of Y.*

Proof. It suffices to construct for each p a strong deformation retraction of $\mathcal{H}_p$ onto $D(p,v) \cup (\mathcal{H}_p \cap \partial_0 W)$. By the very definition of handles there is a homeomorphism of pairs

$$\Big(\mathcal{H}_p,\ D(p,v) \cup (\mathcal{H}_p \cap \partial_0 W)\Big) \approx \Big(D^k \times D^{m-k},\ D^k \times 0 \cup S^{k-1} \times D^{m-k}\Big).$$

It is an easy exercise in elementary algebraic topology to construct a deformation retraction of $D^k \times D^{m-k}$ onto $D^k \times 0 \cup S^{k-1} \times D^{m-k}$. □

Theorem 1.7. (1) *The inclusion $X \hookrightarrow W$ is a homotopy equivalence.*

(2) *The inclusion of pairs $(X, \partial_0 W) \hookrightarrow (W, \partial_0 W)$ is a homology equivalence.*†

Proof. The second assertion follows from the first one by an easy application of the Five-Lemma to the exact homology sequences of the two pairs. Let us prove the first assertion. In view of the preceding proposition it suffices to prove that the inclusion $Y \hookrightarrow W$ is a homotopy equivalence. Recall the continuous map $L : W \times \mathbf{R}_+ \to W$ from Subsection 3.2 of Chapter 3 (page 105):

$$\begin{aligned} L(x,t) &= \gamma(x,t;-v) && \text{if } t \leqslant \tau(x,-v),\\ L(x,t) &= \gamma(x,\tau(x,-v);-v) && \text{if } t \geqslant \tau(x,-v), \end{aligned}$$

where $\tau(x,-v)$ is the moment when the $(-v)$-trajectory $\gamma(x,\cdot,-v)$ quits W (by a certain abuse of notation we set $\tau(x,-v) = \infty$ if this trajectory *does not* quit W). Put $L_t(x) = L(x,t)$, then for every t the map L_t is a continuous map of W to itself. Every handle $\mathcal{H}_p$ is $(-v)$-invariant, therefore for every t the set Y is invariant with respect to L_t. Note moreover, that for T sufficiently large we have $L_T(W) \subset Y$ (by Proposition 3.12 of Chapter 3, page 106). Therefore the proof is completed by the application of the next lemma.

† A map of pairs is called a *homology equivalence* if it induces an isomorphism in the homology groups of the pairs.

Lemma 1.8. *Let X be a topological space and $Y \subset X$. Assume that there is a homotopy $\Phi_t : X \to X$, $t \in [0, T]$ with the following properties:*

(i) $\Phi_0 = id$, *and* $\Phi_T(X) \subset Y$.
(ii) Y *is invariant with respect to* Φ_t *for all* t.

Then the inclusion $i : Y \hookrightarrow X$ *is a homotopy equivalence.*

Proof. The map $\Phi_T : X \to Y$ is the homotopy inverse for i. Indeed, the composition $\Phi_T \circ i : Y \to Y$ is equal to $\Phi_T \mid Y$ which is homotopic to the identity map $Y \to Y$ via the homotopy $\Phi_t \mid Y$. The composition $i \circ \Phi_T : X \to X$ is homotopic to the identity map $X \to X$ via the homotopy Φ_t. □

Corollary 1.9. *The homology $H_*(W, \partial_0 W)$ is a free abelian group which has one free generator in degree k for each critical point of f of index k.*

Proof. We have $H_*(W, \partial_0 W) \approx H_*(X, \partial_0 W)$, and the homology of $(X, \partial_0 W)$ is easily deduced from Lemma 1.5. □

We shall need in the sequel a relative version of these results (the proof is completely similar to the above).

Proposition 1.10. *Let $f : W \to [a, b]$ be a Morse function on a cobordism W and v an f-gradient. Let c be a regular value of f; put*

$$W' = f^{-1}([a, c]), \quad W'' = f^{-1}([c, b]).$$

Assume that the pair $(f|W'', v|W'')$ is elementary. Let X denote the union of all descending discs of critical points of $f|W''$. Then the inclusion $W' \cup X \hookrightarrow W$ is a homotopy equivalence. The inclusion of pairs

$$(W' \cup X, W') \hookrightarrow (W, W')$$

is a homology equivalence. □

2. Morse functions and the cellular structure of manifolds

It turns out that a delicate task of constructing a CW decomposition from the Morse stratification becomes a lot easier if we limit ourselves to the *homotopy type* of the underlying manifold.

Theorem 2.1. *Let M be a closed manifold, and $f : M \to \mathbf{R}$ be a Morse function. Then M is homotopy equivalent to a CW-complex having exactly $m_k(f)$ cells in dimension k.*

Proof. Let v be an f-gradient satisfying the almost transversality condition. Pick an ordered Morse function $\phi : M \to \mathbf{R}$ adjusted to (f, v), and let $a = a_0 < \cdots < a_{m+1} = b$ be the ordering sequence of ϕ. Consider the corresponding Morse-Smale filtration $M^{(k)} = \phi^{-1}([a_0, a_{k+1}])$ of M. Proposition 1.10 implies that for each k the space $M^{(k)}$ is homotopy equivalent to the space obtained from $M^{(k-1)}$ by attaching $m_k(f)$ cells of dimension k. Now our theorem follows by an induction argument applying the following result from elementary algebraic topology.

Lemma 2.2 ([**94**], Lemma 3.7). *Let X, Y be two homotopy equivalent topological spaces. Let X' be the space obtained from X by attaching one k-dimensional cell. Then X' is homotopy equivalent to a space Y' obtained from Y by attaching one k-dimensional cell.* □

This theorem is widely used in algebraic topology. It makes applicable to manifolds the usual tools of the homotopy theory of CW complexes, like cellular approximation theorems, cellular homology and others. However the theorem is only about the *homotopy type* of the manifold. Moreover it gives information only about the number of cells, but says nothing about the characteristic maps.

3. Homology of elementary cobordisms

Let $f : W \to [a, b]$ be a Morse function on a cobordism W and v an f-gradient. Assume that the pair (f, v) is elementary. Corollary 1.9 says that the homology of the pair $(W, \partial_0 W)$ is a free abelian group. The main aim of the present section is the detailed study of this group and of naturally arising bases in it. The first two subsections mostly contain preliminaries, so the reader may wish to begin with the third subsection and return to the first two when necessary.

3.1. Orientations of gradients. Let $f : W \to [a, b]$ be a Morse function on a cobordism W and v an f-gradient. We do not assume in this subsection that the pair (f, v) is elementary.

Definition 3.1. A choice of orientation of the manifold $D(p, v)$ for every $p \in S(f)$ is called an *orientation of v*. An f-gradient endowed with an orientation is called *oriented.*

Let $\mathcal{O}$ be an orientation of v. Let p be a critical point of f of index k. We have an isomorphism

$$H_k\big(D(p, v), D(p, v) \setminus p\big) \approx \mathbf{Z}.$$

The chosen orientation of $D(p,v)$ determines a fundamental class

$$\omega_p \in H_k\big(D(p,v), D(p,v)\setminus p\big)$$

which is a free generator of this group. Put

$$W_p = \{x \in W \mid f(x) \leqslant f(p)\}.$$

The image of ω_p in the group $H_*(W_p, W_p \setminus p)$ will be denoted $\mathcal{O}_p$.

Definition 3.2. Let v, w be f-gradients, let $\mathcal{O}$ be an orientation of v, and $\mathcal{O}'$ be an orientation of w. We say that $\mathcal{O}$ and $\mathcal{O}'$ are *equivalent at* p if $\mathcal{O}_p = \mathcal{O}'_p$. Orientations of two f-gradients v, w are called *equivalent* if they are equivalent at every $p \in S(f)$. Oriented gradients are called *similarly oriented* if their orientations are equivalent.

Proposition 3.3. *Let $f : M \to \mathbf{R}$ be a Morse function (where M is a manifold without boundary or a cobordism), and let p be a critical point of f of index k. Let $L \subset M$ be a k-dimensional submanifold (maybe with non-empty boundary), such that*

$$L \subset M_p, \quad p \in L \setminus \partial L.$$

Then the inclusion

$$(L, L\setminus p) \overset{j}{\hookrightarrow} (M_p, M_p \setminus p)$$

is a homology equivalence.

Proof. Applying the excision isomorphism and the Morse lemma the assertion is easily reduced to the following particular case:

$$M = B^k(0,R) \times B^{m-k}(0,R), \quad p = (0,0),$$
$$f(x,y) = -||x||^2 + ||y||^2.$$

In this case

$$M_p = \{(x,y) \mid ||x|| \geqslant ||y||\}.$$

Put

$$B = B^k(0,R), \quad B' = B^{m-k}(0,R),$$

and let π denote the first coordinate projection

$$B \times B' \to B, \quad (x,y) \mapsto x.$$

The restriction

$$\pi|(M_p, M_p\setminus p) : (M_p, M_p \setminus p) \to (B, B\setminus p)$$

is a homology equivalence. (See Figure 24 below, where the manifold $M = B \times B'$ is light shaded, and the set M_p is dark shaded.)

Let $T_pL \subset \mathbf{R}^m$ denote the vector subspace of all vectors in $\mathbf{R}^m$ tangent to L at $p = (0,0)$. Since $L \subset M_p$, we have $T_pL \subset M_p$, and therefore

$T_pL \pitchfork \mathbf{R}^{m-k}$, and the differential of the map $\pi|L$ at the point $p = (0,0)$ is an isomorphism. Diminishing R if necessary, we can assume that the restriction $\pi|L$ is a diffeomorphism of L onto B. In this case the restriction induces a homology equivalence

$$\pi|(L, L \setminus p) : (L, L \setminus p) \to (B, B \setminus p)$$

and therefore j also is a homology equivalence. □

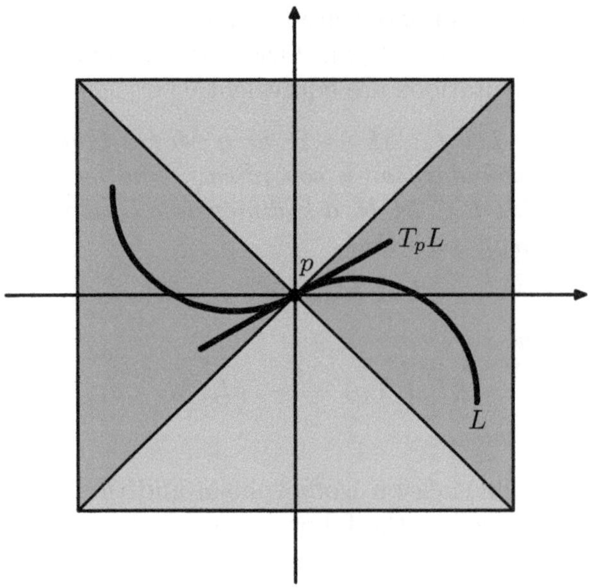

FIGURE 24.

Applying the proposition above to the submanifold $L = D(p, v)$ we deduce that for every orientation $\mathcal{O}$ of an f-gradient v the element $\mathcal{O}_p$ is a generator of the group $H_*(W_p, W_p \setminus \{p\}) \approx \mathbf{Z}$. There are exactly two classes of equivalence of orientations of gradients at a given $p \in S(f)$. On the whole there are exactly $2^{m(f)}$ classes of equivalence of orientations.

3.2. Homology of elementary cobordisms: preliminaries. Let $f : W \to [a, b]$ be a Morse function on a cobordism W and v an f-gradient.

Proposition 3.4. *The inclusion*

$$(W, \partial_0 W) \xrightarrow{j} (W, W \setminus D(-v))$$

is a homology equivalence.

Proof. It suffices to check that the inclusion $\partial_0 W \hookrightarrow W \setminus D(-v)$ is a homology equivalence. This is proved by the usual gradient descent argument: the formula

$$H(x,t) = \gamma\big(x, t\cdot\tau(x,-v); -v\big), \quad \text{with} \quad t\in[0,1]$$

defines a deformation retraction of $W\setminus D(-v)$ onto $\partial_0 W$ (recall that $\tau(x,-v)$ is the moment when the curve $\gamma(x,\cdot;-v)$ reaches $\partial_0 W$). □

Now we proceed to elementary cobordisms.

Proposition 3.5. *Let $f: W \to [a,b]$ be a Morse function on a cobordism W and v an f-gradient. Assume that the critical points of f are not linked to each other. Let $\Sigma = D(v)\cap\partial_0 W$. The inclusion*

$$(\partial_0 W, \partial_0 W\setminus\Sigma) \overset{i}{\hookrightarrow} (W, W\setminus D(v))$$

is a homology equivalence.

Proof. For each critical point $p\in S(f)$ choose a handle $\mathcal{H}_p$ for p, in such a way that $\mathcal{H}_p$ and $\mathcal{H}_q$ are disjoint whenever $p\neq q$. Let

$$\mathcal{H} = \cup_p \mathcal{H}_p, \quad A_p = \mathcal{H}_p\cap\partial_0 W, \quad A = \cup_p A_p,$$

so that A_p is the "sole" of the handle $\mathcal{H}_p$, and A is the union of these soles. Consider the commutative diagram of inclusions:

$$\begin{array}{ccc} (\mathcal{H}, \mathcal{H}\setminus D(v)) & \hookrightarrow & (W, W\setminus D(v)) \\ \uparrow & & \uparrow \\ (A, A\setminus\Sigma) & \hookrightarrow & (\partial_0 W, \partial_0 W\setminus\Sigma). \end{array}$$

The horizontal arrows are homology equivalences (by excision) so it remains to prove that the left vertical arrow is a homology equivalence. Since

$$H_*(A, A\setminus\Sigma) \approx \bigoplus_{p\in S(f)} H_*(A_p, A_p\setminus\Sigma),$$

$$H_*(\mathcal{H}, \mathcal{H}\setminus D(v)) \approx \bigoplus_{p\in S(f)} H_*(\mathcal{H}_p, \mathcal{H}_p\setminus D(v)),$$

it suffices to prove that for each p the inclusion

$$(A_p, A_p\setminus\Sigma) \hookrightarrow (\mathcal{H}_p, \mathcal{H}_p\setminus D(v))$$

is a homology equivalence. And this follows immediately from the definition of handles. □

Applying the preceding proposition to the $(-f)$-gradient $(-v)$ we obtain the following corollary.

Corollary 3.6. *Let $f : W \to [a, b]$ be a Morse function on a cobordism W and v an f-gradient. Assume that the critical points of f are not linked to each other. Let $S = D(-v) \cap \partial_1 W$. Then the inclusion*

$$(\partial_1 W, \partial_1 W \setminus S) \xrightarrow{i} (W, W \setminus D(-v))$$

is a homology equivalence. □

3.3. The basis in the homology of $(W, \partial_0 W)$. We continue our study of elementary cobordisms. Let $f : W \to [a, b]$ be a Morse function on a cobordism W and v an f-gradient. We assume that the critical points of f are not linked to each other, and moreover, that they all have the same index k. Corollary 1.9 says that the homology $H_*(W, \partial_0 W)$ is a free abelian group, and suggests a natural basis in this group. Namely, choose an orientation of the f-gradient v. For every critical point $p \in S(f)$ we have a diffeomorphism of the pairs

$$\big(D(p, v), \partial D(p, v)\big) \approx (D^k, S^{k-1}) \quad \text{where} \quad k = \operatorname{ind} p,$$

and the chosen orientation determines a generator in the group $H_k(D(p, v), \partial D(p, v)) \approx \mathbf{Z}$. The image of this generator with respect to the homomorphism

$$H_k\big(D(p, v), \partial D(p, v)\big) \to H_k(W, \partial_0 W)$$

induced by the inclusion will be denoted by d_p. We obtain the following proposition.

Proposition 3.7. *Assume that the critical points of f are not linked to each other and the index of every critical point of f equals k. Then the homology $H_k(W, \partial_0 W)$ is a free abelian group and the family $\{d_p\}$ is a basis of this group. If $i \neq k$, then $H_i(W, \partial_0 W) = 0$.* □

The rest of this section is devoted to the detailed study of this basis.

3.4. Multiplicities of cycles at critical points. As in the previous subsection let $f : W \to [a, b]$ be a Morse function and v an f-gradient, such that (f, v) is an elementary pair, and all critical points have the same index k. Choose an orientation of v. For $x \in H_k(W, \partial_0 W)$ write

$$x = \sum_{p \in S(f)} m(x, p) d_p \quad \text{where} \quad m(x, p) \in \mathbf{Z}.$$

The integer $m(x, p)$ will be called the *multiplicity of x at p*. The geometric interpretation of numbers $m(x, p)$ is related to one of the very basic intuitive ideas in Morse theory, namely the idea of *cycles hanging on critical points.*

Assuming for simplicity that x is represented by a closed oriented k-dimensional submanifold N of $\partial_1 W$, let us allow N to descend along the flow lines of the gradient. If $x \neq 0$, the manifold N will never land on $\partial_0 W$, and therefore it will *hang* on one or more critical points, approaching the descending discs in a neighbourhood of critical points. To see how it works, let us consider the simplest possible example. In Figure 25 we have depicted the "pants" manifold endowed with the usual height function having the unique critical point p. Let N be one of the connected components of $\partial_0 W$. The fat lines on the figure represent the subsets $L_t(N)$ for several values of t. As $t \to \infty$ locally in a neighbourhood of the critical point p the subset $L_t(N)$ becomes closer and closer to the descending disc $D(p, v)$. The homology class of $L_t(N)$ in the relative homology of the pair $(W, \partial_0 W)$ is therefore equal to that of the descending disc modulo its boundary, and the multiplicity of the homology class $[N]$ with respect to p is equal to 1. Observe that the intersection index of N with $D(p, -v)$ is also equal to 1.

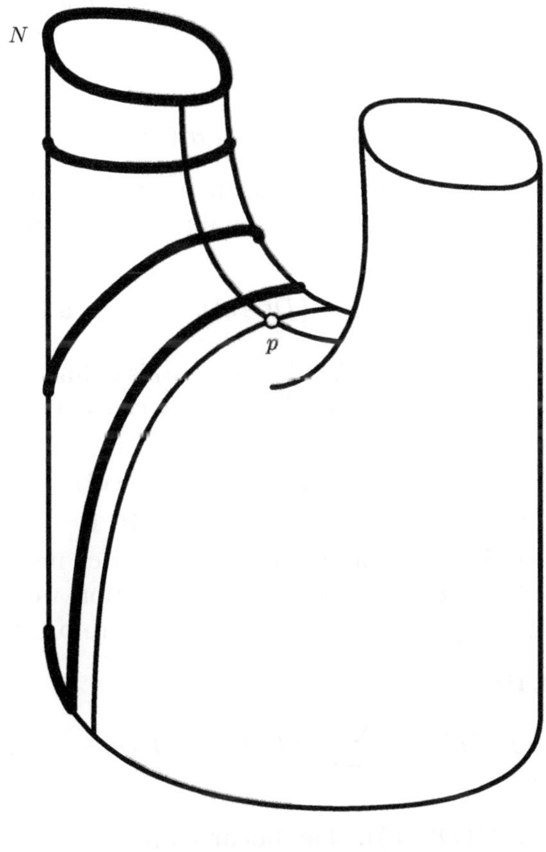

FIGURE 25.

In the case of arbitrary cobordism W the situation is more complicated: in general a submanifold $N \subset \partial_1 W$ descending along the gradient flow falls into *several copies* of $D(p,v)$ in a neighbourhood of a critical point p. Each of these copies inherits an orientation from N. Comparing these orientations with the orientation of $D(p,v)$ and summing up the signs we obtain an integer, which turns out to be equal to $m(x,p)$, and is also equal to the algebraic intersection number of $D(p,-v)$ with N. This last result and its generalizations form the subject of the present section.

Exercise 3.8. Consider the case when W is the "pants" manifold and $N = \partial_0 W$. Distinguish between the cases corresponding to different orientations of N (four possibilities in total).

Theorem 3.9. *Let $f : W \to [a,b]$ be a Morse function on a cobordism W and v an f-gradient. Assume that (f,v) is elementary and all the critical points have the same index k. Let $N \subset W$ be a closed oriented k-dimensional submanifold transverse to $D(-v)$. Choose any orientation of v. Then the algebraic intersection number of N and $D(p,-v)$ is equal to the multiplicity of $[N]$ at p:*

$$N \mathbin{\#} D(p,-v) = m\big([N],p\big). \tag{1}$$

Proof. We shall deduce the theorem from a somewhat more general proposition. Let L be an oriented k-dimensional manifold without boundary, and $F : L \to W$ a C^∞ map, such that

$$F \pitchfork D(-v) \quad \text{and} \quad F^{-1}(D(-v)) \subset L \quad \text{is a finite set.}$$

Then the intersection index $F(L) \mathbin{\#} D(p,-v)$ is defined for every $p \in S_k(f)$ (see Subsection 4.3 of the Appendix to this chapter). Also we have the fundamental class

$$F_*([L]) \in H_k(W, W \setminus D(-v))$$

(see Subsection 4.1, page 189). The group $H_k(W, W \setminus D(-v))$ is generated by the fundamental classes of the descending discs modulo the boundary (by Proposition 3.4). By certain abuse of notation we shall denote these fundamental classes by the same symbol d_p (where $p \in S(f)$).

Proposition 3.10.

$$F_*([L]) = \sum_{p \in S_k(f)} \big(F(L) \mathbin{\#} D(p,-v)\big) d_p.$$

Proof. Let $a \in F^{-1}(D(-v))$. The linear map

$$F'(a) : T_a L \to T_{F(a)} W$$

sends T_aL to a linear subspace transverse to the $(m-k)$-dimensional vector subspace $T_{F(a)}D(-v)$, thus the dimension of $F'(a)(T_aL)$ equals k, and the map $F'(a)$ is injective. Therefore the map F restricted to a small neighbourhood of a in L is a C^∞ embedding. For every $a \in F^{-1}(D(-v))$ choose an embedding $\phi_a : \mathcal{B} \to L$ of the open k-dimensional disc $\mathcal{B} = B^k(0,r)$, such that $\phi_a(0) = a$. Restricting this embedding to a smaller disc if necessary, we can assume that the map $F \circ \phi_a$ is a C^∞ embedding such that

$$(F \circ \phi_a)(\mathcal{B}) \cap D(-v) = \{a\}.$$

We can also assume that the closures of the images $\operatorname{Im} \phi_a$ are pairwise disjoint for different points a. Applying Lemma 4.2 of the Appendix it is not difficult to show that the fundamental class $F_*([L])$ satisfies

$$F_*([L]) \;=\; \sum_a (F \circ \phi_a)_*([\mathcal{B}]) \;\in\; H_*(W, W \setminus D(-v)),$$

where $[\mathcal{B}]$ is the fundamental class of $\mathcal{B}$ in $H_*(\mathcal{B}, \mathcal{B} \setminus \{0\})$. By linearity it suffices to prove our proposition for the case when F is the embedding of one k-dimensional disc, $F : \mathcal{B} \hookrightarrow W$ which intersects $D(-v)$ at one point $b = F(a) \in D(p,-v)$, the intersection is transversal and the sign of intersection equals 1. To do the proof in this particular case, we shall allow the disc $F(\mathcal{B})$ to descend along the flow lines of $(-v)$ to a small neighbourhood of p, where the computations in homology are immediate. Choose any diffeomorphism

$$\psi : B^k(0,r_1) \times B^{m-k}(0,r_2) \to U$$

where U is a neighbourhood of p, such that

$$\psi(0,0) = p,$$

$$\psi(B^k(0,r_1)) = D(p,v) \cap U, \quad \psi(B^{m-k}(0,r_2)) = D(p,-v) \cap U,$$

$$\text{and} \quad \psi | B^k(0,r_1) \quad \text{preserves the orientation.}$$

We shall use the following abbreviation:

$$B = B^k(0,r_1), \quad B' = B^{m-k}(0,r_2),$$

and we shall identify B, respectively B' with the subset $B \times 0$, respectively $0 \times B'$ of $B \times B'$. Let $T > 0$ be sufficiently large, so that $\gamma(F(a),T;-v) \in U$. Restricting if necessary the embedding ϕ_a to a smaller disc, we can assume that $\gamma(F(\mathcal{B}),T;-v) \subset U$. Define a map $g : \mathcal{B} \to U$ by the following formula:

$$g(z) = \gamma(F(z),T;-v).$$

We have now a diagram of embeddings

$$
\begin{array}{ccccccc}
(\mathcal{B}, \mathcal{B} \setminus 0) & \xhookrightarrow{g} & (U, U \setminus \psi(B')) & \xleftarrow[\approx]{\psi} & (B, B \setminus 0) \times B' & \xhookleftarrow{\alpha} & (B, B \setminus 0) \\
 & & \Big\downarrow J & & & & \\
 & & (W, W \setminus D(-v)), & & & &
\end{array}
$$

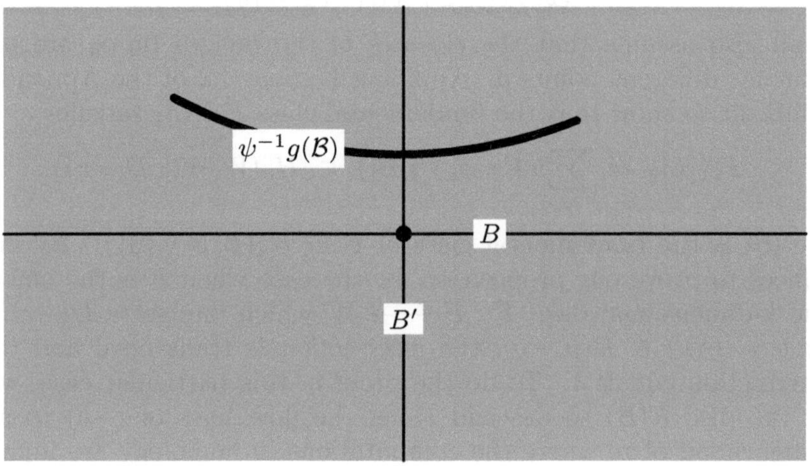

FIGURE 26.

where J is the inclusion map, and α is the embedding

$$\alpha(x, y) = (x \times 0, y \times 0).$$

The fundamental class of the disc $D(p, v)$ in $H_*(W, W \setminus D(-v))$ is equal to the $(J \circ \psi \circ \alpha)_*$-image of the fundamental class of the pair $(B, B \setminus 0)$.

Thus the proof of our proposition is reduced to the following lemma.

Lemma 3.11.

$$(\psi^{-1} \circ g)_* \big(\, [\mathcal{B}, \mathcal{B} \setminus 0] \, \big) = \alpha_* \big(\, [B, B \setminus 0] \, \big).$$

Proof. The map

$$h = \psi^{-1} \circ g$$

is a C^∞ embedding of $\mathcal{B}$ into $B \times B'$, which intersects B' transversally at one point, and the sign of intersection is positive. Let $\pi : B \times B' \to B$ denote the projection onto the first factor. The transversality condition $h \pitchfork B'$ implies that the composition $\pi \circ h$ is a local diffeomorphism in a neighbourhood of the point 0. This local diffeomorphism preserves orientations. Restricting ψ

to the product $B^k(0,\rho_1)\times B^{m-k}(0,\rho_2)$ with appropriate $\rho_1 < r_1, \rho_2 < r_2$ we can assume that the map $\pi\circ h$ is a diffeomorphism preserving orientations. Then

$$(\pi\circ h)_*\big([\mathcal{B},\mathcal{B}\setminus 0]\big) = [B, B\setminus 0] = (\pi\circ\alpha)_*\big([B,B\setminus 0]\big)$$

and since the map

$$\pi : \big(B, B\setminus 0\big)\times B' \xrightarrow{\ \pi\ } (B, B\setminus 0)$$

is a homotopy equivalence of pairs the lemma follows, and Proposition 3.10 is proved. □

Now we can complete the proof of Theorem 3.9. The formula (1) which we are proving is equivalent to the following one:

$$[N] = \sum_{p\in S_k(f)} \big(N \mathbin{\#} D(p,-v)\big)d_p \in H_*(W, \partial_0 W). \tag{2}$$

The proposition above implies that the images of both sides of (2) in the group $H_*(W, W\setminus D(-v))$ coincide. Since the inclusion

$$(W,\partial_0 W) \hookrightarrow (W, W\setminus D(-v))$$

is a homology equivalence (by Proposition 3.4) the theorem follows. □

Let us single out one particular case, important for future applications. For $p\in S_k(f)$ put

$$S_p = D(p,-v)\cap \partial_1 W,$$

then S_p is an embedded $(m-k-1)$-dimensional sphere.

Corollary 3.12. *Let $f : W \to [a,b]$ be a Morse function on a cobordism W and v an f-gradient. Assume that (f,v) is elementary and all the critical points have the same index k. Let $N\subset \partial_1 W$ be a closed k-dimensional oriented submanifold of $\partial_1 W$, which is transverse to S_p for every p. Then for every $p\in S(f)$ we have:*

$$N \mathbin{\#} S_p = m([N],p), \qquad \textit{and} \qquad [N] = \sum_{p\in S_k(f)} (N \mathbin{\#} S_p)d_p,$$

where $N \mathbin{\#} S_p$ stands for the intersection index of the submanifolds in $\partial_1 W$, and $[N]\in H_k(W,\partial_0 W)$ is the fundamental class of N.

Proof. Observe that the intersection index of N and S_p in $\partial_1 W$ equals the intersection index of N and $D_p(-v)$ in W. □

We end this subsection with one more variation on the same theme. Let $f : W \to [a,b]$ be a Morse function on a cobordism W and v an f-gradient. Let $c\in[a,b]$ be a regular value of f; put

$$W_1 = f^{-1}([a,c]), \quad W_2 = f^{-1}([c,b]).$$

Assume that the pair $(f|W_2, v|W_2)$ is elementary, and all critical points in W_2 have the same index k. Let N be an oriented k-dimensional C^∞ manifold without boundary and let $F : N \to W$ be a C^∞ map, such that

$$F^{-1}(W_2) \quad \text{is compact and} \quad F \pitchfork D(-v|W_2).$$

The homology of the pair (W, W_1) is a free abelian group generated by the fundamental classes d_p of the descending discs of the critical points of $f|W_2$. Therefore the fundamental class $F_*([N]) \in H_k(W, W_1)$ is a finite linear combination of the elements d_p with integral coefficients. The next proposition is deduced without difficulty from Proposition 3.10. We leave the proof to the reader as an exercise.

Proposition 3.13.

$$F_*([N]) = \sum_{p \in S_k(f)} \big(F(N) \# D(p, -v)\big) d_p.$$

3.5. C^0-small perturbations of the gradient. We continue our study of elementary pairs (f, v). In this subsection $f : W \to [a, b]$ is a Morse function on a cobordism W, and v is an f-gradient, such that the critical points of f are not linked to each other. We assume, moreover, that all critical points have the same index k.

The descending discs $D(p, v)$ depend on the vector field v, and the same is true for the homology classes $d_p = d_p(v)$ of these discs. To show that the classes $d_p(v)$ can be different for different choices of v let us consider an example depicted in Figure 27. The manifold in question is a 2-dimensional cobordism $W \subset \mathbf{R}^3$. The height function $f : W \to \mathbf{R}$ is a Morse function with two critical points: P and Q, both of index 1. For any almost transverse gradient v the discs $D(P, v)$ and $D(Q, v)$ are diffeomorphic to closed intervals, and the homology $H_1(W, \partial_0 W)$ is the free abelian group generated by the homology classes of these two intervals. The pictures (1) and (2) below show the descending discs corresponding to the point P and two different choices of f-gradient (the arrows show the orientation of the discs, and *not* the direction of the gradient). The difference of the corresponding cycles is homological to the cycle depicted on the picture (3).

The aim of the present subsection is to show that the classes $d_p(v)$ are stable with respect to C^0-small perturbations of v. Let us first study the simplest case.

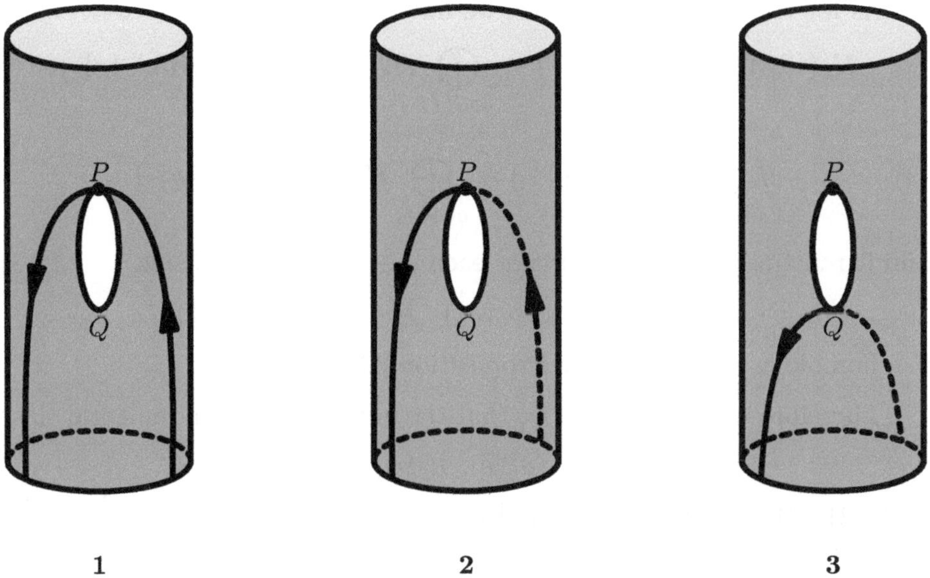

FIGURE 27.

Proposition 3.14. *Let $f : W \to [a,b]$ be a Morse function having only one critical value. Let $p \in S(f)$. Then $d_p(v) = d_p(w)$ for any two similarly oriented f-gradients v, w.*

Proof. Put

$$W' = \{x \in W \mid f(x) \leqslant f(p)\}.$$

Lemma 3.15. *In the commutative diagram*

$$\begin{array}{ccc} (\partial_0 W \cup D(v),\ \partial_0 W) & \hookrightarrow & (W,\ \partial_0 W) \\ \big\downarrow & & \big\downarrow \\ (W',\ W' \setminus S(f)) & \hookrightarrow & (W,\ W' \setminus S(f)) \end{array}$$

all the arrows are homology equivalences.

Proof. The upper horizontal arrow is a homology equivalence by Theorem 1.7. There is a strong deformation retraction of $W' \setminus S(f)$ onto $\partial_0 W$:

$$H(x,t) = \gamma(x, t \cdot \tau(x, -v); -v), \qquad \text{where} \qquad \theta \in [0,1],$$

and therefore the right vertical arrow is also a homology equivalence.

As for the left vertical arrow, we have isomorphisms

$$H_*(\ \partial_0 W \cup D(v),\ \partial_0 W\ ) \approx \bigoplus_{q \in S(f)} H_*(\ D(q,v),\ D(q,v) \setminus \{q\}\ ),$$

$$\text{(3)} \qquad H_*(\ W',\ W' \setminus S(f)\ ) \approx \bigoplus_{q \in S(f)} H_*(\ W',\ W' \setminus \{q\}\ ),$$

and it remains to observe that for each $q \in S(f)$ the inclusion

$$(D(q,v), D(q,v) \setminus \{q\}) \hookrightarrow (W', W' \setminus \{q\})$$

is a homology equivalence by Proposition 3.3. □

Therefore it remains to prove that the images of the fundamental classes

$$d_p(v),\ d_p(w)$$

in $H_*(W', W' \setminus S(f))$ are equal. In view of the isomorphism (3) it suffices to check that for every critical point $q \in S(f)$ the images of the classes $d_p(v)$, $d_p(w)$ in $H_*(W', W' \setminus q)$ are equal. If $p \neq q$, both images vanish. If $p = q$ they are equal since the orientations of w and v are equivalent to each other. □

For the case of Morse functions with more than one critical value, we have the following proposition.

Proposition 3.16. *Let $f : W \to [a,b]$ be a Morse function on a cobordism W, and v an oriented f-gradient. Assume that the pair (f,v) is elementary, and all critical points of f have the same index. Then there is $\delta > 0$, such that for every oriented f-gradient w with $||v - w|| < \delta$ we have $d_p(v) = d_p(w)$ for every $p \in S(f)$ if v and w are similarly oriented.*

Proof. Observe that every f-gradient w, which is sufficiently C^0-close to v is almost transverse (Theorem 3.41 of Chapter 4, page 150), and since all the critical points of f have the same index, the pair (f,w) is elementary. By Theorem 3.17 of Chapter 4 (page 142) there is a Morse function $\phi : W \to \mathbf{R}$ adjusted to (f,v) and having only one critical level. By Proposition 3.40 of Chapter 4 (page 150) there is $\delta > 0$ such that every f-gradient w with $||w - v|| < \delta$ is also a ϕ-gradient. Apply now Proposition 3.14, and the proof is complete. □

Let us denote the intersection $\partial_0 W \cap D(p,v)$ by $M_p(v)$. Proposition 3.16 implies that the homology class of $M_p(v)$ in $\partial_0 W$ is invariant with respect to C^0-small perturbations of v. For our applications we shall need a slightly stronger property:

Proposition 3.17. *Let $f : W \to [a, b]$ be a Morse function on a cobordism W, and v an oriented f-gradient. Assume that the pair (f, v) is elementary, and all critical points of f have the same index k. Let p be a critical point of f. Let $A \subset \partial_0 W$ be an open subset such that $M_p(v) \subset A$. Then there is $\delta > 0$ such that for every f-gradient w with $||w - v|| < \delta$ and oriented similarly to v we have*

$$[M_p(v)] = [M_p(w)] \in H_*(A).$$

Proof. Choose a handle $\mathcal{H}$ for p with respect to v in such a way that $\mathcal{H} \cap \partial_0 W \subset A$. Let

$$\mathcal{H}_p = \{x \in \mathcal{H} \mid f(x) \leqslant f(p)\}.$$

For every f-gradient w sufficiently C^0-close to v we have

$$D(p, w) \subset \mathcal{H}_p.$$

Let $\Delta_p(w)$ denote the fundamental class of the pair $\big(D(p, w), \partial D(p, w)\big)$ in $H_*(\mathcal{H}_p, \mathcal{H}_p \cap \partial_0 W)$. We would like to show that

$$\Delta_p(w) = \Delta_p(v)$$

for every w sufficiently C^0-close to v. Observe that both inclusions

$$(\mathcal{H}_p, \mathcal{H}_p \cap \partial_0 W) \hookrightarrow (\mathcal{H}_p, \mathcal{H}_p \setminus \{p\}) \hookrightarrow (W_p, W_p \setminus \{p\})$$

are homology equivalences (the case of the first inclusion is dealt with the usual gradient descent argument, and the second one with the argument similar to the proof of Proposition 3.3, page 173.) Thus it remains to prove that the fundamental classes of $D(p, v)$ and $D(p, w)$ in $H_*(W_p, W_p \setminus \{p\})$ are equal, and this follows since w and v are similarly oriented. □

Now let us return to multiplicities of cycles with respect to critical points. Recall that for $x \in H_k(W, \partial_0 W)$ the multiplicities $m(x, p)$ are defined by the following formula:

$$x = \sum_{p \in S(f)} m(x, p) d_p(v).$$

Since the classes $d_p(v)$ depend on v, the multiplicities $m(x, p)$ depend also on v, and to stress this dependence we shall now denote them by $m(x, p; v)$. The intersection $D(p, -v) \cap \partial_1 W$ will be denoted $S_p(v)$.

Proposition 3.18. *Let $f : W \to [a, b]$ be a Morse function on a cobordism W, and v an oriented f-gradient. Assume that the pair (f, v) is elementary, and all critical points of f have the same index. Then there is $\delta > 0$ such that for every f-gradient w which is oriented similarly to v and satisfies $||w - v|| < \delta$ the following holds:*

(1) *For every* $x \in H_*(W, \partial_0 W)$ *we have*

$$m(x, p; w) = m(x, p; v).$$

(2) *For every closed oriented* k*-dimensional submanifold* $N \subset W$ *which is transverse to* $D(p, -v)$ *and to* $D(p, -w)$ *for every* p *we have*

$$N \mathbin{\#\!\!\!\!\#} D(p, -v) = N \mathbin{\#\!\!\!\!\#} D(p, -w)$$

(here $\mathbin{\#\!\!\!\!\#}$ *denotes the algebraic intersection index of two submanifolds in* W*).*

(3) *For every closed oriented* k*-dimensional submanifold* $L \subset \partial_1 W$ *which is transverse to* $S_p(v)$ *and to* $S_p(w)$ *for every* p *we have*

$$L \mathbin{\#\!\!\!\!\#} S_p(v) = L \mathbin{\#\!\!\!\!\#} S_p(w)$$

(here $\mathbin{\#\!\!\!\!\#}$ *denotes the algebraic intersection index of two submanifolds in* $\partial_1 W$*).*

Proof. The first point follows immediately from definition of the numbers $m(x, p; v)$ and Proposition 3.16. The second point follows from Theorem 3.9 together with Proposition 3.16. The third point follows as in the proof of Corollary 3.12. □

3.6. Relative case. Let $f : W \to [a, b]$ be a Morse function on a cobordism W, and v an oriented f-gradient such that the pair (f, v) is elementary, and all the critical points of f have the same index k. Let $K \subset \partial_1 W$ be a subset such that

$$K \quad \text{is compact and} \quad K \cap D(-v) = \varnothing.$$

Then every $(-v)$-trajectory starting at a point of K reaches $\partial_0 W$ and the set $T(K, -v)$ is homeomorphic to the product $K \times [0, 1]$. Put

$$Q = T(K, -v) \cup \partial_0 W.$$

In this subsection we suggest a version "mod Q" of the results of the two previous subsections. The homology groups $H_*(W, \partial_0 W)$ and $H_*(\partial_1 W)$ with which we worked in the previous section will be replaced now by the groups $H_*(W, Q)$ and, respectively, $H_*(\partial_1 W, K)$.

Proposition 3.19. *The inclusion*

$$(W, \partial_0 W) \xhookrightarrow{i} (W, Q)$$

is a homology equivalence.

Proof. Observe that Q can be deformed onto its subspace $\partial_0 W \subset Q$ by the downward gradient descent, therefore the inclusion $\partial_0 W \hookrightarrow Q$ is a homology equivalence. Applying the exact sequence of the pair and the five lemma, we deduce the assertion of the proposition. □

Therefore the group $H_*(W, Q)$ is free abelian, and its basis is formed by the homology classes d_p of the descending discs of the critical points $p \in S(f)$. Let $x \in H_*(W, Q)$. Write

$$x = \sum_p m(x,p) d_p \quad \text{with} \quad m(x,p) \in \mathbf{Z};$$

the number $m(x,p)$ will be called the *multiplicity of x at p*. These numbers depend on v and K, so the most careful notation here would be $m(x,p;v,K)$. However we shall omit K from the notation, since usually the value of K is clear from the context. The dependence of these numbers on v is more important, and we shall write $m(x,p;v)$ if we work simultaneously with more than one f-gradient.

We proceed to a generalization of Corollary 3.12. We shall need only a special case when $N \subset \partial_1 W$. Recall the notation

$$S(v) = D(-v) \cap \partial_1 W, \quad S_p(v) = D(p, -v) \cap \partial_1 W.$$

Proposition 3.20. *Let $N \subset \partial_1 W$ be a k-dimensional oriented submanifold without boundary of $\partial_1 W$ such that:*

(1) *The submanifolds $N \subset \partial_1 W$ and $S(v) \subset \partial_1 W$ are transverse to each other in $\partial_1 W$.*

(2) *$N \setminus \operatorname{Int} K$ is compact.*

Then

$$N \mathbin{\#} S_p(v) = m(p, [N])$$

where $\#$ stands for the algebraic intersection number of submanifolds in $\partial_1 W$, and $[N]$ denotes the fundamental class of N in $H_*(W, Q)$.

Proof. We have to prove the following equality:

$$[N] = \sum_{p \in S_k(f)} (N \mathbin{\#} D(p, -v)) d_p \in H_*(W, Q).$$

The inclusion $(W, Q) \hookrightarrow (W, W \setminus D(-v))$ is a homology equivalence, therefore it suffices to prove that the images in $H_*(W, W \setminus D(-v))$ of the both sides of the above formula are equal. And this follows from Proposition 3.10. □

Now we wish to investigate the stability of the multiplicities with respect to C^0-small perturbations of the gradient. Let $x \in H_k(\partial_1 W, K)$. Let x' denote the image of x in $H_k(W, Q)$. By a certain abuse of notation we denote the integer $m(x',p;v)$ by $m(x,p;v)$. The compact $Q = T(K,-v) \cup \partial_0 W$ depends on v, and we shall therefore denote it by $Q(v)$.

Proposition 3.21. *There is $\delta > 0$ such that for every f-gradient w which satisfies $||w-v|| < \delta$ and is oriented similarly to v:*

(1) *For every $x \in H_k(\partial_1 W, K)$ and every $p \in S(f)$ we have*

$$m(x,p;v) = m(x,p;w).$$

(2) *For every oriented k-dimensional submanifold $N \subset \partial_1 W$ such that $N \setminus \operatorname{Int} K$ is compact, we have*

$$N \,\#\, S_p(v) = N \,\#\, S_p(w).$$

Proof. Let K' be any compact subset of $\partial_0 W$ such that

$$K \subset \operatorname{Int} K' \quad \text{and} \quad D(-v) \cap K' = \varnothing.$$

Let $Q'(v) = T(K',-v) \cup \partial_0 W$. For any f-gradient w sufficiently C^0-close to v, we have:

$$Q(w) = T(K,-w) \cup \partial_0 W \subset Q'(v) = T(K',-v) \cup \partial_0 W.$$

With such w we consider the following commutative diagram of inclusions.

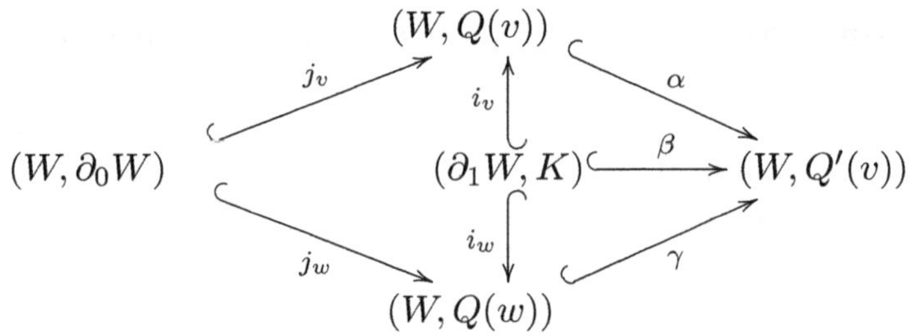

Observe that every diagonal arrow is a homology equivalence. The integers $m(x,p;v)$ are equal to the coefficients of the decomposition of the element $(\alpha \circ i_v)_*(x)$ with respect to the basis

$$E_v = \{(\alpha \circ j_v)_*(d_p(v))\}_{p \in S(f)}.$$

Similarly, the integers $m(x,p;w)$ are equal to the coefficients of the decomposition of the element $(\gamma \circ i_w)_*(x)$ with respect to the basis

$$E_w = \{(\gamma \circ i_w)_*(d_p(w))\}_{p \in S(f)}.$$

It follows from Proposition 3.16 that the bases E_v and E_w are equal if w is sufficiently C^0-close to v. Since $\alpha \circ i_v = \gamma \circ i_w$, the proof of the first part of the proposition is complete.

The second part of the proposition is the consequence of the first part together with Proposition 3.20. □

4. Appendix: Orientations, coorientations, fundamental classes etc.

This appendix is about fundamental classes, Thom classes and related notions. We do not aim at a complete treatment of the subject (the reader can find the details in numerous textbooks, for example [**29**], Ch. 7, [**96**], Appendix, [**63**], Ch. 17, [**155**], Ch. 3). Our aim is rather to list the results and definitions which we use in this and the following chapters.

4.1. Orientations and fundamental classes. Let M be an oriented C^∞ manifold without boundary, $m = \dim M$, and let $U \subset M$.

Definition 4.1. An element $\omega \in H_m(M, U)$ is called the *fundamental class* or the *orientation class* of the pair (M, U) if for every $p \in M \setminus U$ the image of ω in $H_n(M, M \setminus p)$ is the generator of the group $H_m(M, M \setminus p) \approx \mathbf{Z}$ induced by the given orientation of M.

If the set $M \setminus U$ is compact, then such fundamental class exists and is unique (see for example, [**63**], Ch. 17, Th. 4.4 or [**29**],Ch. VII, §3 and §4). Let (A, B) be a pair of topological spaces, and

$$f : (M, U) \to (A, B)$$

be a continuous map of pairs. If $M \setminus U$ is compact, then we have the fundamental class of M in the group $H_m(M, U)$, and its image in the group $H_m(A, B)$ with respect to the map f will be called *the image of the fundamental class of M in (A, B)*. It will be denoted $f_*([M])$ (or simply $[M]$ when no confusion is possible).

The following functoriality property is immediate:

$$(f \circ g)_*([M]) = f_*\big(g_*([M])\big).$$

In some cases the fundamental class of a map $f : M \to A$ can be computed from the restriction of f to a submanifold of M. This is the contents of the next lemma (the proof is easy and will be omitted).

Lemma 4.2. *Let M be an oriented manifold without boundary, and $U \subset M$ a subset, such that $M \setminus U$ is compact. Let $f : (M, U) \to (A, B)$ be a continuous map. Let $N \subset M$ be a submanifold with* $\dim N = \dim M$ *and such that $M \setminus U \subset N$. Then*

$$f_*([M]) = (f|N)_*([N]) \in H_*(A, B).$$

The fundamental classes have some natural multiplicative properties. Namely, let (A, B) and (A', B') be pairs of topological spaces, and

$$\phi : (M, U) \to (A, B), \quad \phi : (M', U') \to (A', B')$$

be continuous maps of pairs, where M, M' are manifolds, and $M \backslash U$, $M' \backslash U'$ are compact. Consider the product map of pairs

$$F = f \times f' : (M, U) \times (M', U') \to (A, B) \times (A', B'),$$

where the product of pairs is defined by

$$(M, U) \times (M', U') = (M \times M', \ M \times U' \cup U \times M').$$

The set

$$M \times M' \ \setminus \ \big(M \times U' \cup U \times M'\big)$$

is compact, and it is easy to check the formula

$$F_*([M \times M']) = f_*([M]) \ \times \ f'_*([M'])$$

where in the right-hand side of the formula $\times$ denotes the exterior product in homology.

In the rest of this subsection we discuss a generalization of the notion of fundamental class to the case of manifolds with boundary. We work here in a rather restricted category of spaces and manifolds, sufficient for our applications.

Definition 4.3. Let (X, K) be a pair of compact sets. We say that K is an *H-retract* of X if for every neighbourhood V of K there exists a neighbourhood $V' \subset V$ of K such that the embedding $K \hookrightarrow V'$ induces an isomorphism in homology.

Example 4.4. For any cobordism W the subsets $\partial_0 W, \partial_1 W, \partial W$ are H-retracts of W.

Definition 4.5. Let (X, K) be a pair of compact sets and $f : A \to X$ a continuous map of topological spaces. We say that f is *proper rel* K if $f^{-1}(Q)$ is compact whenever Q is a compact subset of $X \setminus K$.

For example if $A \setminus f^{-1}(\text{Int } K)$ is compact, then f is proper rel K. Let (X, K) be a pair of compact sets, such that K is an H-retract of X. Let M be an oriented manifold with boundary (we do not assume that M or ∂M is compact), and $m = \dim M$. Let $f : M \to X$ be a continuous map, such that

(1) f is proper rel K.
(2) $f(\partial M) \subset K$.

We shall now define an element of $H_m(X, K)$ which will be called the fundamental class of f. Pick any open neighbourhood V of K in X such that the inclusion $K \hookrightarrow V$ is a homology equivalence. The set $f^{-1}(V) \setminus \partial M$ is an open subset with compact complement in the manifold $M \setminus \partial M$. Thus the fundamental class

$$\mu_1 \in H_m\Big(M \setminus \partial M, f^{-1}(V) \setminus \partial M\Big)$$

is defined. Let μ_2 be its image in $H_m(X, V)$ via the map f; there is a unique element $\mu_3 \in H_m(X, K)$ which is sent to μ_2 by the isomorphism $(X, K) \hookrightarrow (X, V)$. This element μ_3 is called the *fundamental class* of the singular manifold $f : M \to X$ and denoted by

$$f_*([M]) \in H_m(X, K)$$

(it is not difficult to show that this homology class does not depend on the particular choice of the open subset V used in the definition). Fundamental classes of singular manifolds have a natural functoriality property: if $\phi : (X, K) \to (X', K')$ is a continuous map of compact pairs, where K, K' are H-retracts of X, respectively X' and $f : M \to X$ is proper rel K, then $\phi \circ f$ is proper rel K' and

$$(\phi \circ f)_*([M]) = \phi_*\Big(f_*([M])\Big).$$

Example 4.6. If M is a compact oriented ∂-manifold, then ∂M is an H-retract of M and the identity map $M \to M$ is proper rel ∂M. The resulting element in $H_m(M, \partial M)$ is the usual fundamental class of the pair $(M, \partial M)$.

Here is a generalization of Lemma 4.2 to the case of singular manifolds with boundary:

Lemma 4.7. *Let (X, K) be a pair of compact sets, such that K is an H-retract of X. Let M be an oriented manifold without boundary of dimension m. Let $f : M \to X$ be a continuous map, which is proper rel K. Let $N \subset M$ be a submanifold with boundary of M such that* $\dim M = \dim N$ *and*

$$f(\partial N) \subset K, \quad f(M \setminus N) \subset K.$$

Then the map $f|N : N \to X$ is proper rel K, and

$$(f|N)_*([N]) = f_*([M]) \in H_m(X, K).$$

Proof. Let V be a neighbourhood of K in X such that the inclusion $K \hookrightarrow V$ is a homology equivalence. It suffices to prove that the images of the classes

$$(f|N)_*([N]), \ f_*([M])$$

in $H_*(X, V)$ are equal. The class $f_*([N])$ in $H_*(X, V)$ equals by definition the class $f_*([N \setminus \partial N])$, which is equal to $f_*([M])$ by Lemma 4.2. □

4.2. Coorientations. Let M be a ∂-manifold, $X \subset M$ a ∂-submanifold. The normal bundle $\nu(X)$ to X in M is a vector bundle which fiber at $a \in X$ is defined as follows:

$$\nu_a(X) = T_a(M)/T_a(X).$$

Definition 4.8. An orientation of the normal bundle $\nu(X)$ is called a *coorientation* of X. A submanifold $X \subset M$ endowed with a coorientation is called *cooriented.*

Example 4.9. Let M be an oriented manifold with a non-empty boundary. Then $X = \partial M$ acquires a canonical coorientation as follows: a vector $e \in T_a M, e \notin T_a X$ is called *positive*, if it points outward M.

Definition 4.10. Let M be an oriented ∂-manifold, $X \subset M$ a ∂-submanifold. Put $\dim X = k$, $\dim M = m$. For $a \in X$ a family $N = (e_1, \dots, e_{m-k})$ of vectors in $T_a(M)$ is called a *normal frame* for X at p if the image of N in $\nu_a(X)$ is a basis of this vector space.

An orientation $\mathcal{O}$ of X and a coorientation $\mathcal{C}$ of X are called *coherent* if for every $a \in X$, every positively oriented normal frame N for X at a and every positively oriented basis T of $T_a X$, the basis (N, T) of $T_a M$ is positive.

It is clear that for every orientation $\mathcal{O}$ of X there is a unique coorientation of X which is coherent with $\mathcal{O}$; we say that this coorientation is *induced by* $\mathcal{O}$. Similarly, for every coorientation C of X there is a unique orientation of X which is coherent with C; we say that this orientation is *induced by* C.

Example 4.11. The standard orientation of the boundary ∂M of an oriented ∂-manifold M is coherent with the canonical coorientation of ∂M defined in Example 4.9.

A basis E of the tangent space $T_a \partial M$ is positive with respect to this orientation, if and only if the frame (N, E) of $T_a M$ is positive, where N is any vector in $T_a M$ which points outward.

For a closed submanifold X of a manifold M without boundary there is another approach to coorientability arising from the theory of the Thom classes. Let $\mathscr{D}$ be the total space of the disc bundle $D\nu(X)$ associated to the normal bundle of X, and $\mathscr{S}$ be the total space of the sphere bundle $S\nu(X)$. The fiber D_a of $D\nu(X)$ over a point $a \in X$ is diffeomorphic to a closed disc of dimension $\dim M - \dim X$, and the fiber S_a of $S\nu(X)$ over a point $a \in X$ is diffeomorphic to a sphere of dimension $\dim M - \dim X - 1$. A choice of orientation of the normal space $\nu_a(X)$ is equivalent to a choice of one of two generators in the homology $H_*(D_a, S_a) \approx \mathbf{Z}$. Intuitively, a coorientation of X can be thought of as a family of generators $\mathcal{O}_x \in H_*(D_x, S_x)$ depending continuously on x. Here is a precise statement: the manifold X is cooriented if and only if there exists a cohomology class $U \in H^k(\mathscr{D}, \mathscr{S})$ (the so-called *Thom class* of the normal bundle) such that for every $a \in X$ the image of U in the cohomology $H^*(D_a, S_a)$ is a generator of this group. The image of the Thom class in the group $H^*(M, M \setminus X)$ via the excision isomorphism

$$H^*(\mathscr{D}, \mathscr{S}) \approx H^*(M, M \setminus X)$$

will be called *the coorientation class* and denoted by ω_X. The space $\mathscr{D}/\mathscr{S}$ obtained by shrinking the subspace $\mathscr{S}$ to a point is called the *Thom complex* of the normal bundle ν_X, and denoted $T\nu(X)$.

4.3. Intersection indices. Let M be a manifold without boundary and Y be a ∂-submanifold of M. Let X be a ∂-manifold such that

$$\dim X + \dim Y = \dim M.$$

Assume that X is oriented, and Y is cooriented. Let $F : X \to M$ be a C^∞ map, transverse to Y, that is, for every $p \in F^{-1}(Y)$ we have

$$F'(p)(T_pX) \oplus T_{F(p)}Y = T_{F(p)}M.$$

This direct sum decomposition yields a linear bijection

$$\phi : T_pX \xrightarrow{\approx} \nu_{F(p)}(Y).$$

Put $\varepsilon(p) = 1$ if ϕ conserves the orientation and $\varepsilon(p) = -1$ if ϕ reverses the orientation. If the set $F^{-1}(Y)$ is finite, the number

$$F(X) \mathbin{\#} Y = \sum_{p \in X \cap F^{-1}(Y)} \varepsilon(p)$$

is called the *intersection index* of $F(X)$ and Y, or *the algebraic number of the points of intersection.*

Assume now that both X, Y are oriented submanifolds of M, and the map $F : X \hookrightarrow M$ is the inclusion map. Then the orientation of Y induces a coorientation of Y and we can define the intersection index $X \mathbin{\#} Y$. Permuting X and Y leads to the following sign change:

$$Y \mathbin{\#} X = (-1)^{\dim X \cdot \dim Y} X \mathbin{\#} Y.$$

Now we will give a homological interpretation of intersection indices. Assume that M is a manifold without boundary, and $X, Y \subset M$ are closed submanifolds of complementary dimensions in M, such that X is oriented, Y is cooriented and $X \pitchfork Y$. Then the set $X \cap Y$ is finite, and the intersection index $X \mathbin{\#} Y$ is defined. We have the fundamental class

$$[X] \in H_k(M), \quad \text{where} \quad k = \dim X,$$

and the coorientation class

$$\omega_Y \in H^k(M, M \setminus Y).$$

An easy argument based on the excision isomorphism proves the following formula:

$$X \mathbin{\#\!\!\!\!\!\!\#} Y = \Big\langle \omega_Y, [X] \Big\rangle,$$

where the brackets $\langle \cdot, \cdot \rangle$ denote the canonical pairing

$$H^k(M, M \setminus Y) \otimes H_k(M) \to \mathbf{Z}.$$

CHAPTER 6

The Morse complex of a Morse function

One of the basic tools of algebraic topology is the CW homology theory. For every CW complex X one constructs *the cellular chain complex* $C_*(X)$, which is freely generated (as a graded $\mathbf{Z}$-module) by the cells of X, and the boundary operator is computable in terms of attaching maps of the cells (see [**29**], Ch. 5). The homology of $C_*(X)$ is isomorphic to the singular homology of X, and this isomorphism provides one of the most efficient tools for computing the homology of topological spaces. Morse stratifications have a lot in common with CW decompositions, and one can expect a similar construction in the Morse-theoretic framework. The most natural way to proceed would be to endow the Morse stratification with a structure of a CW decomposition. However we have seen in Chapter 4 that this is a delicate task, and we shall not pursue this direction.

Instead, we use a method introduced by E. Witten in [**163**]. Given a Morse function $f : W \to [a, b]$ and a transverse f-gradient v, Witten constructs a chain complex $\mathcal{M}_*(f, v)$ with

$$H_*(\mathcal{M}_*(f, v)) \approx H_*(W, \partial_0 W).$$

The group $\mathcal{M}_k(f, v)$ is freely generated by the critical points of f of index k, and the boundary operator is defined via counting the flow lines of v joining critical points. Thus the homology of the manifold can be computed directly from the Morse-theoretic data. This chain complex will be called *the Morse complex* for brevity, although the term *the Morse-Thom-Smale-Witten complex* would be certainly more appropriate from the historical point of view.

The Morse complex is the main subject of the present chapter. We begin by proving that its homology is isomorphic to the homology of the underlying manifold. In Section 2 we give a generalization of the Witten construction to the case when the f-gradient v is almost transverse (following the paper [**120**]). We prove that the resulting Morse complex is stable with respect to small perturbations of the f-gradient v.

Then we construct a natural homotopy equivalence between the Morse complex and the singular chain complex of the underlying manifold (Section 3). For this purpose we develop in Subsection 3.1 a relevant algebraic technique, which allows us to work with *cellular filtrations* of chain complexes.

1. The Morse complex for transverse gradients

1.1. Definition. The input data for the construction is:

(1) A Morse function $f : W \to [a, b]$ on the cobordism W.
(2) An oriented f-gradient v satisfying the transversality condition.

(Recall that *oriented f-gradient* means f-gradient together with a choice of orientation for all descending discs of v.) We will construct from this data a free based chain complex $\mathcal{M}_*$ of abelian groups, and we will show that its homology is isomorphic to $H_*(W, \partial_0 W)$. By definition the group $\mathcal{M}_k$ is a free $\mathbf{Z}$-module, generated by $S_k(f)$. Thus the groups $\mathcal{M}_k$ come along with natural bases, and the boundary operator, which we are going to construct, is given by the coefficients of its matrix with respect to these bases. These matrix coefficients are defined via counting the flow lines of $(-v)$. Recall that a flow line of $(-v)$ joining $p \in S(f)$ with $q \in S(f)$ is an integral curve γ of $(-v)$ satisfying

$$\lim_{t \to -\infty} \gamma(t) = p, \lim_{t \to +\infty} \gamma(t) = q,$$

and we identify γ with all the curves obtained from γ by reparametrization. Denote the set of all flow lines of $(-v)$ from p to q by $\Gamma(p, q; v)$.

Lemma 1.1. *If* $\operatorname{ind} p = \operatorname{ind} q + 1$, *then* $\Gamma(p, q; v)$ *is a finite set.*

Proof. The vector field v satisfies the transversality condition, therefore it is a ϕ-gradient for some ordered Morse function ϕ (see Proposition 3.38 of Chapter 4, page 149). Let $a_0 = a < a_1 < \cdots < a_{m+1} = b$ be the ordering sequence of ϕ, where $m = \dim W$. Put

$$k = \operatorname{ind} p, \quad V = \phi^{-1}(a_k).$$

Let

$$\Sigma_p = D(p, v) \cap V, \quad S_q = D(q, -v) \cap V.$$

These are compact submanifolds of V, diffeomorphic to spheres of dimensions respectively $k-1$ and $m-k$. Each flow line joining p with q intersects V at exactly one point, thus we obtain a bijective correspondence

$$\Gamma(p, q; v) \xrightarrow[\approx]{I} \Sigma_p \cap S_q. \tag{1}$$

The transversality condition implies that $\Sigma_p \pitchfork S_q$, thus the intersection $\Sigma_p \cap S_q$ is a finite set, since $\dim \Sigma_p + \dim S_q = \dim V - 1$. □

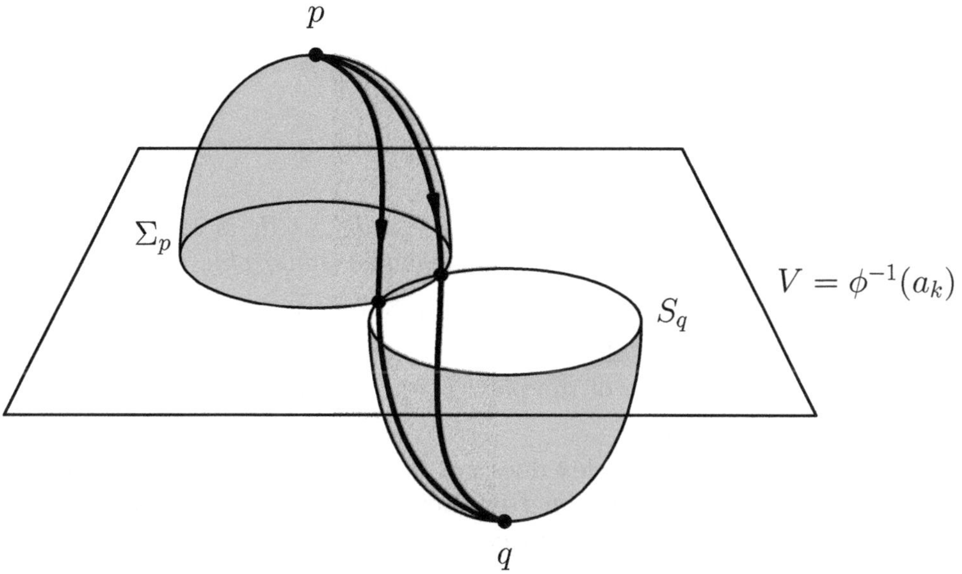

FIGURE 28.

Now we shall use the chosen orientations of $D(p, v)$ and $D(q, v)$ to associate a sign $\varepsilon(\gamma) \in \{-1, +1\}$ to every $\gamma \in \Gamma(p, q; v)$. We have:

$$\Sigma_p = \partial\Big(D(p, v) \cap \phi^{-1}([a_k, a_{m+1}])\Big),$$

and the chosen orientation of $D(p, v)$ induces an orientation of Σ_p. The chosen orientation of $D(q, v)$ induces a *coorientation* of $D(q, -v)$ and therefore a coorientation in V of the transversal intersection

$$D(q, -v) \cap V = S_q.$$

Thus each point of the intersection $\Sigma_p \cap S_q$ acquires a sign. For a flow line $\gamma \in \Gamma(p, q; v)$ denote the sign corresponding to $\gamma \cap V$ by $\varepsilon(\gamma)$. (It is not difficult to check that $\varepsilon(\gamma)$ does not depend on the choice of the ordered function ϕ adjusted to (f, v).)

Definition 1.2. The integer

$$n(p, q; v) = \sum_{\gamma \in \Gamma(p,q;v)} \varepsilon(\gamma) \tag{2}$$

is called *the incidence coefficient of* p *and* q with respect to v.

In other words $n(p,q;v)$ is equal to the intersection index $\Sigma_p \mathrel{\#} S_q$ of the two spheres in V. Define a homomorphism $\partial_k : \mathcal{M}_k \to \mathcal{M}_{k-1}$ by the following formula:

$$\partial_k p = \sum_{q\in S_{k-1}(f)} n(p,q;v)\cdot q.$$

Theorem 1.3. *For every k the composition $\partial_{k-1}\circ\partial_k$ is equal to zero and the homology of the resulting chain complex is isomorphic to $H_*(M)$.*

The proof of the theorem will be given in Subsection 1.3.

Definition 1.4. The resulting chain complex will be denoted $\mathcal{M}_*(f,v)$ and called *the Morse complex* of the pair (f,v). The critical points of f form a natural basis in $\mathcal{M}_*(f,v)$.

Observe that this complex does not depend on the particular choice of the function f for which v is an f-gradient. Indeed, the boundary operator depends only on the flow lines of v.

1.2. Homological version. In this subsection we suggest another construction of a chain complex associated with an ordered Morse function. It is defined in purely homological terms, and provides an intermediate step between the homology of the Morse complex, defined in the preceding subsection and the singular homology of $(W,\partial_0 W)$.

Let us first recall some terminology from the theory of cellular filtrations (cf. [**29**], Ch. 5). *A filtration* of a topological space W is a sequence $W^{(k)}$ of subspaces of W such that

$$W^{(-1)}\subset W^{(0)}\subset\cdots\subset W^{(k)}\subset\cdots \qquad\text{and}\qquad \bigcup_k W^{(k)}=W.$$

To each filtration one associates a chain complex as follows. Put

$$C_k = H_k(W^{(k)},W^{(k-1)})\quad\text{for}\quad k\geqslant 0,$$

and $C_k=0$ for $k<0$, and endow the graded group C_* with the boundary operator

$$\widetilde{\partial}_k : C_k = H_k\big(W^{(k)},W^{(k-1)}\big)\to C_{k-1}=H_{k-1}\big(W^{(k-1)},W^{(k-2)}\big)$$

which equals the composition

$$H_k\big(W^{(k)},W^{(k-1)}\big)\longrightarrow H_k\big(W^{(k-1)}\big)\longrightarrow H_{k-1}\big(W^{(k-1)},W^{(k-2)}\big)$$

(where the first homomorphism is the boundary operator in the exact sequence of the pair $\big(W^{(k)},W^{(k-1)}\big)$ and the second one is the natural projection). †

† The homomorphism $\widetilde{\partial}_k$ can be also defined as the boundary operator in the exact sequence of the triple $\big(W^{(k)},W^{(k-1)},W^{(k-2)}\big)$.

The resulting chain complex is called the *adjoint chain complex* associated with the filtration $W^{(k)}$.

Definition 1.5. A filtration $\{W^{(k)}\}_{k\geqslant -1}$ of a topological space is called *cellular* if

$$H_i\big(W^{(k)}, W^{(k-1)}\big) = 0 \quad \text{whenever} \quad i \neq k.$$

For every cellular filtration the homology of the adjoint chain complex is isomorphic to the homology $H_*(W, W^{(-1)})$:

$$H_*(C_*) \approx H_*(W, W^{(-1)}) \tag{3}$$

(see [**29**], Ch. 5, Prop. 1.3).

Now let us return to the Morse theory. Let $\phi : W \to [a, b]$ be an ordered Morse function on a cobordism W and let $(a_0, \dots, a_{m+1})$ be its ordering sequence. Consider the corresponding Morse-Smale filtration

$$W^{(-1)} = \partial_0 W \subset W^{(0)} \subset \cdots \subset W^{(m)} = W,$$

$$\text{where} \quad W^{(k)} = \phi^{-1}([a, a_{k+1}]);$$

The corresponding adjoint complex will be denoted by $C_*(\phi)$, so that by definition

$$C_k(\phi) = H_k(W^{(k)}, W^{(k-1)})$$

and the boundary operator of this complex is the boundary operator of the exact sequence of the triple $\big(W^{(k)}, W^{(k-1)}, W^{(k-2)}\big)$.

Lemma 1.6. *The Morse-Smale filtration is a cellular filtration, and $C_k(\phi)$ is a free abelian group, freely generated by the set of critical points of index k of ϕ.*

Proof. By excision we have:

$$H_k(W^{(k)}, W^{(k-1)}) \approx H_k(W_k, \partial_0 W_k) \quad \text{where} \quad W_k = \phi^{-1}([a_k, a_{k+1}]).$$

Then W_k is an elementary cobordism, and Proposition 3.7 of Chapter 5 (page 176) implies that the group $H_s(W_k, \partial_0 W_k) = 0$ vanishes for $s \neq k$, and for $s = k$ this group is the free abelian group generated by the homology classes of the descending discs of the points in $S_k(\phi)$. □

Applying the isomorphism (3) we obtain the following theorem:

Theorem 1.7. *The homology of the chain complex $C_*(\phi)$ is isomorphic to $H_*(W, \partial_0 W)$.* □

The chain complex $C_*(\phi)$ has no preferred basis. We can endow this complex with a free basis if we pick an oriented ϕ-gradient v, satisfying the transversality condition. Namely, for $p \in S_k(f)$ let $d_p(v)$ denote the homology class of the descending disc of p in the cobordism $\phi^{-1}([a_k, a_{k+1}])$ with respect to the given orientation $\mathcal{O}$. Then the classes $d_p(v)$ for $p \in S_k(\phi)$ form a base in $C_k(\phi)$. The resulting *based* chain complex will be denoted by $C_*(\phi, v)$ and called *the homological version of the Morse complex.*

1.3. Proof of Theorem 1.3. We are going to compare the constructions from Subsections 1.1 and 1.2. Let $f : W \to [a, b]$ be a Morse function and v an oriented transverse f-gradient. We have the graded based free group $\mathcal{M}_*(f, v)$ endowed with homomorphisms $\partial_k : \mathcal{M}_k(f, v) \to \mathcal{M}_{k-1}(f, v)$ constructed in Subsection 1.1. Pick an ordered Morse function ϕ such that v is a ϕ-gradient. Then we have also the free based chain complex $C_*(\phi, v)$ from Subsection 1.2. Define an isomorphism of graded abelian groups

$$I : \mathcal{M}_*(f, v) \to C_*(\phi, v), \quad I(p) = d_p(v) \quad \text{for} \quad p \in S(f).$$

Theorem 1.3 follows immediately from the next proposition.

Proposition 1.8. *The isomorphism I commutes with the boundary operators: $I \circ \partial_k = \widetilde{\partial}_k \circ I$. In particular, $\mathcal{M}_*(f, v)$ is a chain complex and I is an isomorphism of chain complexes.*

Proof. For $p \in S_k(f)$ write

$$\partial(p) = \sum_{q \in S_{k-1}(f)} n(p, q) q, \qquad \widetilde{\partial}(d_p) = \sum_{q \in S_{k-1}(f)} \nu(p, q) d_q,$$

where we abbreviate $d_p(v)$ to d_p. We are going to prove that $\nu(p, q) = n(p, q)$. Let

$$V = \phi^{-1}(a_k), \quad \Sigma_p = D(p, v) \cap V, \quad S_q = D(q, -v) \cap V.$$

Then Σ_p is a closed $(k-1)$-dimensional oriented submanifold of V and its fundamental class $[\Sigma_p] \in H_*(W^{(k)}, W^{(k-1)})$ equals $\widetilde{\partial}(d_p)$. Therefore it remains to show:

$$[\Sigma_p] = \sum_{q \in S_{k-1}} n(p, q) d_q, \qquad \text{where} \quad n(p, q) = \Sigma_p \mathbin{\#} S_q,$$

and this formula follows immediately from Corollary 3.12 of Chapter 5 (page 181). □

1.4. Morse inequalities. The existence of the Morse complex implies immediately the classical *Morse inequalities*:

Theorem 1.9 (M. Morse). *Let $f : W \to [a, b]$ be a Morse function on a cobordism W. The number $m_k(f)$ of critical points of f of index k is not less than the Betti number $b_k(W, \partial_0 W)$.*

Proof. Consider the Morse complex $\mathcal{M}_*(f,v)$ (associated to any oriented transverse f-gradient v) and let

$$\mathcal{M}'_* = \mathcal{M}_*(f,v) \underset{\mathbf{Z}}{\otimes} \mathbf{Q}.$$

Then $\mathcal{M}'_*$ is a finitely generated chain complex of vector spaces over the field $\mathbf{Q}$, and

$$H_*(\mathcal{M}'_*) \approx H_*(W, \partial_0 W) \underset{\mathbf{Z}}{\otimes} \mathbf{Q}.$$

Thus the vector space $H_k(W, \partial_0 W; \mathbf{Q})$ is isomorphic to a factor space of a subspace of $\mathcal{M}'_k$, and therefore

$$m_k(f) = \dim_{\mathbf{Q}} \mathcal{M}'_k \geqslant \dim_{\mathbf{Q}} H_k(W, \partial_0 W; \mathbf{Q}) = b_k(W, \partial_0 W). \qquad \square$$

These inequalities can be improved if we take into account the torsion in the homology. Let us first recall the corresponding algebraic notions. Let R be a principal ideal domain, and L be a finitely generated module over R. There is an isomorphism:

$$L \approx R^a \oplus \bigoplus_{1\leqslant i\leqslant b} R/\alpha_i R \quad \text{where} \quad \alpha_i \in R \quad \text{are}$$
$$\text{non-zero, non-invertible and} \quad \alpha_i | \alpha_{i+1}, \text{ for every } i.$$

(See for example [**81**], Ch. **XV**, §2). The numbers a, b are determined by the module L. They are called the *rank* and the *torsion number* of L, and denoted respectively rk L and t.n.L. It is easy to prove that the number t.n.L is equal to the minimal possible number of generators of the torsion submodule

$$Tors\ L \approx \bigoplus_{1\leqslant i\leqslant b} R/\alpha_i R.$$

Proposition 1.10. *Let*

$$C_* = \{\cdots \longleftarrow C_{k-1} \xleftarrow{\partial_k} C_k \longleftarrow \cdots\}$$

be a chain complex of finitely generated free R-modules. Then for every k we have

$$(\mathcal{MP}) \qquad \text{rk } C_k \geqslant \text{rk } H_k(C_*) + \text{t.n.}H_k(C_*) + \text{t.n.}H_{k-1}(C_*).$$

Proof. A chain complex isomorphic to

$$0 \longleftarrow R \longleftarrow R \longleftarrow 0.$$

will be called *elementary*. It is clear that the inequalities $(\mathcal{MP})$ hold for any elementary complex. It is also easy to check that, if these inequalities hold for two complexes C_* and D_*, they hold also for the chain complex $C_* \oplus D_*$. Observe finally that it suffices to prove our proposition for *finite* *complexes*, that is, the complexes having only a finite number of non-zero terms. Now our proposition follows from the next lemma.

Lemma 1.11. *Any finite chain complex of finitely generated free modules over R is isomorphic to a finite direct sum of elementary complexes.*

Proof. Put

$$Z_k = \mathrm{Ker}\,\Big(\partial_k : C_k \to C_{k-1}\Big), \quad B_k = \mathrm{Im}\,\Big(\partial_{k+1} : C_{k+1} \to C_k\Big).$$

Then both Z_k, B_k are submodules of C_k, therefore they are free finitely generated modules over R. The homomorphism $\partial_{k+1} : C_{k+1} \to B_k$ is surjective by definition, therefore C_{k+1} splits:

$$C_{k+1} = Z_{k+1} \oplus D_{k+1} \quad \text{where} \quad \partial_{k+1} | Z_{k+1} = 0.$$

Thus the chain complex C_* splits into the finite direct sum of the chain complexes

$$C_*^{(k)} = \{0 \longleftarrow Z_k \longleftarrow D_{k+1} \longleftarrow 0\}$$

and it suffices to prove that each $C_*^{(k)}$ is isomorphic to a finite direct sum of elementary complexes. This assertion follows from the fact that every homomorphism of finitely generated free R-modules can be diagonalized in appropriate bases. □

The next proposition (due to E. Pitcher [**125**]) follows.

Proposition 1.12. *Let $f : W \to [a, b]$ be a Morse function on a cobordism W. Then*

$$m_k(f) \geqslant \mathrm{rk}\; H_k(W, \partial_0 W) + \mathrm{t.n.} H_k(W, \partial_0 W) + \mathrm{t.n.} H_{k-1}(W, \partial_0 W).$$

□

1.5. Examples.

Quadratic forms on S^m

Recall the Morse function $f : S^m \to \mathbf{R}$ defined by

$$f = g \mid S^m, \quad \text{where} \quad g(x_0, \ldots, x_m) = \alpha_0 x_0^2 + \cdots + \alpha_m x_m^2$$

(see page 40). Here α_i are pairwise distinct non-zero real numbers; for notational convenience we shall assume

$$\alpha_0 < \alpha_1 < \cdots < \alpha_m.$$

For each $i, 0 \leqslant i \leqslant m$, this function has two critical points of index i, namely

$$N_i = (0, \ldots, \underset{i}{1}, \ldots, 0), \qquad \text{and} \quad S_i = (0, \ldots, \underset{i}{-1}, \ldots, 0).$$

Let v be the Riemannian gradient of the function f with respect to the restriction to S^m of the Euclidean metric in $\mathbf{R}^{m+1}$. We are going to compute the Morse complex $\mathcal{M}_*(f,v)$. Let us first recall the description of the descending and ascending discs of the critical points of f. Let

$$L_i = \{(x_0,\dots,x_m) \mid x_{i+1} = x_{i+2} = \cdots = x_m = 0\}, \quad \Sigma_i = L_i \cap S^m.$$

The critical points N_i, S_i are both in the sphere Σ_i. Let

$$\Sigma_i^+ = \{x \in \Sigma_i \mid x_i > 0\}, \quad \Sigma_i^- = \{x \in \Sigma_i \mid x_i < 0\}.$$

Proposition 2.26 of Chapter 3 (page 93) says:

$$D(N_i,v) = \Sigma_i^+, \quad D(S_i,v) = \Sigma_i^-.$$

Further, put

$$\widehat{L}_i = \{(x_0,\dots,x_m) \mid x_0 = \cdots = x_{i-1} = 0\}, \quad \widehat{\Sigma}_i = S^m \cap \widehat{L}_i,$$

and

$$\widehat{\Sigma}_i^+ = \{x \in \widehat{\Sigma}_i \mid x_i > 0\}, \quad \widehat{\Sigma}_i^- = \{x \in \widehat{\Sigma}_i \mid x_i < 0\}.$$

Proposition 2.27 of Chapter 3 (page 94) says

$$D(N_i,-v) = \widehat{\Sigma}_i^+, \quad D(S_i,-v) = \widehat{\Sigma}_i^-.$$

It is easy to check that

$$\Sigma_i \pitchfork \widehat{\Sigma}_j \quad \text{for every} \quad i,j.$$

Therefore the f-gradient v satisfies the transversality condition. Now let us describe the Morse complex $\mathcal{M}_*(f,v)$. The $(-v)$-trajectories joining the critical points of index i with critical points of index $(i-1)$ belong to $C = \Sigma_i \cap \widehat{\Sigma}_{i-1}$. This intersection is a big circle of the sphere S^m:

$$C = \{(x_0,\dots,x_m) \mid x_j = 0 \quad \text{for} \quad j < i-1,\ j > i\},$$

and it is easy to see that the complement in C to the four critical points $N_i, S_i, N_{i-1}, S_{i-1}$ is the union of four flow lines of $(-v)$, depicted in Figure 29.

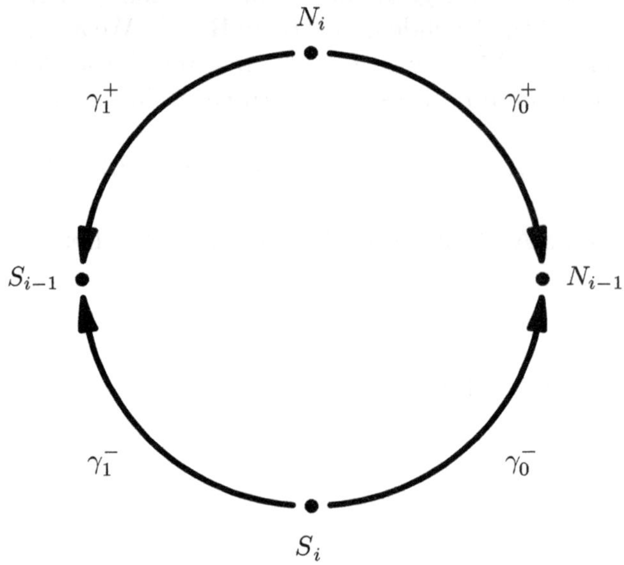

FIGURE 29.

To associate signs to each trajectory we must choose orientations for the descending discs. Let us endow the Euclidean discs

$$D_i = \{x \in \mathbf{R}^m \mid \langle x, x\rangle \leqslant 1, \quad x_k = 0 \quad \text{for} \quad k > i\}$$

with their standard orientations. Every sphere Σ_i being the boundary of the oriented manifold D_i acquires the canonical orientation (see Section 4.2 of Chapter 5, page 191), and therefore the descending discs

$$D(N_i, v) = \Sigma_i^+, \quad D(S_i, v) = \Sigma_i^-$$

also acquire an orientation.

We leave to the reader the computation of the signs corresponding to the four trajectories depicted in Figure 29, and give only the final result:

$$\varepsilon(\gamma_0^+) = (-1)^i = \varepsilon(\gamma_1^+); \quad \varepsilon(\gamma_0^-) = (-1)^{i-1} = \varepsilon(\gamma_1^-).$$

The boundary operator in the Morse complex satisfies therefore

$$\partial N_i = (-1)^i (N_{i-1} + S_{i-1}), \quad \partial S_i = (-1)^{i-1} (N_{i-1} + S_{i-1}).$$

Exercise 1.13. Deduce the isomorphisms

$$H_i(\mathcal{M}_*(f, v)) = \begin{cases} 0 & \text{if} \quad i \neq 0,\ i \neq n, \\ \mathbf{Z} & \text{if} \quad i = 0 \quad \text{or} \quad i = n. \end{cases}$$

Now let us proceed to the real projective space $\mathbf{R}P^m$. Recall that we have the Morse function $h : \mathbf{R}P^m \to \mathbf{R}$ obtained from the quadratic form $f : S^m \to \mathbf{R}$ considered above by factorization with respect to the involution

$$I : S^m \to S^m, \quad I(x) = -x.$$

Similarly we obtain the h-gradient w. For each $i : 0 \leqslant i \leqslant m$ the function h has one critical point E_i of index i. To compute the boundary operator in $\mathcal{M}_*(h, w)$ observe that the involution I being restricted to S^i preserves the orientation for i odd and reverses the orientation for i even. Thus

$$\partial E_i = \begin{cases} (-1)^i \cdot 2E_{i-1} & \text{if} \quad i \quad \text{is even,} \\ 0 & \text{if} \quad i \quad \text{is odd.} \end{cases}$$

The next proposition follows.

Proposition 1.14. *Let m be a positive integer. For $i < m$ we have*

$$H_i(\mathbf{R}P^m) \approx \begin{cases} 0 & \textit{if} \quad i \quad \textit{is even,} \\ \mathbf{Z}/2\mathbf{Z} & \textit{if} \quad i \quad \textit{is odd.} \end{cases}$$

The homology in degree m is given by the following formula:

$$H_m(\mathbf{R}P^m) \approx \begin{cases} 0 & \textit{if} \quad m \quad \textit{is even,} \\ \mathbf{Z} & \textit{if} \quad m \quad \textit{is odd.} \end{cases}$$

□

1.6. The Poincaré duality in Morse complexes. The Poincaré duality theorem for cobordisms says (in its simplest version):

$(\mathcal{P})$ $\qquad H_*(W, \partial_0 W) \approx H^{m-*}(W, \partial_1 W),$

where W is an oriented cobordism and $m = \dim W$.

In this subsection we discuss the Morse-theoretical version of the duality phenomenon, which leads to a short proof of the isomorphism above.

Let $f : W \to [a, b]$ be a Morse function on an oriented cobordism W, and v an f-gradient satisfying the transversality condition. Let W' denote the opposite cobordism (that is, the cobordism which differs from W only by interchanging the components of the boundary, so that $\partial_0 W' = \partial_1 W$, $\partial_1 W' = \partial_0 W$). The cobordism W' acquires the Morse function $-f$ and its gradient $-v$, satisfying the transversality condition.

Definition 1.15. Assume that v is oriented, so that at every point $p \in S(f)$ an orientation of $T_pD(p, v)$ is fixed. Let us orient the tangent space $T_pD(p, -v)$ in such a way that the direct sum of the orientations in

$$T_pD(p, -v) \oplus T_pD(p, v) \approx T_pW$$

(in this order) is the positive orientation of T_pW. This orientation of the $(-f)$-gradient $-v$ will be called *induced* from the orientation of v.

Now let us recall the definition of the Poincaré-dual of a chain complex. Let m be a positive integer, A a commutative ring, and C_* be a chain complex of A-modules such that $C_k = 0$ vanishes for $k > m$ and $k < 0$:

$$C_* = \{\, 0 \longleftarrow C_0 \xleftarrow{\partial_1} C_1 \longleftarrow \cdots \xleftarrow{\partial_m} C_m \longleftarrow 0 \,\}$$

The *m-dimensional Poincaré-dual chain complex* is defined as follows:

$$C_*^\vee = \{\, 0 \longleftarrow D_0 \xleftarrow{d_1} D_1 \longleftarrow \cdots \xleftarrow{d_m} D_m \longleftarrow 0 \,\}$$

where $D_k = \operatorname{Hom}_A(C_{m-k}, A)$ and $d_k = (-1)^{m-k+1}\partial^*_{m-k+1}$. By definition

$$H_k(C_*^\vee) \approx H^{m-k}(C_*) \quad \text{for every} \quad k.$$

If C_* is a free module endowed with a finite free base e, then $C_*^\vee$ acquires the dual base e^*, and the matrix of the homomorphism d_k with respect to this base is the conjugate of the matrix of $(-1)^{m-k+1}\partial_{m-k+1}$.

Theorem 1.16. *Let $m = \dim W$. Endow the vector field $(-v)$ with the orientation induced from v. Then there is a basis-preserving isomorphism between the chain complexes $\mathcal{M}_*^\vee(-f,-v)$ and $\mathcal{M}_*(f,v)$.*

Proof. The boundary operator

$$\partial_k : \mathcal{M}_k(f,v) \to \mathcal{M}_{k-1}(f,v)$$

is given by the following formula:

$$\partial_k p = \sum_{q\in S_{k-1}(f)} n(p,q;v)\cdot q \quad \text{where} \quad p \in S_k(f) \quad \text{and}$$

$$n(p,q;v) = \sum_{\gamma\in\Gamma(p,q;v)} \varepsilon(\gamma).$$

Here $\Gamma(p,q;v)$ is the set of all flow lines of $(-v)$ joining p with q, and $\varepsilon(\gamma)$ is the sign associated to the trajectory γ according to the chosen orientations of the descending discs $D(p,v)$ and $D(q,v)$.

Similarly the boundary operator

$$\bar\partial_k : \mathcal{M}_{m-k+1}(-f,-v) \to \mathcal{M}_{m-k}(-f,-v)$$

reads as follows:

$$\bar\partial_k q = \sum_{p\in S_k(f)} n(q,p;-v)\cdot p \quad \text{where} \quad q \in S_{k-1}(f) \quad \text{and}$$

$$n(q,p;-v) = \sum_{\theta\in\Gamma(q,p;-v)} \varepsilon(\theta).$$

Our theorem follows from the next lemma.

Lemma 1.17. *Let* $p \in S_k(f)$, $q \in S_{k-1}(f)$. *Then*

$$n(p,q;v) = (-1)^{m-k+1} \cdot n(q,p\,;-v).$$

Proof. Pick any ordered Morse function ϕ, such that v is a ϕ-gradient, let $\{a_i\}$ be the corresponding ordering sequence, and let $V = \phi^{-1}(a_k)$, so that the manifold V separates the points p and q. We orient the manifold V in such a way that for any $x \in V$ the pair $(v(x), E(x))$ (where $E(x)$ is a positive frame in T_xV) is a positive frame in T_xW. Let

$$\Sigma_p = D(p,v) \cap V, \quad S_q = D(q,-v) \cap V.$$

Recall from Subsection 1.1 that there are bijective correspondences

$$\Gamma(p,q;v) \xleftarrow[\approx]{I} \Sigma_p \cap S_q \xrightarrow[\approx]{J} \Gamma(q,p;-v)$$

(see Figure 28 on page 197). For $A \in \Sigma_p \cap S_q$ put

$$\varepsilon(A) = \varepsilon(I(A)), \ \varepsilon'(A) = \varepsilon(J(A)).$$

Our task is to prove that

$$\varepsilon(A) = (-1)^{m-k+1}\varepsilon'(A).$$

The chosen orientation of v induces an orientation of Σ_p and a coorientation of S_q. Let τ_p be a positively oriented basis in $T_A\Sigma_p$. Let ν_q be a basis in $T_A\Sigma_p$ inducing a positive basis in T_AV/T_AS_q (that is, ν_q has the same sign in this space as the positive tangent basis to $D(q,v)$ translated to A). Then $\varepsilon(A)$ equals the sign of the determinant $\det(\nu_q, \tau_p)$ of the transition matrix from ν_q to τ_p. Applying the same procedure to obtain a formula for $\varepsilon'(A)$ let us choose a positively oriented basis τ_q in T_AS_q with respect to the orientation induced from $D(q,-v)$ and a basis ν_p in T_AS_q yielding a positive basis un T_AV/T_AS_p (with respect to the coorientation of $D(p,v)$). Then $\varepsilon'(A) = \operatorname{sgn}\det(\nu_p, \tau_q)$.

Observe that $(-v(A), \tau_p)$ is a positive basis of $T_AD(p,v)$. Furthermore, the sign in $T_*W/T_*D(p,v)$ of ν_p is equal to the sign of the positive basis of $T_*D(p,-v)$, and applying Definition 1.15 we deduce that the basis $\mathcal{B}_1 = (\nu_p, -v(A), \tau_p)$ of T_AW is positive. The basis $(v(A), \tau_q)$ of $T_AD(q,-v)$ is positive, therefore applying Definition 1.15 we deduce that the basis $\mathcal{B}_2 = (v(A), \tau_q, \nu_q)$ is positive. The sign of the determinant of the transition matrix between $\mathcal{B}_1$ and $\mathcal{B}_2$ equals 1; on the other hand this sign equals

$$(-1)^{m-k} \cdot (-1) \cdot \operatorname{sgn}\det\Big(\big(v(A), \nu_p, \tau_p\big), \big(v(A), \tau_q, \nu_q\big)\Big).$$

Therefore $\varepsilon(A)\varepsilon'(A) = (-1)^{m-k+1}$. □

To deduce the Poincaré duality isomorphism $(\mathcal{P})$ cited in the beginning of this section we need a lemma which follows immediately from the Universal Coefficient Theorem (see [**153**], Ch 5, §5):

Lemma 1.18. *Let C_*, D_* be chain complexes of free R-modules where R is a principal ring. If $H_k(C_*) \approx H_k(D_*)$ for every k, then $H^k(C_*) \approx H^k(D_*)$ for every k.* □

Let $f : W \to [a, b]$ be a Morse function on an oriented cobordism W and v a transverse f-gradient. We have then

$$H_k(W, \partial_0 W) \approx H_k(\mathcal{M}_*(f, v))$$
$$\approx H_k\big(\mathcal{M}_*^{\vee}(-f, -v)\big) \approx H^{m-k}\big(\mathcal{M}_*(-f, -v)\big) \approx H^{m-k}(W, \partial_1 W),$$

and the formula $(\mathcal{P})$ is established.

2. The Morse complex for almost transverse gradients

The construction of the chain complex $\mathcal{M}_*(f, v)$ given in the previous section uses in an essential way the transversality property of v. If the gradient does not satisfy the transversality condition, then the set of flow lines joining the critical points of adjacent indices is not finite in general, and the construction does not work. In this section we show that the homological version of the Morse complex (Subsection 1.2) can be generalized to the case when the gradient is only almost transverse.

We begin with ordered Morse functions in Subsection 2.1 and proceed to the general case in Subsection 2.2.

2.1. The case of ordered Morse functions. Let $\phi : W \to [a, b]$ be an ordered Morse function on a cobordism W, and v be an oriented almost transverse ϕ-gradient. Recall from Subsection 1.2 the based free chain complex $C_*(\phi, v)$, associated with ϕ and v. Our first aim is to show that this chain complex is determined by the vector field v and does not depend on the particular choice of the ordered Morse function ϕ.

Proposition 2.1. *Let ϕ_1, ϕ_2 be ordered Morse functions such that v is a gradient for both. Then there is a basis-preserving isomorphism*

$$\alpha(\phi_1, \phi_2) : C_*(\phi_1, v) \xrightarrow{\approx} C_*(\phi_2, v). \tag{4}$$

Proof. Let us first consider the case when the Morse-Smale filtrations $W_1^{(k)}, W_2^{(k)}$ of W induced by ϕ_1, and ϕ_2 satisfy

$$W_1^{(k)} \subset W_2^{(k)}.$$

Then we have the inclusion $\mathcal{S}_*(W_1^{(k)}, \partial_0 W) \subset \mathcal{S}_*(W_2^{(k)}, \partial_0 W)$ of the corresponding filtrations in the singular chain complex of $(W, \partial_0 W)$, and we obtain the corresponding chain map of the adjoint complexes:

$$\alpha(\phi_1, \phi_2) : C_*(\phi_1, v) \to C_*(\phi_2, v).$$

It remains to prove that $\alpha(\phi_1, \phi_2)$ preserves the bases. The set

$$D(p, v) \setminus \text{Int } W_1^{(k-1)} = D(p, v) \cap \phi_1^{-1}([a_k, a_{k+1}])$$

is compact, therefore the manifold $D(p, v)$ has the fundamental class

$$\Delta_p^{(1)} \in H_k\big(D(p, v),\ D(p, v) \cap \text{Int } W_1^{(k-1)}\big),$$

and the generator

$$d_p^{(1)}(v) \in H_k(W_1^{(k)}, W_1^{(k-1)})$$

is the image of Δ_p with respect to the homomorphism induced by the inclusion. Similarly, the generator

$$d_p^{(2)}(v) \in H_k(W_2^{(k)}, W_2^{(k-1)})$$

is the image of the fundamental class

$$\Delta_p^{(2)} \in H_k\big(D(p, v),\ D(p, v) \cap \text{Int } W_2^{(k-1)}\big).$$

By the uniqueness property of the fundamental classes the element $\Delta_p^{(2)}$ is the image of the $\Delta_p^{(1)}$ with respect to the inclusion. The commutativity of the next diagram, where all the homomorphisms are induced by inclusions, completes the proof of our proposition in the case $W_1^{(k)} \subset W_2^{(k)}$.

$$\begin{array}{ccc} H_k\big(D(p, v),\ D(p, v) \cap \text{Int } W_1^{k-1}\big) & \longrightarrow & H_k\big(D(p, v),\ D(p, v) \cap \text{Int } W_2^{k-1}\big) \\ \big\downarrow & & \big\downarrow \\ H_k\big(W_1^{(k)}, W_1^{(k-1)}\big) & \longrightarrow & H_k\big(W_2^{(k)}, W_2^{(k-1)}\big) \end{array}$$

As for the general case, it suffices to show that for every two ordered functions ϕ_1, ϕ_2 such that v is a gradient for both, there is a third function ϕ_3 such that the corresponding filtrations of W satisfy

$$W_1^{(k)} \supset W_3^{(k)} \subset W_2^{(k)}.$$

And this follows immediately from Theorem 3.58 of Chapter 4 (page 158). □

Before we proceed to the case of arbitrary Morse functions, let us mention a C^0-stability property for the Morse complexes of ordered functions.

Proposition 2.2. *Let $\phi : W \to [a,b]$ be an ordered Morse function, and v an oriented almost transverse ϕ-gradient. Then there is $\delta > 0$, such that for every ϕ-gradient w which satisfies $||w-v|| < \delta$ and is oriented similarly to v, the Morse complexes $C_*(\phi, v)$ and $C_*(\phi, w)$ are basis-preserving isomorphic.*

Proof. Recall that every f-gradient w sufficiently C^0-close to v is almost transverse. Let a_i be the ordering sequence for ϕ, and let $W_k = \phi^{-1}([a_k, a_{k+1}])$. It suffices to show that for every ϕ-gradient w sufficiently close to v in C^0 topology and every $p \in S_k(\phi)$ the homology classes

$$d_p(v),\ d_p(w) \in H_*(W_k, \partial_0 W_k)$$

are equal. And this follows immediately from Proposition 3.16 of Chapter 5 (page 184). □

2.2. General case. Let $f : W \to [a,b]$ be a Morse function on a cobordism W, and let v be an oriented almost transverse f-gradient. Pick any ordered Morse function ϕ on W, such that v is also a ϕ-gradient. Write

$$\partial(d_p(v)) = \sum_{q \in S_{k-1}(\phi)} n(p,q;v) \cdot d_q(v) \qquad \text{with} \qquad n(p,q;v) \in \mathbf{Z}$$

(here ∂ is the boundary operator in the homological Morse complex $C_*(\phi, v)$). The numbers $n(p,q;v)$ are called *incidence coefficients* corresponding to p, q, and v.

By Proposition 2.1 these numbers are determined by the vector field v, and do not depend on the particular choice of an ordered Morse function ϕ, such that v is also a ϕ-gradient. Note also, that when v satisfies the transversality condition, the number $n(p,q;v)$ coincides with the incidence coefficient from Definition 1.2 obtained by counting the flow lines (this follows from Proposition 1.8, page 200). We shall keep the same notation $n(p,q;v)$ for both types of the incidence coefficients since no confusion is possible.

Definition 2.3. Let $f : W \to [a,b]$ be a Morse function and v be an oriented almost transverse f-gradient. Let $\mathcal{M}_k(f,v)$ be the free abelian group freely generated by the set $S_k(f)$. Define the homomorphism $\partial_k : \mathcal{M}_k(f,v) \to \mathcal{M}_{k-1}(f,v)$ by the following formula:

$$\partial_k(p) = \sum_{q \in S_{k-1}(f)} n(p,q;v) q,$$

where $n(p,q;v) \in \mathbf{Z}$ are the incidence coefficients defined above. Then $\partial_k \circ \partial_{k+1} = 0$ for every k and the resulting chain complex $\mathcal{M}_*(f,v)$ is called *the Morse complex* of the pair (f,v).

Thus by the very definition the Morse complex $\mathcal{M}_*(f,v)$ is isomorphic to the Morse complex $C_*(\phi,v)$ where ϕ is any ordered Morse function such that v is a ϕ-gradient. In particular

$$H_k(\mathcal{M}_*(f,v)) \approx H_k(M) \quad \text{for every} \quad k.$$

In the rest of this subsection we study the basic properties of the Morse complex constructed above. Let $c \in [a,b]$ be any regular value of f. The cobordism W is the union of two cobordisms

$$W' = f^{-1}([a,c]), \quad W'' = f^{-1}([c,b]),$$

and along with the Morse complex $\mathcal{M}_*(f,v)$ we have two Morse complexes

$$\mathcal{M}_*(f|W', v|W'), \quad \mathcal{M}_*(f|W'', v|W'').$$

There is a natural inclusion of graded groups

$$\mathcal{M}_*(f|W', v|W') \xhookrightarrow{I} \mathcal{M}_*(f,v)$$

which sends each critical point $p \in W'$ of f to itself, and similarly a natural projection

$$\mathcal{M}_*(f,v) \xrightarrow{\pi} \mathcal{M}_*(f|W'', v|W'').$$

These two maps form a short exact sequence of graded groups:

$$0 \longrightarrow \mathcal{M}_*(f|W', v|W') \xrightarrow{I} \mathcal{M}_*(f,v) \xrightarrow{\pi} \mathcal{M}_*(f|W'', v|W'') \longrightarrow 0.$$

Proposition 2.4. *The homomorphisms I, π are chain maps.*

Proof. The Morse complexes associated to $f, f|W', f|W''$ are obtained from the filtrations corresponding to ordered Morse functions on W, respectively W', W''. Denote Y_i, respectively Y_i', Y_i'' these filtrations. By Theorem 3.58 of Chapter 4 (page 158) we can choose the ordered Morse functions in such a way that

$$Y_i' \subset Y_i, \quad Y_i \subset W' \cup Y_i''.$$

We obtain therefore chain maps

$$\mathcal{M}_*(f|W', v|W') \xrightarrow{I'} \mathcal{M}_*(f,v), \quad \mathcal{M}_*(f,v) \xrightarrow{\pi'} \mathcal{M}_*(f|W'', v|W'')$$

associated with these inclusions. An argument similar to the proof of Proposition 2.1 shows that $I = I'$ and $\pi = \pi'$. The lemma is proved. $\square$

Our next aim is to show that the incidence coefficients $n(p,q;v)$ can be computed from any smaller cobordism

$$W_0 = f^{-1}([\lambda,\mu]) \quad \text{where} \quad \lambda < f(q) < f(p) < \mu.$$

Proposition 2.5.

$$n(p, q; v) = n(p, q; v|W_0).$$

Proof. The particular case when $\lambda = a$ follows from the previous proposition. Similar is the case $\mu = b$. The general case is easily deduced from these two particular ones. □

Let us now prove the C^0-stability of the Morse complexes.

Theorem 2.6. *Let $f : W \to [a, b]$ be a Morse function on a cobordism W and v an almost transverse f-gradient. There is $\delta > 0$, such that for every f-gradient w with $||w - v|| < \delta$ and oriented similarly to v we have*

$$n(p, q; w) = n(p, q; v)$$

for every p, q. We have therefore a basis-preserving isomorphism between the Morse complexes $\mathcal{M}_*(f, v)$ and $\mathcal{M}_*(f, w)$.

Proof. Let ϕ be an ordered Morse function adjusted to (f, v); then $\mathcal{M}_*(f, v) \approx C_*(\phi, v)$. Proposition 3.40 of Chapter 4 (page 150) implies that there is $\delta > 0$ such that for every f-gradient w with $||w - v|| < \delta$ the function ϕ is adjusted to (f, w). In particular w is an almost transverse ϕ-gradient and $\mathcal{M}_*(f, w) \approx C_*(\phi, w)$. Now our assertion follows from Proposition 2.2. □

Exercise 2.7. Deduce Propositions 2.4 and 2.5 from Theorem 2.6 approximating almost transverse gradients by transverse gradients.

3. The Morse chain equivalence

Let $f : W \to [a, b]$ be a Morse function on a cobordism W, and v an oriented almost transverse f-gradient. In Section 2 we associated to this data the Morse complex $\mathcal{M}_*(f, v)$ with homology isomorphic to $H_*(W, \partial_0 W)$. Now we are going to construct a natural chain equivalence between the Morse complex and the singular chain complex $\mathcal{S}_*(W, \partial_0 W)$ of the pair $(W, \partial_0 W)$. The construction will be given in Subsection 3.3, after developing the necessary algebraic technique.

3.1. Cellular filtrations of chain complexes. The aim of the present subsection is to develop a chain complex version of the theory of cellular filtrations of topological spaces. In view of subsequent applications we shall work in the general framework of chain complexes defined over any ring A. We consider only chain complexes C_* of left A-modules, such that $C_i = 0$ for $i < 0$. We shall often omit the adjective "chain", so that "complex" means "chain complex", "homotopy" means "chain homotopy" etc. The term "homotopy equivalence" in this framework is synonymous with "chain equivalence".

Definition 3.1. A *filtration* of a chain complex C_* is a sequence of subcomplexes $C_*^{(k)}, -1 \leqslant k$, such that

$$0 = C_*^{(-1)} \subset C_*^{(0)} \subset \cdots \subset C_*^{(k)} \subset \cdots \quad \text{and} \quad \bigcup_k C_*^{(k)} = C_*.$$

To any filtration $\{C_*^{(k)}\}$ of a complex C_* we associate a complex C_*^{gr} as follows. Set

$$C_k^{gr} = H_k(C_*^{(k)} \,/\, C_*^{(k-1)})$$

and define the boundary operator $\partial_k : C_k^{gr} \to C_{k-1}^{gr}$ to be the composition

$$H_k(C_*^{(k)} \,/\, C_*^{(k-1)}) \longrightarrow H_{k-1}(C_*^{(k-1)}) \longrightarrow H_{k-1}(C_*^{(k-1)}, C_*^{(k-2)})$$

where the first arrow is the boundary operator in the exact sequence of the pair $(C_*^{(k)}, C_*^{(k-1)})$ and the second is induced by the inclusion. †

Definition 3.2. The chain complex C_*^{gr} with the boundary operator defined as above, is called the *adjoint complex* associated to the filtration of C_*.

Definition 3.3. A filtration

$$0 = C_*^{(-1)} \subset C_*^{(0)} \subset \cdots \subset C_*^{(k)} \subset \cdots$$

of C_* is called *cellular* if

$$H_i(C_*^{(k)} \,/\, C_*^{(k-1)}) = 0 \text{ for } i \neq k.$$

Remark 3.4. For a cellular filtration $C_*^{(k)}$ of a complex C_* the homology groups $H_n(C_*^{(k)})$ vanish for $k < n$. (The proof is an easy induction on k, using the exact sequences of the pairs $(C_*^{(k)}, C_*^{(k-1)})$.)

Example 3.5. Let D_* be any complex. The filtration by the chain complexes

$$D_*^{(k)} = \{0 \leftarrow D_0 \leftarrow \cdots \leftarrow D_k \leftarrow 0\}$$

is called *trivial.* This is a cellular filtration and $D_*^{gr} = D_*$.

Theorem 3.6. *Let $C_*^{(k)}$ be a cellular filtration of a complex C_*. Let D_* be a chain complex of free left A-modules and $\varphi : D_* \to C_*^{gr}$ be a chain map. Endow D_* with the trivial filtration. Then there exists a chain map $f : D_* \to C_*$, which preserves filtrations and induces the map φ in the*

† The homomorphism ∂_k can be described also as the boundary operator in the exact sequence of the triple $(C_*^{(k)}, C_*^{(k-1)}, C_*^{(k-2)})$.

adjoint complexes. This chain map f is unique up to a chain homotopy, preserving filtrations.

Proof. The proof is in a diagram chasing. Suppose by induction that for every $i \leqslant k-1$ we have constructed the maps $f_i : D_i \to C_i$, commuting with the boundary operators in C_* and D_*, preserving filtrations (i.e., Im $f_i \subset C_i^{(i)}$) and inducing the given homomorphisms $\varphi_i : D_i \to C_i^{gr}$ in the graded groups. Our aim is to construct the map $f_k : D_k \to C_k$ such that the properties above hold for all $i \leqslant k$.

It suffices to define f_k on the free generators of D_k. Let e be such a generator. We have then an element

$$\phi(e) \in H_k(C_*^{(k)}/C_*^{(k-1)}).$$

Choose any cycle X of the complex $C_*^{(k)}/C_*^{(k-1)}$ representing $\phi(e)$, and choose any element $x \in C_k^{(k)}$ representing X. We would like to set $f_k(e) = x$, but the required equality $\partial x = f_{k-1}(\delta e)$ may not be true (here ∂ is the boundary operator in C_* and δ is the boundary operator in D_*). Let us consider the element

$$\partial x - f_{k-1}(\delta e) \in C_{k-1}^{(k-1)}.$$

This is a cycle of C_* and its projection to the group $H_{k-1}(C_*^{(k-1)}/C_*^{(k-2)})$ vanishes, since this projection is equal to $\partial\phi(e) - \phi(\delta(e))$. Therefore we can write

$$\partial x - f_{k-1}(\delta e) = \partial z + u \quad \text{with} \quad z \in C_k^{(k-1)}, u \in C_{k-1}^{(k-2)}.$$

The element u is a cycle of degree k in $C_*^{(k-2)}$, therefore it is homologous to zero in this complex (see Remark 3.4):

$$u = \partial v, \quad \text{with} \quad v \in C_k^{(k-2)}.$$

Now define the map f_k on the element e by:

$$f_k(e) = x - (z + v);$$

the required properties are easy to check.

Now let us proceed to the homotopical uniqueness. Let $f, g : D_* \to C_*$ be chain maps, preserving filtrations and inducing in the adjoint complexes the same map $\phi : D_* \to C_*^{gr}$. Assume by induction that we have the maps

$$H_i : D_i \to C_{i+1}^{(i)} \quad \text{with } \partial H_i + H_{i-1}\delta = f_i - g_i, \quad \text{for} \quad i < k;$$

we shall now construct the map H_k. As before, it suffices to define it on the free generators of D_k. Let e be such a generator; our task is to construct an element $z \in C_{k+1}^{(k)}$, such that

$$f_k(e) - g_k(e) = H_{k-1}(\delta e) + \partial z. \tag{5}$$

Put

$$\mathcal{A} = f_k(e) - g_k(e) - H_{k-1}(\delta e) \in C_k^{(k)}.$$

It follows from the induction assumption that $\partial \mathcal{A} = 0$. Both f and g preserve filtrations, therefore the images of the elements $f_k(e), g_k(e)$ in the quotient chain complex $C_*^{(k)}/C_*^{(k-1)}$ are cycles. By our hypotheses we have the equality of the homology classes

$$[f_k(e)] = [g_k(e)] \in H_k\big(C_*^{(k)}/C_*^{(k-1)}\big).$$

Therefore

$$f_k(e) - g_k(e) = \partial\xi + v \quad \text{with} \quad \xi \in C_{k+1}^{(k)},\ v \in C_k^{(k-1)}.$$

Thus $\mathcal{A} = \partial\xi + v - H_{k-1}(\delta e)$, and the element $v - H_{k-1}(\delta e)$ is a cycle in $C_k^{(k-1)}$; applying Remark 3.4 write $v - H_{k-1}(\delta e) = \partial u$ with $u \in C_{k+1}^{(k-1)}$. Then

$$\mathcal{A} = \partial z \quad \text{with} \quad z = \xi + u \in C_{k+1}^{(k)}$$

so that z satisfies the condition (5). The induction step is now complete. □

Definition 3.7. A cellular filtration $C_*^{(k)}$ of a complex C_* is called *free cellular* if every module $H_k(C_*^{(k)} \,/\, C_*^{(k-1)})$ is a free left A-module.

Recall that a chain map $f : C_* \to D_*$ of complexes C_*, D_* is called a *homology equivalence* if it induces an isomorphism in homology.

Corollary 3.8. *For a free cellular filtration $C_*^{(k)}$ of a complex C_* there exists a homology equivalence $\chi : C_*^{gr} \to C_*$, preserving filtrations and inducing the identity map in the adjoint complexes. (Here C_*^{gr} is endowed with the trivial filtration.) The equivalence χ is functorial up to chain homotopy preserving filtrations. If C_* is a chain complex of free modules, then χ is a homotopy equivalence.*

Proof. Apply Theorem 3.6 to obtain a chain map $\chi : C_*^{gr} \to C_*$ such that the adjoint map $\chi^{gr} : C_*^{gr} \to C_*^{gr}$ is the identity map. It is easy to check by induction (using the five-lemma) that χ is a homology equivalence. If C_* is free, then every homology equivalence $C_*^{gr} \to C_*$ is a homotopy equivalence.

Now let us prove the functoriality. Let C_* and D_* be chain complexes endowed with cellular filtrations $C_*^{(k)}$, $D_*^{(k)}$; let $f : C_* \to D_*$ be a chain map, preserving filtrations. Denote by

$$\varphi : C_*^{gr} \to D_*^{gr}$$

the chain map, induced by f, and by

$$g : C_*^{gr} \to C_*, \quad h : D_*^{gr} \to D_*$$

the corresponding maps, preserving filtrations (C_*^{gr}, D_*^{gr} are endowed with the trivial filtrations). The chain maps

$$f \circ g,\ h \circ \varphi : C_*^{gr} \to D_*$$

preserve filtrations and induce the same map φ in the graded homology. Hence they are chain homotopic via a homotopy preserving filtrations. □

Let us mention one particular case of the preceding corollary. Assume that a complex C_* is endowed with two cellular filtrations $C_*^{(k)}$ and $\widetilde{C}_*^{(k)}$, such that for every k we have

$$C_*^{(k)} \subset \widetilde{C}_*^{(k)}.$$

This inclusion induces a chain map

$$\lambda : C_*^{gr} \to \widetilde{C}_*^{gr}$$

of the adjoint complexes.

Corollary 3.9. *The diagram*

$$\begin{array}{ccc} C_*^{gr} & \xrightarrow{\ \lambda\ } & \widetilde{C}_*^{gr} \\ & {\scriptstyle \chi}\searrow \quad \swarrow{\scriptstyle \widetilde{\chi}} & \\ & C_* & \end{array}$$

is homotopy commutative (where $\chi, \widetilde{\chi}$ are the chain equivalences provided by Corollary 3.8).

Proof. Both maps $\widetilde{\chi} \circ \lambda$ and χ preserve the filtrations (where the complexes $C_*^{gr}, \widetilde{C}_*^{gr}$ are filtered trivially, and the complex C_* is endowed with the filtration $\widetilde{C}_*^{(k)}$). Further, they both induce the same homomorphism λ in the adjoint complexes. □

3.2. An example: CW homology. Let X be a CW complex, and let $X^{(k)}$ denote the k-th skeleton of X. Then we have the filtration

$$\varnothing = X^{(-1)} \subset X^{(0)} \subset \cdots \subset X^{(k)} \subset \cdots$$

which is obviously a cellular filtration of the space X. The corresponding filtration in the group of the singular chains:

$$\mathcal{S}_*(\varnothing) = 0 \subset \mathcal{S}_*(X^{(0)}) \subset \cdots \subset \mathcal{S}_*(X^{(k)}) \subset \cdots$$

is therefore a cellular filtration of chain complexes. The adjoint complex $C_*^{gr}(X)$ is by definition *the cellular chain complex* of X. Let us denote it by $\mathcal{CW}_*(X)$. Applying Corollary 3.8 we deduce the following result:

Proposition 3.10. *Let X be a CW complex. There exists a chain equivalence*

$$C_X : \mathcal{CW}_*(X) \longrightarrow \mathcal{S}_*(X)$$

preserving filtrations and inducing the identity map in the adjoint complexes. This chain equivalence is functorial up to a chain homotopy preserving filtrations, that is, for any cellular map $f : X \to Y$ we have a homotopy commutative diagram

$$\begin{array}{ccc} \mathcal{CW}_*(X) & \xrightarrow{f_\sharp} & \mathcal{CW}_*(Y) \\ {\scriptstyle C_X}\downarrow & & \downarrow{\scriptstyle C_Y} \\ \mathcal{S}_*(X) & \xrightarrow{f_*} & \mathcal{S}_*(Y) \end{array}$$

where $f_\sharp$, f_* are the homomorphisms induced by f. □

3.3. Definition of the Morse chain equivalence. Let us return to the Morse theory. In this subsection we apply the algebraic technique developed above to construct a natural chain equivalence between the Morse complex and the singular chain complex of the underlying manifold. We will first consider the case of ordered Morse functions. Let W be a cobordism, $\phi : W \to [a, b]$ an ordered Morse function on W, and v an almost transverse oriented ϕ-gradient. Let $W^{(k)}$ denote the corresponding Morse-Smale filtration of W. The induced filtration

$$\mathcal{S}_*^{(k)}(\phi) = \mathcal{S}_*(W^{(k)}, \partial_0 W)$$

in the singular chain complex is a cellular filtration. By Corollary 3.8 we have a chain equivalence

$$\chi(\phi, v) : C_*(\phi, v) \to \mathcal{S}_*(W, \partial_0 W),$$

which preserves the filtrations, induces the identity map in the adjoint chain complexes, and is uniquely determined by these properties. It turns out that this chain equivalence is in a sense independent of the particular choice of the ordered Morse function ϕ, such that v is a ϕ-gradient:

Proposition 3.11. *Let ϕ_1, ϕ_2 be ordered Morse functions, such that v is a gradient for both. Then there is a homotopy commutative diagram*

$$\begin{array}{ccc} C_*(\phi_1, v) & \xrightarrow{\alpha(\phi_1,\phi_2)} & C_*(\phi_2, v) \\ & {\scriptstyle \chi(\phi_1,v)}\searrow \quad \swarrow{\scriptstyle \chi(\phi_2,v)} & \\ & \mathcal{S}_*(W, \partial_0 W) & \end{array}$$

where $\alpha(\phi_1, \phi_2)$ is the basis-preserving isomorphism (4) from Proposition 2.1 (page 208).

Proof. We shall first prove our assertion in the particular case when the filtrations $W_1^{(k)}, W_2^{(k)}$ corresponding to ϕ_1, and ϕ_2 satisfy $W_1^{(k)} \subset W_2^{(k)}$. In this case the inclusions $\mathcal{S}_*(W_1^{(k)}) \hookrightarrow \mathcal{S}_*(W_2^{(k)})$ induce the isomorphism $\alpha(\phi_1, \phi_2) : C_*(\phi_1, v) \to C_*(\phi_2, v)$ in the adjoint complexes (see Proposition 2.1), and the commutativity of our diagram follows now from Corollary 3.9.

The proof of the general case is similar to the proof of Proposition 2.1. Our argument will be completed once we show that for every pair ϕ_1, ϕ_2 of ordered functions such that v is a ϕ_i-gradient for $i = 1, 2$, there is a third one ϕ_3 such that the corresponding Morse-Smale filtrations of W satisfy

$$W_1^{(k)} \supset W_3^{(k)} \subset W_2^{(k)}.$$

And this follows from Theorem 3.58 of Chapter 4 (page 158). □

Now let $f : W \to \mathbf{R}$ be any Morse function, and v be an oriented almost transverse f-gradient. Pick any ordered Morse function ϕ on W such that v is also a ϕ-gradient. Composing the chain equivalence $\chi(\phi, v)$ with the basis-preserving isomorphism $\mathcal{M}_*(f, v) \to C_*(\phi, v)$ we obtain a chain equivalence

$$\mathcal{E} = \mathcal{E}(f, v) : \mathcal{M}_*(f, v) \xrightarrow{\sim} \mathcal{S}_*(W, \partial_0 W);$$

its homotopy class does not depend on the particular choice of an ordered Morse function.

Definition 3.12. The chain equivalence $\mathcal{E}(f, v)$ will be called *the Morse chain equivalence.*

4. More about the Morse complex

We discuss here some further developments and results related to the Morse complex. The first subsection is about the functoriality properties of the Morse complex and the Morse chain equivalence. In the second subsection we outline a formula for the inverse of the Morse chain equivalence, due to M. Hutchings and Y.-J. Lee. In the last subsection we discuss an equivariant version of the Morse complex.

4.1. Functoriality properties. Let $f_1 : M_1 \to \mathbf{R}$, $f_2 : M_2 \to \mathbf{R}$ be Morse functions on closed manifolds. Let v_1 be an oriented almost transverse f_1-gradient, and v_2 an oriented almost transverse f_2-gradient. Let $A : M_1 \to M_2$ be a continuous map. Our first aim is to associate to A a chain map

$$A_\sharp : \mathcal{M}_*(f_1, v_1) \to \mathcal{M}_*(f_2, v_2)$$

of Morse complexes. We shall construct the homomorphism $A_\sharp$ only when A satisfies the following condition:

$$(\mathcal{I})\quad A\big(D(p,v_1)\big)\ \cap\ D(q,-v_2)=\varnothing$$
$$\text{for every}\quad p\in S(f_1),\ q\in S(f_2)\quad\text{with}\quad \operatorname{ind}p<\operatorname{ind}q.$$

The subset $\mathcal{C}(v_1,v_2)$ of the maps satisfying $(\mathcal{I})$ is open in the space $C^0(M_1,M_2)$ of all the continuous maps from M_1 to M_2. Indeed, the condition $(\mathcal{I})$ is equivalent to the following one:

$$A\big(D(\text{ind}{\leqslant}k\ ,v_1)\big)\cap D(\text{ind}{\leqslant}m_2-k-1\ ,-v_2)=\varnothing\qquad\text{for every}\quad k$$

(where $m_2 = \dim M_2$) and this condition is open, since the sets $D(\text{ind}{\leqslant}k\ ,v_1)$, $D(\text{ind}{\leqslant}m_2-k-1\ ,-v_2)$ are compact. The set $\mathcal{C}(v_1,v_2)$ is dense in $C^0(M_1,M_2)$; we postpone the proof of this result until the end of the subsection, and now we will construct the homomorphism $A_\sharp$ for the maps satisfying $(\mathcal{I})$. As usual, let $\Phi(-v_2,t)$ denote the diffeomorphism of M_2 induced by the gradient descent.

Lemma 4.1. *There is $T\geqslant 0$ and ordered Morse functions $\phi_i : M_i\to\mathbf{R}$ (where $i=1,2$) such that v_i is a ϕ_i-gradient, and*

$$\big(\Phi(-v_2,T)\circ A\big)\big(M_1^{(k)}\big)\subset M_2^{(k)}\qquad\textit{for every}\quad k,$$

where $M_1^{(k)}, M_2^{(k)}$ are the Morse-Smale filtrations induced by ϕ_1, respectively ϕ_2.

Proof. Pick any ordered Morse function ϕ_2, such that v_2 is a ϕ_2-gradient, and let $M_2^{(k)}$ be the filtration corresponding to ϕ_2. Condition $(\mathcal{I})$ implies that for every $T>0$ sufficiently large we have

$$\big(\Phi(-v_2,T)\circ A\big)\big(D(\text{ind}{\leqslant}k\ ,v_1)\big)\ \subset\ \operatorname{Int} M_2^{(k)}\quad\text{for every}\quad k.$$

Thus the subset

$$U_k=(\Phi(-v_2,T)\circ A)^{-1}(\operatorname{Int} M_2^{(k)})$$

is an open neighbourhood of $D(\text{ind}{\leqslant}k\ ;v_1)$ for every k. Applying Theorem 3.58 of Chapter 4 (page 158) we find a Morse function ϕ_1 such that the corresponding Morse-Smale filtration satisfies $M_1^{(k)}\subset U_k$. □

For every ϕ_1,ϕ_2 and T satisfying the conclusions of the preceding lemma, the map

$$\Phi(-v_2,T)\circ A : M_1\to M_2$$

preserves the filtrations, and therefore determines a chain map

$$A_\sharp(\phi_1,\phi_2,T):\mathcal{M}_*(f_1,v_1)\to\mathcal{M}_*(f_2,v_2)$$

between the adjoint complexes.

Proposition and definition 4.2. The chain map $A_\sharp(\phi_1, \phi_2, T)$ does not depend on the particular choice of the functions ϕ_i and the number T. This chain map will be denoted $A_\sharp$.

Proof. Given ϕ_1 and ϕ_2 the homomorphism $A_\sharp$ does not depend on the choice of T, since the maps $\Phi(-v_2, T) \circ A$ and $\Phi(-v_2, T') \circ A$ are homotopic via the homotopy $\Phi(-v_2, \tau) \circ A$ with $\tau \in [T, T']$.

Now let us prove independence of the map from the choice of ϕ_1. Let $\widetilde{\phi_1}$ be another ordered Morse function satisfying the hypotheses of Subsection 4.1. If the corresponding filtration $\widetilde{M_1^{(k)}}$ satisfies $\widetilde{M_1^{(k)}} \subset M_1^{(k)}$ for every k, the assertion follows immediately from the functoriality of the adjoint chain maps. The general case is reduced to this particular one similarly to Proposition 2.1 (page 208). The independence of $A_\sharp$ on the choice of ϕ_2 is proved in the same way. $\square$

Proposition 4.3. *The following diagram is chain homotopy commutative:*

$$\begin{array}{ccc} \mathcal{M}_*(f_1, v_1) & \xrightarrow{A_\sharp} & \mathcal{M}_*(f_2, v_2) \\ \downarrow{\scriptstyle \mathcal{E}_1} & & \downarrow{\scriptstyle \mathcal{E}_2} \\ \mathcal{S}_*(M_1) & \xrightarrow{A_*} & \mathcal{S}_*(M_2) \end{array}$$

(where A_* is the chain map in the singular chain complexes induced by A, and $\mathcal{E}_i$ are the Morse chain equivalences).

Proof. Pick ϕ_1, ϕ_2 and T as in Lemma 4.1. If we replace in the diagram above the chain map A_* (the bottom horizontal arrow) by the chain map $(\Phi(-v_2, T) \circ A)_*$, the commutativity of the resulting diagram follows immediately from the functoriality of the Morse chain equivalence. Observe now that the chain maps A_* and $(\Phi(-v_2, T) \circ A)_*$ are chain homotopic, and the proof is over. $\square$

Exercise 4.4. Let M_0, M_1, M_2 be closed manifolds, $f_i : M_i \to \mathbf{R}$ be Morse functions (where $i = 0, 1, 2$), and v_i oriented almost transverse f_i-gradients. Let

$$B : M_0 \to M_1, \ A : M_1 \to M_2$$

be continuous maps, such that A satisfies the condition $(\mathcal{I})$. Assume that

$$B(D(\mathrm{ind}{\leqslant}k\ ; v_0)) \subset D(\mathrm{ind}{\leqslant}k\ ; v_1) \quad \text{for every} \quad k.$$

Show that

$$(A \circ B)_\sharp = A_\sharp \circ B_\sharp.$$

Now we shall obtain a formula for the map $A_\sharp$ in terms of intersection numbers. We shall assume that A is of class C^∞ and that

$(\mathcal{J})$ $A|D(p,v) : D(p,v) \to M_2$ is transverse to $D(q,-v)$

for every $p \in S(f_1), q \in S(f_2)$.

(Observe that the condition $(\mathcal{J})$ implies $(\mathcal{I})$.) In the end of this section we shall show that the subset of all maps satisfying the condition $(\mathcal{J})$ is residual in $C^\infty(M_1, M_2)$ (which implies in particular, that this set is dense, since $C^\infty(M_1, M_2)$ is a Baire space, see [**61**], Ch.2, Sect. 4). Let $\dim M_1 = m_1$, $\dim M_2 = m_2$.

Lemma 4.5. *If A satisfies the condition $(\mathcal{J})$, then for every $p \in S_k(f_1)$ and $q \in S_k(f_2)$ the set*

$$T(p,q) = D(p,v_1) \cap A^{-1}\big(D(q,-v_2)\big)$$

is finite.

Proof. Let

$$R = D(\mathrm{ind}{\leqslant}k{-}1\ ; v_1) \cup D(p,v_1), \qquad Q = D(\mathrm{ind}{\leqslant}m_2{-}k{-}1\ ; -v_2) \cup D(q,-v_2).$$

Since v_i are almost transverse, both R and Q are compact. The condition $(\mathcal{J})$ implies that

$$R \cap A^{-1}(Q) = T(p,q)$$

therefore $T(p,q)$ is a compact subset of R. By transversality it consists of isolated points, and therefore it is finite. □

Thus the intersection index

$$n(p,q;A) = A(D(p,v_1)) \mathrel{\#} D(q,-v_2) \in \mathbf{Z}$$

is defined (see page 193). Introduce a homomorphism of graded groups

$$A_\flat : \mathcal{M}_*(f_1,v_1) \to \mathcal{M}_*(f_2,v_2)$$

by the formula

$$A_\flat(p) = \sum_q n(p,q;A)\cdot q,$$

where $p \in S_k(f_1)$ and the summation in the right-hand side of the formula is over $q \in S_k(f_2)$.

Theorem 4.6. $A_\sharp = A_\flat$.

Proof. Pick any ordered Morse functions $\phi_1 : M_1 \to \mathbf{R}, \phi_2 : M_2 \to \mathbf{R}$ and a positive integer T such that

$$\big(\Phi(T,-v_2)\circ A\big)(M_1^{(k)}) \subset M_2^{(k)} \quad \text{for every} \quad k,$$

where $M_i^{(k)}$ denote the filtrations induced by the functions ϕ_i. Let $\widetilde{A} = \Phi(T, -v_2) \circ A$. It is clear that $\widetilde{A}_\flat = A_\flat$. By the definition the map $A_\sharp$ is the homomorphism induced by $\widetilde{A}$ in the adjoint complexes. Therefore it remains to prove that for every $p \in S_k(f_1)$ we have the following formula for the $\widetilde{A}_*$-image of the homology class $d_p(v_1)$:

$$\widetilde{A}_*(d_p(v_1)) = \sum_q n(p, q; A) \cdot d_q(v_2),$$

where the summation in the formula is over $q \in S_k(f_2)$. And this follows from Proposition 3.13 of Chapter 5 (page 182). □

It remains to show that the condition $(\mathcal{J})$ is residual. The space $C^\infty(M_1, M_2)$ is endowed with the Whitney topology (see [**61**]).

Proposition 4.7. *The set of all C^∞ maps $\Phi : M_1 \to M_2$ satisfying $(\mathcal{J})$ is residual in $C^\infty(M_1, M_2)$.*

Before proceeding to the proof let us deduce from this proposition the property promised in the beginning of the subsection.

Corollary 4.8. *The set $\mathcal{C}(v_1, v_2)$ is dense in $C^0(M_1, M_2)$.*

Proof. Let $g : M_1 \to M_2$ be any continuous map. Approximate it (in a C^0 topology) by a C^∞ map $g_1 : M_1 \to M_2$, and then approximate the map g_1 (in a C^∞ topology) by a C^∞ map g_2 satisfying $(\mathcal{J})$. Since $(\mathcal{J}) \Rightarrow (\mathcal{I})$, the proof of the corollary is complete. □

Proof of Proposition 4.7. It suffices to prove that for two given points $p \in S(f_1), q \in S(f_2)$ the set of maps Φ satisfying the condition

$(\mathcal{J}_{pq})$ $\qquad \Phi | D(p, v) : D(p, v) \to M_2$ is transverse to $D(q, -v)$

is residual. Cover $D(p, v)$ by a countable family of closed embedded discs D_i of dimension $\operatorname{ind} p$, such that every disc D_i is in the domain of definition of a chart of M_1. Similarly, cover $D(q, -v)$ by a countable family of closed embedded discs Δ_j of dimension $m_2 - \operatorname{ind} q$ such that every disc Δ_j is in the domain of definition of a chart of M_2. The transversality condition $(\mathcal{J})$ is equivalent to the transversality of each restriction $\Phi | D_i$ to every disc Δ_j. Therefore the proposition follows from the next lemma.

Lemma 4.9. *Let $D_1 \subset M_1, D_2 \subset M_2$ be embedded closed discs such that each of the two discs is in the domain of definition of a chart of the corresponding manifold. The set of all C^∞ maps $f : M_1 \to M_2$ such that the map $f | D_1$ is transverse to D_2 is open and dense in $C^\infty(M_1, M_2)$.*

Proof. The lemma follows without much difficulty from general transversality theorems. Let us start with the openness property. Assume that $f|D_1 \pitchfork D_2$. For every C^∞ map $g : M_1 \to M_2$ sufficiently C^∞ close to f, the restriction $g|D_1 : D_1 \to D_2$ is C^∞-close to $f|D_1$, and therefore transverse to D_2, since both D_1, D_2 are compact (see [**61**], Th. 2.1).

The density follows from the next lemma, which deals with embedded discs in Euclidean spaces.

Lemma 4.10. *Let $D \subset U$ be an embedded closed disc in an open subset of $\mathbf{R}^n$, let $U_0 \subset U$ be a compact neighbourhood of D in U. Let $\phi : U \to V \subset \mathbf{R}^k \times \mathbf{R}^l$ be a C^∞ map. Let r be a positive integer.*

For every $\epsilon > 0$ there is a C^∞ map $\psi : U \to V$, such that the restriction $\psi|D$ is transverse to $\mathbf{R}^k$ and such that ψ equals ϕ in $U \setminus U_0$ and the C^r-norm of $\psi - \phi$ is $< \epsilon$.

Proof. Let $\bar{\phi}$ denote the composition of ϕ and the projection $\mathbf{R}^k \times \mathbf{R}^l \to \mathbf{R}^l$. Let $\lambda \in \mathbf{R}^l$ be a regular value of $\bar{\phi}|D$, so small that $\phi(U_0) - \lambda \subset V$. Let $h : U \to \mathbf{R}_+$ be a C^∞ function such that $h(x) = 1$ for $x \in U_0$ and $\operatorname{supp} h \subset U$. Then the map

$$x \mapsto h(x)(\phi(x) - \lambda) + (1 - h(x))\phi(x)$$

satisfies the requirements of the lemma, if λ is small enough. □

4.2. The Hutchings-Lee formula for the inverse of the Morse chain equivalence. The Morse chain equivalence $\mathcal{E} : \mathcal{M}_*(f, v) \to \mathcal{S}_*(M)$ is defined only up to chain homotopy. It turns out that there is in a sense an explicit formula for the inverse equivalence. In this subsection we briefly discuss this formula, due to M. Hutchings and Y.-J. Lee [**64**].

Let M be a closed C^∞ manifold, $f : M \to \mathbf{R}$ a Morse function, and v an oriented f-gradient. Let us say that a C^∞ singular simplex $\sigma : \Delta^k \to M$ is transverse to v if the map σ is transverse to the ascending disc of every critical point p of f, and the same is true also for the restriction of f to every face of Δ^k. The free abelian group generated by singular simplices transverse to v will be denoted by $\mathcal{S}_*(M, \pitchfork v)$. One can prove that the inclusion

$$\mathcal{S}_*(M, \pitchfork v) \hookrightarrow \mathcal{S}_*(M)$$

is a chain equivalence (see [**64**]). We are going to define a chain map

$$\xi : \mathcal{S}_*(M, \pitchfork v) \to \mathcal{M}_*(f, v).$$

Lemma 4.11. *For every singular simplex $\sigma \in \mathcal{S}_k(M, \pitchfork v)$ and every critical point $p \in S_k(F)$ the intersection $\sigma(\Delta^k) \cap D(p, -v)$ is a finite set.*

Proof. Let ϕ be an ordered Morse function adjusted to (f, v), let $M^{(k)}$ be the corresponding Morse-Smale filtration. Applying the gradient descent if necessary, we can assume that

$$\sigma(\Delta^k) \subset M^{(k)}$$

and then

$$\sigma(\Delta^k) \cap D(p, -v) = \sigma(\Delta^k) \cap \big(D(p, -v) \cap M^{(k)}\big).$$

The map σ is transverse to the compact $(\dim M - k)$-dimensional manifold $D(p, -v) \cap M^{(k)}$ and therefore their intersection is a finite set. □

The simplex Δ^k inherits the orientation of $\mathbf{R}^k$, and the manifold $D(p, -v)$ is cooriented, so each point of $\sigma(\Delta^k) \cap D(p, -v)$ acquires a sign. Put

$$[\sigma : p] = \sigma(\Delta^k) \# D(p, -v),$$

and

$$\xi(\sigma) = \sum_{p \in S_k(f)} [\sigma : p]p.$$

Extend ξ to a homomorphism $\mathcal{S}_*(M, \pitchfork v) \to \mathcal{M}_*(f, v)$ of abelian groups.

Theorem 4.12 (M. Hutchings, Y.J. Lee). *ξ is a chain map and the following diagram is homotopy commutative*

$$\begin{array}{ccc} \mathcal{S}_*(M, \pitchfork v) & \xrightarrow{\quad \xi \quad} & \mathcal{M}_*(f, v) \\ & {}_{I}\searrow \quad \swarrow_{\mathcal{E}} & \\ & \mathcal{S}_*(M) & \end{array}$$

where the left diagonal arrow I is the inclusion, and the right diagonal arrow is the canonical chain equivalence.

Outline of the proof. The Morse-Smale filtration $\{M^{(k)}\}$ of M induces a filtration in $\mathcal{S}_*(M, \pitchfork v)$ and in $\mathcal{S}_*(M)$. Both chain maps $\mathcal{E} \circ \xi$ and I preserve the filtrations and it suffices to show that they induce the same map in the adjoint complexes. Any cycle in $\mathcal{S}_*\big(M^{(k)}, M^{(k-1)}\big)$ can be represented by a linear combination

$$\lambda = \sum_i n_i \sigma_i$$

of singular simplices such that the boundary of each simplex σ_i is in $M^{(k-1)}$. Now it suffices to apply Proposition 3.13 of Chapter 5 (page 182) and the proof is complete. □

If we replace the singular chain complex by the simplicial chain complex we can define a chain equivalence

$$\Delta_*(M) \xrightarrow{\xi_\Delta} \mathcal{M}_*(f,v)$$

from the simplicial chain complex of M associated to a C^∞ triangulation of M to the Morse complex of (f,v). One must assume that the simplices of the triangulation are transverse to the ascending discs of the critical points. In this assumption let σ be a k-dimensional simplex of the triangulation. For $p \in S_k(f)$ denote by $\langle \sigma, p\rangle$ the algebraic intersection number

$$\sigma \mathbin{\#} D(p,-v)$$

and put

$$\xi_\Delta(\sigma) = \sum_{p\in S_k(f)} [\sigma : p]p.$$

The next theorem is due to D. Schütz [**140**].

Theorem 4.13. *ξ_Δ is a chain map and the following diagram is homotopy commutative:*

$$\begin{array}{ccc} \Delta_*(M) & \xrightarrow{\quad\xi_\Delta\quad} & \mathcal{M}_*(f,v) \\ & \searrow \qquad \swarrow_{\mathcal{E}} & \\ & \mathcal{S}_*(M) & \end{array}$$

where the left diagonal arrow is the natural chain equivalence between the simplicial and singular chain complexes.

4.3. The universal Morse complex. Let $f : M \to \mathbf{R}$ be a Morse function on a closed connected C^∞ manifold, and v an oriented transverse f-gradient. The construction of the Morse complex $\mathcal{M}_*(f,v)$, associated to these data, uses the procedure of counting of flow lines joining critical points of f; each flow line is counted with the sign corresponding to the choice of orientations of the descending discs. One can refine this method, taking into account the homotopy classes of the flow lines. The resulting chain complex encodes much more information about the manifold and the Morse function. Let $G = \pi_1(M)$ denote the fundamental group of M. Consider the universal covering $\Pi : \widetilde{M} \to M$. We have a left action of G on $\widetilde{M}$ and the quotient equals M. We will now define a chain complex $\widetilde{\mathcal{M}}_*(f,v)$ such that

(1) $\widetilde{\mathcal{M}}_k(f,v)$ is a free left module over $\mathbf{Z}G$ freely generated by the critical points of f of index k;
(2) $H_*(\widetilde{\mathcal{M}}_*(f,v)) \approx H_*(\widetilde{M})$.

Let $F = f \circ \Pi$, so that F is a G-invariant Morse function on $\widetilde{M}$. Let $\widetilde{\mathcal{M}}_k$ denote the free abelan group generated by the critical points of index k of F, then $\widetilde{\mathcal{M}}_k$ has a natural structure of a free left $\mathbf{Z}G$-module, any lift to $\widetilde{M}$ of the set $S_k(f)$ being a free $\mathbf{Z}G$-basis for $\widetilde{\mathcal{M}}_k$. Let us define the incidence coefficient $\widetilde{n}(p,q;v) \in \mathbf{Z}G$ for every pair $p,q \in \underline{S(F)}$ of critical points, such that $\operatorname{ind} p = \operatorname{ind} q + 1$. Lift v to a vector field $\widetilde{v}$ on $\widetilde{M}$, so that $\widetilde{v}$ is a G-invariant F-gradient. Consider the set $\Gamma(p,q;\widetilde{v})$ of flow lines of $(-\widetilde{v})$ joining p and q. Let $p' = \Pi(p), q' = \Pi(q)$. In view of the unique lifting property of the covering, the map

$$\Gamma(p,q;\widetilde{v}) \to \Gamma(p',q';v)$$

induced by the projection Π is injective. In particular the set $\Gamma(p,q;\widetilde{v})$ is finite. Since v is oriented, every flow line of $(-v)$ joining p' with q' acquires a sign $\varepsilon(\gamma)$. For $\gamma \in \Gamma(p,q,\widetilde{v})$ let $\varepsilon(\gamma) = \varepsilon(\Pi(\gamma))$, and put

$$\widetilde{n}(p,q;v) = \sum_{\gamma \in \widetilde{\Gamma}(p,q;v)} \varepsilon(\gamma) \in \mathbf{Z}.$$

Define a homomorphism $\widetilde{\partial}_k : \widetilde{\mathcal{M}}_k \to \widetilde{\mathcal{M}}_{k-1}$ by the following formula (here $p \in S_k(F)$):

$$\widetilde{\partial}_k p = \sum_{q \in S_{k-1}(F)} \widetilde{n}(p,q;v) \cdot q.$$

Theorem 4.14. *The composition $\widetilde{\partial}_{k-1} \circ \widetilde{\partial}_k$ is equal to zero for every k and the homology of the resulting chain complex is isomorphic to $H_*(\widetilde{M})$ as a $\mathbf{Z}G$-module.*

The proof repeats the proof of Theorem 1.3 (page 198) with corresponding modifications, see for example [**110**], Appendix. The chain complex $\widetilde{\mathcal{M}}_*(f,v)$ will be called *the universal Morse complex.* Similarly to the case of the ordinary Morse complex (see Subsection 3.3, page 217) we have *the universal Morse equivalence*:

Theorem 4.15. *There is a chain equivalence*

$$\widetilde{\mathcal{M}}_*(f,v) \xrightarrow{\widetilde{\mathcal{E}}} \mathcal{S}_*(\widetilde{M})$$

of free chain complexes over $\mathbf{Z}G$.

The assertion of Theorem 4.15 is in general stronger than the assertion of Theorem 4.14, since two chain complexes over $\mathbf{Z}G$ with isomorphic homology are not necessarily homotopy equivalent. One can strengthen Theorem 4.15 still further, see Subsection 3.3 of Chapter 13 (page 405).

History and Sources

The seminal article [**98**] of Marston Morse starts with *the Morse Lemma* (see Theorem 1.6 of Chapter 2, page 36), and continues with the investigation of the subsets

$$M_a = \{x \in M \mid f(x) \leqslant a\}$$

where $f : M \to \mathbf{R}$ is a Morse function on a manifold M. It is proved that the topology of the set M_a does not change while a varies without crossing a critical value of f, and for a critical value c the set $M_{c+\epsilon}$ is obtained from $M_{c-\epsilon}$ by attaching the handles of indices equal to the indices of the critical points on the critical level c. Then Morse deduces the inequalities which now bear his name (see Theorem 1.9).

A stronger version of the Morse inequalities (see Proposition 1.12) was found much later by E. Pitcher. These inequalities are in a sense optimal, since it was proved by S. Smale [**148**] that on a simply-connected closed manifold of dimension greater than 5 there is a Morse function for which the Morse-Pitcher inequalities become equalities. This result is one of the by-products of Smale's proof of the generalized Poincaré Conjecture in dimensions $\geqslant 5$ ([**149**]). This proof uses the Morse theory in an essential way. The main idea can be outlined as follows. Let M be a closed manifold of the homotopy type of S^n. Starting with an arbitrary Morse function on M one shows that it is possible to cancel the critical points so that one ends up with two critical points, the maximum and the minimum (and therefore M is homeomorphic to S^n, as the Poincaré Conjecture claims). Among the technical innovations of his work one finds the Rearrangement Lemma (see Subsection 3.3 of Chapter 4, page 137), the systematic use of the notion of handles (see Chapter 5) and many others. A detailed exposition of this work can be found in J. Milnor's book [**92**].

The notion of the Morse complex (see Section 1 of this chapter) did not appear explicitly in Smale's works, although he used the procedure of counting the flow lines joining the critical points. It was introduced by E. Witten in his paper [**163**], along with yet another approach to the Morse theory, based on the de Rham cohomology (see Section 1 of Chapter 14, page 413 for discussion of this subject).

Part 3

Cellular gradients.

We show that the gradient descent for a C^0-generic gradient of a Morse function on a cobordism gives rise to a structure similar to a cellular map between CW complexes. This structure enables us to perform a detailed study of the properties of the gradient descent in the C^0-generic case.

CHAPTER 7

Condition ($\mathfrak{C}$)

Let $f : W \to [a, b]$ be a Morse function on a cobordism W and v an f-gradient. The gradient descent from $\partial_1 W$ to $\partial_0 W$ yields a diffeomorphism

$$(-v)^{\rightsquigarrow} : \partial_1 W \setminus D(-v) \to \partial_0 W \setminus D(v);$$

by definition $(-v)^{\rightsquigarrow}(x)$ is the point where the trajectory $\gamma\big(x, \tau(x, -v); -v\big)$ intersects $\partial_0 W$. The map $(-v)^{\rightsquigarrow}$ does not admit a continuous extension to $\partial_1 W$ and its properties in general are rather difficult to study. In this part we show that for a C^0-generic gradient v one can derive from $(-v)^{\rightsquigarrow}$ a geometric structure (which can be called *a graded-continuous map*) closely resembling a cellular map between CW complexes. The components of this structure are: a Morse-Smale filtration $\partial_1 W^k$ of $\partial_1 W$,[†] a Morse-Smale filtration $\partial_0 W^k$ of $\partial_0 W$, and a sequence of continuous maps

$$\phi_k : \partial_1 W^k / \partial_1 W^{k-1} \to \partial_0 W^k / \partial_0 W^{k-1}$$

such that for every k:

(A) $(-v)^{\rightsquigarrow}(\partial_1 W^k) \subset \partial_0 W^k$ and
(B) the following diagram is commutative (where the horizontal arrows are the projection maps).

$$\begin{array}{ccc} \partial_1 W^k & \longrightarrow & \partial_1 W^k / \partial_1 W^{k-1} \\ \Big\downarrow {\scriptstyle (-v)^{\rightsquigarrow}} & & \Big\downarrow {\scriptstyle \phi_k} \\ \partial_0 W^k & \longrightarrow & \partial_0 W^k / \partial_0 W^{k-1} \end{array}$$

Each ϕ_k is a continuous map between spaces homotopy equivalent to wedges of k-dimensional spheres, and the invariants of these maps are quite useful for the study of the gradient descent.

[†] To abbreviate the notation we write $\partial_1 W^k$ instead of $\partial_1 W^{(k)}$ in this part.

1. The gradient descent revisited

Before constructing the graded-continuous map, mentioned in the introduction, let us discuss the question in a somewhat greater generality. Let $f : W \to [a, b]$ be a Morse function on a cobordism W and v an f-gradient. The subset $D(-v) \cap \partial_1 W$ can be considered as the set of singularities of the partially defined map $(-v)^{\leadsto}$. Let us *quotient out* this singularity set. Namely, put

$$S_1 = D(-v) \cap \partial_1 W, \quad S_0 = D(v) \cap \partial_0 W.$$

We will construct a continuous map

$$\phi : \partial_1 W / S_1 \to \partial_0 W / S_0$$

between the quotient spaces. Let ω_1 denote the point of $\partial_1 W / S_1$ corresponding to the collapsed subset S_1, and ω_0 denote the point of $\partial_0 W / S_0$ corresponding to the collapsed subset S_0. Put

$$\phi(\omega_1) = \omega_0.$$

For every point

$$x \in (\partial_1 W / S_1) \setminus \{\omega_1\} = \partial_1 W \setminus S_1$$

define $\phi(x)$ to be the image of $(-v)^{\leadsto}(x) \in \partial_0 W$ via the quotient map $\partial_0 W \to \partial_0 W / S_0$.

Proposition 1.1. *The map*

$$\phi : \partial_1 W / S_1 \to \partial_0 W / S_0 \tag{1}$$

is continuous.

Proof. Continuity on the complement to the point ω_1 follows from the continuity of $(-v)^{\leadsto}$. As for the point ω_1, it suffices to show that for every open neighbourhood U of $\partial_0 W \cap D(v)$ there is a neighbourhood V of $\partial_1 W \cap D(-v)$ such that $(-v)^{\leadsto}(V) \subset U$. Pick $\delta > 0$ such that $B_\delta(v) \cap \partial_0 W \subset U$ (this is possible since $D_\delta(v)_{\delta > 0}$ form a fundamental system of neighbourhoods for $D(v)$). Then put $V = B_\delta(-v) \cap \partial_1 W$ and the result follows. □

Exercise 1.2. (1) Show that for the "pants" manifold (see page 41) the map (1) is a homeomorphism between two copies of a wedge of two circles.

(2) Prove that in general the map (1) is a homeomorphism.

1.1. The map $(-v)^{\rightsquigarrow}$. The central part of this chapter is Section 2 where we give the definition of cellular gradients. The aim of the present subsection is to develop necessary techniques for working with these gradients, so the reader may wish to proceed directly to Section 2 and return to this subsection when necessary.

Let $f : W \to [a, b]$ be a Morse function on a cobordism W and v an f-gradient. Let

$$B_1 \subset A_1 \subset \partial_1 W, \quad B_0 \subset A_0 \subset \partial_0 W$$

be compact sets. Under some restrictions on v, A_i, B_i we shall construct a continuous map $A_1/B_1 \to A_0/B_0$ derived from the gradient descent.

Recall that we write $q \lhd p$ if $p, q \in S(f)$ and there is a broken flow line of v joining q with p (see Subsection 3.4 of Chapter 4, page 142). The relation $q \lhd p$ is a partial order on the set $S(f)$, and a subset $\Sigma \subset S(f)$ is called an *initial segment* of $S(f)$ if

$$\Big(p \in \Sigma \ \&\ q \lhd p\Big) \Rightarrow \Big(q \in \Sigma\Big).$$

Definition 1.3. Let

$$B_1 \subset A_1 \subset \partial_1 W, \quad B_0 \subset A_0 \subset \partial_0 W$$

be compact sets. We say that the condition $(\mathcal{D})$ is fulfilled for f, v and the pairs (A_1, B_1), (A_0, B_0), if

$(\mathcal{D}1)$ $\qquad (-v)^{\rightsquigarrow}(A_1) \subset \operatorname{Int} A_0, \quad (-v)^{\rightsquigarrow}(B_1) \subset \operatorname{Int} B_0,$ †

and there is an initial segment Σ of $S(f)$ such that:

$(\mathcal{D}2)$ $\qquad$ If $D(p, -v) \cap A_1 \neq \varnothing$, then $p \in \Sigma$.

$(\mathcal{D}3)$ $\qquad$ For every $p \in \Sigma$ we have: $D(p, v) \cap \partial_0 W \subset \operatorname{Int} B_0$.

The geometric sense of the conditions $(\mathcal{D}2)$, $(\mathcal{D}3)$ is that the set of points in A_1 where $(-v)^{\rightsquigarrow}$ is *not defined* corresponds to the ascending discs of the critical points whose soles belong to B_0, so when we quotient out the subset B_0 we obtain a continuous map $A_1/B_1 \to A_0/B_0$.

Let us proceed to the construction of this continuous map. Let ω_1 denote the point in A_1/B_1 corresponding to the collapsed set B_1, and let ω_0 denote the point in A_0/B_0 corresponding to the collapsed set B_0. The image of an element $x \in A_1$ (respectively of a subset $X \subset A_1$) via the projection $A_1 \to A_1/B_1$ will be denoted by $\overline{x}$ (respectively $\overline{X}$); similarly for the images of subsets with respect to the projection $A_0 \to A_0/B_0$.

† Recall that $(-v)^{\rightsquigarrow}(X)$ is an abbreviation for $(-v)^{\rightsquigarrow}(X \setminus D(-v))$.

Definition 1.4. Assuming that $(\mathcal{D})$ is satisfied, define a map

$$(-v)^{\twoheadrightarrow} : A_1/B_1 \to A_0/B_0$$

as follows:

$$\begin{aligned}
&(-v)^{\twoheadrightarrow}(\omega_1) = \omega_0; && \\
&(-v)^{\twoheadrightarrow}(\overline{x}) \ = \ \omega_0 && \text{if} \quad x \in D(-v); \\
&(-v)^{\twoheadrightarrow}(\overline{x}) = \overline{(-v)^{\rightsquigarrow}(x)} && \text{if} \quad x \in A_1 \setminus (B_1 \cup D(-v)).
\end{aligned}$$

Proposition 1.5. *The map $(-v)^{\twoheadrightarrow}$ is continuous.*

Proof. For any point x in the domain of definition of $(-v)^{\rightsquigarrow}$ the continuity of $(-v)^{\twoheadrightarrow}$ at $\overline{x}$ follows immediately from continuity of $(-v)^{\rightsquigarrow}$. It remains therefore to prove continuity of $(-v)^{\twoheadrightarrow}$ at any point

$$x \in \{\omega_1\} \cup \Big(D(-v) \cap A_1\Big) = \{\omega_1\} \cup \Big(D(\Sigma, -v) \cap A_1\Big),$$

and this follows from the next lemma.

Lemma 1.6. *Let x_n be a sequence of points in A_1, such that*

$$x_n \to x, \qquad \textit{and} \qquad x \in D(\Sigma, -v), \quad x_n \notin D(\Sigma, -v).$$

Then

$$(-v)^{\twoheadrightarrow}(\overline{x_n}) \to \omega_0 \in A_0/B_0.$$

Proof. For any $\delta > 0$ the point x_n is in the set $B_\delta(\Sigma, -v)$ if n is sufficiently large. Therefore the points $z_n = (-v)^{\rightsquigarrow}(x_n)$ are in $B_\delta(\Sigma, v)$. These sets form a fundamental system of neighbourhoods of $D(\Sigma, v)$ (see Proposition 3.24 of Chapter 4, page 144), and since

$$D(\Sigma, v) \cap \partial_0 W \subset \text{Int } B_0,$$

the point z_n is in B_0 for every n sufficiently large, and we have

$$(-v)^{\twoheadrightarrow}(\overline{x_n}) = \overline{z_n} = \omega_0.$$

□

Here are some examples when the condition $(\mathcal{D})$ is fulfilled (the main example provided by cellular gradients will be constructed in Section 2).

Examples 1.7. 1). If the set $D(-v)$ does not intersect A_1, the condition $(\mathcal{D})$ is fulfilled with any B_1, $\Sigma = \varnothing$, and B_0, A_0 satisfying $(\mathcal{D}1)$. The map $(-v)^{\twoheadrightarrow}$ is the quotient map of $(-v)^{\rightsquigarrow}$, so that the next diagram of continuous maps is commutative:

$$\begin{array}{ccc}
(A_1, B_1) & \longrightarrow & (A_1/B_1, *) \\
{\scriptstyle (-v)^{\rightsquigarrow}} \big\downarrow & & {\scriptstyle (-v)^{\twoheadrightarrow}} \big\downarrow \\
(A_0, B_0) & \longrightarrow & (A_0/B_0, *).
\end{array}$$

2) Put $\Sigma = S(f)$; then the condition $(\mathcal{D})$ is fulfilled for $B_1 = \varnothing$, and any set B_0 containing $D(v) \cap \partial_0 W$ in its interior. Here $(-v)^{\twoheadrightarrow}$ is a continuous map from A_1 to A_0/B_0.

The next proposition will be referred to as *the Composition Lemma.* It gives a precise sense to the intuitively obvious idea that for three regular values $\lambda_0 < \lambda_1 < \lambda_2$ of a function f the gradient descent from the level surface $f^{-1}(\lambda_2)$ to $f^{-1}(\lambda_0)$ is the composition of the gradient descent from $f^{-1}(\lambda_2)$ to $f^{-1}(\lambda_1)$ with the gradient descent from $f^{-1}(\lambda_1)$ to $f^{-1}(\lambda_0)$. Let us first develop the necessary terminology.

Denote $f^{-1}([\lambda_0, \lambda_1])$ by W'. Let

$$A_1 \subset B_1 \subset f^{-1}(\lambda_1), \quad A_0 \subset B_0 \subset f^{-1}(\lambda_0)$$

be compact subsets. Let $\Sigma' \subset S(f|W')$ be an initial segment of $S(f|W')$ such that the condition $(\mathcal{D})$ holds for the restriction of f and v to W' and the pairs (A_1, B_1), (A_0, B_0). The corresponding map, constructed in Definition 1.4, will be denoted by

$$(-v)^{\twoheadrightarrow}_{[\lambda_1,\lambda_0]} : A_1/B_1 \to A_0/B_0.$$

Denote $f^{-1}([\lambda_1, \lambda_2])$ by W''. Let

$$A_2 \subset B_2 \subset f^{-1}(\lambda_2), \quad A_1' \subset B_1' \subset f^{-1}(\lambda_1)$$

be compact subsets. Let $\Sigma'' \subset S(f|W'')$ be an initial segment of $S(f|W'')$ such that the condition $(\mathcal{D})$ holds for the pairs (A_2, B_2), (A_1', B_1') and the restriction of f and v to W''. The corresponding continuous map constructed in Definition 1.4 will be denoted by

$$(-v)^{\twoheadrightarrow}_{[\lambda_2,\lambda_1]} : A_2/B_2 \to A_1'/B_1'.$$

Proposition 1.8. *In the notation above, assume that*

$$A_1' \subset A_1, \ B_1' \subset B_1.$$

Then the condition $(\mathcal{D})$ holds for the cobordism $\widetilde{W} = f^{-1}([\lambda_0, \lambda_2])$, the restrictions of f and v to $\widetilde{W}$, the set $\Sigma = \Sigma' \cup \Sigma'' \subset S(f|\widetilde{W})$ and the subsets

$$A_2 \subset B_2 \subset \partial_1 \widetilde{W}, \ A_0 \subset B_0 \subset \partial_0 \widetilde{W}.$$

The map

$$(-v)^{\twoheadrightarrow}_{[\lambda_2,\lambda_0]} : A_2/B_2 \to A_0/B_0$$

equals the composition of the following maps:

$$A_2/B_2 \xrightarrow{(-v)^{\twoheadrightarrow}_{[\lambda_2,\lambda_1]}} A_1'/B_1' \xrightarrow{I} A_1/B_1 \xrightarrow{(-v)^{\twoheadrightarrow}_{[\lambda_1,\lambda_0]}} A_0/B_0$$

where I denotes the map induced by the inclusion of pairs $(A_1', B_1') \subset (A_1, B_1)$.

Proof. To prove that the condition $(\mathcal{D})$ is satisfied for the pairs (A_2, B_2) and (A_0, B_0), it suffices to show that Σ is an initial segment of $S(f|\widetilde{W})$; the other requirements of Definition 1.3 are easy to check. Let $p \in \Sigma$, $q \in S(f|\widetilde{W})$ and $q \lhd p$. We must prove that $q \in \Sigma$. Assume that $p \in \Sigma''$, $q \in \Sigma'$ (the other cases are obvious). There is a sequence of critical points $q_0 = q, q_1, \ldots, q_k = p$ and for every j there is a v-link from q_{j-1} to q_j. There is a unique j, such that $q_{j-1} \in W'$ and $q_j \in W''$. Then $q_j \in \Sigma''$ and by condition $(\mathcal{D}3)$ the point of intersection of the $(-v)$-trajectory from q_j to q_{j-1} is in $B_1' \subset B_1$. Condition $(\mathcal{D}2)$ applied to the cobordism W' implies the point q_{j-1} must be in Σ' and therefore $q_i \in \Sigma$ for every $i \leqslant j-1$. The verification of the equality

$$(-v)^{\twoheadrightarrow}_{[\lambda_2,\ \lambda_0]} = (-v)^{\twoheadrightarrow}_{[\lambda_1,\ \lambda_0]} \circ I \circ (-v)^{\twoheadrightarrow}_{[\lambda_2,\ \lambda_1]}$$

is not difficult and will be omitted. □

The advantage of considering domains of definition other than $\partial_0 W/S_1$ is that now we can compare the maps corresponding to different gradients. Our next aim is to prove that the map $(-v)^{\twoheadrightarrow}$ is homotopically stable with respect to C^0-small perturbations of v.

Theorem 1.9. *Let $f : W \to [a,b]$ be a Morse function on a cobordism W and v an f-gradient. Let*

$$B_1 \subset A_1 \subset \partial_1 W, \quad B_0 \subset A_0 \subset \partial_0 W$$

be compact subsets. Let $\Sigma \subset S(f)$ be an initial segment of $S(f)$ and assume that the condition $(\mathcal{D})$ holds with respect to the pairs (A_i, B_i) and $\Sigma \subset S(f)$ (where $i = 1, 2$).

Then there is $\delta > 0$ such that for every f-gradient w with $||v - w|| < \delta$ we have:

(1) *The condition $(\mathcal{D})$ is also satisfied with respect to the same subset $\Sigma \subset S(f)$ and the pairs (A_i, B_i).*
(2) *The maps*

$$(-v)^{\twoheadrightarrow}, (-w)^{\twoheadrightarrow} : A_1/B_1 \to A_0/B_0$$

are homotopic.

Proof. α) The easiest case is when the set $D(-v)$ does not intersect A_1 and $\Sigma = \varnothing$. In this case the map $(-v)^{\twoheadrightarrow}$ can be lifted to a continuous map

$$(-v)^{\leadsto} : A_1 \to \text{Int } A_0.$$

Our assertion follows immediately from Proposition 3.5 of Chapter 3 (page 102).

β) The next particular case is when $\Sigma = S(f)$. The condition $(\mathcal{D}2)$ is clearly satisfied for every f-gradient w. The conjunction of the conditions $(\mathcal{D}1)$, $(\mathcal{D}3)$ is equivalent to the conjunction of the following two conditions:

$$\Big(T(B_1, -v) \cup D(v)\Big) \cap \partial_0 W \subset \text{Int } B_0,$$

$$\Big(T(A_1, -v) \cup D(v)\Big) \cap \partial_0 W \subset \text{Int } A_0,$$

which will still hold if we replace the f-gradient v by any f-gradient w, sufficiently C^0-close to v (see Proposition 3.7 of Chapter 3, page 103). As for homotopy of the two gradient descent maps corresponding to v and w, set $w_t = (1-t)v + tw$; this is a path of f-gradients satisfying condition $(\mathcal{D})$, therefore for every $t \in [0,1]$ we have the maps

$$(-w_t)^{\twoheadrightarrow} : A_1/B_1 \to A_0/B_0.$$

It remains to show that the map

$$(x,t) \mapsto (-w_t)^{\twoheadrightarrow}(x) : A_1/B_1 \times [0,1] \to A_0/B_0$$

is continuous. This argument which is completely similar to the proof of Proposition 1.5 will be omitted.

γ) The general case will be reduced to the two particular cases above with the help of the Composition Lemma. Pick a Morse function $g : W \to [a,b]$ adjusted to (f,v) and having a regular value λ such that

$$\Sigma = S(f) \cap g^{-1}([a,\lambda])$$

(this is possible since Σ is an initial segment of $S(f)$, see Proposition 3.23 of Chapter 4, page 143). Put

$$W' = g^{-1}([a,\lambda]), \quad W'' = g^{-1}([\lambda,b]).$$

According to condition $(\mathcal{D})$ the ascending discs of the critical points of $f|W''$ do not intersect A_1. Therefore the sets

$$A' = (-v)^{\leadsto}_{[b,\lambda]}(A_1), \quad B' = (-v)^{\leadsto}_{[b,\lambda]}(B_1)$$

are compact subsets of $V = g^{-1}(\lambda)$. For every pair (K,M) of compact subsets of V such that

$$A' \subset \text{Int } K, \ B' \subset \text{Int } M, \tag{2}$$

the condition $(\mathcal{D})$ is satisfied with respect to the pairs (A_1, B_1) and (K,M) and $\Sigma'' = \varnothing$.

Now let us turn to cobordism W'. This cobordism satisfies condition $(\mathcal{D})$ with respect to the pairs (A', B') and (A_0, B_0) and the initial segment $\Sigma = S(f) \cap W'$. It is easy to check that it is possible to choose the pair (K,M) satisfying (2) and such that the condition $(\mathcal{D})$ is still verified with respect to the pairs (K,M) and (A_0,B_0).

Now we are ready to check the point (1). For every w sufficiently C^0-close to v the condition $(\mathcal{D})$ is still verified with respect to the pairs (A_1, B_1) and (K, M). Also it is verified with respect to the pairs (K, M) and (A_0, B_0). It remains to apply Composition Lemma, Proposition 1.8 and the first point is proved.

As for the second point, observe that (again by Proposition 1.8) the map

$$(-v)^{\twoheadrightarrow}_{[b,a]} : A_1/B_1 \to A_0/B_0$$

is equal to the composition

$$A_1/B_1 \xrightarrow{(-v)^{\twoheadrightarrow}_{[b,\lambda]}} K/M \xrightarrow{(-v)^{\twoheadrightarrow}_{[\lambda,a]}} A_0/B_0.$$

Now just apply α) and β) above and the proof of the theorem is complete. □

2. Definition and first properties of cellular gradients

Let V be a closed m-dimensional manifold and $\{V^k\}_{0 \leqslant k \leqslant m}$ be a Morse-Smale filtration of V associated to an ordered Morse function ϕ with the ordering sequence $\{a_k\}$. The Morse-Smale filtration of V corresponding to the function $-\phi$ and the ordering sequence $b_k = -a_{m+1-k}$ will be called the *dual filtration* of V and denoted $\{\widehat{V}^k\}$, so that we have $\widehat{V}^{m-k-1} = \overline{V \setminus V^k}$.

Definition 2.1. Let $f : W \to [a, b]$ be a Morse function on a cobordism W and v an almost transverse f-gradient. We say that v satisfies condition $(\mathfrak{C})$ if there is a Morse-Smale filtration $\{\partial_1 W^k\}$ of $\partial_1 W$ and a Morse-Smale filtration $\{\partial_0 W^k\}$ of $\partial_0 W$ such that

$$(\mathfrak{C}1) \qquad (-v)^{\leadsto}(\partial_1 W^k) \subset \operatorname{Int} \partial_0 W^k \supset D(\mathrm{ind}{\leqslant}k{+}1, v) \cap \partial_0 W,$$

$$(\mathfrak{C}2) \qquad \overset{\leadsto}{v}\left(\widehat{\partial_0 W}^k\right) \subset \operatorname{Int} \widehat{\partial_1 W}^k \supset D(\mathrm{ind}{\leqslant}k{+}1, -v) \cap \partial_1 W.$$

The gradients satisfying condition $(\mathfrak{C})$ will be also called *cellular gradients*, or $\mathfrak{C}$-gradients. The set of all cellular gradients of f will be denoted by $G_C(f)$.

The successive terms of a Morse-Smale filtration can be considered as thickenings of the skeleta of the corresponding Morse stratification. Observe an analogy between the gradient descent map associated to a cellular gradient and a cellular map between CW complexes: $(\mathfrak{C}1)$ implies that the thickened k-th skeleton of $\partial_1 W$ is carried by the gradient descent map to the thickened k-th skeleton of $\partial_0 W$, similarly for $(\mathfrak{C}2)$.

The existence of $\mathfrak{C}$-gradients for every given Morse function f will be established in the next chapter, where we prove that the property ($\mathfrak{C}$) is C^0-generic, that is, the set $G_C(f)$ is open and dense in $G(f)$ with respect to C^0-topology. And now let us discuss the basic properties of the gradient descent map associated with a cellular gradient.

2.1. Cellular gradient descent maps. We shall now associate to each cellular gradient a graded-continuous map, derived from the gradient descent. Let $f : W \to [a, b]$ be a Morse function on a cobordism W, and v be an f-gradient satisfying the condition ($\mathfrak{C}$). Define a map

$$(-v)^{\twoheadrightarrow} : \partial_1 W^k/\partial_1 W^{k-1} \to \partial_0 W^k/\partial_0 W^{k-1}$$

between the successive quotients of the corresponding Morse-Smale filtrations as follows. Let

$$\omega_1 \in \partial_1 W^k/\partial_1 W^{k-1}, \quad \omega_0 \in \partial_0 W^k/\partial_0 W^{k-1}$$

be the points corresponding to the collapsed subsets $\partial_1 W^{k-1}$, respectively $\partial_0 W^{k-1}$. For a point $x \in \partial_1 W^k$ we denote by $\overline{x}$ its image in the quotient space $\partial_1 W^k/\partial_1 W^{k-1}$; similar notation will be used for the points of $\partial_0 W^k$. For $x \in \partial_1 W^k$ put

$$\begin{aligned} (-v)^{\twoheadrightarrow}(\bar{x}) &= \omega_0 && \text{if} \quad \bar{x} = \omega_1 \quad \text{or} \quad x \in D(-v), \\ (-v)^{\twoheadrightarrow}(\bar{x}) &= \overline{(-v)^{\rightsquigarrow}(x)} && \text{if} \quad x \notin \partial_1 W^{k-1} \cup D(-v). \end{aligned}$$

Intuitively, to obtain the image of a point $x \in \partial_0 W^k$ we apply the map $(-v)^{\rightsquigarrow}$; when it is not possible we just send x to the point ω_0.

This definition is a particular case of Definition 1.4 from Section 1 (page 234). Indeed, let Σ denote the set of all critical points of f of indices $\leqslant k$. Condition ($\mathfrak{C}2$) implies that only the ascending discs $D(p, -v)$ with $p \in \Sigma$ can intersect $\partial_1 W^k$. Put

$$A_1 = \partial_1 W^k, \ B_1 = \partial_1 W^{k-1}, \tag{3}$$

$$A_0 = \partial_0 W^k, \ B_0 = \partial_0 W^{k-1}, \tag{4}$$

and observe that the condition ($\mathcal{D}$) from Definition 1.3 holds for the pairs (A_1, B_1) and (A_0, B_0). Proposition 1.5 implies the following:

Proposition 2.2. *The map*

$$(-v)^{\twoheadrightarrow} : \partial_1 W^k/\partial_1 W^{k-1} \to \partial_0 W^k/\partial_0 W^{k-1}$$

is continuous. □

Let us recall the main geometric reason for the continuity of this map. The map $(-v)^{\rightsquigarrow}$ being continuous, only the points in $D(-v)$ could make a problem. If $x \in D(p, -v)$ for some critical point $p \in S(f)$, then the condition ($\mathfrak{C}2$) implies that $\operatorname{ind} p \leqslant k$. Therefore the sole of the descending disc $D(p, v)$ is in $\partial_0 W^{k-1}$ (by condition ($\mathfrak{C}1$)). Observe that in the space $\partial_0 W^k / \partial_0 W^{k-1}$ all the subspace $\partial_0 W^{k-1}$ is collapsed to one point, and this leads to the continuity of the map $(-v)^{\twoheadrightarrow}$.

2.2. An example. In this subsection we give an example of a cellular gradient. The cobordism W in question is the "pants" manifold (see page 42), which is diffeomorphic to a 2-dimensional closed disc, with two smaller open discs removed. Embed this cobordism in the Euclidean plane as it is shown in Figure 30 where W is shaded. The set $\partial_1 W$ (the interior boundary of the shaded domain) is the disjoint union of two circles. The set $\partial_0 W$ (the exterior boundary) is an ellipse. The function f is the usual height function, see page 42. It has one critical point p of index 1 (the center of the ellipse). The two descending discs of v are shown by the vertical segments, and the two ascending discs are shown by the horizontal segments.† The Morse-Smale filtration of $\partial_1 W$, required by the condition ($\mathfrak{C}$), arises from the Morse function whose restriction to each of the circles is the standard height function on the circle with two critical points. The critical points of index 1 are depicted by small white circles, and the points of index 0 by small black circles. The set $\partial_1 W^0$ is the union of two black half-circles, and the set $\widehat{\partial_1 W}^0 = \overline{\partial_1 W \setminus \partial_1 W^0}$ is the union of two white half-circles. As for the lower component $\partial_0 W$, the corresponding Morse function has four critical points: two maxima and two minima. Four curves with arrows on the picture show four flow lines of $(-v)$; it is clear from the picture that

$$(-v)^{\rightsquigarrow}(\partial_1 W^0) \subset \operatorname{Int} \partial_0 W^0 \supset D(p, -v) \cap \partial_0 W.$$

† The arrows show the direction of the vector field $(-v)$.

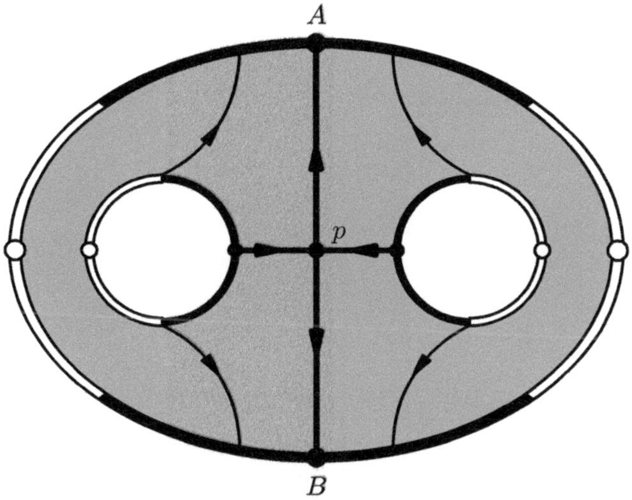

FIGURE 30.

The set $\partial_1 W^1/\partial_1 W^0$ is homeomorphic to the wedge of two circles $S_l \vee S_r$ corresponding to the components of $\partial_1 W$. The set $\partial_0 W^1/\partial_0 W^0$ is also homeomorphic to the wedge of two circles $S_l' \vee S_r'$ corresponding to the left and the right white half-circles on the picture. The map $(-v)^{\twoheadrightarrow}$ sends S_l to S_l', and this map is of degree 1. Similarly, $(-v)^{\twoheadrightarrow}(S_r) = S_r'$ and this map is of degree 1.

This picture illustrates the general principle of definition of the map $(-v)^{\twoheadrightarrow}$. Let us see what happens with the left component of $\partial_1 W$ (the left circle on the figure), while it descends to $\partial_0 W$ along the gradient flow lines. This circle is cut by the ascending disc of p, and its image in $\partial_0 W$ with respect to $(-v)^{\leadsto}$ is diffeomorphic to an open interval. The ends of this interval (the fat black point A on the top and the fat black point B in the bottom) belong to the set $\partial_0 W^0$, and after collapsing $\partial_0 W^0$ to a point, A and B are identified, and the map $(-v)^{\twoheadrightarrow}$ is continuous.

CHAPTER 8

Cellular gradients are C^0-generic

1. Introduction

Let $f : W \to [a,b]$ be a Morse function on a cobordism W; we denote by $G(f)$ the set of all f-gradients. As usual the C^0 norm in the space of all vector fields is denoted by $|| \cdot ||$. The next result will be called the *$\mathfrak{C}$-approximation theorem*; it is one of the central topics of the book.

Theorem 1.1. *The subset $G_C(f)$ of all cellular f-gradients is open and dense in $G(f)$ with respect to C^0 topology.*

Moreover, let u be any almost transverse f-gradient, $\epsilon > 0$, and U be a neighbourhood of ∂W. Then there is $v \in G_C(f)$, such that

$$||v - u|| \leqslant \epsilon \quad \textit{and} \quad \text{supp}\,(v-u) \subset U \setminus \partial W.$$

Before embarking on the proof which will occupy the rest of the present chapter, let us deduce some corollaries. We shall often work with sets which are open and dense with respect to C^0 topology, and it is convenient to reserve a special term for this property.

Definition 1.2. Let $\xi : E \to W$ be a C^∞ Riemannian vector bundle over a cobordism W. Let $\mathcal{E}$ denote the vector space of all the C^∞ sections of ξ, and let $X \subset \mathcal{E}$. A subset $A \subset \mathcal{E}$ is called *C^0-generic* in X if $A \cap X$ is open and dense in X with respect to C^0 topology. A property $\mathcal{A}$ is called *C^0-generic* in X if the set of all elements of X with the property $\mathcal{A}$ is C^0-generic in X.

Theorem 1.1 above says that the condition ($\mathfrak{C}$) is C^0-generic in $G(f)$. Observe that by definition every cellular gradient is almost transverse, therefore the property ($\mathfrak{C}$) is C^0-generic in the set $G_A(f)$ of all almost transverse f-gradients. The case of transverse gradients is only slightly more difficult:

Corollary 1.3. *The property ($\mathfrak{C}$) is C^0-generic in the set $G_T(f)$ of all transverse f-gradients.*

Proof. The openness is guaranteed by Theorem 1.1. Let us prove that

$$G_C(f) \cap G_T(f)$$

is C^0-dense in $G_T(f)$. Let $\delta > 0$, and let $u \in G_T(f)$. Pick a cellular f-gradient w with $||w - u|| < \delta/2$. The set $G_C(f)$ being C^0-open, there is $\epsilon > 0$ such that every f-gradient v with $||w - v|| < \epsilon$ is in $G_C(f)$. Now pick a transverse f-gradient v with $||w - v|| < \min(\epsilon, \delta/2)$ (see Corollary 3.32 of Chapter 4 (page 147)), then $||v - u|| < \delta$, and the proof is finished. □

2. The stratified gradient descent

In this subsection we begin our preparations for the proof of Theorem 1.1. Let $f : W \to [a, b]$ be a Morse function on a cobordism W and v an f-gradient. Recall the transport map

$$(-v)^{\rightsquigarrow} : \partial_1 W \setminus D(-v) \xrightarrow{\approx} \partial_0 W \setminus D(v)$$

(see Section 3 of Chapter 3, page 99). For a compact submanifold $L \subset \partial_1 W$ the image $(-v)^{\rightsquigarrow}(L \setminus D(-v))$ is a submanifold of $\partial_0 W$ which is not compact in general. To compactify it we can add to it the soles of the descending discs of v. The resulting object is not a submanifold any longer, but can be considered as a stratified submanifold. Thus one of the possible ways of regularizing the gradient descent is to regard it as a map between stratified manifolds. This approach to the gradient descent is the aim of the present chapter. Our notion of a stratified submanifold differs from the classical notion of Whitney-stratified submanifolds. The objects which we introduce and study in the next subsection (we call them s-submanifolds) are simpler to define and carry less structure than the Whitney-stratified spaces.

2.1. s-submanifolds. Let $\mathbb{A} = \{A_0, \ldots, A_k\}$ be a finite sequence of subsets of a topological space X. We denote A_s also by $\mathbb{A}_{(s)}$, and we denote $A_0 \cup \cdots \cup A_s$ by $A_{\leqslant s}$ and also by $\mathbb{A}_{(\leqslant s)}$. We say that $\mathbb{A}$ is a *compact family* if $\mathbb{A}_{(\leqslant s)}$ is compact for every s.

Definition 2.1. Let M be a manifold without boundary. A finite sequence $\mathbb{A} = \{A_0, \ldots, A_k\}$ of subsets of M is called an *s-submanifold of M* (s for *stratified*) if:

(1) The sets $A_0, \ldots, A_k$ are pairwise disjoint and each A_i is a submanifold of M of dimension i.
(2) $\mathbb{A}$ is a compact family.

A compact C^∞ submanifold $A \subset M$ can be regarded as an s-submanifold $(A_0, \ldots, A_k)$ where $k = \dim A$ and $A_i = \emptyset$ for $i \neq k$, and $A_k = A$. A compact ∂-submanifold also gives rise to an s-submanifold.

For an s-submanifold $\mathbb{A} = \{A_0, \ldots, A_k\}$ the largest s such that $A_s \neq \emptyset$ is called the *dimension* of $\mathbb{A}$, and denoted by $\dim \mathbb{A}$. For a diffeomorphism

$\Phi : M \to N$ and an s-submanifold $\mathbb{A}$ of M we denote by $\Phi(\mathbb{A})$ the s-submanifold of N defined by $\Phi(\mathbb{A})_{(i)} = \Phi(\mathbb{A}_{(i)})$.

Here is a basic example of an s-submanifold. Let $f : W \to [a, b]$ be a Morse function on a cobordism W, v be an f-gradient. Recall the notation

$$D(\mathrm{ind}{=}i\ ; v) = \bigcup_{\mathrm{ind}\, p=i} D(p, v).$$

Define a family of subsets of W:

$$\mathbb{D}(v) = \{D(\mathrm{ind}{=}i\ ; v)\}_{0 \leqslant i \leqslant \dim W}.$$

For $\lambda \in [a, b]$ define a family of subsets of $f^{-1}(\lambda)$:

$$\mathbb{D}_\lambda(v) = \{D(\mathrm{ind}{=}i{+}1\ ; v) \cap f^{-1}(\lambda)\}_{0 \leqslant i \leqslant \dim W - 1}.$$

Lemma 2.2. *Assume that v satisfies the almost transversality condition. Then:*

(1) *$\mathbb{D}(v)$ is a compact family.*
(2) *For every regular value λ of f the family $\mathbb{D}_\lambda(v)$ is an s-submanifold of $f^{-1}(\lambda)$.*
(3) *If $\partial W = \emptyset$, then $\mathbb{D}(v)$ is an s-submanifold of W, transverse to $f^{-1}(\lambda)$ for every regular value λ of f.*

Proof. Follows immediately from Proposition 3.43 of Chapter 4 (page 151). □

Now we proceed to the transversality theory for s-submanifolds.

Definition 2.3. Let $\mathbb{A}, \mathbb{B}$ be s-submanifolds of M.

(1) We say, that $\mathbb{A}$ is *transverse to* $\mathbb{B}$ (notation: $\mathbb{A} \pitchfork \mathbb{B}$) if $\mathbb{A}_{(i)} \pitchfork \mathbb{B}_{(j)}$ for every i, j.
(2) We say that $\mathbb{A}$ is *almost transverse to* $\mathbb{B}$ (notation: $\mathbb{A} \nmid \mathbb{B}$) if $\mathbb{A}_{(i)} \pitchfork \mathbb{B}_{(j)}$ for $i + j < \dim M$.

If V is a compact submanifold of M, transverse to an s-submanifold $\mathbb{A}$, then the family $\{\mathbb{A}_{(i)} \cap V\}$ is an s-submanifold of V which will be denoted by $\mathbb{A} \cap V$.

Remark 2.4. Note, that $\mathbb{A} \nmid \mathbb{B}$ if and only if $\mathbb{A}_{(\leqslant i)} \cap \mathbb{B}_{(\leqslant j)} = \emptyset$ whenever $i + j < \dim M$.

R. Thom's Transversality Theory implies that two compact submanifolds of a given manifold can be made transverse by a small perturbation of one of them.

We are going to prove that two s-submanifolds of a given manifold can be made almost transverse by a small perturbation of one of them. The perturbations with which we shall be working, are the isotopies defined by time-dependent vector fields (see Subsection 1.3 of Chapter 4, page 119).

Let I, I' be a pair of intervals of $\mathbf{R}$ such that $\overline{I'} \subset \operatorname{Int} I$. Recall that for a compact subset $K \subset M$ we denote by $\operatorname{Vect}^\infty(M \times I; K, I')$ the subset of $\operatorname{Vect}^\infty(M \times I)$ formed by all vector fields v such that:

(1) supp $(v - v_0)$ is a compact subset of $K \times I'$, where $v_0 = (0, 1)$,
(2) the second coordinate of v is equal to 1 everywhere.

To each vector field $v \in \operatorname{Vect}^\infty(M \times I; K, I')$ we associated the *shift diffeomorphism* $\mathscr{T}_v : M \to M$ (see page 120).

Proposition 2.5. *Let $\mathbb{A}, \mathbb{B}$ be s-submanifolds of a closed manifold M. Let I, I' be intervals satisfying $\overline{I'} \subset \operatorname{Int} I$. Put $\mathcal{V} = \operatorname{Vect}^\infty(M \times I; M, I')$. Then the subset of vector fields $v \in \mathcal{V}$, such that*

$$\mathscr{T}_v(\mathbb{A}) \pitchfork \mathbb{B}$$

is open in $\mathcal{V}$ with respect to C^0-topology and dense in $\mathcal{V}$ with respect to C^∞ topology.

Proof. C^0-openness follows from the next lemma.

Lemma 2.6. *Let $X, Y \subset M$ be two compact sets. Then the set of all $v \in \mathcal{V}$ such that*

$$\mathscr{T}_v(X) \cap Y = \varnothing$$

is an open subset in $\mathcal{V}$ with respect to C^0-topology.

Proof. The diffeomorphism $\mathscr{T}_v : M \to M$ can be considered as the exit function for the vector field v on the cobordism $W = M \times I$. Now the result is an immediate consequence of Proposition 3.5 of Chapter 3 (page 102). □

Let us proceed to the density property. Let A_k be one of the submanifolds in the family $\mathbb{A}$. Let us call a *patch* for A_k a chart

$$\Phi : U \to B^k(0, 3R) \subset \mathbf{R}^k$$

for A_k such that the normal bundle to A_k restricted to U is trivial. For a given patch Φ let us denote by Δ_Φ the set $\Phi^{-1}(D^k(0, R))$. The next lemma asserts that by a small perturbation of v we can put Δ_Φ in a general position with respect to $\mathbb{B}$.

Lemma 2.7. *Let $\Phi : U \to B^k(0, 3R) \subset \mathbf{R}^k$ be a patch for A_k. Let $\mathcal{L}_\Phi$ be the set of all $v \in \mathcal{V}$ such that*

$$\mathscr{T}_v(\Delta_\Phi) \cap B_j = \emptyset$$

for every $B_j \in \mathbb{B}$ with $j + k < \dim M$. Then $\mathcal{L}_\Phi$ is C^0-open and C^∞-dense in $\mathcal{V}$.

Before proving the lemma we shall deduce from it our theorem. Choose a countable family of patches $\Phi_i : U_i \to \mathbf{R}^k$ (each for its own A_k) such that the sets $\Phi_i^{-1}(B(0,R))$ form a covering of $\cup_i A_i$. This is possible since each A_i is a topological space with countable base. For each patch Φ_i the set $\mathcal{L}_{\Phi_i}$ is open and dense in $\mathcal{V}$ with respect to C^∞ topology, therefore by the Baire theorem the intersection of these sets is dense in $\mathcal{V}$.

Proof of Lemma 2.7. Lemma 2.6 implies that $\mathcal{L}_\Phi$ is C^0-open. As for the C^∞ density, it suffices to prove that the vector field $v_0 = (0,1)$ is in the closure of $\mathcal{L}_\Phi$. This last assertion is easily reduced to the proof of the following lemma about submanifolds and vector fields in the Euclidean space.

Lemma 2.8. *Let*

$$N = B^k(0,3R) \times \mathbf{R}^{m-k} \quad \text{and} \quad Q = D^k(0,2R) \times 0 \subset N.$$

Let $S \subset N$ be any submanifold of dimension $< m-k$ and

$$\mathcal{U} \subset \mathrm{Vect}^\infty(N \times I; Q, I')$$

be a neighbourhood of $v_0 = (0,1)$. Then there is $w \in \mathcal{U}$ such that

$$\mathcal{T}_w(D^k(0,R)) \cap S = \varnothing.$$

Proof. Let $S' \subset \mathbf{R}^{m-k}$ be the projection of S to the second component. We shall construct a vector field $w \in \mathcal{U}$ and a vector $h \in \mathbf{R}^{m-k} \setminus S'$ such that

$$\mathcal{T}_w(D^k(0,R)) = D^k(0,R) + h,$$

then the sets $\mathcal{T}_w(D^k(0,R))$ and S are obviously disjoint.

Pick any numbers $0 < \rho < r$ and recall that by Lemma 1.8 of Chapter 4 (page 120) there is a continuous map

$$\sigma : B^{m-k}(0,\rho/2) \to \mathrm{Vect}^\infty\Big(B^{m-k}(0,r) \times I; D^{m-k}(0,\rho), I'\Big)$$

such that $\mathcal{T}_{\sigma(h)}(0) = h$. Pick a smooth function λ on $B^k(0,3R)$ such that $\lambda(y) = 1$ for $|y| \leqslant R$ and $\lambda(y) = 0$ for $|y| \geqslant 2R$. Then the time-dependent vector field

$$w_h(y,x,t) = \big(0, \lambda(y)\sigma(h)(x,t)\big), \quad \text{where} \quad y \in B^k(0,3R),\ x \in \mathbf{R}^{m-k},\ t \in I$$

satisfies

$$\mathcal{T}_w(D^k(0,R)) = D^k(0,R) + h.$$

For a sufficiently small vector h the vector field w_h is in the neighbourhood $\mathcal{U}$. It remains to observe that by the Sard Lemma there are arbitrarily small vectors $h \in \mathbf{R}^{m-k} \setminus S'$, and the proof of our lemma is complete. □

2.2. Tracks of s-submanifolds. Let $f : W \to [a, b]$ be a Morse function on a cobordism W and v an f-gradient. Recall the definition of the track of a subset $X \subset \partial_0 W$:

$$T(X, -v) = \{\gamma(x, t; -v) \mid x \in X, \quad t \geqslant 0\}.$$

In this subsection we extend the notion of track to the category of s-submanifolds. Let $\mathbb{A} = \{A_0, A_1, \ldots, A_k\}$ be an s-submanifold of $\partial_1 W$, and v an almost transverse f-gradient.

Definition 2.9. For $0 \leqslant i \leqslant k+1$ set

$$TA_i(-v) = T(A_{i-1}, -v) \cup D(\mathrm{ind}{=}i\ ; v). \qquad \dagger$$

The family

$$\mathbb{T}(\mathbb{A}, -v) = \{TA_i(-v)\}_{0 \leqslant i \leqslant k+1}$$

is called the *track of* $\mathbb{A}$ *with respect to* $(-v)$.

Lemma 2.10. *If*

(1) $$\mathbb{A} \nmid \mathbb{D}_b(-v),$$

then $\mathbb{T}(\mathbb{A}, -v)$ *is a compact family.*

Proof. Our task is to prove that the set

$$Q_s = \bigcup_{0 \leqslant i \leqslant s} TA_i(-v)$$

is compact for every s. Let $\phi : W \to [a, b]$ be an ordered Morse function, such that v is a ϕ-gradient, and let $(a_0, \ldots, a_{m+1})$ be the ordering sequence for ϕ (here $m = \dim W$). Let

$$W' = \phi^{-1}\big([a, a_s]\big), \quad W'' = \phi^{-1}\big([a_s, b]\big).$$

Then all the critical points of ϕ of indices $< s$ are in W' and those of indices $\geqslant s$ are in W''. The condition (1) implies that

$$A_{\leqslant s-1} \cap D(\mathrm{ind}{\leqslant}m{-}s\ ; -v) = \varnothing.$$

Thus every $(-v)$-trajectory, starting at a point of $A_{\leqslant s-1}$ reaches $\partial_0 W'' = \phi^{-1}(a_s)$. Then

$$Q_s \cap W'' = T(A_{\leqslant s-1}, -v) \cap W''$$

is compact by Lemma 3.2 of Chapter 3 (page 100). Let

$$K_s = T(A_{\leqslant s-1}, -v) \cap \partial_1 W'.$$

Then

$$Q_s \cap W' = T(K_s, -v) \cup D(v|W'),$$

and this set is compact by one more application of the same lemma. □

† For notational convenience we set $A_{-1} = \varnothing$.

In general the subsets of W, which form the family $\mathbb{T}(\mathbb{A}, -v)$ are manifolds with non-empty boundary. For example, if v has no zeros, then $\mathbb{T}(\mathbb{A}, -v)$ is homeomorphic to $\mathbb{A} \times [0,1]$.

Definition 2.11. The family

$$(-v)^{\rightsquigarrow}(\mathbb{A}) = \{TA_{i+1}(-v) \cap \partial_0 W)\}_{0 \leqslant i \leqslant k}$$

is called the $(-v)^{\rightsquigarrow}$-*image of* $\mathbb{A}$.

The next lemma is obvious.

Lemma 2.12. *The family* $(-v)^{\rightsquigarrow}(\mathbb{A})$ *is an* s*-submanifold of* $\partial_0 W$.

Similarly, for every regular value λ we have an s-submanifold

$$(-v)^{\rightsquigarrow}_{[b,\lambda]}(\mathbb{A}) = \{TA_{i+1}(-v) \cap f^{-1}(\lambda)\}_{0 \leqslant i \leqslant k}$$

of $f^{-1}(\lambda)$.

Now let us return to the condition (1) which is required for the definition of $\mathbb{T}(\mathbb{A}, -v)$. It turns out that this condition can always be achieved by a small perturbation of v. This will be proved in Subsection 2.4 after we develop the necessary techniques in Subsection 2.3.

2.3. Horizontal drift constructions. Let $f : W \to [a,b]$ be a Morse function on a cobordism W, and v be an f-gradient. Let

$$\overset{\rightsquigarrow}{v} : \partial_0 W \setminus D(v) \to \partial_1 W \setminus D(-v)$$

denote the transport map. Let ξ be a C^∞ vector field on $\partial_1 W$, and let $\Phi(\xi, 1)$ be the corresponding diffeomorphism of $\partial_1 W$ to itself. We will modify v in a neighbourhood of $\partial_1 W$ in such a way that the transport map for the resulting gradient w satisfies:

$$\overset{\rightsquigarrow}{w} = \Phi(\xi, 1) \circ \overset{\rightsquigarrow}{v}.$$

The construction is geometrically quite obvious. In a small neighbourhood of $\partial_1 W$ we add to the vector field v a component "parallel to ξ" so that the trajectories of the new vector field w will drift in the horizontal direction along the vector field ξ while approaching $\partial_1 W$.

The aim of the present subsection is to describe this construction in detail; the results will be used many times in the sequel.

A. Horizontal drift construction nearby $\partial_1 W$.

We shall impose an additional restriction on the pair (f, v), which makes the constructions easier.

Definition 2.13. Let $f : W \to [a, b]$ be a Morse function on a cobordism W and v an f-gradient. We say that the pair (f, v) has *collars of height* α if $a + \alpha < b - \alpha$ and

$$f'(x)(v(x)) = 1 \quad \text{for every} \quad x \in f^{-1}\Big([a, a+\alpha] \cup [b-\alpha, b]\Big).$$

Not every pair (f, v) has collars, but it is always possible to find a vector field w, equivalent to v in such a way that the pair (f, w) has collars of some height $\alpha > 0$. If (f, v) has collars of height α, then we have C^∞ maps

$$\Phi_1 : \partial_1 W \times [-\alpha, 0] \to W; \qquad \Phi_1(x, t) = \gamma(x, t; v),$$
$$\Phi_0 : \partial_0 W \times [0, \alpha] \to W; \qquad \Phi_0(x, t) = \gamma(x, t; v)$$

(so that $\partial_0 W$ and $\partial_1 W$ have collars of height α with respect to v, see Definition 2.5 of Chapter 1, page 25).

Proposition 2.14. *Assume that the pair* (f, v) *has collars of height* α*. Let* $I = [-\alpha, 0]$*, and choose any closed interval* $I' \subset \operatorname{Int} I$*. Then for every time-dependent vector field*

$$\eta \in \operatorname{Vect}^\infty\Big(\partial_1 W \times I;\ \partial_1 W,\ I'\Big)$$

on $\partial_1 W$ *there exists an* f*-gradient* Z_η *such that:*

(1) $\operatorname{supp}(Z_\eta - v) \subset \Phi_1(\partial_1 W \times I')$.
(2) $\overset{\leadsto}{Z_\eta} = \mathscr{T}_\eta \circ \overset{\leadsto}{v}$, *where* $\mathscr{T}_\eta$ *is the shift diffeomorphism associated with* η.
(3) *The map*

$$\eta \mapsto Z_\eta; \quad \operatorname{Vect}^\infty\Big(\partial_1 W \times I;\ \partial_1 W,\ I'\Big) \to G(f)$$

is continuous (with respect to the C^∞ *topologies), and* $Z_0 = v$.

Proof. Define a vector field Z_η on W by

$$Z_\eta(y) = v(y) \qquad \text{if} \quad y \notin \operatorname{Im} \Phi_1,$$
$$Z_\eta(y) = \big((\Phi_1)_* \eta\big)(y) \qquad \text{if} \quad y \in \operatorname{Im} \Phi_1.$$

Observe that Z_η is again an f-gradient, since the vector field $Z_\eta - v$ is tangent to the level surfaces of f. It is clear that

$$\overset{\leadsto}{Z_\eta} = \mathscr{T}_\eta \circ \overset{\leadsto}{v};$$

the property (3) is also obvious. □

We shall use in the sequel a particular case of this construction.

Definition 2.15. Assume that W is endowed with a Riemannian metric. We shall say that the pair (f, v) has *framed collars of height* α if (f, v) has collars of height α and the maps Φ_0, Φ_1 from Definition 2.13 are isometries when the manifolds $\partial_1 W \times [-\alpha, 0]$, $\partial_0 W \times [0, \alpha]$ are endowed with the product metric.

Proposition 2.16. *Assume that (f, v) has framed collars of height α. For every $\xi \in \mathrm{Vect}^\infty(\partial_1 W)$ there is an f-gradient v_ξ such that:*

(1) $(v_\xi)^{\rightsquigarrow} = \Phi(\xi, 1) \circ \overset{\rightsquigarrow}{v}$.
(2) $\mathrm{supp}\,(v_\xi - v) \subset \mathrm{Int}\ \mathrm{Im}\ \Phi_1 \setminus \partial_1 W$.
(3) $||v_\xi - v|| \leqslant \frac{3}{\alpha}||\xi||$.
(4) *The map*

$$\xi \mapsto v_\xi; \quad \mathrm{Vect}^\infty(\partial_1 W) \to G(f)$$

is continuous in C^∞ topology and also in C^0 topology.

Proof. Pick a C^∞ function $h : [0, 1] \to \mathbf{R}_+$ satisfying the following conditions:

A) the support of h is in $]\frac{1}{4}, \frac{3}{4}[$;
B) h is constant on the interval $]\frac{1}{3}, \frac{2}{3}[$, where it takes its maximal value H;
C) the integral of h is equal to 1.

The maximal value H of the function h is then between 2 and 3. Define a vector field

$$\eta \in \mathrm{Vect}\Big(\partial_1 W \times I,\ \partial_1 W,\ I'\Big), \quad \text{with} \quad I = [-\alpha, 0],\ I' = \left[-\frac{3}{4}\alpha, -\frac{1}{4}\alpha\right]$$

as follows:

$$\eta(x, t) = v_0(x, t) + h_\alpha(t) \cdot \xi(x), \quad \text{where}$$
$$v_0(x, t) = (0, 1) \quad \text{and} \quad h_\alpha(t) = \frac{1}{\alpha} h(-t/\alpha)$$

(the graph of the function $h_\alpha(t)$ is depicted in the next picture).

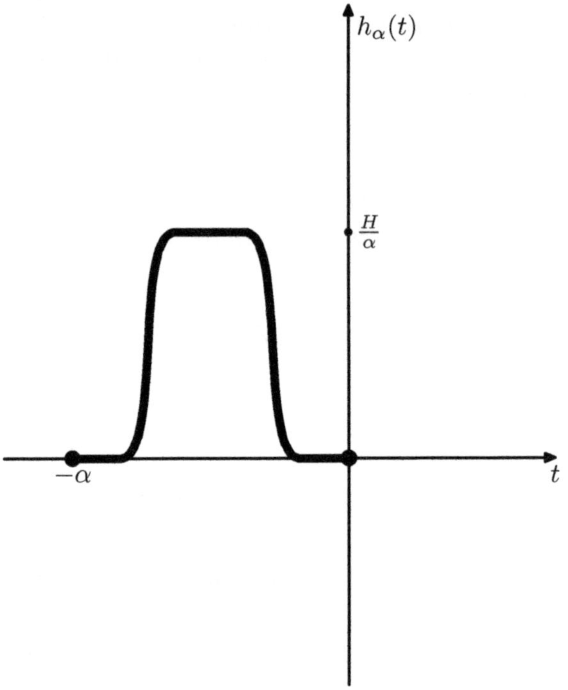

Observe that the shift diffeomorphism $\mathscr{T}_\eta$ associated to η equals $\Phi(\xi, 1)$. Now define the vector field v_ξ as the result of the horizontal drift construction applied to v and η, that is, put

$$v_\xi = Z_\eta,$$

where Z_η was defined during the proof of Proposition 2.14. The properties (1), (2) and (4) of the statement of our proposition follow from Proposition 2.14. As for (3), recall that Φ_1 is an isometry, and write

$$||v - v_\xi|| = \frac{1}{\alpha} \cdot H \cdot ||\xi|| \leqslant \frac{3}{\alpha} \cdot ||\xi||. \qquad \square$$

B. Horizontal drift construction with respect to both components of the boundary.

A similar construction can be applied to the lower component of the boundary $\partial_0 W$. We can also apply simultaneously the two constructions nearby both components of the boundary. The properties of the resulting gradient are described in the proposition below (we work with the gradient descent $(-v)^{\rightsquigarrow}$ as it is necessary for subsequent applications.)

Proposition 2.17. *Assume that (f, v) has framed collars of height α. Let*

$$\xi_0 \in \mathrm{Vect}^\infty(\partial_0 W), \quad \xi_1 \in \mathrm{Vect}^\infty(\partial_1 W).$$

There is an f-gradient $w = w_{\xi_0,\xi_1}$ such that:

(1) $(-w)^{\leadsto} = \Phi(-\xi_0, 1) \circ (-v)^{\leadsto} \circ \Phi(-\xi_1, 1)$.

(2) $\mathrm{supp}\,(w - v) \subset \mathrm{Int}\,\big(\mathrm{Im}\,\Phi_0 \cup \mathrm{Im}\,\Phi_1\big) \setminus \partial_0 W$.

(3) $||w - v|| \leqslant \frac{3}{\alpha} \cdot \max\big(||\xi_0||, ||\xi_1||\big)$.

(4) *The map*

$$(\xi_0, \xi_1) \mapsto w_{\xi_0,\xi_1}; \quad \mathrm{Vect}^\infty(\partial_0 W) \times \mathrm{Vect}^\infty(\partial_1 W) \to G(f)$$

is continuous in C^∞ topology and also in C^0 topology. □

2.4. A lemma on small perturbations. Now we can prove the property mentioned in the end of Subsection 2.2.

Proposition 2.18. *Let $f : W \to [a, b]$ be a Morse function on a cobordism W and v an almost transverse f-gradient. Let $\mathbb{A}$ be an s-submanifold of $\partial_1 W$, $\mathbb{B}$ be an s-submanifold of $\partial_0 W$. Let U be a neighbourhood of ∂W, and $\mathcal{W}$ be a neighbourhood of v in the space $G(f)$ (endowed with C^∞ topology). Then there is an almost transverse f-gradient $w \in \mathcal{W}$, such that:*

(1) $\mathbb{D}_b(-w) \pitchfork \mathbb{A}, \quad \mathbb{D}_a(w) \pitchfork \mathbb{B}$;

(2) $(-w)^{\leadsto}(\mathbb{A}) \pitchfork \mathbb{B}; \quad \overset{\leadsto}{w}(\mathbb{B}) \pitchfork \mathbb{A}$;

(3) $\mathrm{supp}\,(v - w) \subset U \setminus \partial W$.

Proof. Let us first consider the case when the pair (f, v) has collars of height α. Let $C_0 = \mathrm{Im}\,\Phi_0$, $C_1 = \mathrm{Im}\,\Phi_1$ where Φ_1, Φ_0 are defined as follows:

$$\Phi_1 : \partial_1 W \times [-\alpha, 0] \to W; \quad \Phi_1(x, t) = \gamma(x, t; v),$$
$$\Phi_0 : \partial_0 W \times [0, \alpha] \to W, \quad \Phi_0(x, t) = \gamma(x, t; v).$$

Let $I = [-\alpha, 0]$, pick any closed interval $I' \subset \mathrm{Int}\, I$. Choose a time-dependent vector field

$$\eta \in \mathrm{Vect}^\infty(\partial_1 W \times I;\ \partial_1 W,\ I')$$

such that

$$\mathscr{T}_\eta(\mathbb{D}_b(-v)) \pitchfork \mathbb{A}$$

(this is possible by Proposition 2.5, page 246). Applying Proposition 2.14 we obtain an f-gradient $w_0 = Z_\eta$ which satisfies

$$\mathbb{D}_b(-w_0) = \mathscr{T}_\eta\Big(\mathbb{D}_b(-v)\Big) \pitchfork \mathbb{A}, \quad \mathrm{supp}\,(v - w_0) \subset \mathrm{Int}\, C_1 \setminus \partial_1 W.$$

Therefore the $(-w_0)^{\leadsto}$-image of the s-submanifold $\mathbb{A}$ is defined (see Definition 2.11). Choosing the vector field η sufficiently small we can achieve that $w_0 \in \mathcal{W}$ (see point (3) of Proposition 2.14).

Now apply a similar construction to the s-submanifold $(-w_0)^{\leadsto}(\mathbb{A})$, and the s-submanifold $\mathbb{B}$. We obtain a vector field $w \in \mathcal{W}$ such that

$$(-w)^{\leadsto}(\mathbb{A}) \pitchfork \mathbb{B},$$
$$\text{supp}\,(w - w_0) \subset \text{Int}\, C_0 \setminus \partial_0 W.$$

It is easy to deduce that $w^{\leadsto}(\mathbb{B}) \pitchfork \mathbb{A}$. Moreover, $D_b(-w) = D_b(-w_0) \pitchfork \mathbb{A}$. Therefore the vector field w satisfies all the requirements of our proposition. □

Now let us consider the general case when we do not assume that the pair (f, v) has collars. Let f be a Morse function and v an f-gradient. Let ϕ be a strictly positive C^∞ function on W such that supp $(1-\phi)$ is in a small neighbourhood of ∂W, and such that in a neighbourhood of ∂W we have $\phi(x) = f'(x)(v(x))$. The vector field $v_1(x) = \phi(x)^{-1} \cdot v(x)$ is an f-gradient, and the pair (f, v_1) has collars. Choose a vector field w_1 which satisfies the conclusions of the proposition with respect to (f, v_1), and set $w(x) = \phi(x) \cdot w_1(x)$. Then w satisfies (1)–(3) and is close to v in C^∞ topology, if w_1 is sufficiently close to v_1 in C^∞ topology. □

3. Quick flows

Let $f : M \to \mathbf{R}$ be a Morse function on a closed manifold M, and v a transverse f-gradient. The family

$$\mathbb{D}(v) = \{D(\text{ind}{=}i\ ; v)\}$$

is a cellular decomposition of M. The family $\mathbb{D}(-v)$ is a cellular decomposition, which is in a sense dual to $\mathbb{D}(v)$. Recall in particular that the Morse complex $\mathcal{M}_*(-f, -v)$ is dual to $\mathcal{M}_*(f, v)$ (see Theorem 1.16 of Chapter 6, page 206). We begin this section with a discussion of another manifestation of the duality phenomenon. Namely, we prove that for a given k the δ-thickening

$$B_\delta(\text{ind}{\leqslant}k\ ; v)$$

of the k-skeleton of the Morse stratification is homotopy equivalent to the complement

$$M \setminus B_\delta(\text{ind}{\leqslant}m{-}k{-}1\ ; -v)$$

of the δ-thickened $(m-k-1)$-skeleton of the dual Morse stratification (where $m = \dim M$). Observe first of all that for every $\delta > 0$ sufficiently small we have an inclusion

$$(2) \qquad B_\delta(\text{ind}{\leqslant}k\ ; v) \subset M \setminus B_\delta(\text{ind}{\leqslant}m{-}k{-}1\ ; -v) \quad \text{for every} \quad k.$$

Indeed, pick any ordered Morse function $\phi : M \to \mathbf{R}$ adjusted to (f, v), and let M^k and $\widehat{M}^k$ be the Morse-Smale filtrations corresponding to ϕ and, respectively, $-\phi$ (see Subsection 3.8 of Chapter 4, page 157). Every $\delta > 0$ for which we have

$$\begin{aligned} B_\delta(p) &\subset M^k && \text{for} \quad \operatorname{ind} p \leqslant k; \\ B_\delta(q) &\subset \widehat{M}^{m-k-1} && \text{for} \quad \operatorname{ind} q \geqslant k+1 \end{aligned}$$

satisfies (2).

Proposition 3.1. *Let $f : M \to \mathbf{R}$ be a Morse function on a closed m-dimensional manifold M, and v an almost transverse f-gradient. Let $\delta > 0$ be a real number such that (2) holds. Then for every k we have:*

(A) *The inclusion*

$$I : B_\delta(\text{ind}{\leqslant}k\ ; v) \hookrightarrow M \setminus B_\delta(\text{ind}{\leqslant}m-k-1\ ; -v)$$

is a homotopy equivalence.

(B) *For every $T > 0$ sufficiently large we have:*

$$\Phi(-v, T)\big(M \setminus B_\delta(\text{ind}{\leqslant}m-k-1\ ; -v)\big) \subset B_\delta(\text{ind}{\leqslant}k\ ; v)$$

and the restriction of $\Phi(-v, T)$ to $M \setminus B_\delta(\text{ind}{\leqslant}m-k-1\ ; -v)$ is the homotopy inverse for the inclusion I.

Proof. Let us begin with (B). Put

$$B_\delta = \bigcup_{p \in S(f)} B_\delta(p), \qquad \text{and} \qquad K = M \setminus B_\delta.$$

There are no critical points of f in K, therefore there is $C > 0$ such that $f'(x)(v(x)) > C$ for $x \in K$. For any v-trajectory $\gamma(t)$ we have

$$\frac{d}{dt} f(\gamma(t)) = f'(\gamma(t))(\gamma'(t)) = f'(\gamma(t))\big(v(\gamma(t))\big),$$

therefore

$$\frac{d}{dt} f(\gamma(t)) > C \quad \text{if} \quad \gamma(t) \in K.$$

Thus the time during which a v-trajectory can stay in K is not more then $T_0 = (\max f - \min f)/C$. Now let $y \in M \setminus B_\delta(\text{ind}{\leqslant}m-k-1\ ; -v)$. The $(-v)$-trajectory $\gamma(y, \cdot; -v)$ intersects the set B_δ at some moment $t_0 \leqslant T = T_0 + 1$. The point $p \in S(f)$ such that $\gamma(y, t_0; -v) \in B_\delta(p)$ is necessarily of index less than k, since the set $M \setminus B_\delta(\text{ind}{\leqslant}m-k-1\ ; -v)$ is $(-v)$-invariant. The set $B_\delta(\text{ind}{\leqslant}k\ ; v)$ is also $(-v)$-invariant, therefore $\gamma(y, t; -v) \in B_\delta(\text{ind}{\leqslant}k\ ; v)$ for every $t \geqslant T$, and the first point of (B) is proved.

The rest follows by an application of Lemma 1.8 of Chapter 5 (page 171) to the homotopy

$$\Phi(-v, t), \quad t \in [0, T]$$

(observe that the subspaces $B_\delta(\text{ind}{\leqslant}k\ ; v)$ and $M \setminus B_\delta(\text{ind}{\leqslant}m-k-1\ ; -v)$ of M are both invariant with respect to this homotopy). □

The point (B) of the above proposition implies that for a given Morse function $f : M \to \mathbf{R}$ and an almost transverse f-gradient v the downward gradient descent along the flow lines of v will push the complement $M \setminus B_\delta(\mathrm{ind}{\leqslant}m{-}k{-}1\ ; -v)$ to $B_\delta(\mathrm{ind}{\leqslant}k\ ; v)$ if the time T of the descent is sufficiently large. Given $\delta > 0$ one can ask, what is the *infimum* $T(v, \delta)$ of the values of T such that

$$\Phi(-v, T)\Big(M \setminus B_\delta(\mathrm{ind}{\leqslant}m{-}k{-}1\ ; -v)\Big) \subset B_\delta(\mathrm{ind}{\leqslant}k\ ; v).$$

The number $T(v, \delta)$ is an interesting geometric characteristic of the dynamical system associated to the vector field v. It is clear that for a given vector field v we have

$$T(v, \delta) \to \infty \quad \text{when} \quad \delta \to 0,$$

so that $T(v, \delta)$ tends to be very large when δ is small. The main result of the present section in a sense contradicts this intuition. Namely, we will show that for a given closed Riemannian manifold and given $\epsilon > 0, \delta > 0$ there is a Morse function $f : M \to \mathbf{R}$ and an almost transverse f-gradient v, such that

$$||v|| \leqslant 1, \quad T(v, \delta) \leqslant \epsilon.$$

The critical points of such Morse functions are spread very densely in the manifold, so that a v-trajectory starting at a given point of M hits almost immediately the δ-ball around one of critical points.[†]

Since $T(Cv, \delta) = C^{-1}T(v, \delta)$, we can reformulate this result as follows: for a given closed Riemannian manifold and a given $C > 0, \delta > 0$ there is always a Morse function $f : M \to \mathbf{R}$ and an almost transverse f-gradient v, such that

$$||v|| \leqslant C, \quad T(v, \delta) \leqslant 1.$$

This is the contents of Theorem 3.6 of the present section. These results will be used in Chapter 12.

Here is the outline of the section. Our first step is to show that the properties of the gradient flow discussed above are directly related to the properties of the maximal length of its integral curves (Proposition 3.4). Theorem 3.6 says that on every cobordism for every $C > 0$ there is a Morse function f and its gradient v such that the maximal length of the integral curves of v is less than C. The proof is done in Subsection 3.6 by induction on the dimension of the cobordism. We show that the essential case is that of the product cobordism $V \times [a, b]$ and we study this case in Subsections 3.3–3.5. The main part of the argument is the construction of some special

[†] The Morse stratification corresponding to such a vector field is a Morse-theoretic counterpart of a triangulation whose simplices have small diameter; such triangulations can be obtained by subdividing any given triangulation of a manifold.

Morse function on the cobordism $V \times [a, b]$. Let us give here a brief outline of this construction in the particular case $V = S^1$, which can be easily visualized (see Figure 31).

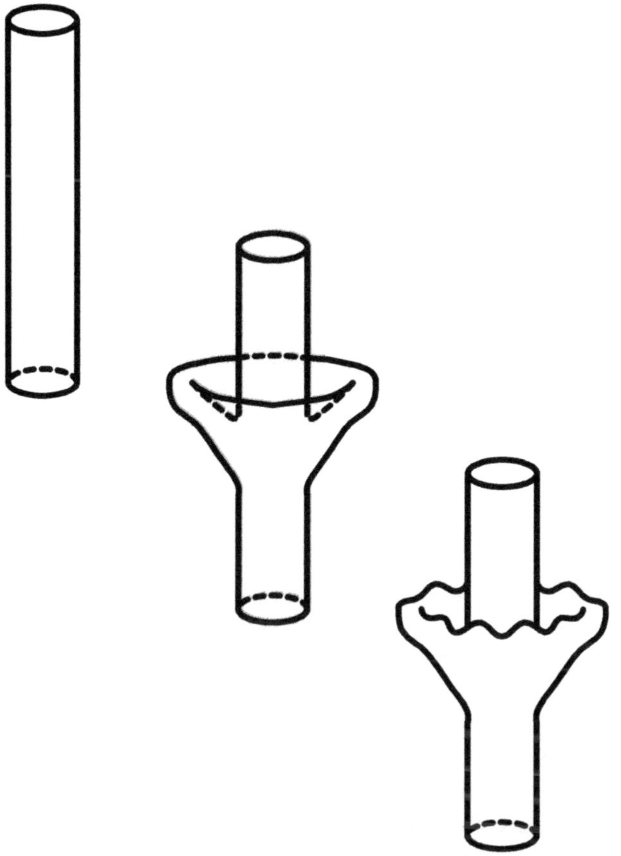

FIGURE 31.

We start with the projection $\phi_0 : V \times [a, b] \to [a, b]$, which is the height function corresponding to the embedding of the cylinder $V \times [a, b]$ to $\mathbf{R}^3$ shown on the left of the figure. Now let us "fold" the function ϕ_0 †) and obtain a function $\phi_1 : V \times [a, b] \to [a, b]$ having two critical submanifolds each diffeomorphic to a circle (see the second picture in the figure). Perturbing ϕ_1 in a neighbourhood of these circles we obtain a Morse function

† The letter "S" in the term "*S*-construction" refers to the Russian word "skladka" which means "fold" in English.

$\phi_2 : V \times [a,b] \to [a,b]$ (see the right hand side of the figure), which is by definition the result of the S-construction, applied to the cobordism $V \times [a,b]$.

The S-construction will be used as follows. Let us start with the ϕ_0-gradient $v_0 = (0,1)$. The integral curves of v_0 are vertical straight lines. It is not difficult to construct from it a ϕ_2-gradient v_2, whose integral curves are shorter than those of v_0. Consider for example a $(-v_2)$-trajectory starting at $\partial_1 W \approx V$. It descends to the valley which it can not leave; in particular, it never reaches $\partial_0 W$.

The basic geometric idea of the proof of Theorem 3.6 for the product cobordisms is as follows. Start with a product cobordism $V \times [a,b]$ and cut it into very thin slices of the form $V \times [a_i, b_i]$. Applying to each of these slices the S-construction, we obtain at the end a Morse function on $V \times [a,b]$ and its gradient, whose integral curves have very small length. And this leads to the proof of our theorem.

3.1. Theorem on the existence of quick flows. Let W be a Riemannian cobordism, $f : W \to [a,b]$ a Morse function, v a gradient-like vector field for f.

Exercise 3.2. Show that every integral curve of v has finite length.

Definition 3.3. We say that v is *C-quick* if the length of every integral curve of v is not greater than C. The minimum of the numbers C such that v is C-quick is called the *quickness constant* of v and denoted $Q(v)$.

The following proposition relates the quickness constant to the properties of the gradient descent discussed in the beginning of this section. Recall that two vector fields v, w are called *equivalent* if $w = \phi \cdot v$ where ϕ is a strictly positive C^∞ function $f : W \to \mathbf{R}$ such that $\phi(x) = 1$ for every x in a neighbourhood of $S(v)$.

Proposition 3.4. *Let $\partial W = \varnothing$ and let v be a C-quick gradient-like vector field for f. Then for every $\delta > 0$ there is a gradient-like vector field w for f such that w is equivalent to v, $||w|| \leqslant C$ and for every k we have:*

(A) $$\Phi(-w,1)(W \setminus B_\delta(\mathrm{ind}_{\leqslant m-k-1}\ ; -w)) \subset B_\delta(\mathrm{ind}_{\leqslant k}\ ; w).$$

(B) $$\Phi(w,1)(W \setminus B_\delta(\mathrm{ind}_{\leqslant m-k-1}\ ; w)) \subset B_\delta(\mathrm{ind}_{\leqslant k}\ ; -w).$$

Proof. Let $B_\delta = \cup_{p \in S(f)} B_\delta(p)$. Pick a C^∞ function $h : W \to \mathbf{R}$ such that

$$h(x) = 1 \quad \text{in a neighbourhood of} \quad S(f),$$
$$||h(x)v(x)|| = C \quad \text{for every} \quad x \notin B_\delta,$$
$$||h(x)v(x)|| \leqslant C \quad \text{for every} \quad x.$$

Put

$$w(x) = h(x)v(x),$$

then the vector field w is equivalent to v, and $||w|| \leqslant C$. Let us prove that the properties (A) and (B) hold for w.

Lemma 3.5. *Let $y \in W$. Then*

(1) *There is $t_0 \leqslant 1$ such that $\gamma(y, t_0; w) \in B_\delta$.*
(2) *There is $t_0 \leqslant 1$ such that $\gamma(y, t_0; -w) \in B_\delta$.*

Proof. Observe that each maximal integral curve of w is defined on $\mathbf{R}$, and its length is not more than C. If the property 1) does not hold, then $\gamma(y, [0,1]; w) \subset W \setminus B_\delta$. By the construction of the function h we have $||w(x)|| = C$ for every $x \notin B_\delta$ therefore the length of the curve $\gamma(y, [0,1]; w)$ is equal to C. We deduce that the length of the maximal integral curve of w which passes through y is strictly greater than C and obtain a contradiction with the condition of C-quickness of v. Thus we have proved the property (1), the proof of (2) is similar. □

Now let us return to the proof of the proposition. To prove the property (B) let $y \in W \setminus B_\delta(\mathrm{ind}_{\leqslant m-k-1}\, ; w)$. By the lemma above we have:

$$\gamma(y, t_0; w) \in B_\delta(p) \quad \text{for some} \quad p \in S(f) \quad \text{and} \quad t_0 \leqslant 1.$$

The set $W \setminus B_\delta(\mathrm{ind}_{\leqslant m-k-1}\, ; w)$ is $(-w)$-invariant, therefore $\mathrm{ind}\, p \geqslant m - k$. Thus

$$\gamma(y, t_0; w) \in B_\delta(\mathrm{ind}_{\leqslant k}\, ; -w)$$

and since the last set is w-invariant, we have

$$\gamma(y, 1; w) \in B_\delta(\mathrm{ind}_{\leqslant k}\, ; -w).$$

The property (A) is proved similarly. □

The main aim of the present section is the following theorem.

Theorem 3.6. *Let W be a Riemannian cobordism and $C > 0$. There is a Morse function f on W and an almost transverse C-quick gradient-like vector field for f.*

The proof occupies the rest of the section.

Definition 3.7. Let us call a Riemannian cobordism W *regular* if for every $C > 0$ there is a Morse function $f : W \to [a, b]$ having an almost transverse C-quick gradient-like vector field for f.

Theorem 3.6 says every cobordism is regular. Observe that if W, W' are two diffeomorphic Riemannian cobordisms, then W is regular if and only if W' is regular, so that regularity of a cobordism does not depend on the particular choice of the Riemannian metric on it. Let us start our study of regular cobordisms with a simple example.

Proposition 3.8. *S^1 is regular.*

Proof. Identifying $\mathbf{R}^2$ with $\mathbf{C}$ write

$$S^1 = \{z \in \mathbf{C} \mid |z| = 1\}.$$

Let $n \in \mathbf{N}$ and define a Morse function $\phi_n : S^1 \to \mathbf{R}$ by the formula $\phi_n(z) = \mathrm{Re}\,(z^n)$, so that

$$\phi_n(e^{it}) = \cos(nt) \quad \text{for} \quad t \in \mathbf{R}.$$

The Euclidean length of every non-constant integral curve of any ϕ_n-gradient equals π/n. Thus for any $C > 0$ any ϕ_n-gradient is C-quick, if n is sufficiently large. □

Exercise 3.9. Prove that a gradient-like vector field equivalent to a C-quick one, is C-quick itself.

Exercise 3.10. Prove that the manifold $\mathbf{T}^2 = S^1 \times S^1$ is regular.

3.2. Properties of C-quick flows. Our aim in this subsection is to develop the techniques necessary for the sequel. We begin by investigating the influence of C^0-small perturbations of the gradient on its quickness constant.

Lemma 3.11. *Let $f : W \to [a, b]$ be a Morse function without critical points. Let w be an f-gradient. For any $x \in W$ let $L(x, w)$ denote the length of the maximal integral curve starting at x.*

Then for every $\epsilon > 0$ there is $\delta > 0$ such that for every f-gradient v with $||w - v|| < \delta$ we have

$$L(x, v) \leqslant L(x, w) + \epsilon \qquad \textit{for every} \quad x \in W.$$

Proof. In view of the compactness of W our proposition follows from the next lemma.

Lemma 3.12. *The function*

$$L : W \times G(f) \to \mathbf{R}$$

is continuous with respect to the C^0-topology on $G(f)$.

Proof. Let $w_n \to w$, $x_n \to x$; we have to show that $L(x_n, w_n) \to L(x, w)$. Replacing the vector fields by equivalent ones, we can assume that for every $x \in W$ we have $f'(x)(w_n(x)) = f'(x)(w(x)) = 1$. Then every maximal integral curve of the vector fields in question starting at x is defined on the interval $[0, b - a]$ and applying Theorem 1.1 of Chapter 1 (page 19) we deduce that the curves $\gamma_n = \gamma(x_n, \cdot; w_n)$ converge to the curve $\gamma = \gamma(x, \cdot; w)$ pointwise. Since the C^0-norms of the vector fields w_n, w are bounded from the above, the Arzela theorem implies that the convergence is uniform. The length $L(\gamma_n)$ of the curve γ_n is given by the following formula:

$$L(\gamma_n) = \int_0^{b-a} ||\gamma_n'(t)||dt = \int_0^{b-a} ||w_n(\gamma_n(t))||dt,$$

and therefore $L(x_n, w_n) \to L(x, w)$. □

Proposition 3.13. *Let $f : W \to [a, b]$ be a Morse function on a cobordism W and v a C-quick gradient-like vector field for f. Let $[c, b] \subset [a, b]$ be a regular interval for f and $\epsilon > 0$. Then there is $\delta > 0$ such that every gradient-like vector field w for f with $||w - v|| < \delta$ and* supp $(v - w) \subset f^{-1}([c, b])$ *is $(C + \epsilon)$-quick.*

Proof. Let w be an f-gradient such that supp $(v - w) \subset f^{-1}\big([c, b]\big)$. Let $\gamma = \gamma(x, \cdot; w)$ be any maximal integral curve of w, where $x \in f^{-1}\big([a, c[\big)$. If γ is entirely in $f^{-1}\big([a, c[\big)$, then it is also an integral curve of v and $L(\gamma) \leqslant C$. If γ intersects $f^{-1}(c)$, let us denote by γ_0 the part of γ lying in $f^{-1}\big([a, c]\big)$ and by γ_1 the part of γ lying in $f^{-1}\big([c, b]\big)$. Let y be the intersection of γ with $f^{-1}(c)$ and denote by $\widetilde{\gamma}_1$ the v-trajectory starting at y. Then we have $L(\gamma_1) \leqslant L(\widetilde{\gamma}_1) + \epsilon$ if $||w - v||$ is small enough (by the previous lemma) and

$$L(\gamma) = L(\gamma_0) + L(\gamma_1) \leqslant L(\gamma_0) + L(\widetilde{\gamma}_1) + \epsilon \leqslant C + \epsilon. \qquad \square$$

Proposition 3.14. *Let $f : W \to [a, b]$ be a Morse function on a cobordism W and v a C-quick f-gradient. Let w be an upward normal vector field defined in a neighbourhood of $\partial_1 W$. Let $\epsilon > 0$, and U be a neighbourhood of $\partial_1 W$. Then there is a gradient-like vector field u for f with the following properties:*

(1) *u is $(C + \epsilon)$-quick,*
(2) supp $(v - u) \subset U$,
(3) *u equals w in a neighbourhood of $\partial_1 W$.*

If v is almost transverse, and $\mathbb{A}$ is an s-submanifold of $\partial_1 W$, then we can choose the vector field u so that it is almost transverse and satisfies $\mathbb{A} \pitchfork \mathbb{D}(-u) \cap \partial_1 W$.

Proof. Diminishing U is necessary; we can assume that w is defined everywhere in U and that it is a gradient-like vector field for $f|U$. Let $h : U \to [0, 1]$ be a C^∞ function such that supp $h \subset U$ and $h(x) = 1$ for every x in some neighbourhood of $\partial_1 W$. Define the vector field u as follows:

$$u(x) = \big(1 - h(x)\big)v(x) + h(x)w(x) \quad \text{for} \quad x \in U,$$
$$u(x) = v(x) \quad \text{for} \quad x \in W \setminus U.$$

The vector field u is a gradient-like vector field for f. In the next lemma we prove that u is $(C+\epsilon)$-quick if the support of h is chosen sufficiently small.

Lemma 3.15. *There is $\delta > 0$ such that for every h with* supp $h \subset f^{-1}\big([b-\delta, b]\big)$ *the gradient-like vector field u defined above is $(C+\epsilon)$-quick.*

Proof. Let γ be any u-trajectory in W. The part of γ contained in $f^{-1}([a, b-\delta])$ is not longer than C. As for the part contained in $W' = f^{-1}([b-\delta, b])$, an easy computation shows that its length is not more than

$$\delta \cdot \frac{\max_{x\in W'} ||u(x)||}{\min_{x\in W'} f'(x)(u(x))}. \tag{3}$$

We have

$$\max_{x\in W'} ||u(x)|| \leqslant ||v|| + ||w||.$$

Further,

$$f'(x)(u(x)) \geqslant \min\Big(f'(x)(v(x)), \quad f'(x)(w(x))\Big) \quad \text{for every} \quad x.$$

Thus the fraction in (3) is bounded from above by a constant which does not depend on δ. The lemma follows. □

Thus we have constructed a vector field u satisfying the properties (1)–(3). Now let us proceed to the second part of the proposition. If

$$\text{supp } (u - v) \subset f^{-1}\big([b-\delta, b]\big)$$

where δ is sufficiently small so that $f^{-1}([b-\delta, b])$ does not contain critical points of f, the field u is necessarily almost transverse. To make the s-submanifold $\mathbb{D}(-u) \cap \partial_1 W$ almost transverse to $\mathbb{A}$, it suffices by Proposition 2.18 to make a small perturbation of u nearby $\partial_1 W$. Applying Proposition 3.13 we complete the proof. □

Proposition 3.16. *Let W be a Riemannian cobordism, and $g : W \to [a, b]$ a C^∞ function. Let $c \in]a, b[$ be a regular value of g. Put*

$$W_0 = g^{-1}([a, c]), \;\; W_1 = g^{-1}([c, b]),$$

Let

$$f_0 : W_0 \to [a, c], \;\; f_1 : W_1 \to [c, b]$$

be Morse functions and v_0, v_1 *be gradient-like vector fields for* f_0*, resp.* f_1*. Assume that* v_i *are* C_i*-quick (for* $i = 0, 1$*). Let* $\epsilon > 0$*.*

Then there is a Morse function $f : W \to [a, b]$ *and a* $(C_0+C_1+\epsilon)$*-quick gradient-like vector field* v *for* f*. If both* v_0, v_1 *are almost transverse, then* v *can be chosen almost transverse.*

Proof. The idea of the proof is quite obvious: we will glue together the vector fields v_0, v_1 and the functions f_0, f_1 in a neighbourhood of the manifold

$$V = g^{-1}(c).$$

Let w be any C^∞ vector field defined in a neighbourhood U of the manifold V, pointing inward W_1. We can assume that w is a gradient-like vector field for functions

$$f_0|W_0 \cap U, \;\; f_1|W_1 \cap U, \;\; g|U.$$

Proposition 3.14 yields a $(C_1 + \epsilon/2)$-quick f_1-gradient u_1 which is equal to w in a neighbourhood of $V = \partial_0 W_1$. Let

$$W_1' = f_1^{-1}([c, c']),$$

where $c' > c$ is so close to c that the interval $[c, c']$ is regular. Applying Proposition 3.7 of Chapter 2 (page 64) we obtain a Morse function

$$\phi : W_1' \to [c, c'] \quad \text{such that}$$

$$\phi = g \quad \text{nearby} \quad \partial_0 W_1' \quad \text{and}$$

$$\phi = f_1 \quad \text{nearby} \quad \partial_1 W_1',$$

and u_1 is a ϕ-gradient. Glue the functions ϕ and $f_1|(W_1 \setminus W_1')$ and obtain a Morse function

$$\widetilde{f_1} : W_1 \to [c, b] \qquad \text{with} \quad \widetilde{f_1} = g \quad \text{nearby} \quad V = \partial_0 W_1$$

and a $(C_1+\epsilon/2)$-quick $\widetilde{f_1}$-gradient u_1 such that $u_1 = w$ nearby V. Observe that if u_1 is almost transverse, then the same is true for u_1 if U is chosen sufficiently small. Similarly we construct a Morse function

$$\widetilde{f_0} : W_0 \to [a, c] \qquad \text{with} \quad \widetilde{f_0} = g \quad \text{nearby} \quad V = \partial_1 W_0$$

and a $(C_0 + \epsilon/2)$-quick $\widetilde{f_0}$-gradient u_0 such that $u_0 = w$ nearby V. If u_1, u_0 are almost transverse, then we can choose u_0 to be almost transverse. Applying Proposition 2.18 we can also assume that

$$\text{(AT)} \qquad \mathbb{D}(-u_0) \cap V \pitchfork \mathbb{D}(u_1) \cap V.$$

Now let us glue together the functions $\widetilde{f_1}, \widetilde{f_0}$ and the gradients u_1, u_0 to obtain a Morse function $f : W \to [a, b]$ and an f-gradient v, which is obviously $(C_1 + C_2 + \epsilon)$-quick.

If v_0, v_1 are almost transverse, then v also is almost transverse as it follows from (AT). □

Here is an immediate corollary of Proposition 3.16.

Corollary 3.17. *Let W be a Riemannian cobordism, and $g : W \to [a,b]$ be a C^∞ function. Let $c \in\,]a,b[$ be a regular value of g. Put*

$$W_0 = g^{-1}([a,c]), \ W_1 = g^{-1}([c,b]).$$

If W_0, W_1 are both regular, then W is regular. □

For a cobordism W let $\mathcal{M}(W)$ denote the minimal possible number of critical points of a Morse function on W. The preceding corollary implies that it suffices to prove Theorem 3.6 for the case of cobordisms with $\mathcal{M}(W) \leqslant 1$. The next proposition shows, moreover, that it suffices to do the case $\mathcal{M}(W) = 0$.

Proposition 3.18. *Let m be a positive integer. If any cobordism W of dimension m with $\mathcal{M}(W) = 0$ is regular, then any cobordism of dimension m with $\mathcal{M}(W) = 1$ is regular.*

Proof. Let W be a cobordism with

$$\dim W = m, \ \mathcal{M}(W) = 1,$$

and let $g : W \to [a,b]$ be a Morse function having exactly one critical point p. Let

$$c = g(p), \quad W_\delta = g^{-1}\big([c-\delta, c+\delta]\big).$$

Pick a gradient-like vector field v for g. Denote by $\lambda(\delta)$ the supremum of the lengths of integral curves of $v|W_\delta$.

Lemma 3.19. $\lambda(\delta) \to 0$ *as* $\delta \to 0$.

Proof. Let $\Psi : U \to V$ be a Morse chart for (g,v) at p. Let G be the Riemannian metric on W, and G' be the Riemannian metric on U, induced by Ψ from the Euclidean metric on V. Diminishing the domain U if necessary, we can assume that there is $K > 0$ such that the length of any curve in U with respect to G is not more than K times the length of this curve with respect to G'. Choose any $r > 0$ sufficiently small so that $B^k(0,r) \times B^{m-k}(0,r) \subset V$, and put

$$U_r = \Psi^{-1}\big(B^k(0,r) \times B^{m-k}(0,r)\big).$$

We are going to prove that

$$\lim_{\delta\to 0} \lambda(\delta) \leqslant 3Kr \tag{4}$$

which implies obviously the assertion of our lemma (observe that this limit exists, since $\lambda(\delta)$ is an increasing function of δ). Let

$$C_r = \min_{W \setminus U_r} g'(x)(v(x)).$$

Let γ be any v-trajectory. The maximal time which γ can spend in $W_\delta \setminus U_r$ is not more than $2\delta/C_r$. The length of the part of γ which is in $W_\delta \setminus U_r$ is therefore not more than $2||v||\delta/C_r$.

Now let us estimate the length of the part of γ contained in U_r. The Ψ-image η of γ is an integral curve of the vector field

$$v_0(x,y) = (-x, y),$$

where (x,y) is a generic point in $\mathbf{R}^m = \mathbf{R}^k \times \mathbf{R}^{m-k}$. It is easy to show that the Euclidean length of η is not greater than $2r$. Therefore the length of the Ψ^{-1}-image of η is not more than $2r \cdot K$. Thus for every given r we have:

$$\lambda(\delta) < 2Kr + 2||v||\delta/C_r,$$

and we obtain (4). □

Now we can complete the proof of the proposition. Let $C > 0$. Pick $\delta > 0$ sufficiently small, so that $\lambda(\delta) < C/4$. The cobordisms $g^{-1}([a, c-\delta])$ and $g^{-1}([c+\delta, b])$ are both regular by our assumption, therefore there are Morse functions ϕ_0, ϕ_1 on W_0, resp. W_1 and their gradients u_0, u_1 which are both $C/4$-quick. Applying Proposition 3.16 we obtain a Morse function $f : W \to [a,b]$ and a C-quick gradient-like vector field v for f. □

Thus we have reduced the proof of Theorem 3.6 to the case of product cobordisms. In the next three subsections we develop some geometric constructions which will allow us to deal with the product cobordisms and to complete the proof in Subsection 3.6.

3.3. Digging in the bottom of a channel. Let V be a closed Riemannian manifold, and $\alpha > 0$; put

$$Z = V \times [-\alpha, \alpha].$$

Consider Z as a cobordism with

$$\partial_1 Z = V \times \{-\alpha, \alpha\}, \quad \partial_0 Z = \varnothing.$$

Consider the function $F_0(x,t) = c + t^2$ on the cobordism Z. The number c is a critical value of F_0; it is degenerate if $\dim V > 0$, and $V \times 0$ is the corresponding critical level. In this subsection we construct a perturbation of F_0 having only non-degenerate critical points. The construction is illustrated in Figure 32, which suggested the title of the subsection.

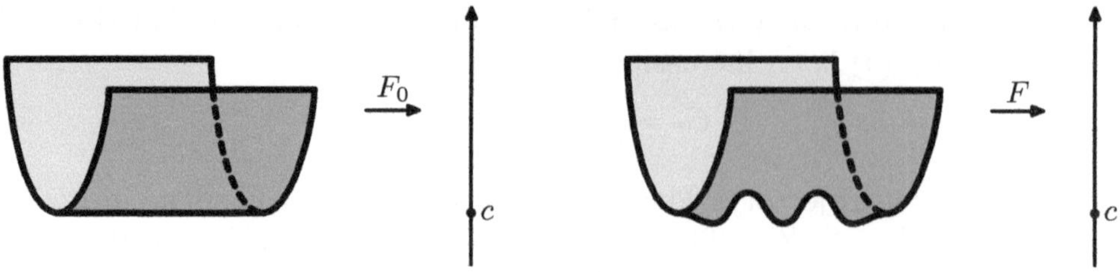

FIGURE 32.

Let $\phi : V \to \mathbf{R}$ be a Morse function and u a gradient-like vector field for ϕ. Pick a C^∞ function $B : \mathbf{R} \to [0,1]$ with the following properties:

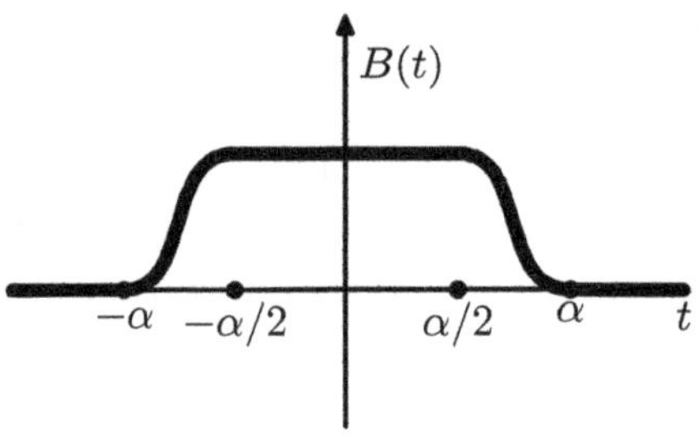

$$B(t) = 1$$
in a neighborhood of $[-\alpha/2, \alpha/2]$,
$$B(t) = 0$$
in a neighborhood of $[-\infty, -\alpha] \cup [\alpha, \infty]$

FIGURE 33.

Set
$$F(x,t) = \lambda B(t)\phi(x) + c + t^2.$$
It is clear that if $\lambda > 0$ is sufficiently small, then $F : Z \to \mathbf{R}$ is a Morse function which equals F_0 in a neighbourhood of $V \times \{-\alpha, \alpha\}$. The critical points of F are all of the form $(p, 0)$ where $p \in S(\phi)$; and $\operatorname{ind}(p,0) = \operatorname{ind} p$. Now we shall introduce a gradient-like vector field for F. Set
$$\xi(x,t) = \big(B(t)u(x),\ \theta(t)\big), \quad \text{where} \quad \theta(t) = tB(t) + (1 - B(t)) \operatorname{sgn} t$$
(so that in a neighbourhood of $V \times 0$ the vector field ξ equals $(u(x), t)$ and in a neighbourhood of $V \times \{-\alpha, \alpha\}$ it equals $(0, \operatorname{sgn} t)$).

Lemma 3.20. *For every $\lambda > 0$ sufficiently small the vector field ξ is a gradient-like vector field for F.*

Proof. Let us show that

(5) $$F'(x,t)\big(\xi(x,t)\big) > 0 \quad \text{for every} \quad (x,t) \notin S(F).$$

We have

$$F'(x,t)\big(\xi(x,t)\big) = \frac{\partial F}{\partial x}(x,t)\big(B(t)u(x)\big) + \frac{\partial F}{\partial t}(x,t)\cdot\theta(t).$$

The first term in the right-hand side is always non-negative and is strictly positive if $t = 0$ and $x \notin S(f)$. For the second term we have:

$$\frac{\partial F}{\partial t}(x,t)\cdot\theta(t) = \Big(\lambda B'(t)\phi(x) + 2t\Big)\cdot\theta(t);$$

this vanishes for $t = 0$ and is strictly positive for every $t \neq 0$ if $\lambda > 0$ is sufficiently small. The inequality (5) follows.

Now let us investigate the local structure of the pair (F, ξ) around critical points of F. Let $p \in S_k(\phi)$, and let

$$\Psi : U \to \mathcal{B}(r) = B^k(0,r) \times B^{m-k}(0,r)$$

be a Morse chart for (ϕ, u) at p. Let $I =]-\alpha/2, \alpha/2[$. The diffeomorphism

$$\widetilde{\Psi} : U \times I \to \mathcal{B}(\sqrt{\lambda}\cdot r)\times I; \quad \widetilde{\Psi}(x,t) = \Big(\sqrt{\lambda}\cdot\Psi(x), t\Big)$$

is a Morse chart for (F, ξ) (cf. Exercise 2.7 of Chapter 2, page 52). □

Lemma 3.21. *Assume that the Riemannian metric on $Z = V \times [-\alpha, \alpha]$ is the product metric. If u is C-quick, then ξ is $(C+\alpha)$-quick.*

Proof. We will give an explicit formula for the integral curve of ξ with the initial condition (x_0, t_0) where $0 < t_0 \leqslant \alpha$. Let μ denote the solution of the differential equation

$$\mu'(\tau) = \theta\big(\mu(\tau)\big), \quad \mu(0) = t_0.$$

Then μ is a positive increasing C^∞ function defined on the whole of $\mathbf{R}$ and $\mu(\tau) \to 0$ as $\tau \to -\infty$. There is a unique τ_0 such that $\mu(\tau_0) = \alpha$, and $\mu(]-\infty, \tau_0]) =]0, \alpha]$. Put

$$\varkappa(\tau) = \int_0^\tau B(\mu(\tau))d\tau,$$

and let γ denote the maximal integral curve of u with the initial value $\gamma(0) = x_0$. Put

$$\Gamma(\tau) = \big(\gamma(\varkappa(\tau)), \mu(\tau)\big).$$

An easy computation shows that

$$\Gamma'(\tau) = \xi\big(\Gamma(\tau)\big) \quad \text{for all} \quad \tau.$$

The function Γ is defined on the whole of $\mathbf{R}$, and takes its values in $V \times \mathbf{R}$; we shall be interested only in the values of the argument $\tau \in]-\infty, \tau_0]$, since at $\tau = \tau_0$ the curve Γ quits $V \times [-\alpha, \alpha]$. For the length of this part of Γ we have the following inequality:

$$L(\Gamma) = \int_{-\infty}^{\tau_0} ||\Gamma'(\tau)|| d\tau \leqslant L(\gamma) + L(\mu) = L(\gamma) + \alpha.$$

A similar estimate holds for the integral curves of ξ with the initial condition (x_0, t_0) where $t_0 < 0$. Finally, any integral curve of ξ with an initial condition in $V \times 0$ equals an integral curve of u, and its length therefore is not more than C. □

Our next task is to describe the structure of the descending and ascending discs of the vector field ξ. We identify V with the submanifold $V \times 0 \subset Z$.

Lemma 3.22. *For every $p \in S(\phi)$ we have*

$$D(p, \xi) = D(p, u) \subset V.$$

Proof. Let p be a critical point of ϕ, and Γ an integral curve of ξ. We have seen during the proof of the preceding lemma that if Γ passes through a point (x, t) with $t \neq 0$, then Γ quits the cobordism Z, and therefore it can not belong to the descending disc $D(p, \xi)$. Thus we have $D(p, \xi) \subset D(p, u)$, and the inverse inclusion follows from the fact that V is a $(\pm\xi)$-invariant submanifold of Z. □

Exercise 3.23. Prove that for every $p \in S(\phi)$ we have

$$D(p, -\xi) = D(p, -u) \times [-\alpha, \alpha].$$

Lemma 3.24. *If u is almost transverse, then ξ is also almost transverse.*

Proof. If γ is a ξ-link from q to p, then it is in $D(p, u)$ by Lemma 3.22. Therefore γ is also a u-link, which implies $\operatorname{ind}_\phi q < \operatorname{ind}_\phi p$. Since $\operatorname{ind}_\phi q = \operatorname{ind}_F q$, $\operatorname{ind}_\phi p = \operatorname{ind}_F p$, our lemma follows. □

Let us summarize the properties of the constructed objects.

(1) $\max F = c + \alpha^2$, and $F^{-1}(c + \alpha^2) = \partial_1 Z = V \times \{-\alpha, \alpha\}$.
(2) $S(F) = S(\phi)$, and for every $p \in S(F)$ we have $\operatorname{ind}_F p = \operatorname{ind}_\phi p$.
(3) $D(p, \xi) = D(p, u)$ for every $p \in S(F)$.
(4) If u is almost transverse, then ξ is almost transverse.
(5) ξ is $(C + \alpha)$-quick.
(6) The subsets $V \times 0$, $V \times]0, \alpha]$, $V \times [-\alpha, 0[$ are $(\pm\xi)$-invariant.
(7) In a neighbourhood of $\partial_1 Z = V \times \{-\alpha, \alpha\}$ we have:

$$\xi(x, t) = (0, \operatorname{sgn} t), \ F(x, t) = c + t^2.$$

For a normal vector field v on a cobordism let us call a *boundary trajectory* any v-trajectory intersecting the boundary of the cobordism. The property (6) above implies that any boundary trajectory of ξ can intersect at most one of the two sets $V \times \{\alpha\}, V \times \{-\alpha\}$.

3.4. Digging on the top of a hill. We can perform a similar construction starting with the function $d - t^2$ instead of $c + t^2$. This time we consider the manifold $V \times [-\alpha, \alpha]$ as a cobordism with the empty upper component of the boundary and the lower component $V \times \{-\alpha, \alpha\}$. This cobordism will be denoted by Z'. As in the previous subsection we assume that the Riemannian metric on Z' is the product metric. For a Morse function $\phi : V \to \mathbf{R}$ and its gradient-like vector field u, we construct a Morse function $G : Z' \to \mathbf{R}$ and its gradient-like vector field η, having the following properties:

(1) $\min G = d - \alpha^2$, and $G^{-1}(d - \alpha^2) = \partial_0 Z' = V \times \{-\alpha, \alpha\}$.

(2) $S(G) = S(\phi)$, and for every $p \in S(G)$ we have $\operatorname{ind}_G p = \operatorname{ind}_\phi p + 1$.

(3) $D(p, -\eta) = D(p, -u)$ for every $p \in S(G)$.

(4) If u is almost transverse, then η is almost transverse.

(5) η is $(C + \alpha)$-quick.

(6) The subsets $V \times 0$, $V \times]0, \alpha]$, $V \times [-\alpha, 0[$ are $(\pm\eta)$-invariant.

(7) Any boundary trajectory of η can intersect only one of sets $V \times \{\alpha\}, V \times \{-\alpha\}$.

(8) In a neighbourhood of $\partial_0 Z' = V \times \{-\alpha, \alpha\}$ we have:

$$\eta(x, t) = (0, -\operatorname{sgn} t), \ \ G(x, t) = d - t^2.$$

3.5. *S*-construction. The S-construction is in a sense a superposition of the constructions of the two previous subsections; it is essential for the proof of Theorem 3.6.

Let W be a Riemannian cobordism having a Morse function $\phi : W \to [a, b]$ without critical points. Let u be a ϕ-gradient. Let

$$V_1 = \phi^{-1}(c), \quad V_2 = \phi^{-1}(d)$$

be regular level surfaces of ϕ (where $a < c < d < b$). Let

$$\phi_1 : V_1 \to \mathbf{R}, \quad \phi_2 : V_2 \to \mathbf{R}$$

be Morse functions. Let u_1 be a gradient-like vector field for ϕ_1, and u_2 a gradient-like vector field for ϕ_2. Let $\alpha > 0$ be a real number such that the maps

$$\Psi_1 : V_1 \times [-\alpha, \alpha] \to W, \quad \Psi_1(x, t) = \gamma(x, t; u),$$
$$\Psi_2 : V_2 \times [-\alpha, \alpha] \to W, \quad \Psi_2(x, t) = \gamma(x, t; u)$$

are diffeomorphisms onto their images and these images are disjoint. Let A be a real number such that

$$A \geqslant \max_{x,t} ||\Psi_1'(x,t)||, \quad A \geqslant \max_{x,t} ||\Psi_2'(x,t)||.$$

The main aim of this subsection is the following theorem.

Theorem 3.25. *Assume that u is C-quick, u_1 is C_1-quick, u_2 is C_2-quick. Let $\nu > 0$. There is a Morse function $f : W \to [a,b]$ and a gradient-like vector field v for f, such that:*

(1) *V_1 and V_2 are $\pm v$-invariant.*
(2) *f equals ϕ and v equals u in a neighbourhood of ∂W.*
(3) *No boundary trajectory of v intersects both $\partial_1 W$ and $\partial_0 W$.*
(4) *v is C'-quick, where*

$$C' = C + A(C_1 + C_2) + \nu.$$

If u_1 and u_2 are almost transverse we can choose v to be almost transverse. Moreover, for a given s-submanifold $\mathbb{A}$ of $\partial_0 W$ and a given submanifold $\mathbb{B}$ of $\partial_1 W$ we can choose v so that it satisfies the following conditions:

$$\mathbb{A} \pitchfork \mathbb{D}(v) \cap \partial_0 W, \quad \mathbb{B} \pitchfork \mathbb{D}(-v) \cap \partial_1 W.$$

Proof. Using Proposition 3.7 of Chapter 2 (page 64) we can modify the function ϕ in a neighbourhood of $V_1 \cup V_2$ so that $\phi'(x)(u(x)) = 1$ holds for every x in some neighbourhood R of $V_1 \cup V_2$. Diminishing α if necessary, we can also assume that R contains the image of Ψ_1 and of Ψ_2, so that $2\alpha < d - c$ and

$$(\phi \circ \Psi_1)(x,t) = c + t, \quad (\phi \circ \Psi_2)(x,t) = d + t.$$

For the construction of the function χ below we shall impose additional restrictions on α, namely we require that

$$2\alpha^2 < d - c.$$

Let $\chi : [a,b] \to [a,b]$ be a C^∞ function, such that

$$\begin{aligned} \chi(t) &= t && \text{for every } t \text{ in a neighbourhood of } \{a,b\} \\ \chi(t) &= d - (t-c)^2 && \text{for} \quad t \in [c-\alpha, c+\alpha] \\ \chi(t) &= c + (t-d)^2 && \text{for} \quad t \in [d-\alpha, d+\alpha] \\ \chi'(t) &> 0 && \text{for} \quad t \in [a, c[\cup]d, b] \\ \chi'(t) &< 0 && \text{for} \quad t \in]c, d[. \end{aligned}$$

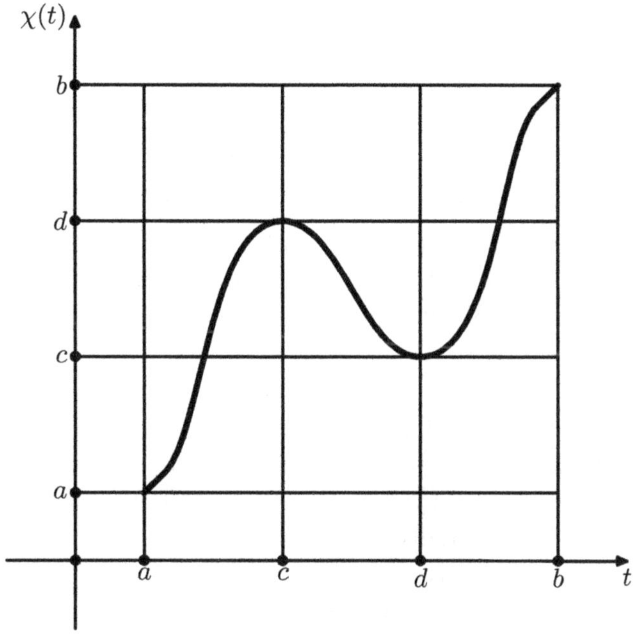

FIGURE 34.

Let $h = \chi \circ \phi$. We have:

$$(h \circ \Psi_1)(x,t) = d - t^2, \quad (h \circ \Psi_2)(x,t) = c + t^2.$$

Both V_1 and V_2 are critical manifolds for h, therefore all the critical points of h are degenerate if $\dim W > 1$. We shall now modify the function h in the subsets Im Ψ_1 and Im Ψ_2 in order to obtain a Morse function. Apply the results of Subsection 3.3 to construct a Morse function F on the cobordism $V_2 \times [-\alpha, \alpha]$ and its gradient-like vector field ξ satisfying (1)–(6) from page 268. We have also a Morse function G on the cobordism $V_1 \times [-\alpha, \alpha]$ and an G-gradient η (see Subsection 3.4). To define the Morse function f, we just glue together the functions $h, F \circ \Psi_2^{-1}$ and $G \circ \Psi_1^{-1}$. Here are the formulae:

$$f(x) = h(x) \quad \text{if} \quad x \notin \text{Im } \Psi_1 \cup \text{Im } \Psi_2,$$
$$f(x) = (F \circ \Psi_2^{-1})(x) \quad \text{if} \quad x \in \text{Im } \Psi_2,$$
$$f(x) = (G \circ \Psi_1^{-1})(x) \quad \text{if} \quad x \in \text{Im } \Psi_1.$$

It follows from the properties of the functions F, G, h that these formulas define indeed a C^∞ function on the whole of W. Let us define a vector field

v by the following formulae:

$$\begin{aligned} v(x) &= u(x) &\text{if}\quad &\phi(x)\in[a,c-\alpha] \quad\text{or}\quad \phi(x)\in[d+\alpha,b],\\ v(x) &= -u(x) &\text{if}\quad &\phi(x)\in[c+\alpha,d-\alpha],\\ v(x) &= (\Psi_1)_*(\eta) &\text{if}\quad &\phi(x)\in[c-\alpha,c+\alpha],\\ v(x) &= (\Psi_2)_*(\xi) &\text{if}\quad &\phi(x)\in[d-\alpha,d+\alpha]. \end{aligned}$$

It is clear that f is a Morse function and v is a gradient-like vector field for f. The points (1) – (3) of Theorem 3.25 clearly follow from the properties of the functions F, G. We will show now that if

$$\alpha < \frac{\nu}{2A},$$

then the property (4) also holds. Indeed, let γ be a v-trajectory, starting at some point of $\phi^{-1}(\,]c,d[\,)$. Then γ stays in this domain forever, and the part of γ, contained in $\phi^{-1}([c+\alpha,d-\alpha])$ is not longer than C. The part of γ contained in $\phi^{-1}([d-\alpha,d])$ is not longer than $A(C_2+\alpha)$ by the results of Subsection 3.4. The part of γ contained in $\phi^{-1}([c,c+\alpha])$ is not longer than $A(C_1+\alpha)$. The total length $L(\gamma)$ of γ satisfies therefore

$$L(\gamma) \leqslant C + A(C_1+C_2) + 2A\alpha. \tag{6}$$

Since $2A\alpha < \nu$ we deduce that

$$L(\gamma) < C + A(C_1+C_2) + \nu.$$

Similarly one proves the estimate (6) for the case of v-trajectories, contained in $\phi^{-1}(\,]d,b[\,)$ or in $\phi^{-1}(\,]a,c[\,)$. The case of trajectories in V_1 or V_2 is obvious.

Let us now investigate the almost transversality property of v assuming that u_1, u_2 are almost transverse. Consider a cobordism

$$W_0 = \phi^{-1}([c-\alpha,\ c+\alpha]),$$
$$\text{with}\quad \partial_0 W_0 = \phi^{-1}(c-\alpha)\ \cup\phi^{-1}(c+\alpha),\quad \partial_1 W_0 = \varnothing.$$

We have a Morse function $f|W_0$ and its almost transverse gradient $v|W_0$. Put $N_1 = \phi^{-1}(c+\alpha)$, and denote by $\mathbb{K}$ the s-submanifold $\mathbb{D}(v|W_0)\cap N_1$. Similarly we have a cobordism

$$W_1 = \phi^{-1}([d-\alpha,d+\alpha]),$$
$$\text{with}\quad \partial_0 W_1 = \varnothing,\quad \partial_1 W_1 = \phi^{-1}(d-\alpha)\ \cup\phi^{-1}(d+\alpha).$$

We have a Morse function $f|W_1$ and an almost transverse $(f|W_1)$-gradient $v|W_1$. Put $N_0 = \phi^{-1}(d-\alpha)$, and denote by $\mathbb{H}$ the s-submanifold $\mathbb{D}(-v|W_1)\cap N_0$. Let

$$W' = \phi^{-1}([c-\alpha,c+\alpha])$$

and endow this cobordism with a Morse function $f|W'$ and its gradient $v|W'$. It is clear that v is almost transverse if and only if

$$(v|W')^{\leadsto}(\mathbb{H}) \pitchfork \mathbb{K}. \tag{7}$$

Let $\epsilon > 0$. Applying Propositions 2.18 and 3.13 it is easy to show that perturbing the vector field $v|W'$ we can achieve the property (7) and moreover the quickness constant of $v|W'$ will not increase more than by ϵ. Then the total length of any integral curve of v belonging to $\phi^{-1}(]c,d[)$ will be not more than $C+\epsilon+A(C_1+C_2)+2A\alpha$ which is less than $C+A(C_1+C_2)+\nu$ if ϵ is chosen to be small enough.

Similarly, we can achieve the condition

$$\mathbb{A} \pitchfork \mathbb{D}(v) \cap \partial_0 W, \quad \mathbb{B} \pitchfork \mathbb{D}(-v) \cap \partial_1 W$$

and this completes the proof of the theorem. □

3.6. Proof of Theorem 3.6. We will proceed by induction on the dimension of the cobordism in question. The 1-dimensional case is trivial (cf. Proposition 3.8, page 260). Assuming that the theorem is proved for dimensions $\leqslant m-1$, we shall now prove it for the dimension m. We know that it suffices to establish the theorem for the product cobordisms (see Corollary 3.17 and Proposition 3.18). Let W be such a cobordism, pick a Morse function $f: W \to [a,b]$ without critical points. Let v be a gradient-like vector field for f. Let $C>0$. Choose a sequence $a=a_0<a_1<\cdots<a_N=b$ of regular values of f such that for every cobordism $W_i = f^{-1}([a_i,a_{i+1}])$ the $(f|W_i)$-gradient $v|W_i$ is $C/4$-quick.

Lemma 3.26. *For every i there is a Morse function $g_i : W_i \to [a_i, a_{i+1}]$ and an almost transverse gradient-like vector field w_i for g_i such that:*

(1) *f equals g_i and v equals w_i in a neighbourhood of ∂W_i.*

(2) *No boundary trajectory of w_i intersects both $\partial_0 W_i$ and $\partial_1 W_i$.*

(3) *w_i is $C/2$-quick.*

(4) $\mathbb{D}(w_i) \cap f^{-1}(a_i) \pitchfork \mathbb{D}(-w_{i-1}) \cap f^{-1}(a_i)$ *for every* i.

Proof. Construct the functions g_i and their gradients w_i by induction on i applying Theorem 3.25 successively to the cobordisms W_i (recall that by the induction assumption for every $\epsilon>0$ there are almost transverse ϵ-quick gradient flows on every regular level surface of g_i). □

Now let us glue together all the Morse functions g_i and their gradients w_i to obtain a Morse function $g: W \to [a,b]$ and its gradient-like vector field w. The property (2) from the lemma above implies that any w-trajectory can intersect at most two of the cobordisms W_i. Therefore the maximal length of a w-trajectory is not more than twice the maximal length of a

w_i-trajectory, hence $\leqslant C$. The points (2) and (4) of the lemma above imply that w is almost transverse. The proof of Theorem 3.6 is complete. □

4. Proof of the $\mathfrak{C}$-approximation theorem

Theorem 1.1 asserts that for a given Morse function $f : W \to [a,b]$ the subset of cellular f-gradients is open and dense in the set of all gradients with respect to C^0 topology.

4.1. Density. Let us first explain the basic idea of the proof. Let u be an almost transverse f-gradient. Pick any ordered Morse functions

$$\phi_0 : \partial_0 W \to \mathbf{R}, \ \phi_1 : \partial_1 W \to \mathbf{R}.$$

Let us see if we can modify the f-gradient u in such a way that the condition ($\mathfrak{C}$) will be fulfilled with respect to the Morse-Smale filtrations defined by the functions ϕ_i. Pick an almost transverse ϕ_1-gradient ξ_1 and an almost transverse ϕ_0-gradient ξ_0.

Let us say that a subset X of a manifold M is *smoothly k-dimensional*, if it is a finite union of submanifolds of M of dimensions $\leqslant k$. For example the set

$$\Sigma_k = D(\mathrm{ind}_{\leqslant k+1}, u) \cap \partial_0 W$$

is smoothly k-dimensional. Also the set $D(\mathrm{ind}_{\leqslant k}; \xi_1)$ is smoothly k-dimensional, and the same is true for its $(-u)^{\rightsquigarrow}$-image in $\partial_0 W$. Therefore perturbing the f-gradient u, we can assume that the set

$$Z_k = \Sigma_k \cup (-u)^{\rightsquigarrow}\Big(D(\mathrm{ind}_{\leqslant k}; \xi_1)\Big)$$

does not intersect the $(m-k-2)$-skeleton of the Morse stratification $\mathbb{D}(-\xi_0)$ (where $m = \dim W$). Then for $\delta > 0$ sufficiently small Z_k does not intersect the δ-thickening $D_\delta(\mathrm{ind}_{\leqslant m-k-2}; -\xi_0)$ of this $(m-k-2)$-skeleton. Let us call this construction *the first modification of the gradient*; the detailed description of this construction is the content of the subsection B below.

Now, assuming that

$$Z_k \cap D_\delta(\mathrm{ind}_{\leqslant m-k-2}; -\xi_0) = \varnothing,$$

observe that the flow corresponding to the vector field $(-\xi_0)$ will push the set Z_k into an arbitrarily small neighbourhood of $D(\mathrm{ind}_{\leqslant k}; \xi_0)$ if we wait sufficiently long. That is, we have

$$\Phi(-\xi_0, T)(Z_k) \subset D_\delta(\mathrm{ind}_{\leqslant k}; \xi_0) \tag{8}$$

if T is sufficiently large. Applying the horizontal drift construction from Subsection 2.3 (page 249) we construct an f-gradient w such that the transport map $(-w)^{\rightsquigarrow}$ sends the set $D(\mathrm{ind}{\leqslant}k\ ; \xi_1)$ inside $D_\delta(\mathrm{ind}{\leqslant}k\ ; \xi_0)$, and similarly

$$D(\mathrm{ind}{\leqslant}k{+}1\ , w) \cap \partial_0 W \subset D_\delta(\mathrm{ind}{\leqslant}k\ ; \xi_0).$$

Some extra work is required to show that a similar argument applies also to the terms of the Morse-Smale filtration $\partial_1 W^k$ (which can be considered as thickenings of the skeleta $D(\mathrm{ind}{\leqslant}k\ ; \xi_1)$). Now we have an f-gradient w such that

$$(-w)^{\rightsquigarrow}(\partial_1 W^k) \quad \subset \quad \mathrm{Int}\ \partial_0 W^k \quad \supset \quad D(\mathrm{ind}{\leqslant}k{+}1\ , w) \cap \partial_0 W$$

so that the property ($\mathfrak{C}$1) from the definition of the cellular gradients holds. We call this construction *the final modification of the gradient*; it is the content of Subsection C below.

The procedure which we just described works for any Morse-Smale filtrations of $\partial_1 W$ and $\partial_0 W$, but it may happen that in order to satisfy the condition (8) we must choose a very large value of T, so that $w - u$ will be also very large. To assure that we can make the C^0 norm $||w - u||$ arbitrarily small, we must choose appropriate Morse-Smale filtrations of $\partial_1 W, \partial_0 W$. Here we apply the main theorem of Section 3. The corresponding choice of ϕ_i and ξ_i is the content of the section A below.

A. Handle-like filtrations of $\partial_0 W$ and $\partial_1 W$.

Proceeding to precise arguments, let us first make some reductions. The case of arbitrary f-gradients is immediately reduced to the case of almost transverse ones, so we shall concentrate on the second part of Theorem 1.1. Let $\epsilon > 0$ be arbitrary number, u an almost transverse gradient of a Morse function $f : W \to [a, b]$, and U a neighbourhood of ∂W. Our task is to construct a cellular f-gradient v with

$$||v - u|| < \epsilon \quad \text{and} \quad \mathrm{supp}\ (u - v) \subset U \setminus \partial W.$$

The same argument as already employed for the proof of Proposition 2.18 shows that it suffices to consider the case when the pair (f, u) has framed collars. This means (see Definition 2.15) that $f'(x)(u(x)) = 1$ for every $x \in C_0 \cup C_1$ where $C_0 = \mathrm{Im}\ \Psi_0$ and $C_1 = \mathrm{Im}\ \Psi_1$, and the maps

$$\Psi_0 : \partial_0 W \times [0, \alpha] \longrightarrow W, \quad \Psi_0(x, t) = \gamma(x, t; u),$$
$$\Psi_1 : \partial_1 W \times [-\alpha, 0] \longrightarrow W, \quad \Psi_1(x, t) = \gamma(x, t; u)$$

are isometries onto their images which are disjoint. Choose an ordered Morse function ϕ, adjusted to (f, u). We can assume that $C_0 \cup C_1 \subset U$ and that f and ϕ coincide in $C_0 \cup C_1$. Let

$$A = \frac{\epsilon\alpha}{6}.$$

Applying Theorem 3.6 let us choose:

1. An ordered Morse function $\phi_1 : \partial_1 W \to \mathbf{R}$ and an A-quick ϕ_1-gradient η_1.
2. An ordered Morse function $\phi_0 : \partial_0 W \to \mathbf{R}$ and an A-quick ϕ_0-gradient η_0.

We will show that the Morse-Smale filtrations associated with these functions satisfy the conditions ($\mathfrak{C}1$), ($\mathfrak{C}2$) with respect to some f-gradient v which will be chosen afterwards. During the proof we shall work constantly with certain handle-like filtrations of $\partial_1 W$, so we introduce special abbreviations:

$$A_k = D(\text{ind}{\leqslant}k\ ; \eta_1), \qquad \boldsymbol{A}_{\delta,k} = D_\delta(\text{ind}{\leqslant}k\ ; \eta_1).$$

For the lower component $\partial_0 W$ we shall work mostly with the handle-like filtrations associated with $-\eta_0$. Set

$$B_k = D(\text{ind}{\leqslant}k\ ; -\eta_0), \qquad \boldsymbol{B}_{\delta,k} = D_\delta(\text{ind}{\leqslant}k\ ; -\eta_0).$$

The number $\delta > 0$ will be always chosen so small that

$$D_\delta(\text{ind}{\leqslant}k\ ; \eta_1) \subset \partial_1 W^k, \qquad D_\delta(\text{ind}{\leqslant}k\ ; -\eta_0) \subset \widehat{\partial_0 W}^k.$$

B. The first modification of the gradient.

Now we begin the construction of v. Our first task is to perturb the given almost transverse gradient u in such a way that the corresponding gradient descent map sends A_k to the subset of $\partial_0 W$ disjoint from B_{m-k-2}. This is done by a direct application of Proposition 2.18, which implies that there is an almost transverse ϕ-gradient w with

$$(-w)^{\leadsto}(\mathbb{D}(\eta_1)) \pitchfork \mathbb{D}(-\eta_0) \quad \text{and} \quad ||w-u|| < \epsilon/2.$$

Moreover we can assume that w equals u in $C_0 \cup C_1$ (to see this just apply Proposition 2.18 to the cobordism $f^{-1}([a+\alpha, b-\alpha])$). The cited proposition implies also that we can choose w so that $\mathbb{D}_b(-w) \pitchfork \mathbb{A}$, that is,

$$(9) \qquad D(\text{ind}{\leqslant}m-k-1\ ; -w) \cap A_k = \varnothing \quad \text{for every} \quad k.$$

The condition of almost transversality of the s-submanifolds $(-w)^{\leadsto}(\mathbb{D}(\eta_1))$ and $\mathbb{D}(-\eta_0)$ is equivalent to the following:

$$(10) \qquad \Big((-w)^{\leadsto}(A_k) \cup D(\text{ind}{\leqslant}k+1\ ; w)\Big) \cap B_{m-k-2} = \varnothing \quad \text{for every} \quad k.$$

In the next proposition we show that a condition stronger than (10) holds, namely, we can replace the sets A_k and B_{m-k-2} by their δ-thickenings.

Proposition 4.1. *For every $\delta > 0$ sufficiently small and every k the sets*

$$(-w)^{\rightsquigarrow}(\boldsymbol{A}_{\delta,k}) \cup D(\text{ind}{\leqslant}k{+}1\ ; w) \qquad \textit{and} \qquad \boldsymbol{B}_{\delta,\, m-k-2}$$

are disjoint.

Proof. Recall that the family $\{\boldsymbol{B}_{\delta,\, m-k-2}\}_{\delta>0}$ is a fundamental system of neighbourhoods of B_{m-k-2} in $\partial_0 W$. Therefore it follows from (10) that for $\delta > 0$ small enough we have

$$D(\text{ind}{\leqslant}k{+}1\ ; w) \cap \boldsymbol{B}_{\delta,m-k-2} = \emptyset. \tag{11}$$

It remains therefore to show that for $\delta > 0$ small enough we have

$$(-w)^{\rightsquigarrow}(\boldsymbol{A}_{\delta,k}) \cap \boldsymbol{B}_{\delta,\, m-k-2} = \emptyset. \tag{12}$$

For this purpose choose a regular value c of ϕ separating the critical points of indices $\leqslant k$ from the critical points with indices $\geqslant k+1$, so that

$$\phi(p) < c \Leftrightarrow \operatorname{ind} p \leqslant k$$

for every critical point p. Let $V = \phi^{-1}(c)$. Let us denote

$$F = \overset{\rightsquigarrow}{w}_{[a,c]}, \quad G = \overset{\rightsquigarrow}{(-w)}_{[b,c]}$$

so that F (respectively G) is a diffeomorphism of an open subset of $\partial_0 W$ (respectively $\partial_1 W$) onto an open subset of V. It follows from (11) that for every $\delta > 0$ small enough the set $\boldsymbol{B}_{\delta,m-k-2}$ is in the domain of definition of the map F. Therefore the family

$$\{F(\boldsymbol{B}_{\delta,m-k-2})\}_{\delta>0}$$

of subsets of V is a fundamental system of neighbourhoods of the set $F(B_{m-k-2})$. The condition (9) implies that for $\delta > 0$ small enough the set $\boldsymbol{A}_{\delta,k}$ is in the domain of definition of the map G and therefore the sets $G(\boldsymbol{A}_{\delta,k})$ form a fundamental system of neighbourhoods of the set $G(A_k)$. We have:

$$G(A_k) \cap F(B_{m-k-2}) = \varnothing$$

as it follows from (10), therefore

$$G(\boldsymbol{A}_{\delta,k}) \cap F(\boldsymbol{B}_{\delta,\, m-k-2}) = \varnothing$$

whenever $\delta > 0$ is sufficiently small, and this last condition is equivalent to (12). The proof of the proposition is over. □

The following proposition is a dual version of Proposition 4.1, and its proof is completely similar.

Proposition 4.2. *For every $\delta > 0$ sufficiently small and every k the sets*

$$\overset{\rightsquigarrow}{w}(\boldsymbol{B}_{\delta,m-k-2}) \cup D(\text{ind}{\leqslant}m{-}k{-}1\ ; -w) \qquad \textit{and} \qquad \boldsymbol{A}_{\delta,k}$$

are disjoint. □

The properties of the vector field w established in Propositions 4.1 and 4.2 are close to the condition ($\mathfrak{C}$) but still somehow weaker.

C. The final modification of the gradient.

Now we can construct the f-gradient v satisfying the condition ($\mathfrak{C}$). Pick a number $\delta > 0$ so as to satisfy the conclusions of Propositions 4.1 and 4.2. By our choice of η_1 there is a ϕ_1-gradient ξ_1, equivalent to η_1 and such that

$$\Phi(-\xi_1, 1)\Big(\partial_1 W \setminus B_\delta(\mathrm{ind}{\leqslant}m{-}2{-}k\ ; -\xi_1)\Big) \subset B_\delta(\mathrm{ind}{\leqslant}k\ ; \xi_1) \quad \text{for every} \quad k$$

$$\text{and} \quad ||\xi_1|| < A = \frac{\epsilon\alpha}{6}.$$

(see Proposition 3.4, page 258). Similarly, choose a ϕ_0-gradient ξ_0, equivalent to η_0 and such that

$$\Phi(-\xi_0, 1)\Big(\partial_1 W \setminus B_\delta(\mathrm{ind}{\leqslant}m{-}2{-}k\ ; -\xi_0)\Big) \subset B_\delta(\mathrm{ind}{\leqslant}k\ ; \xi_0) \quad \text{for every} \quad k$$

$$\text{and} \quad ||\xi_0|| < A.$$

Applying Proposition 2.17 (page 252) to the fields ξ_0, ξ_1 we obtain an f-gradient $v = v_{\xi_0,\xi_1}$, such that

$$(-v)^{\leadsto} = \Phi(-\xi_0, 1) \circ (-w)^{\leadsto} \circ \Phi(-\xi_1, 1), \qquad \overset{\leadsto}{v} = \Phi(\xi_1, 1) \circ \overset{\leadsto}{w} \circ \Phi(\xi_0, 1).$$

By construction the diffeomorphism $\Phi(-\xi_1, 1)$ sends the set

$$\partial_1 W^k \subset \partial_1 W \setminus B_\delta(\mathrm{ind}{\leqslant}m{-}2{-}k\ ; -\xi_1)$$

to the set $B_\delta(\mathrm{ind}{\leqslant}k\ ; \xi_1)$ for every k. The map $(-w)^{\leadsto}$ descends $B_\delta(\mathrm{ind}{\leqslant}k\ ; \xi_1)$ into $\partial_0 W \setminus B_\delta(\mathrm{ind}{\leqslant}m{-}2{-}k\ ; -\xi_0)$. This last set is sent by the diffeomorphism $\Phi(-\xi_0, 1)$ to

$$B_\delta(\mathrm{ind}{\leqslant}k\ ; \xi_0) \subset \mathrm{Int}\ \partial_0 W^k.$$

Therefore we have

$$(-v)^{\leadsto}(\partial_1 W^k) \subset \mathrm{Int}\ \partial_0 W^k$$

for every k. Similarly

$$D(\mathrm{ind}{\leqslant}k{+}1\ ; v) \cap \partial_0 W \subset \mathrm{Int}\ \partial_0 W^k$$

for every k, and the condition ($\mathfrak{C}1$) holds. A similar argument proves that the condition ($\mathfrak{C}2$) holds also.

Observe that (by the point (3) of Proposition 2.17) we have

$$||v - w|| \leqslant \frac{3}{\alpha} \cdot \max\big(||\xi_0||, ||\xi_1||\big) \leqslant \epsilon/2.$$

Recall that by our choice of w we have $||w - u|| < \epsilon/2$ and therefore

$$||v - u|| \leqslant ||v - w|| + ||w - u|| < \epsilon.$$

The proof of C^0-density of the condition ($\mathfrak{C}$) is over.

4.2. Openness. Let v be an almost transverse f-gradient satisfying $(\mathfrak{C})$. Pick an ordered Morse function $\phi : W \to [a,b]$, adjusted to (f,v), and let $a_0 = a, a_1, \ldots, a_{m+1} = b$ be the ordering sequence for ϕ. Set

$$V_k = \phi^{-1}(a_{k+1}), \quad G_k = (-v)^{\rightsquigarrow}_{[b,a_{k+1}]}, \quad H_k = v^{\rightsquigarrow}_{[a,a_{k+1}]}$$

so that G_k is a partially defined map $\partial_1 W \to V_k$ and H_k is a partially defined map $\partial_0 W \to V_k$. Let $\partial_1 W^k$, $\partial_0 W^k$ be Morse-Smale filtrations of the components of ∂W. It is easy to see that the condition $(\mathfrak{C})$ with respect to these filtrations holds for v if and only if for every k the three following conditions hold:

(R1) For every $x \in \partial_1 W^k$ the $(-v)$-trajectory starting at x reaches V_k.

(R2) For every $y \in \widehat{\partial_0 W}^{\,m-k-1}$ the v-trajectory starting at y reaches V_k.

(R3) $G_k(\partial_1 W^k) \cap H_k\Big(\widehat{\partial_0 W}^{\,m-k-2}\Big) = \varnothing$.

Any f-gradient w with $||w - v||$ sufficiently small, is also a ϕ-gradient (Proposition 3.40 of Chapter 4, page 150). We shall now prove that the properties (R1) – (R3) still hold for w if $||w - v||$ is sufficiently small. As for the properties (R1) and (R2) this follows immediately from Proposition 3.5 of Chapter 3 (page 102). Now let us proceed to the property (R3). Let A and B be disjoint open subsets in V_k such that

$$(-v)^{\rightsquigarrow}_{[b,a_{k+1}]}(\partial_1 W^k) \subset A, \quad v^{\rightsquigarrow}_{[a,a_{k+1}]}\Big(\widehat{\partial_0 W}^{\,m-k-2}\Big) \subset B.$$

Proposition 3.5 cited above implies that these conditions still hold if we replace the vector field v by the vector field w if $||w - v||$ is sufficiently small. The proof of the openness of $(\mathfrak{C})$ is now complete. $\square$

CHAPTER 9

Properties of cellular gradients

In this chapter we continue our study of gradient descent for cellular gradients. Let $f : W \to [a, b]$ be a Morse function, v a cellular f-gradient, and $\partial_1 W^k, \partial_0 W^k$ the corresponding Morse-Smale filtrations of $\partial_1 W$ and $\partial_0 W$. Recall the continuous map

$$(-v)^{\twoheadrightarrow} : \partial_1 W^k / \partial_1 W^{k-1} \to \partial_0 W^k / \partial_0 W^{k-1}$$

derived from the gradient descent (see Subsection 2.1 of Chapter 7, page 239). The homomorphism induced by this map in homology contains a lot of information about the gradient flow, and we reserve for it a special term: *the homological gradient descent.* We discuss its properties in Section 1 of this chapter.

The theory of cellular gradients will be applied in Part 4 to circle-valued Morse maps. We will work there with cobordisms W endowed with an isometry

$$I : \partial_0 W \to \partial_1 W.$$

Such cobordisms will be called *cyclic.* For a cyclic cobordism the map

$$\Theta = I \circ (-v)^{\leadsto}$$

is a partially defined dynamical system on $\partial_1 W$, whose properties are important for the circle-valued Morse theory. In Section 2 of the present chapter we introduce the notion of a *cyclic cellular gradient* for cyclic cobordisms. For a cyclic cellular gradient the Morse-Smale filtrations $\partial_1 W^k$ and $\partial_0 W^k$ can be chosen in such a way that

$$\partial_1 W^k = I(\partial_0 W^k) \quad \text{for every} \quad k$$

and therefore we obtain a self-map of the space $\partial_1 W^k / \partial_1 W^{k-1}$:

$$I \circ (-v)^{\twoheadrightarrow} : \partial_1 W^k / \partial_1 W^{k-1} \to \partial_1 W^k / \partial_1 W^{k-1}.$$

The iterations of this map yield a dynamical system whose properties are closely related to the properties of Θ. In particular we show in Section 2 that the Lefschetz zeta function counting the periodic orbits of the map Θ can be computed from the homological gradient descent.

1. Homological gradient descent

1.1. Homotopical stability of the homological gradient descent. Recall from Subsection 4.2 of Chapter 8 (page 279) that every f-gradient w sufficiently C^0-close to a cellular gradient v satisfies also the condition $(\mathfrak{C})$ with respect to the same Morse-Smale filtrations on $\partial_1 W$ and $\partial_0 W$. The next result follows immediately from Theorem 1.9 of Chapter 7 (page 236).

Theorem 1.1. *Let v be a cellular gradient of a Morse function $f : W \to [a,b]$. There is $\delta > 0$ such that for every f-gradient w with $||w - v|| < \delta$ the maps*

$$(-v)^{\twoheadrightarrow}, (-w)^{\twoheadrightarrow} : \partial_1 W^k / \partial_1 W^{k-1} \to \partial_0 W^k / \partial_0 W^{k-1}$$

are homotopic. □

The space $\partial_1 W^{k-1}$ is a deformation retract of its neighbourhood in $\partial_1 W^k$ (Proposition 3.15 of Chapter 3, page 107). Therefore we have a natural isomorphism

$$\widetilde{H}_*(\partial_1 W^k / \partial_1 W^{k-1}) \approx H_*(\partial_1 W^k, \partial_1 W^{k-1})$$

(see [**58**], p. 124).

Definition 1.2. The map induced by $(-v)^{\twoheadrightarrow}$ in homology will be denoted

$$\mathcal{H}_k(-v) : H_k(\partial_1 W^k, \partial_1 W^{k-1}) \to H_k(\partial_0 W^k, \partial_0 W^{k-1})$$

and called the *homological gradient descent.*

Corollary 1.3. *Let v be a cellular gradient of a Morse function $f : W \to [a,b]$. There is $\delta > 0$ such that for every f-gradient w with $||w - v|| < \delta$ the homomorphisms*

$$\mathcal{H}_k(-v),\ \mathcal{H}_k(-w) : H_k(\partial_1 W^k, \partial_1 W^{k-1}) \to H_k(\partial_0 W^k, \partial_0 W^{k-1})$$

are equal. □

1.2. Images of fundamental classes. Now we will show that the $(-v)^{\rightsquigarrow}$-images of fundamental classes of submanifolds of $\partial_1 W$ can be computed with the help of the homological gradient descent. Let N be an oriented k-dimensional submanifold of $\partial_1 W$, such that

$$N \subset \partial_1 W^k \quad \text{and} \quad N \setminus \operatorname{Int} \partial_1 W^{k-1} \quad \text{is compact.}$$

The fundamental class

$$[N] \in H_k(\partial_1 W^k, \partial_1 W^{k-1})$$

is defined (see page 190). The $(-v)^{\rightsquigarrow}$-image N' of N is an oriented k-dimensional submanifold of $\partial_0 W$.

Proposition 1.4. (1) *The manifold* $N' = (-v)^{\rightsquigarrow}(N)$ *is in* $\partial_0 W^k$ *and* $N \setminus \mathrm{Int}\, \partial_0 W^{k-1}$ *is compact.*

(2) *The fundamental class* $[N'] \in H_k(\partial_0 W^k, \partial_0 W^{k-1})$ *is equal to* $\mathcal{H}_k(-v)([N])$.

Proof. We can assume that f is ordered. Let $\{a_k\}$ be the ordering sequence for f. We can present the gradient descent from $\partial_1 W$ to $\partial_0 W$ as the composition of two steps: the first one corresponding to the descent from $\partial_1 W$ to

$$V = f^{-1}(a_{k+1}),$$

and the second one corresponding to the descent from V to $\partial_0 W$. Let us begin with the first step. Put

$$W_1 = f^{-1}([a_{k+1}, b]), \quad v_1 = v|W_1.$$

Let

$$M = (-v_1)^{\rightsquigarrow}(N) \subset V.$$

Consider the following subsets of V:

$$A = (-v_1)^{\rightsquigarrow}(\partial_1 W^k), \quad B = (-v_1)^{\rightsquigarrow}(\partial_1 W^{k-1}).$$

The condition $(\mathfrak{C})$ implies that every $(-v_1)$-trajectory starting at a point of $\partial_1 W^k$ reaches V, therefore we have the homeomorphism of pairs

$$(-v_1)^{\rightsquigarrow} : (\partial_1 W^k, \partial_1 W^{k-1}) \xrightarrow{\approx} (A, B)$$

and also a homeomorphism

$$(-v_1)^{\twoheadrightarrow} : \partial_1 W^k / \partial_1 W^{k-1} \xrightarrow{\approx} A/B.$$

Both A and B are compact, and $B \subset \mathrm{Int}\, A$.

Lemma 1.5. *The manifold* M *is in* A *and* $M \setminus \mathrm{Int}\, B$ *is compact, and*

$$[M] = (-v_1)_*^{\twoheadrightarrow}([N]),$$

where $[M]$ *is the fundamental class of* M *in* $H_*(A, B)$.

Proof. The lemma follows immediately from the functoriality properties of fundamental classes (see Subsection 4.1 of Chapter 5, page 189). □

Now let us proceed to the gradient descent map from $V = f^{-1}(a_{k+1})$ to $\partial_0 W$. Put

$$W_0 = f^{-1}([a, a_{k+1}]), \quad v_0 = v|W_0.$$

Choose δ sufficiently small so that for every critical point p of index $\leqslant k$ we have

$$D_\delta(p, v) \cap \partial_0 W \subset \operatorname{Int} \partial_0 W^{k-1}.$$

Put

$$D_\delta = V \cap \bigcup_{\operatorname{ind} p \leqslant k} D_\delta(p, -v)$$

and

$$A' = A \cup D_\delta, \quad B' = B \cup D_\delta.$$

The gradient descent from V to $\partial_0 W$ yields a continuous map

$$(-v_0)^{\twoheadrightarrow} : A'/B' \to \partial_0 W^k / \partial_0 W^{k-1}.$$

(Indeed, it is easy to check that the condition $(\mathcal{D})$ from Subsection 1.1 of Chapter 7 (page 233) is satisfied with respect to the pairs

$$(A', B') \quad \text{and} \quad (\partial_0 W^k, \partial_0 W^{k-1})$$

and the initial segment Σ containing all the critical points of $f|W_0$.) Recall that we denote $(-v)^{\rightsquigarrow}(N)$ by N'.

Lemma 1.6. *The set* $N' \setminus \operatorname{Int} \partial_0 W^{k-1}$ *is compact, and*

$$[N'] = (-v_0)_*^{\twoheadrightarrow}([M]).$$

Proof. We have

$$M \subset \operatorname{Int} B' \cup (M \setminus \operatorname{Int} B').$$

The $(-v_0)^{\twoheadrightarrow}$-image of $\operatorname{Int} B'$ is in $\operatorname{Int} \partial_0 W^{k-1}$ and the set $M \setminus \operatorname{Int} B'$ is a compact subset of the domain of definition of $(-v_0)^{\twoheadrightarrow}$. Therefore $(-v_0)^{\twoheadrightarrow}(M \setminus \operatorname{Int} B')$ is compact and the first assertion follows. Proceeding to the formula for the fundamental class $[N']$ put

$$C' = \operatorname{Int} B \cup \Big(V \cap \bigcup_{\operatorname{ind} p \leqslant k} B_{\delta/2}(p, -v)\Big).$$

Let $M_0 = M \setminus C'$; then M_0 is a submanifold of $A' \setminus C'$ and the fundamental class

$$[M_0] \in H_*(A' \setminus C', B' \setminus C')$$

is defined. The image of $[M_0]$ in $H_*(A', B')$ equals $[M]$ so it suffices to check that

$$(-v_0)_*^{\twoheadrightarrow}([M_0]) = [N']$$

where

$$(-v_0)^{\twoheadrightarrow} : (A' \setminus C') \big/ (B' \setminus C') \to \partial_0 W^k / \partial_0 W^{k-1}$$

is the map derived from the gradient descent. Since $A' \setminus C'$ is a compact subset of the domain of definition of $(-v_0)^{\twoheadrightarrow}$, we have

$$[(-v_0)^{\rightsquigarrow}(M_0)] = (-v_0)_*^{\rightsquigarrow}([M_0]) = (-v_0)_*^{\twoheadrightarrow}([M_0])$$

by functoriality (cf. 1.7 of Chapter 7, page 234).

Our proposition follows now from the Composition Lemma (page 235):

$$\mathcal{H}_k(-v)([N]) = (-v)_*^{\twoheadrightarrow}([N]) = \Big((-v_0)^{\twoheadrightarrow} \circ (-v_1)^{\twoheadrightarrow}\Big)_*([N]) = (-v_0)_*^{\twoheadrightarrow}([M]) = [N']. \quad \square$$

1.3. Descending discs. Let v be a cellular oriented gradient of a Morse function $f : W \to [a, b]$ and p a critical point of f; put $k = \operatorname{ind} p - 1$. It follows from the condition $(\mathfrak{C})$ that the sole

$$N(v) = D(p, v) \cap \partial_0 W$$

of the descending disc is a k-dimensional oriented submanifold of Int $\partial_0 W^k$, and the subset $N(v) \setminus \text{Int } \partial_0 W^k$ is compact. Therefore we have the fundamental class

$$[N(v)] \in H_k(\partial_0 W^k, \partial_0 W^{k-1}).$$

Proposition 1.7. *There is $\delta > 0$ such that for every f-gradient w with $||w - v|| < \delta$ we have*

$$[N(v)] = [N(w)].$$

Proof. Assuming that f is ordered, let $\{a_k\}$ be the ordering sequence for f and put $V = f^{-1}(a_{k+1})$. Let $M(v) = D(p, v) \cap V$ and let A be a small tubular neighbourhood of $M(v)$ in V. The pairs $(A, \varnothing)$ and $(\partial_0 W^k, \partial_0 W^{k-1})$ satisfy the condition $(\mathcal{D})$ from Definition 1.3 of Chapter 7, page 233, so we have a continuous map

$$(-v)^{\twoheadrightarrow} : A \to \partial_0 W^k / \partial_0 W^{k-1}.$$

An argument similar to the proof of Lemma 1.6, page 284 shows that

$$[N(v)] = (-v)_*^{\twoheadrightarrow}([M(v)]).$$

It follows from Proposition 3.17 of Chapter 5, page 185 that

$$[M(v)] = [M(w)] \in H_*(A)$$

if $||w - v||$ is small enough and w is oriented similarly to v. The homomorphism $(-v)_*^{\twoheadrightarrow}$ is also C^0-stable and our proposition is therefore established. $\square$

1.4. Multiplicities of cycles of $\partial_1 W$ at critical points. Let $f : W \to [a, b]$ be a Morse function on a cobordism W and v be a cellular f-gradient. Let $\partial_1 W^k$, $\partial_0 W^k$ be the corresponding Morse-Smale filtrations of $\partial_1 W$ and $\partial_0 W$. Let

$$x \in H_k(\partial_1 W^k, \partial_1 W^{k-1}).$$

The element

$$\mathcal{H}_k(-v)(x) \in H_k(\partial_0 W^k, \partial_0 W^{k-1})$$

can be considered as "the part of x which reaches $\partial_0 W$". In this subsection we show that it is possible also to describe "the part of x which hangs on the critical points". As in the preceding subsection we assume that the Morse function $f : W \to [a, b]$ is ordered, with the ordering sequence $\{a_k\}$. Put

$$W_1 = f^{-1}([a_{k+1}, b]), \quad v_1 = v|W_1.$$

The gradient descent yields an embedding

$$(-v_1)^{\leadsto} : (\partial_1 W^k, \partial_1 W^{k-1}) \to (V, B)$$

$$\text{where} \quad V = f^{-1}(a_{k+1}), \quad B = (-v_1)^{\leadsto}(\partial_1 W^{k-1}).$$

Let

$$W_0 = f^{-1}([a_k, a_{k+1}]), \quad v_0 = v|W_0.$$

Then W_0 is an elementary cobordism. Condition ($\mathfrak{C}$) implies that every $(-v)$-trajectory starting at a point of B reaches $\partial_0 W_0 = f^{-1}(a_k)$. Let

$$\bar{x} = (-v_1)^{\leadsto}_*(x) \in H_*(V, B).$$

In Subsection 3.6 of Chapter 5, page 186 we have associated to $\bar{x}$ and any point $p \in S_k(F|W_0)$ an integer $m(\bar{x}, p; v)$ (*the multiplicity of* $\bar{x}$ *with respect to* p). Intuitively these multiplicities are the coefficients in the decomposition of the cycle $\bar{x}$ with respect to the basis of the homology $H_k(W_0, \partial_0 W_0)$ formed by the descending discs of critical points.

Definition 1.8. The multiplicity of $\bar{x}$ at $p \in S_k(f)$ will be also called the *multiplicity of* x *at* p, and denoted by $m(x, p; v)$.

We obtain a homomorphism

$$x \mapsto m(x, p; v); \ H_*(\partial_1 W^k, \partial_1 W^{k-1}) \to \mathbf{Z}.$$

The next proposition follows directly from Proposition 3.20 of Chapter 5 (page 187).

Proposition 1.9. *Let* $N \subset \partial_1 W^k$ *be a smooth* k*-dimensional oriented submanifold of* $\partial_1 W$*, such that* $N \setminus \partial_1 W^{k-1}$ *is compact. Let* p *be a critical point of* f *of index* k*. Then*

$$N \mathbin{\#\!\!\!\!\#} (D(p, -v) \cap \partial_1 W) = m([N], p; v). \qquad \square$$

The multiplicities are stable with respect to C^0-small perturbations of the gradient:

Proposition 1.10. *Let $x \in H_k(\partial_1 W^k, \partial_1 W^{k-1})$. There is $\delta > 0$ such that for every f-gradient w with $||w - v|| < \delta$ and for every $p \in S_k(f)$ we have:*

$$m(x, p; v) = m(x, p; w).$$

Proof. Choose any compact $Q \subset V$ such that

$$B \subset \operatorname{Int} Q \quad \text{and} \quad Q \cap D(-v|W') = \varnothing.$$

Choose $\delta > 0$ sufficiently small so that for every f-gradient w with $||w - v|| < \delta$ any $(-w)$-trajectory starting at a point of $\partial_1 W^k$ reaches V and

$$(-w)^{\leadsto}_{[b, a_{k+1}]}(\partial_1 W^{k-1}) \subset \operatorname{Int} Q.$$

Let w be any f-gradient with $||w - v|| < \delta$. Put

$$\widetilde{x}(w) = \left((-w)^{\leadsto}_{[b, a_{k+1}]}\right)_*(x) \in H_*(V, Q).$$

It is easy to see that for every $x \in H_k(\partial_1 W^k, \partial_1 W^{k-1})$ we have

$$m(x, p; w) = m(\widetilde{x}(w), p; w).$$

Proposition 3.6 of Chapter 3 (page 103) implies that $\widetilde{x}(w) = \widetilde{x}(v)$ if w is sufficiently C^0-close to v. Applying Proposition 3.21 of Chapter 5 (page 188) we deduce that

$$m(\widetilde{x}(v), p; w) = m(\widetilde{x}(v), p; v)$$

if w is sufficiently C^0-close to v. The proof of our proposition is now complete. □

2. Cyclic cobordisms and iterations of the gradient descent map

The cobordisms which will appear in our applications are obtained by cutting a closed manifold along a regular level surface of a C^∞ map of the manifold to S^1. Such cobordisms have an additional structure: there is a diffeomorphism of the lower component of the boundary onto the upper component.

Definition 2.1. *A cyclic cobordism* W is a Riemannian cobordism together with a diffeomorphism $I : \partial_0 W \to \partial_1 W$ preserving the Riemannian metric.

Let W be a cyclic cobordism, $f : W \to [a, b]$ a Morse function on W, and v an f-gradient. The composition $\Theta = I \circ (-v)^{\leadsto}$ is a *self-map* of the manifold $\partial_1 W$. The iterations of this map can be considered as a dynamical system. Unfortunately the map Θ and its iterations are not everywhere defined, and the classical theory of dynamical systems can not be applied immediately. However in the case when the gradient v satisfies an appropriate analog of the condition ($\mathfrak{C}$) (see Definition 2.22) the gradient descent gives rise to a dynamical system of *graded-continuous self-maps* of the space $\partial_1 W$.[†] In this section we study this dynamical system, its fixed and periodic points. We obtain a formula expressing its Lefschetz zeta function in terms of the homological gradient descent (Theorem 2.29).

2.1. Fixed points and Lefschetz zeta functions of continuous maps. In this section we recall the definition and basic properties of Lefschetz zeta functions. Our basic reference for the fixed point theory is A. Dold's textbook [**29**], Chapter 7, §4 – 6; we shall use the terminology and results from this source.

Definition 2.2. Let X be a topological space, $U \subset X$ be an open subset, and $f : U \to X$ a continuous map. A point $p \in U$ is a *fixed point* of f if $f(p) = p$. The set of all fixed points of f will be denoted $\mathrm{Fix}(f)$.

A point $p \in U$ is a *periodic point* of f if for some natural number k we have:

$$f(p), f^2(p), \ldots, f^{k-1}(p) \in U \quad \text{and} \quad f^k(p) = p.$$

The minimal number k satisfying this condition is called *the least period* of p, and denoted by $l(p)$. The subset $\{f^k(p) \mid k \in \mathbf{N}\}$ where p is a periodic point is called a *finite orbit* of f. The set of all periodic points of f is denoted $\mathrm{Per}(f)$.

For any finite orbit ω of f the cardinality of ω equals the least period of any of the points $p \in \omega$. If $U = X$ and f is bijective, the finite orbits of f are exactly the finite orbits of the $\mathbf{Z}$-action induced by f.

Now we shall discuss the notion of *the index of a fixed point set.* Let us start with the case when $X = \mathbf{R}^m$ and $U \subset \mathbf{R}^m$ is an open subset. Let $f : U \to \mathbf{R}^m$ be a continuous map, and $p \in U$ an isolated fixed point of f. Put $g(x) = x - f(x)$, we obtain a continuous map of pairs

$$g : (U', U' \setminus p) \longrightarrow (\mathbf{R}^m, \mathbf{R}^m \setminus 0)$$

where U' is a neighbourhood of p, sufficiently small, so that p is a unique fixed point of f in U'. The homology groups

$$H_m(U', U' \setminus p), \; H_m(\mathbf{R}^m, \mathbf{R}^m \setminus 0)$$

are both isomorphic to $\mathbf{Z}$, and a choice of orientation in $\mathbf{R}^m$ yields a generator in each of these groups. The homomorphism in homology induced by g is identified therefore with a homomorphism $\mathbf{Z} \to \mathbf{Z}$, which is the multiplication by an integer called *the index of f at p*, and denoted $I(f, p)$. If f

[†] See the introduction to this part for the definition of graded-continuous maps

is of class C^∞, then $I(f,p)$ can be calculated in terms of $f'(p)$, as stated in the next proposition (we omit the proof).

Proposition 2.3. *Let p be an isolated fixed point of a diffeomorphism $f : U \to V$ where U, V are open sets in $\mathbf{R}^m$. Assume that 1 is not an eigenvalue of the linear map $f'(p) : \mathbf{R}^m \to \mathbf{R}^m$. Then*

$$I(f,p) = \operatorname{sgn}\big(\det(\mathrm{Id} - f'(p))\big). \qquad \square$$

Definition 2.4. Let $U \subset \mathbf{R}^m$ be an open subset and $f : U \to \mathbf{R}^m$ a continuous map such that $\mathrm{Fix}(f)$ is a finite set. The sum of the indices of all fixed points of f is called *the index of f*, or *the algebraic number of fixed points of f* or *the Lefschetz number of f* and denoted $L(f)$.

Now we proceed to a generalization of this construction. Recall that a subset X of a topological space Y is called a *neighbourhood retract* if there is an open neighbourhood $X' \subset Y$ of X such that X is a retract of X'. A topological space U is called a *Euclidean neighbourhood retract* (or *ENR* for brevity) if there is a homeomorphism of U onto a neighbourhood retract in $\mathbf{R}^m$. Let $U \subset Y$ be an open subset of a topological space Y, and assume that U is an ENR. If $h : U \to Y$ is a continuous map, such that the set of the fixed points of h is compact, *the index of h* is defined. This is an integer, which we shall denote $L(h)$. Referring to A. Dold's textbook for the definition (see [**29**], Ch. 7, Def. 5.10), we will only list some natural properties of the index, to be used later on.

(A) The index is *local*: If W is an open subset of U such that $\mathrm{Fix}(h) \subset W$, then $L(h) = L(h|W)$.
(B) The index of a constant map h equals 1 if $h(U) \in U$ and 0 if $h(U) \notin U$.
(C) The index is *additive*: If U is the union of a finite family of open sets U_i such that for every $i \neq j$ we have $U_i \cap U_j \cap \mathrm{Fix}(h) = \varnothing$, then
$$L(h) = \sum_i L(h|U_i).$$
(D) The index is *homotopy invariant*: let $H : U \times [a,b] \to Y$ be a homotopy, and assume that the set $\cup_t \mathrm{Fix}(H_t)$ is compact (where $H_t(x) = H(x,t)$); then $L(H_a) = L(H_b)$.
(E) When $U \subset \mathbf{R}^m$ is an open subset and $h : U \to \mathbf{R}^m$ a continuous map, and the set $\mathrm{Fix}(h)$ is finite, we have:
$$L(h) = \sum_{p \in \mathrm{Fix}(h)} I(h,p),$$

that is, $L(h)$ equals the algebraic number of fixed points of h, defined above.

The index of a continuous map is computable in homological terms. This is the contents of the celebrated *Lefschetz fixed point theorem*; to state it we need a definition.

Definition 2.5. Let $\{F_i\}_{i\in\mathbf{N}}$ be a sequence of vector spaces over some field K. Let $A : F_* \to F_*$ be a graded linear map, that is, a collection of linear maps $A_i : F_i \to F_i$. Assume that each of A_i is a linear map of finite rank (that is, the image of each A_i is finite-dimensional). The element

$$\mathcal{T}r(A) = \sum_i (-1)^i \mathrm{Tr}(A_i) \in K$$

is called *the graded trace* of A.

For example, a continuous map $f : X \to X$ of any topological space such that $H_*(X, \mathbf{Q})$ is finite-dimensional, induces a graded endomorphism f_* of $H_*(X, \mathbf{Q})$, and the graded trace $\mathcal{T}r f_*$ is defined.

Here is the Lefschetz fixed point theorem (the reader will find a more general statement in [**29**], Ch. 7, Prop. 6.6).

Theorem 2.6. *Let Y be a compact ENR and $f : Y \to Y$ a continuous map; let $f_* : H_*(Y, \mathbf{Q}) \to H_*(Y, \mathbf{Q})$ be the graded homomorphism induced by f in homology. Then the set* $\mathrm{Fix}(f)$ *is compact, the homomorphism f_* has a finite rank, and*

$$L(f) = \mathcal{T}r(f_*). \tag{1}$$

This theorem can be applied to iterations of a continuous map $f : Y \to Y$ of a Euclidean neighbourhood retract Y. Put

$$\eta_L(f) = \sum_{n=1}^{\infty} \frac{L(f^n)}{n} t^n,$$

so that $\eta_L(f)$ is a formal power series with rational coefficients. This series will be called the eta-function of f. The sequence $\{L(f^n)\}_{n\in\mathbf{N}}$ and the series $\eta_L(f)$ determine each other. The exponent of $\eta_L(f)$ is a power series with the constant term equal to 1. This exponent

$$\zeta_L(f) = \exp(\eta_L(f))$$

is called the *Lefschetz zeta function* of f. It is a much more known and used invariant than $\eta_L(f)$.

Theorem 2.7.

$$\zeta_L(f) = \prod_i \det(\mathrm{Id} - t \cdot f_{i,*})^{(-1)^{i+1}}$$

where $f_{i,*}$ is the endomorphism of $H_i(Y, \mathbf{Q})$ induced by f.

Proof. The proof consists in linear-algebraic manipulations with the Lefschetz fixed point formula.

Lemma 2.8. *For any square matrix A with coefficients in $\mathbf{C}$ we have the following equality of two power series in t:*

$$\log\det(\mathrm{Id} - tA) = -\sum_{k=1}^{\infty} \frac{1}{k}\,\mathrm{Tr}(tA)^k.$$

Proof. Applying the Jordan normal form theorem the equality above is easily reduced to the case when A is a 1×1-matrix. For this case our equality is just the Taylor decomposition of the function $\log(1 - \lambda t)$ in a neighbourhood of 0. □

Now to deduce our theorem it suffices to apply the previous lemma to the linear map $f_{i,*}$ for each i, add up the results with corresponding signs, and apply the exponential function. □

Exercise 2.9. Check Theorem 2.7 by a direct computation for the following particular cases:

(1) $f : Y \to Y$ is a constant map.
(2) $f : S^1 \to S^1$ is given by the formula $f(z) = \bar{z}$.

Corollary 2.10. *For a continuous map $f : Y \to Y$ of a compact ENR Y, the power series $\zeta_L(f)$ has integral coefficients.*

Proof. Choose any finite basis in the free abelian group $H_i(Y, \mathbf{Z})/Tors H_i(Y, \mathbf{Z})$. It yields a basis in the vector space $H_i(Y, \mathbf{Q})$. The matrix of the linear map $f_{i,*}$ with respect to this basis has integral coefficients. Thus the power series $\zeta_L(f)$ is a quotient of integer polynomials whose constant terms are 1, therefore it has integral coefficients. □

Let us now consider a particular case which will appear in our study of gradient descent maps.

Definition 2.11. Let X be a manifold, U an open subset of X, and $f : U \to X$ a C^∞ map. A fixed point p of f is called *hyperbolic* if the differential $f'(p) : T_pX \to T_pX$ is a hyperbolic linear map, that is, $f'(p)$ has no eigenvalues on the unit circle.

A hyperbolic fixed point is isolated in $\mathrm{Fix}(f)$, and it is also a hyperbolic fixed point of any power f^k of f. We have the direct sum decomposition

$$T_pX = T^e \oplus T^c$$

(T^e is the $f'(p)$-invariant subspace where $f'(p)$ is expanding, and T^c is the $f'(p)$-invariant subspace where $f'(p)$ is contracting). Put

$$\mu(f,p) = \dim T^e.$$

Let

$$\begin{aligned}
\varepsilon(f,p) &= 1 \qquad \text{if} \quad f'(p) \quad \text{conserves the orientation of} \quad T^e,\\
\varepsilon(f,p) &= -1 \qquad \text{if} \quad f'(p) \quad \text{reverses the orientation of} \quad T^e.
\end{aligned}$$

Proposition 2.3 implies that

$$I(f^k,p) = \varepsilon(f,p)^k \cdot (-1)^{\mu(f,p)} \qquad \text{for every} \quad k.$$

Let ω be a finite orbit of f; denote the cardinality of ω by $l(\omega)$. Choose any point $p \in \omega$ and put

$$\nu(\omega) = (-1)^{\mu(f^{l(\omega)},p)}, \quad \varepsilon(\omega) = \varepsilon(f^{l(\omega)},p);$$

it is clear that the numbers $\mu(f^{l(\omega)},p)$ and $\varepsilon(f^{l(\omega)},p)$ do not depend on the particular choice of $p \in \omega$.

The next proposition says that in favorable cases the zeta function can be factorized as an infinite product of terms corresponding to finite orbits.

Proposition 2.12 ([**37**], Prop.5.19). *Let X be a C^∞ manifold without boundary and $f : U \to X$ a C^∞ map where $U \subset X$ is an open subset. Assume that all periodic points of f are hyperbolic, and there is only a finite number of periodic points of any given period. The set of all finite orbits of f will be denoted $\mathcal{O}$; by our hypotheses on f this is a finite or countable set. We have then*

$$\eta_L(f) = \sum_{\omega\in\mathcal{O},k\geqslant 1} \frac{\nu(\omega)\varepsilon(\omega)^k}{k} t^{k\cdot l(\omega)},$$

$$\zeta_L(f) = \prod_{\omega\in\mathcal{O}} \left(1-\varepsilon(\omega)t^{l(\omega)}\right)^{-\nu(\omega)}.$$

Proof. Write

$$\eta_L(f) = \sum_{n=1}^{\infty} \frac{L(f^n)}{n} t^n = \sum_{n=1}^{\infty} \Big(\sum_{p\in\operatorname{Fix}(f^n)} \frac{I(f^n,p)}{n} t^n \Big).$$

The set $\operatorname{Fix}(f^n)$ is the disjoint union of a finite family of orbits of f. For each orbit $\omega \subset \operatorname{Fix}(f^n)$ its length $l(\omega)$ divides n and we have

$$\sum_{p\in\omega} \frac{I(f^n,p)}{n} t^n = l(\omega)\frac{I(f^n,p)}{n} t^n = \frac{1}{k}\varepsilon^k(\omega)\nu(\omega)t^{k\cdot l(\omega)},$$

where we denoted $k = \frac{n}{l(\omega)}$. Therefore

$$\eta_L(f) = \sum_{\omega\in\mathcal{O}} \nu(\omega) \sum_{k=1}^{\infty} \frac{1}{k} \left(\varepsilon(\omega)t^{l(\omega)}\right)^k = -\sum_{\omega\in\mathcal{O}} \nu(\omega) \log\left(1-\varepsilon(\omega)t^{l(\omega)}\right).$$

Thus

$$\zeta_L(f) = \exp(\eta_L(f)) = \prod_{\omega \in \mathcal{O}} \left(1 - \varepsilon(\omega) t^{l(\omega)}\right)^{-\nu(\omega)},$$

and the proof of our formulas is complete. □

Corollary 2.13. *In the assumption of the previous proposition the power series $\zeta_L(f)$ has integral coefficients.* □

The assumptions of the proposition are satisfied for example if X is compact, $U = X$, and all fixed points of f are hyperbolic. The assumptions hold also in the case of the maps derived from the gradient descent flows, (see Chapter 13).

2.2. Gradient descent on cyclic cobordisms. Now we proceed to iterations of the gradient descent map on cyclic cobordisms. Let $f : W \to [a, b]$ be a Morse function on a cyclic cobordism W. Let v be an f-gradient. The composition $\Theta = I \circ (-v)^{\leadsto}$ will be denoted in this subsection by Θ_v, since we shall consider more than one f-gradient at a time. Thus

$$\Theta_v : \partial_1 W \setminus D(-v) \to \partial_1 W$$

is a C^∞ map, which is a diffeomorphism onto its image.

Definition 2.14. An f-gradient v satisfies the *cyclic almost transversality condition*, if

(1) v is almost transverse;

(2) $I(\mathbb{D}(v) \cap \partial_0 W) \pitchfork \mathbb{D}(-v) \cap \partial_1 W$.

We shall abbreviate *f-gradient, satisfying the cyclic almost transversality condition* as *CAT-gradient.*

In this subsection we establish some basic properties of CAT-gradients. In particular we shall see that as far as the fixed point theory is concerned, the map Θ_v associated to a CAT-gradient behaves like a continuous map of a compact space to itself.

Proposition 2.15. *The subset of all CAT-gradients is open in the set $G(f)$ of all f-gradients with respect to C^0 topology.*

Proof. Let v be any CAT-gradient for f. Let m denote the dimension of W. For every k we have:

$$D(\text{ind}_{\leqslant m-k} ; -v) \cap I\Big(D(\text{ind}_{\leqslant k} ; v) \cap \partial_0 W\Big) = \varnothing.$$

Recall from Proposition 3.46 of Chapter 4 (page 151) that for every k the family

$$\{B_\delta(\text{ind}_{\leqslant k} ; v)\}_{\delta > 0}$$

is a fundamental system of neighbourhoods of $D(\text{ind}\leqslant k\ ; v)$. Similarly, the family

$$\{B_\delta(\text{ind}\leqslant m-k\ ; -v)\}_{\delta>0}$$

is a fundamental system of neighbourhoods of $D(\text{ind}\leqslant m-k\ ; -v)$. Therefore there is $\delta > 0$, such that the compact subsets

$$D_\delta(\text{ind}\leqslant m-k\ ; -v) \cap \partial_1 W \ \subset\ \partial_1 W \ \supset\ I\Big(D_\delta(\text{ind}\leqslant k\ ; v) \cap \partial_0 W\Big)$$

are disjoint. Let $U_k, V_k \subset \partial_1 W$ be disjoint open subsets such that

$$D_\delta(\text{ind}\leqslant m-k\ ; -v) \cap \partial_1 W \subset U_k, \quad I\Big(D_\delta(\text{ind}\leqslant k\ ; v) \cap \partial_0 W\Big) \subset V_k.$$

For every f-gradient w such that $||w - v||$ is sufficiently small, we have

$$D_\delta(\text{ind}\leqslant m-k\ ; -w) \cap \partial_1 W \subset U_k, \quad I\Big(D_\delta(\text{ind}\leqslant k\ ; w) \cap \partial_0 W\Big) \subset V_k$$

(see Proposition 3.8 of Chapter 3, page 104), and this establishes the CAT property for w if $||w - v||$ is sufficiently small. □

Exercise 2.16. Prove that the subset of CAT-gradients is dense in the set of all f-gradients with respect to C^∞ topology.

Now we shall study the properties of the fixed point set $\text{Fix}(\Theta_v)$.

Proposition 2.17. *Let v be a CAT-gradient for f. There is $\delta > 0$ such that*

$$\text{Fix}(\Theta_v) \cap D_\delta(-v) = \varnothing.$$

Proof. It suffices to prove that for every integer k with $0 \leqslant k \leqslant m$ we have

$$\text{Fix}(\Theta_v) \cap D_\delta(\text{ind}{=}m-k\ ; -v) = \varnothing.$$

Pick any $\delta > 0$ such that

$$(2) \qquad \Big(D_\delta(\text{ind}\leqslant m-k\ ; -v) \cap \partial_1 W\Big) \cap I\Big(D_\delta(\text{ind}\leqslant k\ ; v) \cap \partial_0 W\Big) = \varnothing.$$

For $x \in D_\delta(\text{ind}{=}m-k\ ; -v)$ the $(-v)$-trajectory starting at x can quit W only through a point in $D_\delta(\text{ind}{=}k\ ; v) \cap \partial_0 W$. Therefore if

$$x \in \text{Fix}(\Theta_v) \cap D_\delta(\text{ind}{=}m-k\ ; -v),$$

then $(-v)^{\leadsto}(x)$ is in $D_\delta(\text{ind}{=}k\ ; v)$, and this contradicts (2). □

Proposition 2.18. *If v is a CAT-gradient, then the set $\text{Fix}(\Theta_v)$ is compact.*

Proof. We must prove that any sequence of points $a_n \in \mathrm{Fix}(\Theta_v)$ contains a subsequence converging to a point in $\mathrm{Fix}(\Theta_v)$. The space $\partial_1 W$ is compact, and extracting a subsequence if necessary, we can assume that $a_n \to a \in \partial_1 W$. Proposition 2.17 implies that $a \notin D(-v)$, so that a is in the domain of definition of $(-v)^{\rightsquigarrow}$, and we have

$$(-v)^{\rightsquigarrow}(a_n) \to (-v)^{\rightsquigarrow}(a),$$

and therefore

$$a_n = I\Big((-v)^{\rightsquigarrow}(a_n)\Big) \to I\Big((-v)^{\rightsquigarrow}(a)\Big).$$

Since $a_n \to a$, we deduce $I \circ (-v)^{\rightsquigarrow}(a) = a$ and the proof of our proposition is complete. □

The set $\partial_1 W \setminus D(-v)$ is an ENR (as it follows from Proposition 8.10 of [**29**], Chapter 4). The set of fixed points of the continuous map $\Theta_v : \partial_1 W \setminus D(-v) \to \partial_1 W$ is compact by the previous proposition, therefore we can consider the *index* of the map $\Theta_v : \partial_1 W \setminus D(-v) : \partial_1 W$ (as discussed on page 289). This index will be denoted by

$$L(v) = L(\Theta_v);$$

it is invariant under C^0-small perturbations of v, as the next proposition shows.

Proposition 2.19. *Let v be a CAT-gradient for f. Then there is $\delta > 0$ such that for every f-gradient w with $||v - w|| < \delta$ we have:*

$$L(v) = L(w).$$

Proof. The main step of the proof is the next lemma.

Lemma 2.20. *There is an open neighbourhood U of* $\mathrm{Fix}(\Theta_v)$ *and real numbers $\delta, \epsilon > 0$ such that for every w with $||w - v|| < \epsilon$ we have:*

(A) $$\overline{U} \cap D_\delta(-w) = \varnothing.$$

(B) $$\mathrm{Fix}(\Theta_w) \subset U.$$

Proof. Pick a $\delta > 0$ such that

$$\mathrm{Fix}(\Theta_v) \cap D_\delta(-v) = \varnothing$$

(see Proposition 2.17). Pick an open neighbourhood $U \subset \partial_1 W$ of the set $\mathrm{Fix}(\Theta_v)$ such that

$$\overline{U} \cap D_\delta(-v) = \varnothing.$$

Proposition 3.8 of Chapter 3 (page 104) implies that for some $\epsilon > 0$ every f-gradient w with $||w - v|| < \epsilon$ has the property (A):

$$\overline{U} \cap D_\delta(-w) = \varnothing.$$

In particular the open set U is in the domain of definition of Θ_w if $||w-v|| < \epsilon$. Diminishing further the numbers δ, ϵ if necessary, we can find, for every k, disjoint open subsets $U_k, V_k \subset \partial_1 W$ such that

$$D_\delta(\text{ind}\leqslant m-k\ ; -w) \cap \partial_1 W \subset U_k, \quad I\Big(D_\delta(\text{ind}\leqslant k\ ; w) \cap \partial_0 W\Big) \subset V_k$$

for every f-gradient w with $||w-v|| < \epsilon$ (see the proof of Proposition 2.15). In particular

$$\text{Fix}(\Theta_v) \cap D_\delta(-v) = \varnothing.$$

We are going to prove that, diminishing ϵ if necessary, we can achieve the property (B). The contrary would mean that there is a sequence w_n of f-gradients and a sequence $x_n \in \partial_1 W \setminus U$ such that

$$||w_n - v|| \to 0 \quad \text{and} \quad \Theta_{w_n}(x_n) = x_n.$$

Since $\partial_1 W$ is compact we can assume that $x_n \to x \in \partial_1 W \setminus U$. Observe that $x \notin D(-v)$. Indeed, suppose that for some k we have

$$x \in D(p, -v) \quad \text{with} \quad \text{ind}\, p = k.$$

Choose any η with $0 < \eta < \delta$. For all n sufficiently large we have

$$x_n \in D_\eta(\text{ind}\leqslant m-k\ ; -v) \cap \partial_1 W;$$

diminishing ϵ we can assume that

$$D_\eta(\text{ind}\leqslant m-k\ ; -v) \subset D_\delta(\text{ind}\leqslant m-k\ ; -w)$$

for every w with $||w - v|| < \epsilon$ (see Corollary 3.10 of Chapter 3, page 105). Thus we have

$$x_n \in D_\delta(\text{ind}\leqslant m-k\ ; -w_n) \cap \partial_1 W \subset U_k$$

and

$$(-w_n)^{\leadsto}(x_n) \in D_\delta(\text{ind}\leqslant k\ ; w_n) \cap \partial_0 W.$$

Therefore

$$\Theta_{w_n}(x_n) = I\big((-w_n)^{\leadsto}(x_n)\big) \subset V_k,$$

and the equality $x_n = \Theta_{w_n}(x_n)$ can not be true since U_k and V_k are disjoint; we obtain a contradiction.

Thus we have proved that $x \notin D(-v)$. Then the equality $\Theta_{w_n}(x_n) = x_n$ implies $\Theta_v(x) = x$ by continuity (see Theorem 2.13 of Chapter 1, page 30), hence $x \in U$, which leads to a contradiction. The proof of our lemma is now complete. □

Now we are ready to prove our proposition. Choose an $\epsilon > 0$ as in the previous lemma and for $||w - v|| < \epsilon$ put $w_t = v + t(w - v)$. Recall that the space of all f-gradients is convex (by Proposition 2.13 of Chapter 2, page 55) so w_t is an f-gradient for every t. We are going to prove that

$$L(w_t) = I(\Theta_{w_t})$$

does not depend on t. For this purpose observe that

$$I(\Theta_{w_t}) = I(\Theta_{w_t}|U)$$

(as it follows from the previous proposition). Therefore it suffices to prove that the number $I(\Theta_{w_t}|U)$ does not depend on t. The family

$$\{\Theta_{w_t}|U\}_{t\in[0,1]}$$

is a one-parameter family of continuous maps $U \to \partial_1 W$ and we are going now to apply the homotopy invariance of the index (property (D), page 289). For this it remains to prove the next lemma.

Lemma 2.21. *The set*

$$\bigcup_t \mathrm{Fix}(\Theta_{w_t})$$

is compact.

Proof. Let $x_n \in \mathrm{Fix}(w_{t_n})$ be a sequence of points, where $t_n \in [0,1]$. Extracting from x_n and t_n converging sequences if necessary, we can assume that $t_n \to \alpha$ and $x_n \to y$; it remains to prove that y is a fixed point of Θ_{w_α}. Observe that $x_n \in U$, therefore $y \in \overline{U}$ and $y \notin D(-w_\alpha)$. Thus y is in the domain of definition of $(-w_\alpha)^{\leadsto}$ and we have

$$(-w_{t_n})^{\leadsto}(x_n) \to (-w_\alpha)^{\leadsto}(y).$$

Therefore y is a fixed point of Θ_{w_α}. The lemma and the proposition are proved. □

2.3. Cyclic cellular gradients. Now we have the notion of index for the gradient descent map associated to CAT gradients. We would like to obtain an expression for it in homological terms, thus generalizing the Lefschetz fixed point formula and extending the analogy with continuous maps of closed manifolds. It is possible if we impose an additional restriction on the gradient v, namely a corresponding version of the condition $(\mathfrak{C})$:

Definition 2.22. Let $f : W \to [a,b]$ be a Morse function on a cyclic cobordism W, and v be an f-gradient. We say that v satisfies condition $(\mathfrak{C}')$ if v satisfies condition $(\mathfrak{C})$ (page 238), and, moreover, the corresponding Morse-Smale filtrations $\partial_1 W^k, \partial_0 W^k$ can be chosen in such a way that $I(\partial_0 W^k) = \partial_1 W^k$ for every k. The gradients satisfying condition $(\mathfrak{C}')$ will be also called *cyclic cellular gradients.*

The next theorem implies that the set of all f-gradients satisfying $(\mathfrak{C}')$ is open and dense in $G(f)$ with respect to C^0 topology.

Theorem 2.23. *Let $f : W \to [a,b]$ be a Morse function on a cyclic cobordism W. The subset of all cyclic cellular f-gradients is open and dense in the space of all f-gradients with respect to C^0 topology. Moreover, for any almost transverse f-gradient v, a number $\epsilon > 0$ and a neighbourhood U of ∂W, there is a cyclic cellular f-gradient w, with $||w - v|| < \epsilon$, and* $\operatorname{supp}(w - v) \subset U \setminus \partial W$.

Proof. The proof of C^0-openness of condition ($\mathfrak{C}'$) is completely similar to the corresponding argument from Subsection 4.2 of Chapter 8 (page 279). For the proof of C^0-density just note that in the proof of C^0-density of the condition ($\mathfrak{C}$) (Subsection 4.1 of Chapter 8, page 274) we can choose $\phi_0 = \phi_1 \circ I,\ u_1 = I_*(u_0)$. □

Any cyclic cellular f-gradient v satisfies the cyclic almost transversality condition, and, therefore a fixed point set of the map $\Theta = I \circ (-v)^{\rightsquigarrow}$ † is a compact subset of $\partial_1 W$. The set $\partial_1 W$ has now a filtration $\partial_1 W^k$, which satisfies the following property:

$$\Theta(\partial_1 W^k) \subset \operatorname{Int} \partial_1 W^k \quad \text{for every} \quad k.$$

Put $U_k = \operatorname{Int} \partial_1 W^k \setminus \partial_1 W^{k-1}$. The fixed point set $\operatorname{Fix}(\Theta)$ falls naturally into a disjoint union of the sets

$$\mathcal{F}_k = \operatorname{Fix}(\Theta) \cap U_k.$$

Proposition 2.24. *For every k the set $\mathcal{F}_k$ is compact.*

Proof. By definition we have $\mathcal{F}_k \subset \operatorname{Int} \partial_1 W^k \setminus \partial_1 W^{k-1}$. The sets $\operatorname{Int} \partial_1 W^k \setminus \partial_1 W^{k-1}$ are open and disjoint and the union of the sets $\mathcal{F}_k$ is compact. The proposition follows. □

Now we shall obtain a formula for $L(\Theta)$ in terms of the homological gradient descent associated to v. Consider the continuous map

$$\Theta_k = I \circ (-v)^{\twoheadrightarrow} : \partial_1 W^k / \partial_1 W^{k-1} \to \partial_1 W^k / \partial_1 W^{k-1}.$$

To apply the usual Lefschetz fixed point theory to this map we shall first of all make sure that the space $\partial_1 W^k / \partial_1 W^{k-1}$ has good geometric properties.

Lemma 2.25. *The space $\partial_1 W^k / \partial_1 W^{k-1}$ is an ENR.*

Proof. The space $\partial_1 W^k$ is a compact manifold, therefore an ENR. The quotient of an ENR space by a compact ENR subset is again an ENR, see [**29**, exercise 5 of Chapter 4]. Thus $\partial_1 W^k / \partial_1 W^{k-1}$ is also an ENR, as we have claimed. □

† The notation Θ_v used in the previous subsection will be replaced now by Θ, since we work here with only one f-gradient, and no confusion is possible.

Therefore we can apply the Lefschetz fixed point formula and the Lefschetz number $L(\Theta_k)$ equals the graded trace of the homomorphism induced by $\Theta_k = I \circ (-v)^{\twoheadrightarrow}$ in homology. The space $\partial_1 W^k / \partial_1 W^{k-1}$ has the homotopy type of the wedge of k-spheres and we have:

$$L(\Theta_k) =$$
$$1 + (-1)^k \mathrm{Tr}\Big((\Theta_k)_* : H_k(\partial_1 W^k / \partial_1 W^{k-1}) \to H_k(\partial_1 W^k / \partial_1 W^{k-1}) \Big)$$
$$= 1 + (-1)^k \mathrm{Tr}\big(\overline{\mathcal{H}}_k(-v)\big),$$
$$\text{where} \quad \overline{\mathcal{H}}_k(-v) = I_* \circ \mathcal{H}_k(-v);$$

recall that $\mathcal{H}_k(-v)$ is the homological gradient descent. The fixed point set of the map Θ_k is easy to identify:

$$\mathrm{Fix}(\Theta_k) = \{\omega_k\} \sqcup \bar{\mathcal{F}}_k,$$

where $\bar{\mathcal{F}}_k$ stands for the image of $\mathcal{F}_k$ under the projection $\partial_1 W^k \to \partial_1 W^k / \partial_1 W^{k-1}$, and ω_k is the point corresponding to the collapsed set $\partial_1 W^{k-1}$.

The point ω_k is an isolated fixed point of Θ_k and it follows from condition $(\mathfrak{C}')$ that a whole neighbourhood of ω_k is sent to ω_k by Θ_k. Therefore the index of ω_k is equal to 1, and applying the formula above for the number $L(\Theta_k)$ we deduce

$$I(\Theta | U_k) = (-1)^k \mathrm{Tr}\, \overline{\mathcal{H}}_k(-v). \tag{3}$$

Now we are ready to establish the Lefschetz fixed point formula for the map Θ. Put

$$\mathcal{Z}_k = H_k(\partial_1 W^k, \partial_1 W^{k-1}; \mathbf{Q}).$$

The maps $\overline{\mathcal{H}}_k(-v)$ form a graded linear map of the graded vector space $\mathcal{Z}_*$ to itself. Adding up the equalities (3) we obtain the next proposition.

Proposition 2.26.

$$L(\Theta) = \mathcal{T}r\big(\overline{\mathcal{H}}_*(-v)\big). \qquad \square$$

2.4. Lefschetz zeta functions for cyclic cellular gradients. For cyclic cellular gradients the iterations of the gradient descent map also have good fixed point properties:

Proposition 2.27. *Let $f : W \to [a, b]$ be a Morse function on a cyclic cobordism W, and v a cyclic cellular f-gradient. Let n be a positive integer. Then:*

(1) *There is $\delta > 0$ such that*

$$\mathrm{Fix}(\Theta^n) \cap B_\delta(-v) = \varnothing, \quad \textit{where} \quad \Theta = I \circ (-v)^{\leadsto}.$$

(2) *The set* $\mathrm{Fix}(\Theta^n)$ *is compact.*

Proof. 1) It suffices to prove that for some $\delta > 0$ and any k the sets $\mathrm{Fix}(\Theta^n)$ and $B_\delta(\mathrm{ind}{=}k\ ; -v)$ are disjoint. The condition $(\mathfrak{C}')$ implies that

$$\partial_1 W \cap D(\mathrm{ind}{\leqslant}k\ ; -v) \subset \mathrm{Int}\ \widehat{\partial_1 W}^{k-1}.$$

The family

$$\{B_\delta(\mathrm{ind}{\leqslant}k\ ; -v)\}_{\delta>0}$$

is a fundamental system of neighbourhoods of $D(\mathrm{ind}{\leqslant}k\ ; -v)$ therefore for some $\delta > 0$ we have

$$\partial_1 W \cap B_\delta(\mathrm{ind}{\leqslant}k\ ; -v) \subset \mathrm{Int}\ \widehat{\partial_1 W}^{k-1}.$$

If a point $x \in \mathrm{Fix}(\Theta^n)$ is in $B_\delta(\mathrm{ind}{=}k\ ; -v)$, then

$$(-v)^{\leadsto}(x) \in B_\delta(\mathrm{ind}{=}m-k\ ; v) \cap \partial_0 W,$$

where $m = \dim W$. Diminishing δ if necessary we can assume that

$$B_\delta(\mathrm{ind}{=}m-k\ ; v) \cap \partial_0 W \subset \mathrm{Int}\ \partial_0 W^{m-k-1},$$

and

$$I\Big(B_\delta(\mathrm{ind}{=}m-k\ ; v) \cap \partial_0 W\Big) \subset \mathrm{Int}\ \partial_1 W^{m-k-1}.$$

Condition $(\mathfrak{C}')$ implies that the set $\mathrm{Int}\ \partial_1 W^{m-k-1}$ is Θ-invariant, therefore $\Theta^n(x) \in \mathrm{Int}\ \partial_1 W^{m-k-1}$, and

$$x = \Theta^n(x) \in \mathrm{Int}\ \partial_1 W^{m-k-1}$$

which is impossible since $x \in \mathrm{Int}\ \widehat{\partial_1 W}^{k-1}$ and by the definition of the dual filtration the sets

$$\mathrm{Int}\ \widehat{\partial_1 W}^{k-1}, \quad \mathrm{Int}\ \partial_1 W^{m-k-1}$$

are disjoint.

2) To prove that $\mathrm{Fix}(\Theta^n)$ is a closed subset of $\partial_1 W$, let $x_l \in \mathrm{Fix}(\Theta^n)$ be a sequence of points converging to $y \in \partial_1 W$. We are going to prove that $y \in \mathrm{Fix}(\Theta^n)$. Choose $\delta > 0$ as in part 1) of the proof. Since $x_l \in \partial_1 W \setminus B_\delta(-v)$, the point $y \in \partial_1 W \setminus B_\delta(-v)$ is in the domain of definition of $(-v)^{\leadsto}$ and

$$\Theta(x_l) \to \Theta(y).$$

Applying the same argument to the sequence $\Theta(x_l) \in \mathrm{Fix}(\Theta^n)$ we deduce that $\Theta(y) \in \partial_1 W \setminus B_\delta(-v)$ and

$$\Theta^2(x_l) \to \Theta^2(y).$$

Arguing by induction we show that for every $s \in \mathbf{N}$ we have:

$$\Theta^{s-1}(y) \in \partial_1 W \setminus B_\delta(-v) \quad \text{and} \quad \Theta^s(x_l) \to \Theta^s(y).$$

Put $s = n$ to deduce $x_l \to \Theta^n(y)$, and the proof is complete. □

The same arguments as in Subsection 2.3 prove the Lefschetz fixed point formula for iterations of the map Θ:

Proposition 2.28. *For every $n \geqslant 0$ we have*

$$L(\Theta^n) = \mathcal{T}r\Big(\overline{\mathcal{H}}_*^n(-v)\Big). \qquad \square$$

Now let us turn to Lefschetz zeta functions. Similarly to the case of continuous maps we define two power series in one variable t:

$$\eta_L(\Theta) = \sum_{n=1}^{\infty} \frac{1}{n}\, L(\Theta^n)t^n, \qquad \zeta_L(\Theta) = \exp\big(\eta_L(\Theta)\big).$$

The next theorem is the main aim of the present subsection. It is deduced from Proposition 2.28 in the same way as Theorem 2.7 was deduced from Theorem 2.6.

Theorem 2.29. *Let $f : W \to [a, b]$ be a Morse function on a cyclic cobordism W of dimension m. Let v be a cyclic cellular gradient for f and put $\Theta = I \circ (-v)^{\leadsto}$. Then we have the following formula for the Lefschetz zeta function of Θ:*

$$\zeta_L(\Theta) = \prod_{i=0}^{m-1} \det\Big(\mathrm{Id} - t\cdot\overline{\mathcal{H}}_i(-v)\Big)^{(-1)^{i+1}}$$

(where $\overline{\mathcal{H}}_i(-v) = I_* \circ \mathcal{H}_i(-v)$ and $\mathcal{H}_i(-v)$ is the homological gradient descent associated to v). $\square$

3. Handle-like filtrations associated with the cellular gradients

The Morse stratification associated with a Morse function and its almost transverse gradient is a perfect substitute for CW decompositions while we stay in the realm of closed manifolds. The union of all Morse cells is the manifold itself, and the Morse complex computes the homology of the manifold. The same construction, applied to a Morse function $f : W \to [a, b]$ on a cobordism W and its almost transverse gradient v, gives an analog of the cellular decomposition of the *pair* $(W, \partial_0 W)$. The Morse complex in this case computes the relative homology $H_*(W, \partial_0 W)$. As for the homotopy type of the space W itself, it can not be deduced from this Morse-theoretic data. (A good example is the product cobordism $V \times [0, 1]$, where V is a closed manifold, and the Morse function is the projection onto the second factor.) One natural way to improve the Morse-theoretic approach to cobordisms would be to include into the picture the Morse stratifications of $\partial_1 W$ and $\partial_0 W$. Intuitively, we would aim at a cellular decomposition of W, with cells of the following four types:

(1) the Morse cells of $\partial_1 W$,
(2) the Morse cells of $\partial_0 W$,
(3) the Morse cells of W,
(4) the $(-v)$-tracks of the Morse cells of $\partial_1 W$.

The boundary of each cell must be in the union of the cells of smaller dimensions. In particular for every critical point p of f the sole $D(p,v) \cap \partial_0 W$ of its descending disc must belong to the $(\operatorname{ind} p - 1)$-skeleton of $\partial_0 W$. Let us write down this condition:

$$D(\operatorname{ind}_{\leqslant k}, v) \cap \partial_0 W \subset \partial_0 W^{[k-1]} \quad \text{for every} \quad k. \tag{4}$$

Here $\partial_0 W^{[k-1]}$ is the union of all Morse cells of $\partial_0 W$ of dimensions $\leqslant k-1$. A similar requirement must hold for the tracks of the Morse cells of $\partial_1 W$. The task of construction of Morse stratifications of W, $\partial_1 W$ and $\partial_0 W$ satisfying the conditions above seems quite a delicate one. I do not know if it is feasible in general.

If in the formula (4) we replace the Morse skeleton $\partial_0 W^{[k-1]}$ by the $(k-1)$-th term $\partial_0 W^{k-1}$ of some Morse-Smale filtration, we obtain the following condition:

$$D_\delta(\operatorname{ind}_{\leqslant k}, v) \cap \partial_0 W \subset \partial_0 W^{k-1}.$$

This is a part of condition $(\mathfrak{C})$ (page 238). Actually condition $(\mathfrak{C})$ can be considered as the "thickened analog" of the four conditions above plus similar conditions for the function $-f$ and its gradient $-v$.

In this chapter we shall construct for any cellular f-gradient v a handle-like filtration of the cobordism W, which allows us in particular to recover the homology of the cobordism from the Morse-theoretic data associated with the Morse stratifications of $\partial_0 W$, $\partial_1 W$ and W (see Corollary 3.4 and Theorem 3.8). These techniques will be used later in the proof of the main theorem of Chapter 13.

3.1. Definition of the filtration. Let $f : W \to [a,b]$ be a Morse function on a cobordism W, and v a cellular f-gradient. Denote by $\partial_1 W^k$, $\partial_0 W^k$ the corresponding Morse-Smale filtrations of the components of the boundary. Let ϕ be any ordered Morse function on W, adjusted to (f,v), and $\{a_k\}$ be the ordering sequence for ϕ. Put

$$Z_k = \Big(T(\partial_0 W^k, v) \cup D(-v)\Big) \cap \phi^{-1}\big([a, a_{k+1}]\big).$$

In other words Z_k consists of all $z \in \phi^{-1}\big([a, a_{k+1}]\big)$ such that the $(-v)$-trajectory $\gamma(z,t;-v)$ either

(1) converges to a critical point of ϕ as $t \to \infty$ or
(2) reaches $\partial_0 W$ and intersects it at a point in $\partial_0 W^k$.

The set Z_k contains $\partial_0 W^k$ and the descending discs $D(p, v)$ for $\operatorname{ind} p \leqslant k$. Set

$$T_k = T(\partial_1 W^{k-1}; -v) \cap \phi^{-1}([a_k, b])$$

In other words T_k is the set of all points $y \in \phi^{-1}([a_k, b])$ such that the v-trajectory starting at y reaches $\partial_1 W$ and intersects it at $\partial_1 W^{k-1}$. In Figures 35 and 36 the exterior rectangles depict the cobordism W, and the sets T_k, Z_k are shaded.

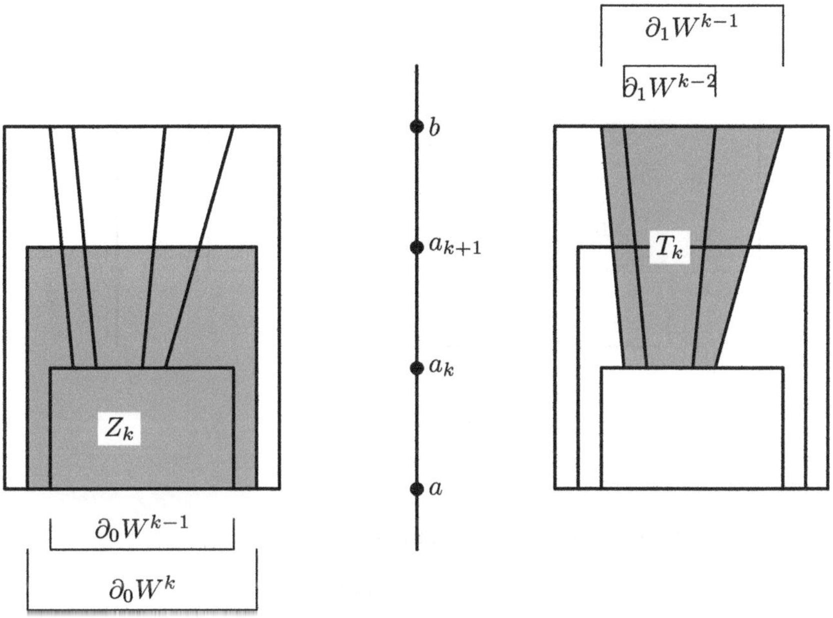

FIGURE 35.

Now we can define the filtration of W: set

$$W^{\langle k \rangle} = \partial_1 W^k \cup Z_k \cup T_k \quad \text{for} \quad k \geqslant 0,$$
$$W^{\langle -1 \rangle} = \varnothing.$$

Thus $W^{\langle k \rangle}$ contains the Morse cells of dimension $\leqslant k$ of the manifolds $\partial_0 W$, $\partial_1 W$, W, and also the tracks of the Morse cells of dimension $\leqslant k-1$ of $\partial_1 W$.

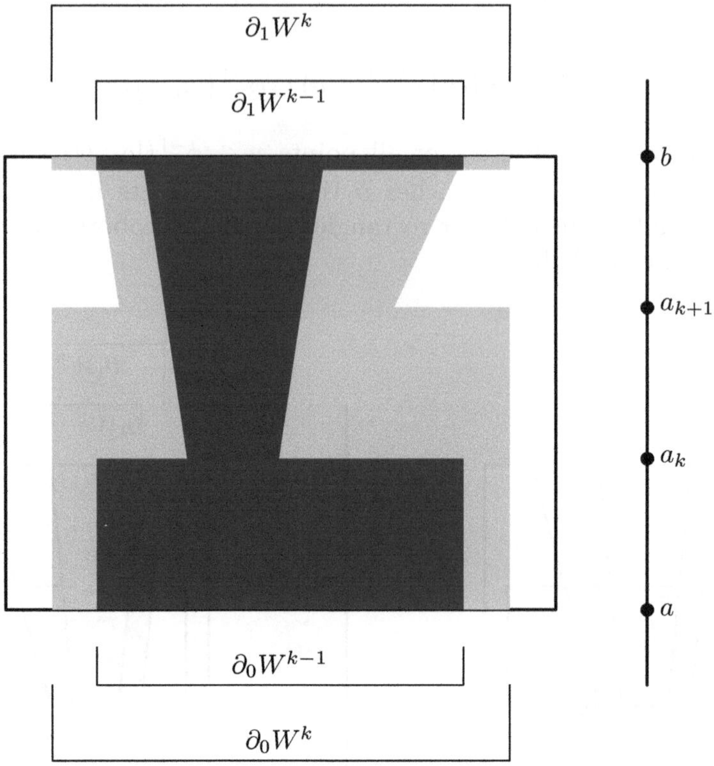

The set $W^{\langle k-1\rangle}$ is shaded black,

and the set $W^{\langle k\rangle} \setminus W^{\langle k-1\rangle}$ is shaded grey.

FIGURE 36.

The main aim of this subsection is to prove that the sets $W^{\langle k\rangle}$ form a cellular filtration of W. During the proof we shall show that the space $W^{\langle k\rangle}$ is homotopy equivalent to a space obtained from $W^{\langle k-1\rangle}$ by attaching four families of k-dimensional cells, corresponding to the families (1)–(4) on page 302. Let

$$E_k = H_k(W^{\langle k\rangle}, W^{\langle k-1\rangle}).$$

Theorem 3.1 below gives a description of the group E_k. To state this theorem we need some preliminaries. The manifolds $\partial_0 W, \partial_1 W$ are endowed with Morse-Smale filtrations satisfying conditions ($\mathfrak{C}1$), ($\mathfrak{C}2$) (page 238). Let us denote

$$\phi_0 : \partial_0 W \to \mathbf{R}, \quad \phi_1 : \partial_1 W \to \mathbf{R}$$

the corresponding ordered Morse functions, and let u_0, u_1 be their gradients. Put

$$\mathcal{M}_*^{(0)} = \mathcal{M}_*(\phi_0, u_0), \quad \mathcal{M}_*^{(1)} = \mathcal{M}_*(\phi_1, u_1).$$

The inclusions

$$\partial_0 W \hookrightarrow W \hookleftarrow \partial_1 W$$

preserve filtrations:

$$\partial_0 W^k \hookrightarrow W^{\langle k \rangle} \hookleftarrow \partial_1 W^k,$$

and yield homomorphisms in the corresponding adjoint chain complexes:

$$\mathcal{M}_*^{(0)} \xrightarrow{\lambda_0} E_* \xleftarrow{\lambda_1} \mathcal{M}_*^{(1)}.$$

Here we use the identification of the Morse complex with the adjoint chain complex of the Morse-Smale filtration, see Proposition 1.8 of Chapter 6 (page 200).

Choose any orientation of v and let $\mathcal{M}_* = \mathcal{M}_*(f, v)$ denote the Morse complex of (f, v). It follows from the condition ($\mathfrak{C}$) that for every $p \in S_k(f)$ the descending disc $D(p, v)$ has a well defined fundamental class $[p]$ in $H_k(W^{\langle k \rangle}, W^{\langle k-1 \rangle})$ and the corresponding map $p \mapsto [p]$ yields a group homomorphism

$$\mathcal{M}_k \xrightarrow{\mu} E_k.$$

The images of λ_1, λ_0 and μ correspond to the components (1)–(3) of our hypothetical cellular decomposition (see page 302). Now let us proceed to the fourth component. Condition ($\mathfrak{C}$) implies that every $(-v)$-trajectory starting at a point of $\partial_1 W^{k-1}$ reaches $\phi^{-1}(a_k)$. Let $I = [0, 1]$. By Proposition 3.3 of Chapter 3 (page 101) we have a canonical homeomorphism

$$(5) \quad F : I \times \partial_1 W^{k-1} \xrightarrow{\approx} T_k = T(\partial_1 W^{k-1}; -v) \cap \phi^{-1}([a_k, b]),$$
$$F(1 \times x) = x; \quad F(0 \times x) \in \phi^{-1}(a_k) \quad \text{for every} \quad x \in \partial_1 W^{k-1}.$$

We obtain therefore a homeomorphism of pairs:

$$(I, \partial I) \times (\partial_1 W^{k-1}, \partial_1 W^{k-2}) \xrightarrow{\approx} (T_k, T_k \cap W^{\langle k-1 \rangle})$$

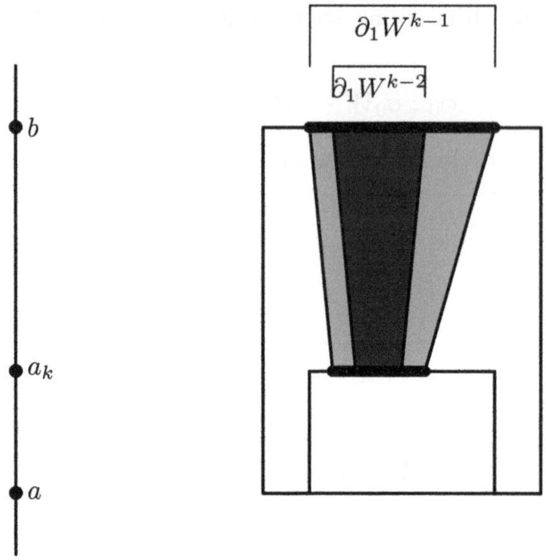

The set $T_k \cap W^{\langle k-1 \rangle}$ is shaded black,

and the set $T_k \setminus T_k \cap W^{\langle k-1 \rangle}$ is shaded grey.

FIGURE 37.

The pair on the left is the suspension of the pair $(\partial_1 W^{k-1}, \partial_1 W^{k-2})$, thus we have the isomorphism

$$H_{k-1}(\partial_1 W^{k-1}, \partial_1 W^{k-2}) = \mathcal{M}^{(1)}_{k-1} \xrightarrow{\approx} H_k\big(T_k, T_k \cap W^{\langle k-1 \rangle}\big).$$

Composing with the inclusion

$$(T_k, T_k \cap W^{\langle k-1 \rangle}) \hookrightarrow (W^{\langle k \rangle}, W^{\langle k-1 \rangle})$$

we obtain a homomorphism which will be denoted

$$\tau : \mathcal{M}^{(1)}_{k-1} \to H_k(W^{\langle k \rangle}, W^{\langle k-1 \rangle}).$$

Theorem 3.1. *The filtration*

$$\varnothing = W^{\langle -1 \rangle} \subset W^{\langle 0 \rangle} \subset \cdots \subset W^{\langle m \rangle} = W, \qquad where \qquad m = \dim W$$

is a cellular filtration of W. *We have:*

$$H_*\big(W^{\langle k \rangle}, W^{\langle k-1 \rangle}\big) = 0 \quad for \quad * \neq k,$$

and the homomorphism

$$\mathcal{L}_k = (\lambda_1, \lambda_0, \mu, \tau) : \mathcal{M}^{(1)}_k \oplus \mathcal{M}^{(0)}_k \oplus \mathcal{M}_k \oplus \mathcal{M}^{(1)}_{k-1} \longrightarrow H_k\big(W^{\langle k \rangle}, W^{\langle k-1 \rangle}\big)$$

is an isomorphism of abelian groups.

Proof. Consider the cobordism

$$W' = \phi^{-1}\big([a_k, a_{k+1}]\big).$$

The pair $(\phi|W', v|W')$ is elementary, and each critical point p of the function $\phi|W'$ has a handle $\mathcal{H}_p$. The disc $D(p,v) \cap W'$ does not intersect the set T_k, and its sole is in the interior of the set $W^{\langle k-1\rangle} \cap \phi^{-1}(a_k)$. Thus we can choose the handle $\mathcal{H}_p$ so close to the disc $D(p,v) \cap W'$ that:

(1) $\mathcal{H}_p \cap T_k = \varnothing$,

(2) $\mathcal{H}_p \cap \phi^{-1}(a_k) \subset W^{\langle k-1\rangle}$.

Let $\mathcal{H}$ be the union of handles $\mathcal{H}_p$ for all critical points $p \in S_k(f)$. Define the subset Q_k as follows:

$$Q_k = \partial_1 W^k \ \cup\ T_k \ \cup\ \mathcal{H} \ \cup\ \big(Z_k \cap \phi^{-1}([a, a_k])\big). \tag{6}$$

The subset Q_k is depicted in Figure 38.

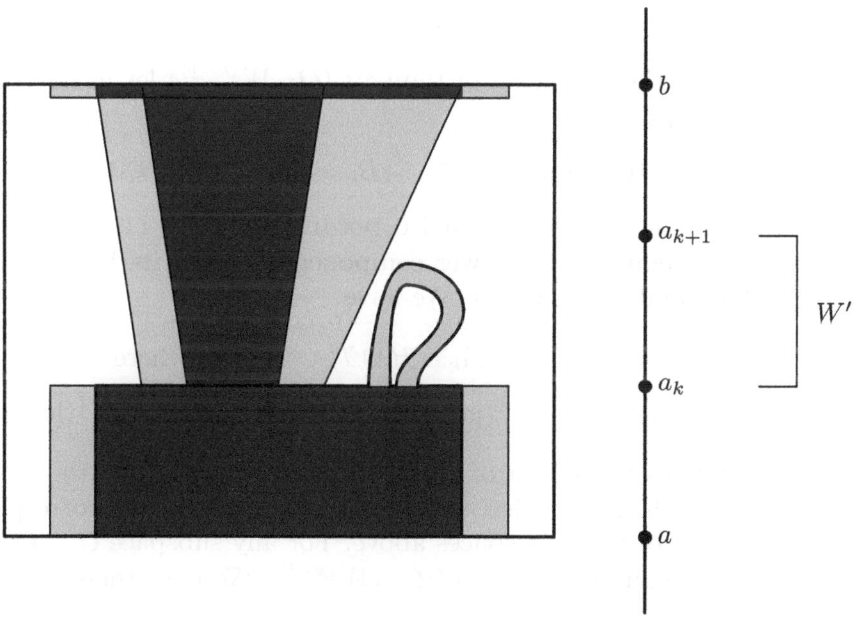

The set $W^{\langle k-1\rangle}$ is shaded black,

and the set Q_k is shaded grey.

The curved component of $Q_k \setminus W^{\langle k-1\rangle}$ depicts a handle.

FIGURE 38.

Lemma 3.2. *The inclusion* $Q_k \subset W^{\langle k \rangle}$ *is a homology equivalence.*

Proof. Observe that the two sets in question differ only inside W', that is,

$$Q_k \cap (W \setminus W') = W^{\langle k \rangle} \cap (W \setminus W').$$

It suffices therefore (by the excision property) to check that the inclusion

$$Q_k \cap W' \hookrightarrow W^{\langle k \rangle} \cap W' = Z_k \cap W'$$

is a homology equivalence. To prove this we apply the usual gradient descent deformation. Recall the homotopy

$$L : W' \times \mathbf{R}_+ \to W'$$

introduced in Subsection 3.2 of Chapter 3 (page 105). The sets

$$Z_k \cap W', \quad Q_k \cap W'$$

are obviously L_t-invariant for each t (where $L_t(x) = L(x,t)$). The image $L_T(Z_k \cap W')$ is in $Q_k \cap W'$ if T is large enough. An application of Lemma 1.8 of Chapter 5 (page 171) completes the proof. □

To compute the homology of the pair $(Q_k, W^{\langle k-1 \rangle})$ let us describe the quotient space $Q_k/W^{\langle k-1 \rangle}$. Put

$$A_k = \partial_1 W^k \setminus \operatorname{Int} \partial_1 W^{k-1}, \quad B_k = \partial_0 W^k \setminus \operatorname{Int} \partial_0 W^{k-1}.$$

Thus A_k is an $(m-1)$-dimensional cobordism, and the intersection of A_k with $W^{\langle k-1 \rangle}$ is equal to the lower component $\partial_0 A_k$ of its boundary. The space B_k has similar properties. We have

$$Q_k = W^{\langle k-1 \rangle} \cup \Big(A_k \cup \mathcal{H} \cup T'_k \cup Z'_k\Big), \quad \text{where}$$

$$T'_k = T(A_{k-1}, -v) \cap \phi^{-1}([a_k, b]), \quad Z'_k = T(B_k, v) \cap \phi^{-1}([a, a_k]).$$

The intersection of any two of these subsets is in $W^{\langle k-1 \rangle}$, therefore the quotient space $Q_k/W^{\langle k-1 \rangle}$ is the wedge of four summands corresponding to the quotients of the four spaces above. For any subspace $C \subset Q_k$ let us denote by $\widehat{C}$ the quotient space $C/C \cap W^{\langle k-1 \rangle}$. We have then:

$$Q_k/W^{\langle k-1 \rangle} \approx \widehat{A}_k \vee \widehat{\mathcal{H}} \vee \widehat{T'_k} \vee \widehat{Z'_k}.$$

Lemma 3.3. *We have*

$$(\mathcal{D}) \qquad H_*(Q_k, W^{\langle k-1 \rangle}) \approx \widetilde{H}_*(\widehat{A}_k) \oplus \widetilde{H}_*(\widehat{\mathcal{H}}) \oplus \widetilde{H}_*(\widehat{T'_k}) \oplus \widetilde{H}_*(\widehat{Z'_k}).^\dagger$$

† Here $\widetilde{H}_*$ denotes reduced homology groups.

Proof. An easy geometric argument shows that $W^{\langle k-1\rangle}$ is a deformation retract of some neighbourhood in Q_k. Therefore

$$H_*(Q_k, W^{\langle k-1\rangle}) \approx \widetilde{H}_*(Q_k/W^{\langle k-1\rangle})$$

(see [**58**], Ch. 2, Proposition 2.22). Also $C \cap W^{\langle k-1\rangle}$ is a deformation retract of some neighbourhood in C, if C is one of the following four spaces: A_k, $\mathcal{H}$, T'_k, Z'_k. Therefore the homology of the wedge of the four spaces $\widehat{A}_k$, $\widehat{\mathcal{H}}$, $\widehat{T}'_k$, $\widehat{Z}'_k$ is isomorphic to the direct sum of the homology of the summands (see [**58**], Ch. 2, Corollary 2.25), and the proof of our lemma is complete. □

We shall now compute the four components of the direct sum decomposition $(\mathcal{D})$ in terms of Morse complexes $\mathcal{M}_*$, $\mathcal{M}_*^{(0)}$, $\mathcal{M}_*^{(1)}$. As for the first summand, we have

$$A_k/(A_k \cap W^{\langle k-1\rangle}) = A_k/\partial_0 A_k = \partial_1 W^k/\partial_1 W^{k-1}.$$

Therefore

$$\widetilde{H}_k(\widehat{A}_k) \approx \mathcal{M}_k^{(1)}.$$

Further

$$\mathcal{H}/(\mathcal{H} \cap W^{\langle k-1\rangle}) = \mathcal{H}/(\mathcal{H} \cap \phi^{-1}(a_k)) \approx \bigvee_{p \in S_k(\phi)} \widehat{\mathcal{H}}_p,$$

where

$$\widehat{\mathcal{H}}_p = \mathcal{H}_p/\mathcal{H}_p \cap \phi^{-1}(a_k)$$

is the quotient of the handle $\mathcal{H}_p$ with respect to its sole. Therefore

$$\widetilde{H}_k(\widehat{\mathcal{H}}) \approx \bigoplus_{p \in S_k(\phi)} \widetilde{H}_k(\widehat{\mathcal{H}}_p) \approx \mathcal{M}_k.$$

Let us proceed to the third summand of $(\mathcal{D})$. It is easy to deduce from (5) that the quotient $T'_k/T'_k \cap W^{\langle k-1\rangle}$ is homeomorphic to the suspension

$$S(\partial_1 W^{k-1}/\partial_1 W^{k-2}).$$

Thus

$$\widetilde{H}_k(\widehat{T}'_k) \approx \mathcal{M}_{k-1}^{(1)}.$$

As for the fourth summand, observe that each v-trajectory starting at a point of B_k reaches $\phi^{-1}(a_k)$ and we have a homeomorphism of pairs

$$\left(Z'_k, Z'_k \cap W^{\langle k-1\rangle}\right) \approx (B_k, \partial_0 B_k) \times [0,1],$$

therefore $Z'_k/Z'_k \cap W^{\langle k-1\rangle}$ is homotopy equivalent to $\partial_0 W^k/\partial_0 W^{k-1}$. We obtain therefore an isomorphism

$$\widetilde{H}_k(\widehat{Z}'_k) \approx \mathcal{M}_k^{(0)}.$$

Form the direct sum of these isomorphisms:

$$\sigma_k : \mathcal{M}_k^{(1)} \oplus \mathcal{M}_k^{(0)} \oplus \mathcal{M}_k \oplus \mathcal{M}_{k-1}^{(1)} \xrightarrow{\approx} H_k(W^{\langle k\rangle}, W^{\langle k-1\rangle}).$$

It is not difficult to check that σ_k is equal to the homomorphism $\mathcal{L}_k$ from the statement of Theorem 3.1, which is now proven. □

The adjoint complex E_* corresponding to the filtration $W^{\langle k\rangle}$ computes therefore the homology of W:

Corollary 3.4. $H_*(E_*) \approx H_*(W)$. □

The groups E_k can be read from the Morse-theoretic data. Now we are going to show that the same is true for the boundary operators in this complex.

3.2. More computations in W'. In the previous subsection we have constructed an isomorphism

$$\mathcal{L}_k = (\lambda_1, \lambda_0, \mu, \tau) : \mathcal{M}_k^{(1)} \oplus \mathcal{M}_k^{(0)} \oplus \mathcal{M}_k \oplus \mathcal{M}_{k-1}^{(1)} \longrightarrow E_k.$$

Let

$$V_1 = \partial_1 W' = \phi^{-1}(a_{k+1}), \quad X_1 = V_1 \cap W^{\langle k\rangle}.$$

The inclusion $X_1 \hookrightarrow W^{\langle k\rangle}$ induces a homomorphism

$$\alpha : H_*(X_1) \to H_*(W^{\langle k\rangle}, W^{\langle k-1\rangle}) = E_k;$$

our aim in this subsection is to compute the homomorphism $\mathcal{L}_k^{-1} \circ \alpha$. The results will be used in Subsection 3.3. The cobordism

$$W' = \phi^{-1}([a_k, a_{k+1}])$$

is elementary. Let $x \in H_k(X_1)$, and put

$$\rho(x) = \sum_{p \in S_k(f)} m(\widetilde{x}, p) p \in \mathcal{M}_k$$

where $\widetilde{x}$ is the image of x in $H_*(W', \partial_0 W')$, and $m(\widetilde{x}, p)$ is the multiplicity of $\widetilde{x}$ with respect to the critical point p (see Subsection 3.4 of Chapter 5, page 176).

For every critical point $p \in S(f)$ of index $\leqslant k$ the sole of the descending disc $D(p, v)$ is in $\partial_0 W^{k-1}$. Therefore the gradient descent yields a continuous map

$$\xi = (-v)^{\twoheadrightarrow}_{[a_{k+1}, a]} : X_1 \to \partial_0 W^k / \partial_0 W^{k-1}$$

(see Definition 1.4 of Chapter 7, page 234), and we obtain an element

$$\xi_*(x) \in H_k(\partial_0 W^k / \partial_0 W^{k-1}) = \mathcal{M}_k^{(0)}.$$

Proposition 3.5. *Let $x \in H_k(X_1)$. We have*

$$\mathcal{L}_k^{-1}(\alpha(x)) = \Big(0,\ \xi_*(x),\ \rho(x),\ 0\Big).$$

Proof. Put

$$V_0 = \phi^{-1}(a_k), \quad Y_0 = W^{\langle k-1\rangle} \cap V_0, \quad X_0 = W^{\langle k\rangle} \cap V_0.$$

Put

$$P = W^{\langle k\rangle} \cap W'.$$

Let us start with the image of x in the homology $H_k(P, Y_0)$. The subsets P, Y_0 are depicted in Figure 39 (compare it with Figures 36, 38).

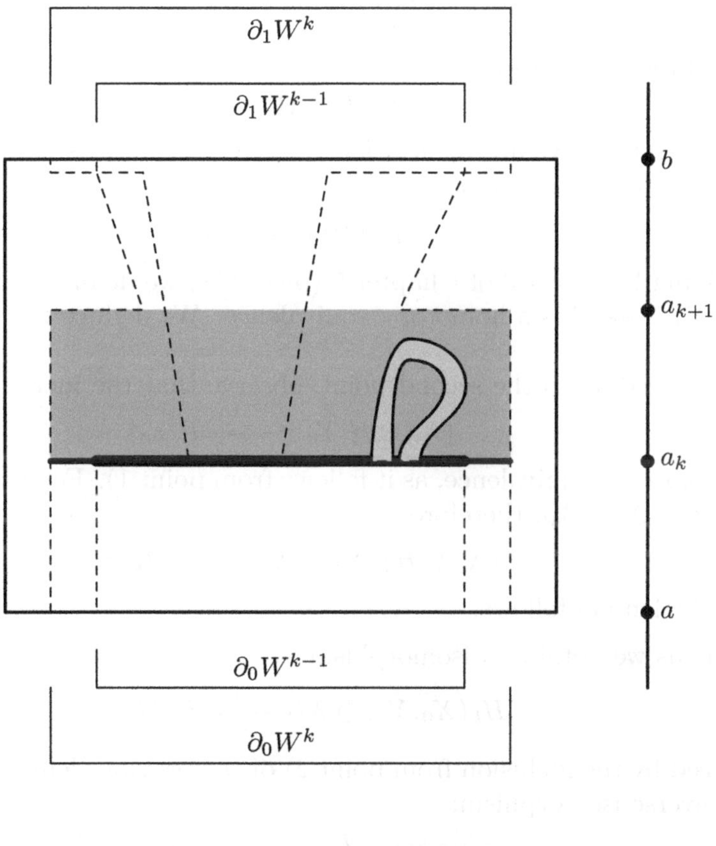

On this figure the set P is shaded, Y_0 is depicted by a fat line.

The sets $W^{\langle k-1\rangle}, W^{\langle k\rangle}$ are shown by dashed lines.

FIGURE 39.

Let $\mathcal{H}$ be the union of all handles of the critical points of $f|W'$; put $\mathcal{H}_0 = \mathcal{H} \cap X_0$, so that $\mathcal{H}_0$ is the union of all the soles of the handles. We choose the handles in such a way that $\mathcal{H}_0 \subset Y_0$.

Lemma 3.6. 1. *The inclusion $X_0 \cup \mathcal{H} \hookrightarrow P$ is a homotopy equivalence.*

2. *The inclusion*

$$X_0/Y_0 \vee \mathcal{H}/\mathcal{H}_0 \hookrightarrow P/Y_0$$

is a homology equivalence.

Proof. Let

$$L : W' \times \mathbf{R}_+ \to W'$$

be the homotopy introduced in Subsection 3.2 of Chapter 3 (page 105). The subset P is L_t-invariant for every t (where $L_t(x) = L(x,t)$). Therefore we obtain a homotopy

$$\widetilde{L} : P \times \mathbf{R}_+ \to P.$$

The subsets $\mathcal{H}, X_0$ are $\widetilde{L}_t$-invariant for every t and it is clear that for every T sufficiently large we have

$$\widetilde{L}_T(P) \subset \mathcal{H} \cup X_0.$$

Applying Lemma 1.8 of Chapter 5 (page 171) we deduce that the inclusion $\mathcal{H} \cup X_0 \hookrightarrow P$ is a homotopy equivalence. We deduce the point 1) of our lemma.

Proceeding to the second point, observe that the inclusion

$$(X_0 \cup \mathcal{H}, Y_0) \hookrightarrow (P, Y_0)$$

is a homology equivalence, as it follows from point 1). Furthermore, observe that $\mathcal{H} \cap X_0 \subset Y_0$, therefore

$$(X_0 \cup \mathcal{H})/Y_0 \approx X_0/Y_0 \vee \mathcal{H}/\mathcal{H}_0$$

and the lemma follows. □

Thus we obtain an isomorphism

$$H_k(X_0, Y_0) \oplus \mathcal{M}_k \xrightarrow[\approx]{J} H_k(P, Y_0)$$

induced by the inclusion from point 2) of the previous lemma. Let I denote the inverse isomorphism:

$$H_k(P, Y_0) \xrightarrow[\approx]{I} H_k(X_0, Y_0) \oplus \mathcal{M}_k.$$

It is now clear how to compute the image $\bar{x}$ of an element $x \in H_k(X_1)$ in $H_*(P, Y_0)$. For the first component of $I(\bar{x})$ we have:

$$I_1(\bar{x}) = (\xi_0)_*(\bar{x})$$

where

(7) $$\xi_0 = (-v)^{\twoheadrightarrow}_{[a_{k+1}, a_k]} : X_1 \to X_0/Y_0$$

is the continuous map induced by the gradient descent (see Definition 1.4 of Chapter 7, page 234). For the second component of $I(\bar{x})$ we have:

$$I_2(\bar{x}) = \rho(x) = \sum_p m(\widetilde{x}, p)p \in \mathcal{M}_k,$$

where $\widetilde{x}$ is the image of x in $H_*(W', \partial_0 W')$.

With this data at hand, let us return to the proof of the proposition. We are going to identify the homomorphism induced by the inclusion $(P, Y_0) \to (W^{\langle k \rangle}, W^{\langle k-1 \rangle})$. It is clear that the following diagram is commutative:

$$\begin{array}{ccc}
H_k(X_1) & \longrightarrow & H_k(W^{\langle k \rangle}, W^{\langle k-1 \rangle}) \\
\downarrow & \nearrow & \uparrow \mathcal{L}_k \\
H_k(P, Y_0) & & \mathcal{M}_k^{(1)} \oplus \mathcal{M}_k^{(0)} \oplus \mathcal{M}_k \oplus \mathcal{M}_{k-1}^{(1)} \\
J \uparrow \approx \downarrow I & & \uparrow i \\
H_k(X_0, Y_0) \oplus \mathcal{M}_k & \xrightarrow{\big((\xi_1)_*,\ \mathrm{Id}\big)} & \mathcal{M}_k^{(0)} \oplus \mathcal{M}_k
\end{array}$$

Here ξ_1 denotes the continuous map

$$\xi_1 = (-v)^{\twoheadrightarrow}_{[a_k, a]} : X_0/Y_0 \to \mathcal{M}_k^{(0)}.$$

The diagonal arrow is induced by the inclusion of the pairs, and i is the natural inclusion.

The image of $x \in H_k(X_1)$ via the composition of the left vertical arrows is equal to the pair

$$y = \Big((\xi_0)_*(\bar{x}), \rho(x)\Big)$$

(recall that $\bar{x}$ stands for the image of x in $H_k(P, Y_0)$ and ξ_0 is defined in (7)). Applying to y the lower horizontal arrow we obtain therefore the element

$$\Big(\xi_*(\bar{x}), \rho(x)\Big),$$

since

$$\xi = (-v)^{\twoheadrightarrow}_{[a_{k+1}, a]} = (-v)^{\twoheadrightarrow}_{[a_k, a]} \circ (-v)^{\twoheadrightarrow}_{[a_{k+1}, a_k]} = \xi_1 \circ \xi_0$$

(by Composition Lemma 1.8 of Chapter 7, page 235). Our proposition follows. □

We shall also need a generalisation of the previous proposition to the case when the absolute cycle x is replaced by a relative one. Put

$$Y_1 = T(\partial_1 W^{k-1}, -v) \cap V_1, \quad R = T(\partial_1 W^{k-1}, -v) \cap W'.$$

We wish to compute the homomorphism

$$H_*(X_1, Y_1) \to H_*(W^{\langle k\rangle}, W^{\langle k-1\rangle} \cup R)$$

induced by the inclusion. The inclusion

$$(W^{\langle k\rangle}, W^{\langle k-1\rangle}) \hookrightarrow (W^{\langle k\rangle}, W^{\langle k-1\rangle} \cup R)$$

is a homological equivalence. Let

$$\mathcal{B} : H_k(X_1, Y_1) \to \mathcal{M}_k^{(1)} \oplus \mathcal{M}_k^{(0)} \oplus \mathcal{M}_k \oplus \mathcal{M}_{k-1}^{(1)}$$

denote the composition of the homomorphism

$$H_k(X_1, Y_1) \to H_*(W^{\langle k\rangle}, W^{\langle k-1\rangle} \cup R) \xrightarrow{\approx} H_k(W^{\langle k\rangle}, W^{\langle k-1\rangle})$$

with the inverse of the isomorphism $\mathcal{L}_k$ from Theorem 3.1. For $x \in H_k(X_1/Y_1)$ let us denote by $\rho(x)$ the element

$$\rho(x) = \sum_{p \in S_k(f)} m(\widetilde{x}, p)p \in \mathcal{M}_k,$$

where $\widetilde{x}$ is the image of x in $H_*(W^{\langle k\rangle}, W^{\langle k-1\rangle} \cup R)$. The proof of the next proposition is similar to the proof of Proposition 3.5 and will be omitted.

Proposition 3.7. *Let $x \in H_*(X_1, Y_1)$. Then*

$$\mathcal{B}(x) = \Big(0, \xi_*(x), \rho(x), 0\Big)$$

where

$$\xi = (-v)^{\twoheadrightarrow}_{[a_{k+1}, a]} : X_1/Y_1 \longrightarrow \partial_0 W^k / \partial_0 W^{k-1}$$

is the continuous map induced by the gradient descent from a_{k+1} to a. □

3.3. The adjoint complex E_*. Now we are ready to compute the boundary operators in the adjoint complex E_*, associated with the filtration $W^{\langle k\rangle}$ of the cobordism W. Theorem 3.1 yields an isomorphism

$$\mathcal{L}_k = (\lambda_1, \lambda_0, \mu, \tau) : \mathcal{M}_k^{(1)} \oplus \mathcal{M}_k^{(0)} \oplus \mathcal{M}_k \oplus \mathcal{M}_{k-1}^{(1)} \longrightarrow E_k.$$

To compute $d_{k+1} : E_{k+1} \to E_k$ it suffices therefore to compute the homomorphism

$$d'_{k+1} = \mathcal{L}_k^{-1} \circ d_{k+1} \circ \mathcal{L}_{k+1} :$$

$$\mathcal{M}_{k+1}^{(1)} \oplus \mathcal{M}_{k+1}^{(0)} \oplus \mathcal{M}_{k+1} \oplus \mathcal{M}_k^{(1)} \longrightarrow \mathcal{M}_k^{(1)} \oplus \mathcal{M}_k^{(0)} \oplus \mathcal{M}_k \oplus \mathcal{M}_{k-1}^{(1)}.$$

The homomorphism d'_{k+1} will be described in terms of its 4×4-matrix which will be displayed *column-wise*, so that for example the terms in the first column are the components of the restriction $d'_{k+1} | \mathcal{M}_{k+1}^{(1)}$. In order to describe the matrix entries we need to introduce two more homomorphisms: η_k and σ_{k+1}.

Define a homomorphism

$$\eta_k : \mathcal{M}_k^{(1)} \to \mathcal{M}_k; \quad \eta_k(x) = \sum_{p \in S_k(f)} m(x, p; v)p,$$

where $m(x, p; v) \in \mathbf{Z}$ is the multiplicity of x with respect to p, see Definition 1.8 (page 286). For the case when x can be represented by a submanifold $X \subset \partial_1 W$ transverse to the ascending discs of critical points of f, the multiplicity $m(x, p; v)$ equals the intersection index of X with $D(p, -v) \cap \partial_1 W$.

To introduce the homomorphism σ_{k+1} consider a critical point $p \in S_{k+1}(f)$ and observe that by condition ($\mathfrak{C}$) the sole $\Sigma_p = D(p, v) \cap \partial_0 W$ of the descending disc of p is a smooth oriented submanifold of $\partial_0 W^k$ and the set $\Sigma_p \setminus \operatorname{Int} \partial_0 W^{k-1}$ is compact. Thus the fundamental class

$$[\Sigma_p] \in H_{k-1}(\partial_0 W^k, \partial_0 W^{k-1}) = \mathcal{M}_k^{(0)}$$

is defined. Denote this class by $\sigma_{k+1}(p)$, and extend the map $p \mapsto \sigma_{k+1}(p)$ to a homomorphism

$$\sigma_{k+1} : \mathcal{M}_{k+1} \to \mathcal{M}_k^{(0)}.$$

Theorem 3.8. *The matrix of d'_{k+1} equals*

$$(8) \qquad \begin{pmatrix} \partial_{k+1}^{(1)} & 0 & 0 & \mathrm{Id} \\ 0 & \partial_{k+1}^{(0)} & \sigma_{k+1} & -\mathcal{H}_k(-v) \\ 0 & 0 & \partial_{k+1} & -\eta_k \\ 0 & 0 & 0 & -\partial_k^{(1)} \end{pmatrix}$$

where $\mathcal{H}_k(-v)$ is the homological gradient descent, and $\partial_*^{(1)}$, $\partial_*^{(0)}$, ∂_* are the boundary operators in the Morse complexes $\mathcal{M}_*^{(1)}$, respectively $\mathcal{M}_*^{(0)}$ and $\mathcal{M}_*$.

Proof. The first two columns which correspond to the components $\mathcal{M}_*^{(0)}$, $\mathcal{M}_*^{(1)}$ are easy to compute. The inclusions

$$\partial_0 W \hookrightarrow W \hookleftarrow \partial_1 W$$

conserve the filtrations and therefore the maps of the corresponding adjoint complexes commute with the boundary operators.

As for the third component, we must compute the value of the boundary operator d_{k+1} on the homology class

$$d_p(v) \in H_k(W^{\langle k+1 \rangle}, W^{\langle k \rangle})$$

corresponding to the critical point $p \in S_{k+1}(f)$. We shall use the notation introduced in the previous subsection:

$$W' = \phi^{-1}([a_k, a_{k+1}]), \quad V_1 = \phi^{-1}(a_{k+1}) = \partial_1 W', \quad X_1 = W^{\langle k \rangle} \cap V_1.$$

The element $d_{k+1}(d_p(v)) \in E_k$ is represented by the oriented embedded sphere

$$\Sigma_p = D(p, v) \cap V_1.$$

According to Proposition 3.5 the $\mathcal{L}_k^{-1}$-image of $[\Sigma_p]$ in $H_*(W^{\langle k \rangle}, W^{\langle k-1 \rangle})$ has the components

$$\Big(0, \ \xi_*([\Sigma_p]), \ \rho([\Sigma_p]), \ 0\Big)$$

with respect to the direct sum decomposition obtained in Theorem 3.1. Here

$$\xi = (-v)^{\twoheadrightarrow}_{[a_{k+1}, a]} : X_1 \to \partial_0 W^k / \partial_0 W^{k-1}$$

is the continuous map induced by the gradient descent. We have by definition:

$$\xi_*([\Sigma_p]) = \sigma_{k+1}(d_p), \quad \rho([\Sigma_p]) = \partial_{k+1}(d_p)$$

and the computation of the third column is complete.

Now let us proceed to the fourth column. This is slightly more complicated than the preceding ones, but also more important for later applications. We must check that

$$d_{k+1}(\tau(x)) = \mathcal{L}_k\Big(x, \ -\mathcal{H}_k(-v)(x), \ -\eta_k(x), \ -\partial_k^{(1)}(x)\Big)$$
$$\text{for every} \quad x \in H_k(\partial_1 W^k, \partial_1 W^{k-1}).$$

Let us recall the definition of $\tau(x)$. Put

$$K = \partial_1 W^k, \quad L = \partial_1 W^{k-1}.$$

Let $I = [0,1]$ and denote by $\iota \in H_1(I, \partial I)$ the fundamental class of the oriented interval I. Denote by $S(x)$ the *suspension* of the class x, that is, the homological exterior product

$$S(x) = \iota \times x \in H_{k+1}\Big((I, \partial I) \times (K, L)\Big),$$

where

$$(I, \partial I) \times (K, L) = (I \times K, I \times L \ \cup \ \partial I \times K).$$

Put

$$W'' = \phi^{-1}([a_{k+1}, b]).$$

Every $(-v)$-trajectory starting at a point of K reaches $V_1 = \phi^{-1}(a_{k+1})$ and we have a natural homeomorphism (see Proposition 3.3 of Chapter 3, page 101)

$$I \times K \xrightarrow{F} T(K, -v) \cap W''$$

with

$$F(1,x) = x, \quad F(0,x) = \gamma\big(x, \tau(x, -v|W''), -v\big) \in V_1.$$

Put

$$TK = F(I \times K), \quad TL = F(I \times L), \quad K' = F(0 \times K), \quad L' = F(0 \times L).$$

It is clear that

$$TL \cup K \cup K' = TK \cap W^{\langle k \rangle},$$

and by definition the element

$$\tau(x) \in H_k(W^{\langle k+1 \rangle}, W^{\langle k \rangle})$$

is the image of

$$\sigma(x) = F_*(S(x)) \in H_*(TK, TK \cap W^{\langle k \rangle})$$

via the inclusion

$$(TK, TK \cap W^{\langle k \rangle}) \hookrightarrow (W^{\langle k+1 \rangle}, W^{\langle k \rangle}).$$

We have to compute the boundary operator of the element $\tau(x)$ in the exact sequence of the triple

(9) $$\Big(W^{\langle k+1 \rangle}, W^{\langle k \rangle}, W^{\langle k-1 \rangle}\Big).$$

Let us first compute the boundary operator of the element $S(x)$ in the exact sequence of the triple

(10) $$\Big(TK,\ TL \cup K \cup K',\ L \cup L'\Big).$$

Lemma 3.9. *Let $x \in H_*(K, L)$ and let*

$$\partial : H_*\big(TK,\ TL \cup K \cup K'\big) \to H_*\big(TL \cup K \cup K',\ L \cup L'\big)$$

denote the boundary operator in the exact sequence of the triple (10). Then

(11) $$\partial(\sigma(x)) = x - \big((-v)^{\leadsto}_{[b, a_{k+1}]}\big)_*(x) - \sigma(\partial(x))$$

where we identify the elements

$$x \in H_*(K, L),$$

$$\big((-v)^{\leadsto}_{[b, a_{k+1}]}\big)_*(x) \in H_*(K', L'),$$

$$\sigma(\partial(x)) \in H_*(TL, L \cup L')$$

with their images in the group $H_*(TL \cup K \cup K', L \cup L')$.

Proof. The boundary operator of the element

$$S(x) \in H_*\Big((I, \partial I) \times (K, L)\Big)$$

is computed immediately:

$$\partial(S(x)) = 1 \times x - 0 \times x - S(\partial x).$$

It remains to identify the three terms in the right-hand side of the preceding formula with the three terms of (11). □

We now proceed to the triple (9). We would like to deduce the required equality by functoriality, using the results already obtained for the triple (10). We have

$$TK \subset W^{\langle k+1 \rangle}, \qquad \text{and} \qquad TL \cup K \cup K' \subset W^{\langle k \rangle}.$$

Observe however that $L \cup L'$ is not a subset of $W^{\langle k-1 \rangle}$, therefore the triple (10) is not a sub-triple of (9) and the functoriality does not apply immediately. We shall now introduce a third triple, which contains both (9) and (10). Put

$$W' = \phi^{-1}([a_k, a_{k+1}]), \quad R = T(\partial_1 W^{k-1}, -v) \cap W'.$$

We have two inclusions of triples:

$$\begin{array}{l}
\Big(W^{\langle k+1 \rangle}, W^{\langle k \rangle}, W^{\langle k-1 \rangle}\Big) \\
\qquad\qquad\qquad \xhookrightarrow{\ I\ } \Big(W^{\langle k+1 \rangle}, W^{\langle k \rangle}, W^{\langle k-1 \rangle} \cup R\Big) \\
\qquad\qquad\qquad \xhookrightarrow{\ J\ } \\
(TK, TL \cup K \cup K', L \cup L')
\end{array}$$

where I is a homology equivalence. The same symbols I, respectively J will also be used to denote the inclusions of pairs

$$(W^{\langle k \rangle}, W^{\langle k-1 \rangle}) \xhookrightarrow{\ I\ } (W^{\langle k \rangle}, W^{\langle k-1 \rangle} \cup R), \qquad \text{respectively}$$

$$(TL \cup K \cup K', L \cup L') \xhookrightarrow{\ J\ } (W^{\langle k \rangle}, W^{\langle k-1 \rangle} \cup R).$$

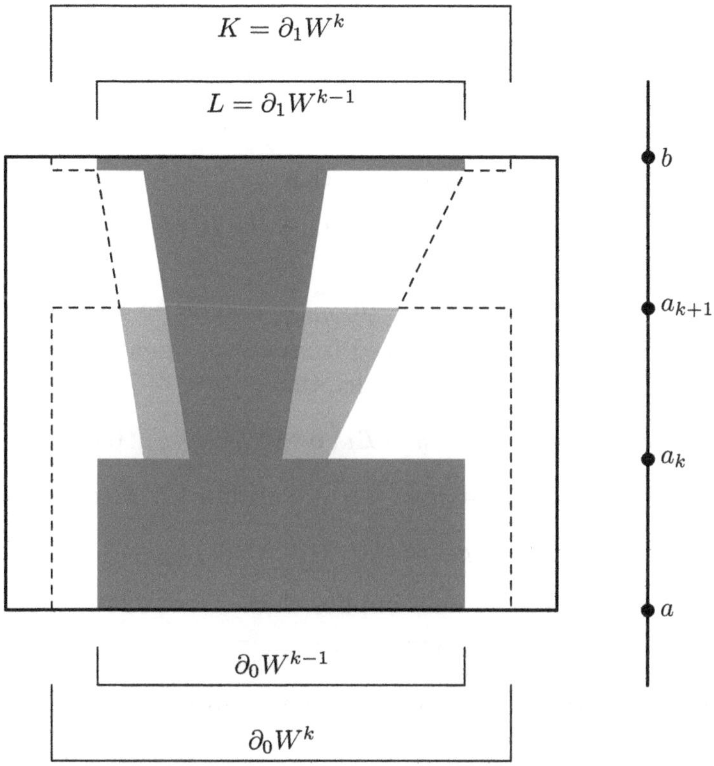

On this figure the set $R \cup W^{\langle k-1 \rangle}$ is shaded light grey,

and the set $W^{\langle k-1 \rangle}$ is shaded dark grey.

The set $W^{\langle k \rangle}$ is shown by dashed lines.

FIGURE 40.

To prove our theorem it suffices to show that for every $x \in H_k(K, L)$ the I_*-image in $H_k(W^{\langle k \rangle}, W^{\langle k-1 \rangle} \cup R)$ of the element

$$\mathcal{L}_k\Big(x,\ -\mathcal{H}_k(-v)(x),\ -\eta_k(x),\ -\partial_k^{(1)}(x)\Big) \in H_k(W^{\langle k \rangle}, W^{\langle k-1 \rangle})$$

equals the J_*-image of the element

$$x - \Big((-v)^{\leadsto}_{[b, a_{k+1}]}\Big)(x) - \sigma(\partial x) \in H_*(TL \cup K \cup K', L \cup L').$$

It follows from the definitions that

$$J_*(x) = I_*\big(\mathcal{L}_k(x,0,0,0)\big), \quad J_*(\sigma(\partial(x))) = I_*\big(\mathcal{L}_k(0,0,0,\partial_k^{(1)}(x))\big).$$

It remains to compute the image in $H_*(W^{\langle k\rangle}, W^{\langle k-1\rangle} \cup R)$ of the element

$$y = \Big(\big(-v\big)^{\rightsquigarrow}_{[b,a_{k+1}]}\Big)_*(x) \in H_*(K', L').$$

We shall identify this element with

$$\lambda_*(x) \in H_*(K'/L')$$

where

$$\lambda = (-v)^{\twoheadrightarrow}_{[b,a_{k+1}]} : K/L \to K'/L'$$

is the continuous map derived from the gradient descent. Applying Proposition 3.7 (page 314) we deduce that

$$y = \mathcal{L}_k\Big(0,\ \xi_*(y),\ \rho(y),\ 0\Big).$$

It is clear that $\rho(y) = \eta_k(x)$, so we obtain the next formula:

$$y = \mathcal{L}_k\Big(0,\ (\xi \circ \lambda)_*(x),\ \eta_k(x),\ 0\Big).$$

The composition formula implies that

$$\xi_* \circ \lambda_* = (-v)^{\twoheadrightarrow}_{[a_{k+1},a]} \circ (-v)^{\twoheadrightarrow}_{[b,a_{k+1}]} = (-v)^{\twoheadrightarrow}_{[b,a]} = \mathcal{H}_k(-v),$$

and the proof of our theorem is now complete. □

Sources

The term *cellular gradient* is new, but the notion itself dates back to 1995, when a similar class of gradients was introduced by the author in order to prove the C^0-generic rationality of the boundary operators in the Novikov complex ([**112**], [**113**]). Several versions of the notion of cellular gradient were discussed in my subsequent articles [**116**], [**121**]. The version chosen for the present exposition aims at maximal simplicity. One of the new features is an emphasis on the continuous map $(-v)^{\twoheadrightarrow}$.

Part 4

Circle-valued Morse maps and Novikov complexes

This part begins with Chapter 10 where we develop the necessary algebraic technique. Our study of circle-valued Morse maps starts in Chapter 11 where we construct the Novikov complex and prove its basic properties. The next two chapters are about the dynamics of the gradient flow associated to a circle-valued Morse function. We apply the theory of cellular gradients to prove that the boundary operators in the Novikov complex are generically rational functions (Chapter 12) and to study the Lefschetz zeta function of the gradient flow (Chapter 13). In Chapter 14 we discuss several selected topics in the Morse-Novikov theory.

CHAPTER 10

Completions of rings, modules and complexes

Let A be a commutative ring. The polynomial ring $A[t]$ will be denoted also by P, the power series ring $A[[t]]$ will be denoted by $\widehat{P}$, and the quotient ring P/t^nP will be denoted by P_n. The Laurent polynomial ring $A[t,t^{-1}]$ will be denoted by L.

1. Chain complexes over $A[[t]]$

Let N be a module over $P = A[t]$. The completion of N is by definition the inverse limit

$$\widehat{N} = \lim_{\leftarrow} N/t^kN$$

of the projective system

$$N/tN \longleftarrow N/t^2N \longleftarrow \cdots \longleftarrow N/t^kN \longleftarrow \cdots$$

of P-modules. Observe that $\widehat{N}$ has a natural structure of a $\widehat{P}$-module.

The natural homomorphism $N \to \widehat{N}$ will be denoted by i_N. This is a homomorphism of P-modules and if N is a $\widehat{P}$-module, then i_N is also a $\widehat{P}$-homomorphism. If N is a free finitely generated $\widehat{P}$-module, then i_N is an isomorphism.

Let C_* be a chain complex over P. Applying to each of the modules C_k the procedure of completion we obtain the *completion* $\widehat{C}_*$ together with a natural chain map

$$i_{C_*} : C_* \to \widehat{C}_*.$$

Definition 1.1. Let R be a commutative ring. A chain complex of R-modules will be also called a *chain complex over* R or an R*-complex.* An R-complex $\{C_i\}$ is called *free* if all modules C_i are free R-modules. An R-complex $\{C_i\}$ is called *finite* if all modules C_i except for a finite number of indices i are equal to zero. A finite R-complex of free finitely generated modules is called *free-finite.* An R-complex is called *homotopically free-finite* if it is homotopy equivalent to a free-finite R-complex.

Lemma 1.2. *If C_* is a free-finite $\widehat{P}$-complex, then the map*

$$i_{C_*} : C_* \to \widehat{C}_*$$

is a chain isomorphism of $\widehat{P}$-complexes.

If C_* is a homotopically free-finite $\widehat{P}$-complex, then i_{C_*} is a homotopy equivalence of $\widehat{P}$-complexes.

Proof. The first assertion is obvious, and the second follows from the first. □

Proposition 1.3. *Let A be a Noetherian ring. If C_* is a homotopically free-finite complex over $P = A[t]$, then the map i_{C_*} induces an isomorphism*

$$H_*(C_*) \underset{P}{\otimes} \widehat{P} \xrightarrow{\approx} H_*(\widehat{C}_*).$$

Proof. It suffices to prove the proposition for the case when C_* is a free-finite P-complex. In this case $\widehat{C}_* \approx C_* \underset{P}{\otimes} \widehat{P}$ and the lemma follows since $\widehat{P}$ is flat over P. □

1.1. Homology of $\widehat{P}$-complexes. Let C_* be a chain complex over $\widehat{P}$. We shall often use the following notation:

$$C_*^{(k)} = C_*/t^k C_*.$$

For every $k \in \mathbf{N}$ the projection

$$C_* \xrightarrow{\pi_k} C_*^{(k)} = C_*/t^k C_*$$

induces a homomorphism in homology

$$(\pi_k)_* : H_*(C_*) \to H_*(C_*/t^k C_*) = H_*(C_*^{(k)}),$$

and the inverse limit of $(\pi_k)_*$ yields a homomorphism

$$\rho_{C_*} : H_*(C_*) \to \varprojlim H_*(C_*/t^k C_*).$$

Theorem 1.4. *Assume that A is Noetherian. Then for every homotopically free-finite chain complex C_* over $\widehat{P} = A[[t]]$ the homomorphism*

$$\rho_{C_*} : H_*(C_*) \to \varprojlim H_*(C_*^{(k)})$$

is an isomorphism.

Proof. It suffices to prove the theorem in the case when C_* is free-finite.

Surjectivity. †

Lemma 1.5. *Let k be a positive integer. Let $x \in H_*(C_*^{(k)})$, $y \in H_*(C_*^{(k+1)})$. Assume that $(p_{k+1})_*(y) = x$, where*

$$p_{k+1} : C_*^{(k+1)} \to C_*^{(k)}$$

is the natural projection. Let X be any cycle in $C_*^{(k)}$ representing x. Then there is a cycle $Y \in C_*^{(k+1)}$ representing y and such that $p_{k+1}(Y) = X$.

Proof. Let Y' be any cycle in $C_*^{(k+1)}$ representing y. We have:

$$p_{k+1}(Y') - X = \partial\xi \quad \text{where} \quad \xi \in C_*^{(k)}.$$

Pick any $\eta \in C_*^{(k+1)}$ with $p_{k+1}(\eta) = \xi$. Then $Y = Y' - \partial\eta$ satisfies our requirements. □

Proceeding to the proof of surjectivity let

$$x = (x_1, x_2, \dots, x_k, \dots) \in \varprojlim H_*(C_*^{(k)})$$

where $x_k \in H_*(C_*^{(k)})$. Applying the previous lemma we can choose cycles $X_k \in C_*^{(k)}$ representing the classes x_k in such a way that

$$p_k(X_k) = X_{k-1} \quad \text{for every} \quad k.$$

Then the inverse sequence

$$X = (X_1, X_2, \dots, X_n, \dots) \in \varprojlim C_*^{(k)}$$

is a cycle in $\widehat{C}_*$ whose image in $\varprojlim H_*(C_*^{(k)})$ equals x. Recall that $C_* \approx \widehat{C}_*$ (since C_* is free-finite) and the proof of the surjectivity of ρ_{C_*} is over.

Injectivity.

Observe that the homology $H_*(C_*)$ is a finitely generated $\widehat{P}$-module. Let $x \in \operatorname{Ker} \rho_{C_*}$; then for every $k \in \mathbf{N}$ the image of x in the module $H_*(C_*^{(k)})$ vanishes. Apply the long exact sequence

$$\cdots \longrightarrow H_*(C_*) \xrightarrow{t^k} H_*(C_*) \longrightarrow H_*(C_*/t^kC_*) \longrightarrow \cdots$$

associated with the pair of chain complexes (C_*, t^kC_*) to deduce that x is divisible by t^k for every k. We complete the proof by applying the next lemma.

† This part of the proof does not use the Noetherian property of A.

Lemma 1.6. *Let H be a finitely generated module over the ring $\widehat{P} = A[[t]]$ where A is Noetherian. Assume that an element $x \in H$ is divisible by t^k for every k. Then $x = 0$.*

Proof. Consider first the case when there is no t-torsion in H, that is,

$$tx = 0 \Rightarrow x = 0.$$

Let $N \subset H$ be the submodule of elements divisible by t^k for any k. Then N satisfies $tN = N$. Indeed, for any $x \in N$ we have

$$x = tx_1 = t^2 x_2 = \cdots = t^n x_n = \cdots$$

and therefore $x_1 = tx_2 = \cdots = t^{n-1}x_n = \cdots$. Since t is in the Jacobson radical of $\widehat{P}$, Nakayama's lemma implies $N = 0$.

Let us now consider the general case. Define the *t-torsion submodule* $T \subset H$ by

$$T = \{x \in H \mid t^m x = 0 \quad \text{for some} \quad m\},$$

and denote H/T by H'. It is easy to see that H' has no t-torsion. Let $x \in H$ be an element divisible by every power of t. Applying to H' the reasoning above, we conclude that $x \in T$. The ring $\widehat{P}$ is Noetherian, therefore the submodule $T \subset H$ is finitely generated, and there is $k \in \mathbf{N}$ such that $t^k y = 0$ for every $y \in T$. Since $x = t^k u$ (with u necessarily in T), we have $x = 0$. □

Corollary 1.7. *Let C_*, D_* be homotopically free-finite complexes over $\widehat{P} = A[[t]]$, where A is a Noetherian ring. Let $f : C_* \to D_*$ be a chain map. The three following properties are equivalent:*

(1) *f is a homotopy equivalence;*
(2) *for every k the map $f/t^k : C_*^{(k)} \to D_*^{(k)}$ is a homotopy equivalence;*
(3) *for every k the induced homomorphism in homology*

$$(f/t^k)_* : H_*(C_*^{(k)}) \to H_*(D_*^{(k)})$$

is an isomorphism.

Proof. The implications (1) ⇒ (2) ⇒ (3) are obvious, therefore it remains to prove that (3) ⇒ (1). Condition (3) implies that $f_* : H_*(C_*) \to H_*(D_*)$ is an isomorphism (Theorem 1.4). Since C_*, D_* are homotopically free-finite $\widehat{P}$-complexes, the result follows. □

1.2. Homotopy of $\widehat{P}$-complexes. Let C_*, D_* be $\widehat{P}$-complexes. We denote by $[C_*,\ D_*]$ the $\widehat{P}$-module of homotopy classes of chain maps $C_* \to D_*$. Any chain map $f : C_* \to D_*$ induces a chain map $f/t^k : C_*^{(k)} \to D_*^{(k)}$ of the truncated complexes, and we obtain a homomorphism

$$\lambda_k : [C_*,\ D_*] \to [C_*^{(k)},\ D_*^{(k)}].$$

Similarly, we have homomorphisms

$$p_k : \big[C_*^{(k+1)},\ D_*^{(k+1)}\big] \to \big[C_*^{(k)},\ D_*^{(k)}\big],$$

which form an inverse system and we have a natural homomorphism of $\widehat{P}$-modules

$$\lambda :\ [C_*,\ D_*] \longrightarrow \varprojlim\ \big[C_*^{(k)},\ D_*^{(k)}\big].$$

Theorem 1.8. *Let C_*, D_* be homotopically free-finite complexes over $\widehat{P} = A[[t]]$, where A is a Noetherian ring. Then the map λ above is an isomorphism.*

Proof. The proof is similar to Theorem 1.4. The injectivity of λ follows from the next proposition.

Proposition 1.9. *Let $f, g : C_* \to D_*$ be chain maps and*

$$f_k, g_k : C_*^{(k)} \to D_*^{(k)}$$

be the corresponding quotient maps. Let

$$\pi_k : C_* \to C_*^{(k)}, \quad \pi_k' : D_* \to D_*^{(k)}$$

denote the natural projections. Then the following four properties are equivalent:

(1) $f \sim g$,
(2) $f_k \sim g_k$ *for every* $k \in \mathbf{N}$,
(3) $\pi_k' \circ f \sim \pi_k' \circ g$ *for every* $k \in \mathbf{N}$,
(4) $f_k \circ \pi_k \sim g_k \circ \pi_k$ *for every* $k \in \mathbf{N}$.

Proof. It is clear that the property (1) implies the others and properties (2)–(4) are equivalent. It remains to prove that (3) implies (1). It suffices to consider the particular case when both C_*, D_* are free-finite over $\widehat{P}$. With this assumption, observe that $[C_*, D_*]$ is a finitely generated $\widehat{P}$-module (it follows from the Noetherian property of $\widehat{P}$). Observe that the maps $\pi_k' \circ f$, $\pi_k' \circ g$ are homotopic if and only if the homotopy class of the map $f - g : C_* \to D_*$ is divisible by t^k in the module $[C_*, D_*]$. Now we apply Lemma 1.6 and the proof is complete. □

Let us move on to the surjectivity property.

Lemma 1.10. *Let C_* be a free complex over $\widehat{P}$, and D_* be any P-complex. Let*

$$\begin{array}{ccc} C_*^{(k)} & \xleftarrow{p_k} & C_*^{(k+1)} \\ \downarrow{\scriptstyle A} & & \downarrow{\scriptstyle B} \\ D_*^{(k)} & \xleftarrow{p'_k} & D_*^{(k+1)} \end{array}$$

be a homotopy commutative square of chain maps of P-complexes (where p_k, p'_k are the natural projections). Then there is a chain map $B' : C_*^{(k+1)} \to D_*^{(k+1)}$ such that $B' \sim B$ and $p'_k \circ B' = A \circ p_k$.

Proof. Let H be a chain homotopy from $p'_k \circ B$ to $A \circ p_k$; that is, $H : C_*^{(k+1)} \to D_{*+1}^{(k)}$ is a graded homomorphism of P-modules, such that

$$\partial H + H\partial = p'_k \circ B - A \circ p_k.$$

Since C_* is a free $\widehat{P}$-module, the module $C_*^{(k+1)}$ is free over the ring $P/t^{k+1}P$ and there is a lift of H to a graded homomorphism

$$H' : C_*^{(k+1)} \to D_{*+1}^{(k+1)}$$

of P-modules. Set

$$B' = B - (\partial H' + H'\partial)$$

and the lemma follows. □

Now we can prove that λ is surjective. It suffices to deal with the case when both C_*, D_* are free-finite. Let $h_k : C_*^{(k)} \to D_*^{(k)}$ be a sequence of chain maps such that every diagram

$$\begin{array}{ccc} C_*^{(k)} & \xleftarrow{p_k} & C_*^{(k+1)} \\ \downarrow{\scriptstyle h_k} & & \downarrow{\scriptstyle h_{k+1}} \\ D_*^{(k)} & \xleftarrow{p'_k} & D_*^{(k+1)} \end{array}$$

is homotopy commutative. An easy inductive argument using Lemma 1.10 shows that there is a sequence of chain maps $h'_k : C_*^{(k)} \to D_*^{(k)}$ such that each h'_k homotopic to h_k and every diagram

$$\begin{array}{ccc} C_*^{(k)} & \xleftarrow{p_k} & C_*^{(k+1)} \\ \Big\downarrow{\scriptstyle h'_k} & & \Big\downarrow{\scriptstyle h'_{k+1}} \\ D_*^{(k)} & \xleftarrow{p'_k} & D_*^{(k+1)} \end{array}$$

is commutative. Put

$$h = \varprojlim h'_k : \widehat{C}_* \to \widehat{D}_*.$$

Then the chain map $C_*^{(k)} \to D_*^{(k)}$ induced by h is homotopic to h_k for every k. Since

$$C_* \approx \widehat{C}_*, \quad D_* \approx \widehat{D}_*$$

the result follows and the proof of our theorem is complete. □

It is convenient to reformulate the theorem above in terms of so-called *virtual maps.*

Definition 1.11. Let C_*, D_* be chain complexes over $\widehat{P}$. A *virtual map* $\mathcal{H} : C_* \dashrightarrow D_*$ is a collection $\{h_k\}_{k\in\mathbf{N}}$ of chain maps

$$h_k : C_*/t^k C_* \longrightarrow D_*/t^k D_*$$

of P-complexes, such that for every k the following diagram is chain homotopy commutative:

$$\begin{array}{ccc} C_*/t^kC_* & \xleftarrow{p_k} & C_*/t^{k+1}C_* \\ \Big\downarrow{\scriptstyle h_k} & & \Big\downarrow{\scriptstyle h_{k+1}} \\ D_*/t^kD_* & \xleftarrow{p'_k} & D_*/t^{k+1}D_* \end{array}$$

(here p_k, p'_k are the natural projections).

A virtual map $\mathcal{H} = \{h_k\}_{k\in\mathbf{N}}$ is called a *virtual homology equivalence* if every h_k is a homology equivalence.

A *realization* of a virtual map $\mathcal{H}$ is a chain map $F : C_* \to D_*$, such that for every k the diagram

$$\begin{array}{ccc} C_*/t^k C_* & \longleftarrow & C_* \\ \Big\downarrow{\scriptstyle h_k} & & \Big\downarrow{\scriptstyle F} \\ D_*/t^k D_* & \longleftarrow & D_* \end{array}$$

is chain homotopy commutative (where horizontal arrows are the natural projections).

Theorem 1.8 and Corollary 1.7 imply the following:

Theorem 1.12. *Let C_*, D_* be homotopically free-finite $\widehat{P}$-complexes, and $\mathcal{H} : C_* \dashrightarrow D_*$ a virtual map. Then $\mathcal{H}$ has a homotopy unique realization, which will be denoted by*

$$|\mathcal{H}| : C_* \longrightarrow D_*;$$

$|\mathcal{H}|$ is a homotopy equivalence if and only if $\mathcal{H}$ is a virtual homology equivalence. □

2. Novikov rings

Let A be a commutative ring with unit.

Definition 2.1. The *Novikov ring* $A((t))$ consists of all Laurent series

$$\lambda = \sum_{i \in \mathbf{Z}} a_i t^i$$

in one variable with coefficients $a_i \in A$, such that the negative part of λ is finite (that is, there is $n = n(\lambda)$ such that $a_k = 0$ if $k < n(\lambda)$). This ring is also denoted by $\overline{L}$.

Exercise 2.2. Show that $A((t))$ has indeed a natural structure of a commutative ring such that the inclusions

$$A[t] \subset A((t)), \ A[t, t^{-1}] \subset A((t)), \ A[[t]] \subset A((t))$$

are ring homomorphisms.

When A is Noetherian, the rings above have nice homological properties:

Theorem 2.3. *Assume that A is Noetherian. Then*

(1) *The ring $A[[t]]$ is flat over $A[t]$.*

(2) *The ring* $A((t))$ *is flat over* $A[t, t^{-1}]$.

Proof. The first item is a classical property of completions in Noetherian commutative algebra, see for example [**9**, Prop. 10.14]. For the proof of (2) note that $A((t))$ is the localization of $A[[t]]$ with respect to the multiplicative subset $S = \{t^n \mid n \geqslant 0\}$, and the flatness of $A[[t]]$ over $A[t]$ implies the flatness of $S^{-1}A[[t]]$ over $S^{-1}A[t] = A[t, t^{-1}]$. □

In our applications we shall work mostly with the case $A = \mathbf{Z}$. This case has a remarkable particularity:

Theorem 2.4. *The ring* $\mathbf{Z}((t))$ *is Euclidean.*

Proof. A *monomial* is an element of $\mathbf{Z}((t))$ of type bt^n with $b, n \in \mathbf{Z}$. For a non-zero Laurent series $A \in \mathbf{Z}((t))$,

$$A = \sum_{k=n}^{\infty} a_k t^k \quad \text{with} \quad n \in \mathbf{Z} \quad \text{and} \quad a_n \neq 0$$

the monomial $a_n t^n$ will be called *the first term* of A and the integer a_n will be called *the first coefficient* of A. The absolute value $|a_n|$ of the first coefficient will be denoted $N(A)$ and the number n will be denoted by $h(A)$. We shall prove that $\mathbf{Z}((t))$ is a Euclidean ring with respect to the map

$$N : \mathbf{Z}((t)) \setminus \{0\} \to \mathbf{N} \setminus \{0\}.$$

It is clear that $N(AB) = N(A)N(B)$ so that $N(AB) \geqslant N(A)$ for any $A,\ B \neq 0$. It remains to prove that our ring admits the Euclidean division algorithm. Let $A, B \in \mathbf{Z}((t))$, assume that $A \neq 0$ and

$$B = \sum_{k=0}^{\infty} b_k t^k, \quad \text{where} \quad b_0 > 0.$$

It suffices to consider the case when B does not divide A. Let

$$\mathcal{R} = \{A - MB \mid M \in \mathbf{Z}((t))\}.$$

A priori two possibilities can occur:

A) There is a series $C \in \mathcal{R}$ with the first coefficient not divisible by b_0.

B) The first coefficient of every element of $\mathcal{R}$ is divisible by b_0.

In the case A) pick some $C \in \mathcal{R}$ with the first term $c_n t^n$ such that

$$A = MB + C, \quad b_0 \nmid c_n.$$

Write

$$c_n = pb_0 + q \quad \text{with} \quad p, q \in \mathbf{Z},\ 0 < q < b_0.$$

Then $C = pt^n B + D$ where the first term of D equals qt^n, so that we have

$$A = (M + pt^n)B + D, \quad \text{where} \quad N(D) = q < b_0 = N(B),$$

and the Euclidean division is over.

Now we will show that the case B) actually does not occur. Indeed, assume that for every M the first coefficient of $A - MB$ is divisible by b_0.

Lemma 2.5. *There is a sequence of non-zero series A_k and monomials M_k such that $A_0 = A$ and for every k we have*

$$A_k = M_k B + A_{k+1},$$
$$h(M_{k+1}) > h(M_k) = h(A_k) < h(A_{k+1}).$$

Proof. We will proceed by induction on k. Assume that we have constructed the monomials $M_0, \ldots, M_k$ and series $A_0, \ldots, A_k, A_{k+1}$ satisfying the cited conditions. Put

$$N_k = \sum_{i=0}^{k} M_i.$$

Observe that

$$A = N_k B + A_{k+1}.$$

Therefore $A_{k+1} \neq 0$ [†] and the first coefficient a_N of $A_{k+1} = a_N t^N + \cdots$ is divisible by b_0. Let $c = a_N / b_0$ then

$$h(A_{k+1} - ct^N B) > h(A_{k+1}).$$

Put

$$M_{k+1} = ct^N, \quad A_{k+2} = A_{k+1} - M_{k+1} B$$

and the induction step is over. □

It is easy to deduce from this lemma that

$$A = NB \quad \text{with} \quad N = \sum_{n=0}^{\infty} M_n$$

and this contradicts our assumption $B \nmid A$. □

[†] Recall that $B \nmid A$.

CHAPTER 11

The Novikov complex of a circle-valued Morse map

Let M be a closed C^∞ manifold and $f : M \to S^1$ a C^∞ map. The manifold S^1 is locally diffeomorphic to $\mathbf{R}$, so all the local notions of Morse theory (non-degenerate critical points and their indices, gradient-like vector fields, etc.) carry over to this situation. On the other hand the usual Morse inequalities are no more valid here, namely, the number of critical points of a circle-valued Morse function is not bounded from below by the Betti numbers. For example, the identity map $\mathrm{id} : S^1 \to S^1$ is a Morse map without critical points. However the homology of S^1 does not vanish, since $H_0(S^1) \approx \mathbf{Z} \approx H_1(S^1)$. In order to apply the Morse-theoretical methods let us consider the infinite cyclic covering $\pi : \bar{M} \to M$ induced by f from the covering

$$\mathrm{Exp} : \mathbf{R} \to S^1, \quad \mathrm{Exp}(t) = e^{2\pi i t}.$$

The function f lifts to a real-valued Morse function F on $\bar{M}$ so that the next diagram is commutative:

(1)
$$\begin{array}{ccc} \bar{M} & \xrightarrow{F} & \mathbf{R} \\ {\scriptstyle \pi}\downarrow & & \downarrow{\scriptstyle \mathrm{Exp}} \\ M & \xrightarrow{f} & S^1 \end{array}$$

However it is not possible to apply the usual Morse theory to F, since $\bar{M}$ is not compact and the number of critical points of F is infinite. The way to overcome this difficulty was indicated by S.P. Novikov in the early 1980s. His approach is based on an ingenious way of counting the critical points and gradient flow lines on the covering $\bar{M}$. We shall present the theory in Section 1; let us first outline the main result.

The ordinary Morse theory associates with each Morse function $f : M \to \mathbf{R}$ and a transverse f-gradient v the Morse complex $\mathcal{M}_*(f, v)$. This complex is freely generated over $\mathbf{Z}$ by the set $S(f)$ of critical points of f and computes the homology of the manifold M.

The Novikov theory associates with each circle-valued map $f : M \to S^1$ and a transverse f-gradient v the Novikov complex $\mathcal{N}_*(f, v)$. This complex is freely generated over $\overline{L} = \mathbf{Z}((t))$ by the set $S(f)$ of critical points of f and computes the homology $H_*(\bar{M}) \underset{L}{\otimes} \overline{L}$, where the structure of $L = \mathbf{Z}[t, t^{-1}] = \mathbf{Z}[\mathbf{Z}]$-module on $H_*(\bar{M})$ is given by the action of $\mathbf{Z}$ on $\bar{M}$.

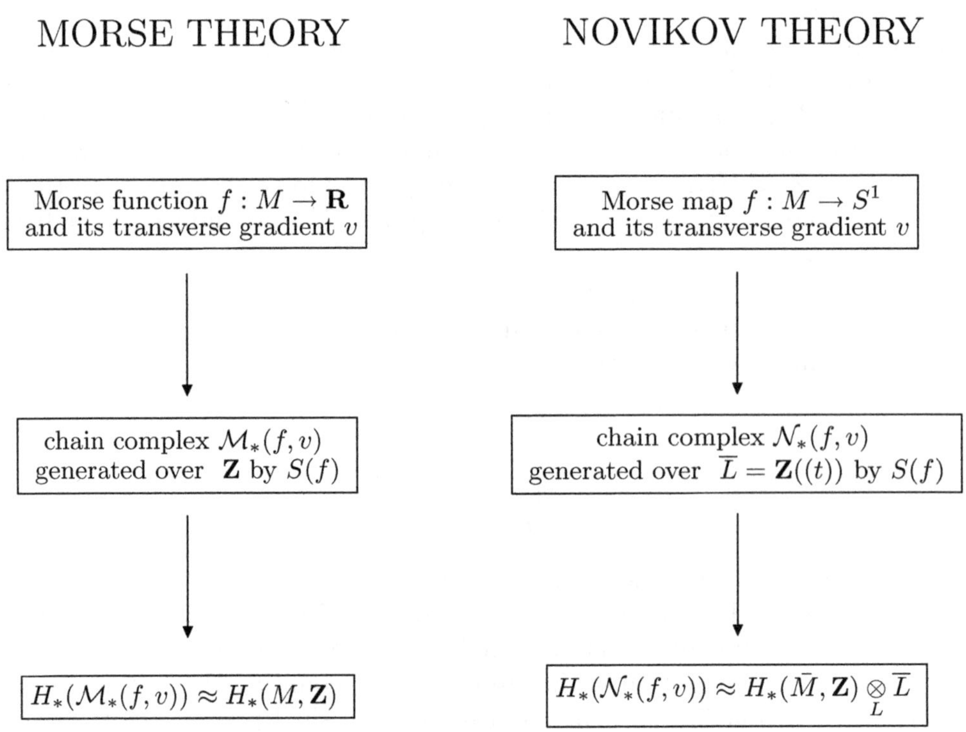

1. The Novikov complex for transverse gradients

Now we proceed to the construction of the Novikov complex. Let $f : M \to S^1$ be a Morse map, where M is a closed manifold. Lift f to a Morse function $F : \bar{M} \to \mathbf{R}$, so that the diagram (1), page 335, is commutative. The free abelian group $\mathcal{M}_k$ generated by the critical points of F of index k, is also a free L-module, any lift to $\bar{M}$ of the set $S_k(f)$ providing a basis for $\mathcal{M}_k$. Let

$$\mathcal{N}_k = \mathcal{M}_k \underset{L}{\otimes} \overline{L}.$$

Here is a geometric description of this module: $\mathcal{N}_k$ is the abelian group of all formal series

$$\sigma = \sum_{p \in S_k(F)} n_p p$$

with integer coefficients n_p, such that for every $C \in \mathbf{R}$ the subset of points p with $n_p \neq 0$ and $F(p) > C$ is finite. Such series will be called *downward.*

We shall introduce boundary operators $\partial_k : \mathcal{N}_k \to \mathcal{N}_{k-1}$, endowing the graded group $\mathcal{N}_*$ with the structure of a chain complex. Similarly to the real-valued Morse theory the construction of the boundary operators uses the count of flow lines of an f-gradient. For an f-gradient v the lift of v to $\bar{M}$ will be denoted by $\bar{v}$; this is an F-gradient.

Lemma 1.1. *For every critical point p of index k of F the stable set $D(p, \bar{v}) = W^{st}(\bar{v}, p)$ is a k-dimensional submanifold of $\bar{M}$.*

Proof. Consider a cobordism $W = F^{-1}([\lambda, \mu])$ where the regular values λ, μ of F are chosen in such a way that $\lambda < F(p) < \mu$. It suffices to prove that for any λ, μ the intersection $D(p, \bar{v}) \cap \operatorname{Int} W$ is a k-dimensional submanifold of $\bar{M}$. Every $\bar{v}$-trajectory converging to p enters W at some moment and never exits. Therefore

$$D(\bar{p}, v) \cap \operatorname{Int} W = D(p, \bar{v}|W) \cap \operatorname{Int} W.$$

It is clear that $F|W$ is a Morse function on W, and $\bar{v}|W$ is its gradient. Applying Proposition 2.6 of Chapter 3 (page 84) we deduce that $D(p, \bar{v}|W) \cap \operatorname{Int} W$ is a k-dimensional submanifold of $\operatorname{Int} W$. □

For any critical point p of F the restriction of the projection $\pi : \bar{M} \to M$ to $D(p, \bar{v})$ is an injective immersion of $D(p, \bar{v})$ to M, and its image equals $W^{st}(p, v)$ (similarly for the unstable manifolds). Therefore the transversality condition holds for $\bar{v}$ if and only if it holds for v.

Definition 1.2. Let v be an f-gradient. An *orientation* of v is a choice of orientation of the local stable manifold $W^{st}_{loc}(v, p)$ for every $p \in S(f)$. An f-gradient endowed with an orientation will be called *oriented.*

Let v be an oriented transverse f-gradient. We shall now define the *incidence coefficient* $n(p, q; \bar{v}) \in \mathbf{Z}$ for any pair p, q of critical points of F with $\operatorname{ind} p = \operatorname{ind} q + 1$. Choose any regular values λ, μ of F with $\lambda < F(q) < F(p) < \mu$. The restriction of F to the cobordism $W = F^{-1}([\lambda, \mu])$

is a Morse function, and $\bar{v}|W$ is a transverse $F|W$-gradient. The chosen orientation of v induces a $\mathbf{Z}$-invariant orientation of $\bar{v}$ and $\bar{v}|W$. We have the incidence coefficient $n(p,q;\bar{v}|W)$ obtained by counting the flow lines from q to p (see Definition 1.2 of Chapter 6, page 197). Clearly $n(p,q;\bar{v}|W)$ does not depend on the auxiliary choice of the regular values λ,μ. It will be denoted by $n(p,q;\bar{v})$.

Proposition 1.3. (1) *There is a unique $\overline{L}$-homomorphism $\partial_k:\mathcal{N}_k\to\mathcal{N}_{k-1}$ which satisfies*

$$(\mathcal{D})\qquad \partial_k(p)=\sum_{q\in S_{k-1}(F)} n(p,q;\bar{v})q$$

for every $p\in S_k(F)$.

(2) *For every k we have $\partial_k\circ\partial_{k+1}=0$.*

Definition 1.4. The graded group $\mathcal{N}_*$ endowed with the boundary operators ∂_*, is therefore a chain complex of $\overline{L}$-modules; it is denoted $\mathcal{N}_*(f,v)$ and called *the Novikov complex* of the pair (f,v).

Exercise 1.5. Prove that a change of the orientation of the f-gradient leads to a chain complex which is basis-preserving isomorphic to the initial one. Thus the choice of the orientation is in a sense less important than the other choices, and we shall not include it in the notation.

Proof of Proposition 1.3. First off observe that the formal sum in the right-hand side of $(\mathcal{D})$ is indeed an element of $\mathcal{N}_{k-1}$, since for $F(p)<F(q)$, we have $n(p,q;\bar{v})=0$. Choose any lift $\mathcal{B}$ of the set $S(f)$ to $\bar{M}$. Then the family $\mathcal{B}$ is an $\overline{L}$-basis for $\mathcal{N}_*$. For any $p\in\mathcal{B}$ define $\partial_k(p)$ by the formula $(\mathcal{D})$, and extend ∂_k to an $\overline{L}$-homomorphism $\mathcal{N}_k\to\mathcal{N}_{k-1}$. Observe that with this definition the formula $(\mathcal{D})$ is valid *for every* $p\in S_k(F)$, since the vector field $\bar{v}$ is $\mathbf{Z}$-invariant, and therefore $n(t^lp,t^lq;v)=n(p,q;v)$ for every $l\in\mathbf{Z}$. The first point of the assertion is proved. Let us proceed to the second point.

Definition 1.6. Let λ be a regular value of F, and let $\mathcal{N}_*^\lambda$ denote the abelian subgroup of $\mathcal{N}_*$ formed by all the downward series whose support is below λ, that is,

$$\mathcal{N}_*^\lambda=\Big\{\sum_{p\in S(F)} n_pp\ \Big|\ n_p=0\quad\text{if}\quad F(p)>\lambda\Big\}.$$

Then $\mathcal{N}_*^\lambda$ is a ∂_*-invariant $\mathbf{Z}[[t]]$-submodule of $\mathcal{N}_*$.

To prove our assertion it suffices to check that for every $k\geqslant 0$ the square of the quotient map

$$\partial_*/t^k:\mathcal{N}_*^\lambda/t^k\mathcal{N}_*^\lambda\to\mathcal{N}_*^\lambda/t^k\mathcal{N}_*^\lambda$$

vanishes. Observe that $t^k\mathcal{N}_*^\lambda = \mathcal{N}_*^{\lambda-k}$; the module $\mathcal{N}_*^\lambda/\mathcal{N}_*^{\lambda-k}$ is the free abelian group generated by the critical points of the function F in the cobordism

$$W_k = F^{-1}\big([\lambda - k, \lambda]\big).$$

We have clearly an isomorphism of two graded abelian groups

$$\mathcal{N}_*^\lambda/\mathcal{N}_*^{\lambda-k} \approx \mathcal{M}_*(F|W_k,\ \bar{v}|W_k)$$

and the homomorphism ∂_*/t^k corresponds via this isomorphism to the boundary operator in the Morse complex $\mathcal{M}_*(F|W_k,\ \bar{v}|W_k)$. Our assertion follows. □

Any lift to $\bar{M}$ of the set $S(f)$ determines a free basis over $\overline{L}$ of the module $\mathcal{N}_*(f,v)$. These bases will be called *geometric bases* for $\mathcal{N}_*(f,v)$. The matrix coefficient of the boundary operator corresponding to generators $p,q \in S(F)$ of a geometric basis is equal to

$$N(p,q;v) = \sum_{k\in\mathbf{Z}} n(p,t^kq;\bar{v})t^k \in \overline{L}.$$

Definition 1.7. The Laurent series $N(p,q;v)$ is called the *Novikov incidence coefficient* corresponding to p,q.

We shall discuss these coefficients later; now let us proceed to the homology of the Novikov complex, which is described in the next theorem.

Theorem 1.8. *Let M be a closed manifold, $f : M \to S^1$ be a Morse map, and v a transverse f-gradient. For every k we have*

$$H_k\big(\mathcal{N}_*(f,v)\big) \approx H_k(\bar{M}) \underset{L}{\otimes} \overline{L}.$$

(The tensor product in the right-hand side of the above formula refers to the structure of the L-module in $H_k(\bar{M})$ induced by the action of the structure group of the covering.)

This theorem will be proved in the present chapter. We shall develop a technique which allows us to obtain a somewhat stronger assertion. First, we shall define the Novikov complex for almost transverse gradients. Then we shall prove (Theorem 6.2) that there is a canonical chain equivalence between the Novikov complex and the completed singular chain complex of $\bar{M}$. We shall see later on that this chain equivalence is an important invariant of the gradient flow related to its dynamical properties.

Example 1.9. Let us return to the map $f = \mathrm{id} : S^1 \to S^1$. It has no critical points, the Novikov complex $\mathcal{N}_*(f,v)$ vanishes, and Theorem 1.8 implies that $H_*(\bar{M}) \underset{L}{\otimes} \overline{L} = 0$. This can be verified independently via a simple computation. The corresponding infinite cyclic covering of S^1 is

$\mathrm{Exp} : \mathbf{R} \to S^1$, and the downward generator t of the group $\mathbf{Z}$ acts on $\mathbf{R}$ as follows:

$$tx = x - 1.$$

The only non-zero homology group of $\mathbf{R}$ is $H_0(\mathbf{R}) \approx \mathbf{Z}$ and every element of the group $\mathbf{Z}$ acts on it as the identity homomorphism. Therefore the element $1 - t$ of the ring $L = \mathbf{Z}[\mathbf{Z}]$ acts trivially in the Novikov completion

$$H_*(\mathbf{R}) \underset{L}{\otimes} \overline{L};$$

on the other hand this element is invertible in the ring $\overline{L}$, since

$$(1-t)^{-1} = 1 + t + t^2 + \cdots .$$

Therefore $H_*(\mathbf{R}) \underset{L}{\otimes} \overline{L} = 0$.

2. Novikov homology

Before proceeding to the proof of Theorem 1.8 let us discuss some of its corollaries. We shall use Theorem 1.8 here; its proof (independent of the material of the present section) will be given later.

The Novikov inequalities

Let X be any topological space, and $f : X \to S^1$ a continuous map. Let $\xi \in [X, S^1]$ denote the homotopy class of f. The infinite cyclic covering induced by f from the covering $\mathrm{Exp} : \mathbf{R} \to S^1$ will be denoted $\bar{X}_\xi$.

Definition 2.1. The graded $\overline{L}$-module

$$\widehat{H}_*(X, \xi) = H_*(\bar{X}_\xi, \mathbf{Z}) \underset{L}{\otimes} \overline{L}$$

is called *the Novikov homology of* X corresponding to the class ξ. In the case when the module $\widehat{H}_*(X, \xi)$ is finitely generated, the integers

$$b_k(X, \xi) = \mathrm{rk}\ _{\overline{L}} \widehat{H}_k(X, \xi)$$

are called the *Novikov Betti numbers*, and the integers

$$q_k(X, \xi) = \mathrm{t.n.}_{\overline{L}} \widehat{H}_k(X, \xi)$$

are called the *Novikov torsion numbers.*†

Example 2.2. If f is homotopic to a constant map, then the corresponding covering is the projection $X \times \mathbf{Z} \to X$; the Novikov Betti and torsion numbers (when defined) are equal to the Betti numbers, respectively torsion numbers of the space X.

† Recall that $\mathrm{t.n.}_{\overline{L}} K$ stands for the *torsion number of the module* K, that is, the minimal possible number of generators of the submodule of torsion elements of K.

Example 2.3. Let $X = Y \times S^1$ and $f : X \to S^1$ be the projection onto S^1. Then $\widehat{H}_*(X, \xi) = 0$. Indeed, the covering $\bar{X}_\xi$ is homeomorphic to $Y \times \mathbf{R}$ and the action of t in the homology of $\bar{X}_\xi$ equals the identity homomorphism. The same argument as in Example 1.9 shows that the Novikov homology $\widehat{H}_*(X, \xi)$ is equal to zero.

The ring $\overline{L}$ is a flat L-module (Theorem 2.3 of Chapter 10, page 332), therefore we have an isomorphism

$$H_*(\mathcal{S}_*(\bar{X}_\xi) \underset{L}{\otimes} \overline{L}) \approx H_*(\bar{X}_\xi) \underset{L}{\otimes} \overline{L}.$$

Thus if the chain complex $\mathcal{S}_*(\bar{X}_\xi)$ is *homotopically free-finite* over L, then the Novikov homology $\widehat{H}_*(X, \xi)$ is a finitely generated $\overline{L}$-module, and the Novikov Betti and torsion numbers are defined. This is the case when X has the homotopy type of a finite CW complex, in particular if X is a closed manifold. Observe that in this case we have also a natural identification of $[X, S^1]$ with $H^1(X, \mathbf{Z})$.

Proposition 2.4 (The Novikov inequalities)**.** *Let M be a closed manifold, and $f : M \to S^1$ be a Morse map belonging to a class $\xi \in H^1(M, \mathbf{Z}) \approx [M, S^1]$. Then*

$$(\mathcal{N}) \qquad m_k(f) \geqslant b_k(M, \xi) + q_k(M, \xi) + q_{k-1}(M, \xi).$$

Proof. Pick any oriented transverse f-gradient v. The homology of the associated Novikov complex $\mathcal{N}_*(f, v)$ is isomorphic to $\widehat{H}_*(M, \xi)$ by Theorem 1.8 and by the purely algebraic Proposition 1.10 of Chapter 6 (page 201) we deduce the conclusion. □

Corollary 2.5. *If $\xi \in H^1(M, \mathbf{Z})$ contains a C^∞ fibration $f : M \to S^1$, then the Novikov numbers $b_k(M, \xi)$, $q_k(M, \xi)$ vanish.* □

Poincaré duality

The Morse theory leads to a quick proof of the Poincaré duality for closed manifolds. Now we shall apply the Morse-Novikov theory to study the Poincaré duality in the Novikov homology.

Let M be a closed oriented m-dimensional manifold, $f : M \to S^1$ a Morse map and v an oriented transverse f-gradient. Consider the Morse map $(-f) : M \to S^1$ (the opposite map to f with respect to the abelian group structure on S^1). The vector field $(-v)$ is a transverse $(-f)$-gradient. We shall endow $(-v)$ with the orientation, induced from v (see Definition 1.15 of Chapter 6, page 205).

Theorem 2.6. *The chain complex of $\overline{L}$-modules $\mathcal{N}_*(-f,-v)$ is basis-preserving isomorphic to the m-dimensional Poincaré-dual of $\mathcal{N}_*(f,v)$.*

Proof. Let $\bar{M} \to M$ denote the infinite cyclic covering induced by f from $\mathrm{Exp} : \mathbf{R} \to S^1$, let $F : \bar{M} \to \mathbf{R}$ be a lift of f, and $\bar{v}$ be the lift to $\bar{M}$ of the vector field v. Let t denote the downward generator of the deck transformation group.

Let us consider similar objects for the map $(-f)$. Let $\bar{M}' \to M$ denote the infinite cyclic covering induced by $(-f)$ from Exp, and let u denote the corresponding downward generator of the deck transformation group.

Thus $\mathcal{N}_*(f,v)$ is a chain complex of $\mathbf{Z}((t))$-modules, and the base ring of $\mathcal{N}_*(-f,-v)$ is $\mathbf{Z}((u))$, so our task is to identify $\mathcal{N}_*(-f,-v)$ with the Poincaré-dual to $\mathcal{N}_*(f,v)$ under the ring isomorphism

$$t \mapsto u, \quad \mathbf{Z}((t)) \xrightarrow{\approx} \mathbf{Z}((u)).$$

It follows from the definition of induced coverings that there is a diffeomorphism $\bar{M} \xrightarrow{\Phi} \bar{M}'$ such that

$$\Phi(tx) = u^{-1}\Phi(x),$$

and the Morse function

$$G = -F \circ \Phi^{-1} : \bar{M}' \to \mathbf{R}$$

is a lift of $(-f)$ to $\bar{M}'$. We have in particular

$$\Phi(S_k(F)) = S_{m-k}(G).$$

The lift of $(-v)$ to $\bar{M}'$ equals to the Φ-image of $-\bar{v}$, and we shall keep the same symbol $-\bar{v}$ for it.

Let p' denote the Φ-image of a critical point $p \in S(F)$. Applying Lemma 1.17 of Chapter 6 (page 207) we deduce easily that for $\operatorname{ind} p = k$, $\operatorname{ind} q = k-1$ we have:

$$n(q', u^n p'; -\bar{v}) = (-1)^{m-k+1} \cdot n(p, t^n q; \bar{v}) \qquad \text{for every} \quad n.$$

Therefore the Novikov incidence coefficient $N(q', p'; -\bar{v}) \in \mathbf{Z}((u))$ is obtained from the Novikov incidence coefficient $N(p, q; \bar{v})$ by changing the variable t to u and multiplying by $(-1)^{m-k+1}$. And the matrix of the operator

$$(-1)^{m-k+1}\partial_{m-k+1} : \mathcal{N}_{m-k+1}(-f,-v) \to \mathcal{N}_{m-k}(-f,-v)$$

is obtained therefore as the conjugate of the matrix of

$$\partial_k : \mathcal{N}_k(f,v) \to \mathcal{N}_{k-1}(f,v)$$

after replacing the variable t by u. The theorem follows. □

To deduce from this result the duality theorem for the Novikov homology, we introduce the notion of the *Novikov cohomology.*

Definition 2.7. Let X be a topological space, and $\xi \in [X, S^1]$ a homotopy class of maps $X \to S^1$. The cohomology of the chain complex

$$\mathrm{Hom}_L\Big(\mathcal{S}_*(\bar{X}_\xi),\ \overline{L}\Big) = \mathrm{Hom}_{\overline{L}}\Big(\mathcal{S}_*(\bar{X}_\xi) \underset{L}{\otimes} \overline{L},\ \overline{L}\Big)$$

of $\overline{L}$-modules is called the *Novikov cohomology* of X with respect to ξ, and denoted $\widehat{H}^*(X, \xi)$.

In other words, $\widehat{H}^*(X, \xi)$ is the cohomology module $H^*(\mathcal{S}_*(\bar{X}_\xi) \underset{L}{\otimes} \overline{L})$.

Proposition 2.8. *Let M be a closed oriented m-dimensional manifold, and $\xi \in H^1(M, \mathbf{Z})$. For every k we have:*

$$\widehat{H}^k(M, \xi) \approx \widehat{H}_{m-k}(M, -\xi).$$

Proof. Observe that $\mathcal{S}_*(\bar{M}_\xi)$ is a free L-complex. Indeed, the structure group $\mathbf{Z}$ of the covering $\bar{M}_\xi \to M$ acts freely on the set of all singular simplices of $\bar{M}_\xi$. Let $\mathcal{O}$ be the set of the orbits of this action. A choice of one singular simplex σ_ω in each orbit $\omega \in \mathcal{O}$ determines an L-basis in $\mathcal{S}_*(\bar{M}_\xi)$.

Therefore $\mathcal{S}_*(\bar{M}_\xi) \underset{L}{\otimes} \overline{L}$ is a free $\overline{L}$-complex and its homology is isomorphic to the homology of $\mathcal{N}_*(f, v)$ (where $f : M \to S^1$ is any Morse function in the homotopy class ξ and v is a transverse f-gradient). Applying Lemma 1.18 of Chapter 6 (page 208) we deduce that the cohomology modules of these two complexes are isomorphic, hence

$$\begin{aligned}\widehat{H}^k(M, \xi) &= H^k(\mathcal{S}_*(\bar{M}_\xi) \underset{L}{\otimes} \overline{L})\\ &\approx H^k\big(\mathcal{N}_*(f, v)\big) \approx H_{m-k}\big(\mathcal{N}_*(-f, -v)\big) \approx \widehat{H}_{m-k}\big(\bar{M}, -\xi\big). \qquad \square\end{aligned}$$

The next corollary is immediate.

Corollary 2.9. *Let M be a closed m-dimensional oriented manifold. Then for every k we have*

$$b_k(M, \xi) = b_{m-k}(M, -\xi), \quad q_k(M, \xi) = q_{m-k-1}(M, -\xi). \qquad \square$$

It is sometimes useful to interpret the Novikov homology with respect to the opposite classes ξ and $-\xi$ as the completion of one and the same L-module $H_*(\bar{M}_\xi)$, but performed with respect to two different "ends" of $\bar{M}_\xi$. Consider a ring isomorphism

$$\chi : \mathbf{Z}((t)) \to \mathbf{Z}((t^{-1})); \quad \chi\Big(\sum_i a_i t^i\Big) = \sum_i a_i t^{-i}.$$

A $\mathbf{Z}((t))$-module N will be called *χ-isomorphic to a $\mathbf{Z}((t^{-1}))$-module K if there is an isomorphism $\phi : N \to K$ of abelian groups, such that for every $\lambda \in \mathbf{Z}((t))$ we have $\phi(\lambda x) = \chi(\lambda)\phi(x)$.*

Lemma 2.10. *The $\mathbf{Z}((t))$-module $\widehat{H}_*(M, -\xi)$ is χ-isomorphic to the $\mathbf{Z}((t^{-1}))$-module $H_*(\bar{M}_\xi) \underset{L}{\otimes} \mathbf{Z}((t^{-1}))$.*

Proof. As we have already seen in the proof of Theorem 2.6 there is a commutative diagram:

$$\begin{array}{ccc} \bar{M}_\xi & \xrightarrow{\quad\Phi\quad} & \bar{M}_{-\xi} \\ & \searrow \quad \swarrow & \\ & M & \end{array}$$

where Φ is a diffeomorphism with $\Phi(tx) = u^{-1}\Phi(x)$ (t and u denote the downward generators of the deck transformation groups of $\bar{M}_\xi$, $\bar{M}_{-\xi}$). The lemma follows immediately. □

Necessary conditions for fibring of manifolds over a circle

Let M be a C^∞ closed manifold and $\xi \in H^1(M, \mathbf{Z}) \approx [M, S^1]$. Corollary 2.5 says that if ξ is represented by a C^∞ fibration then thc Novikov homology $\widehat{H}_*(M, \xi)$ vanishes. Therefore the Novikov homology can be considered as an obstruction for the homotopy class ξ to contain a C^∞ fibration over a circle. Another natural obstruction is provided by the next lemma.

Lemma 2.11. *If a class $\xi \in H^1(M, \mathbf{Z})$ contains a C^∞ fibration $M \to S^1$, then $H_*(\bar{M}_\xi)$ is a finitely generated abelian group.*

Proof. Let $f : M \to S^1$ be a C^∞ fibration belonging to the class ξ. The infinite cyclic covering $\bar{M}_\xi$ has obviously the same homotopy type as the fiber of the map f; therefore its homology is finitely generated. □

It is interesting to note that the two necessary conditions provided by Corollary 2.5 and by the above lemma are equivalent, and this equivalence can be proved algebraically without any reference to the circle-valued Morse theory. The next proposition is a particular case of a theorem of A. Ranicki [**134**].

Proposition 2.12. *Let M be a closed oriented manifold, and $\xi \in H^1(M)$. Then $H_*(\bar{M}_\xi)$ is a finitely generated abelian group if and only if the Novikov homology module $\widehat{H}_*(M, \xi)$ is equal to zero.*

Proof. Assume that $H_*(\bar{M}_\xi)$ is a finitely generated abelian group, and let us prove that the Novikov homology vanishes. The group

$$G = H_*(\bar{M}_\xi) \underset{L}{\otimes} \overline{L} \approx H_*(\bar{M}_\xi) \underset{P}{\otimes} \widehat{P}$$

is a finitely generated module over $\widehat{P}$. Since the action of t on this group is invertible, we have $G = tG$. The element t is in the Jacobson radical of the ring $\widehat{P}$, and, therefore, Nakayama's lemma (see [**9**], Ch. 2, Proposition 2.6) implies that $G = 0$.

To prove the inverse implication we need a lemma.

Lemma 2.13. *Let N be a finitely generated L-module. Assume that $N \underset{L}{\otimes} \overline{L} = 0$. Then there is a polynomial $A(t) \in \mathbf{Z}[t]$ such that $1+tA(t)$ annihilates the module N.*

Proof. We have $N \approx N_0 \underset{P}{\otimes} L$ with N_0 a finitely generated P-module. Let

$$N_0 \xrightarrow{j} N_0 \underset{P}{\otimes} \widehat{P}$$

be the natural map. The kernel of this map is described by the classical Krull theorem ([**9**], Th. 10.17) which implies that there is a polynomial $A(t) \in \mathbf{Z}[t]$ such that every element of the kernel of j is annihilated by $1 + tA(t)$ where $A(t) \in \mathbf{Z}[t]$. Let us denote by $S \subset P$ the multiplicative subset of all elements t^n, with $n \in \mathbf{N}$. Since

$$S^{-1}(N_0 \underset{P}{\otimes} \widehat{P}) = N \underset{L}{\otimes} \overline{L} = 0$$

for each element $x \in N_0$ there is an m such that $j(t^m x) = 0$. Therefore for each $x \in N_0$ we have:

$$(1 + tA(t))t^m x = 0$$

for some $m \in \mathbf{N}$. Since t is invertible in N our lemma is proved. □

Thus the homology $H_*(\bar{M}_\xi)$ is annihilated by some polynomial of the form $1 + tA(t)$ where $A(t) \in \mathbf{Z}[t]$. Observe that

$$H_*(\bar{M}_\xi) \underset{L}{\otimes} \mathbf{Z}((t^{-1})) = 0$$

(by Lemma 2.10 and Proposition 2.8). By the same argument as above there exists a polynomial $B \in \mathbf{Z}[t]$ such that the module $H_*(\bar{M}_\xi)$ is annihilated by $1 + t^{-1}B(t^{-1})$. The module $H_*(\bar{M}_\xi)$ is finitely generated over L, therefore our assertion follows immediately from the next lemma.

Lemma 2.14. *Let $A, B \in \mathbf{Z}[t]$, and let I be the ideal generated by the polynomials $1 + tA(t)$ and $1 + t^{-1}B(t^{-1})$. The quotient L/I is a finitely generated abelian group.*

Proof. We shall prove that L/I is generated over $\mathbf{Z}$ by the monomials t^k with

$$-\deg A \leqslant k \leqslant \deg B.$$

Indeed for every $r \geqslant \deg B + 1$ write $t^r + t^{r-1}B(t^{-1}) = 0$ to express t^r as a $\mathbf{Z}$-linear combination of elements t^i with $0 \leqslant i < r$. Similarly, every monomial t^q with $q \leqslant -1 - \deg A$ is a $\mathbf{Z}$-linear combination of elements t^i with $q < i \leqslant 0$, and we obtain our lemma by a simple induction argument. □

3. The Novikov complex for almost transverse gradients

In this section we work in the same geometric setting as in Section 1, with the only difference that the oriented f-gradient v is no longer assumed to be transverse, but only *almost transverse*, that is, for every $p, q \in S(f)$ with $\operatorname{ind} p \leqslant \operatorname{ind} q$ we have

$$W^{st}(v,p) \pitchfork W^{un}(v,q).$$

As in the case of real-valued Morse functions it is easy to show that v is almost transverse if and only if

$$\Big(p \neq q \quad \text{and} \quad \operatorname{ind} p \leqslant \operatorname{ind} q\Big) \Rightarrow W^{st}(v,p) \cap W^{un}(v,q) = \varnothing.$$

The set of all almost transverse f-gradients will be denoted by $G_A(f)$. It is dense in $G(f)$ with respect to the C^∞ topology since it contains the residual subset $G_T(f)$, but contrarily to the case of real-valued Morse functions, $G_A(f)$ is not in general a C^0-open subset of $G(f)$.

Let $F : \bar{M} \to \mathbf{R}$ be any lift of f (see diagram (1), page 335). The gradient v is almost transverse if and only if the lift $\bar{v}$ of v to $\bar{M}$ is an almost transverse F-gradient. Let us define *the incidence coefficient* $n(p,q;v)$ where $p, q \in S(F)$ are critical points of F with $\operatorname{ind} p = \operatorname{ind} q + 1$. Pick any regular values λ, μ of F such that

$$\lambda < F(p), F(q) < \mu.$$

Consider the cobordism $W = F^{-1}([\lambda,\mu])$. The restriction $F|W$ is a Morse function, and $\bar{v}|W$ is an almost transverse $F|W$-gradient. As we know from Subsection 2.2 of Chapter 6 (page 210), in this situation the incidence coefficient $n(p,q;\bar{v}|W) \in \mathbf{Z}$ is defined. Proposition 2.5 of Chapter 6 (page 212) implies that this coefficient does not depend on the particular choice of regular values λ, μ; so it will be denoted $n(p,q;\bar{v})$. When v is transverse, the coefficient $n(p,q;\bar{v})$ equals the incidence coefficient defined via counting the flow lines joining p and q.

Lemma 3.1. (1) *If $F(p) < F(q)$, then $n(p,q;\bar{v}) = 0$.*

(2) *For every $n \in \mathbf{Z}$ we have*

$$n(p, q; \bar{v}) = n(t^n p, t^n q; \bar{v}).$$

Proof. The first point is obvious. The second point follows easily from the invariance of $\bar{v}$ with respect to the deck transformations of the covering. □

Now we can define the Novikov complex. Recall the module $\mathcal{M}_*$ freely generated over $\mathbf{Z}$ by the critical points of F and the module $\mathcal{N}_* = \mathcal{M}_* \underset{L}{\otimes} \overline{L}$. The proof of the next proposition repeats almost verbatim the proof of Proposition 1.3.

Proposition 3.2. (1) *There is a unique $\overline{L}$-homomorphism $\partial_k : \mathcal{N}_k \to \mathcal{N}_{k-1}$ which satisfies*

$$\partial_k(p) = \sum_{q \in S_{k-1}(f)} n(p, q; \bar{v}) q$$

for every $p \in S_k(F)$.

(2) *For every k we have $\partial_k \circ \partial_{k+1} = 0$.* □

Definition 3.3. The graded group $\mathcal{N}_*$ endowed with the boundary operator ∂_*, is therefore a chain complex of $\overline{L}$-modules, which is denoted $\mathcal{N}_*(f, v)$ and called *the Novikov complex* of the pair (f, v).

Note that if v satisfies the transversality condition, then the Novikov complex just introduced and the Novikov complex from Section 1 are basis-preserving isomorphic, thus using the same term for both does not lead to confusion.

Let λ be any regular value of F. Recall the submodule $\mathcal{N}_*^\lambda \subset \mathcal{N}_*$ formed by all downward series with support below λ (see page 338); it is a free-finite $\mathbf{Z}[[t]]$-complex. We have

$$\mathcal{N}_*^\lambda = \varprojlim \mathcal{N}_*^\lambda / t^n \mathcal{N}_*^\lambda.$$

The *truncated Novikov complexes*

$$\mathcal{N}_*(\lambda, n) = \mathcal{N}_*^\lambda / t^n \mathcal{N}_*^\lambda$$

are important for the proof of Theorems 1.8 and 6.2, and we will study them in detail in the next section.

4. Equivariant Morse equivalences

For a regular value λ of F set

$$W = F^{-1}([\lambda-1,\lambda]), \quad W_n = F^{-1}([\lambda-n,\lambda]),$$
$$V = F^{-1}(\lambda),\ V^- = F^{-1}(]-\infty,\lambda]).$$

Have a look at Figure 41 to visualize these objects.

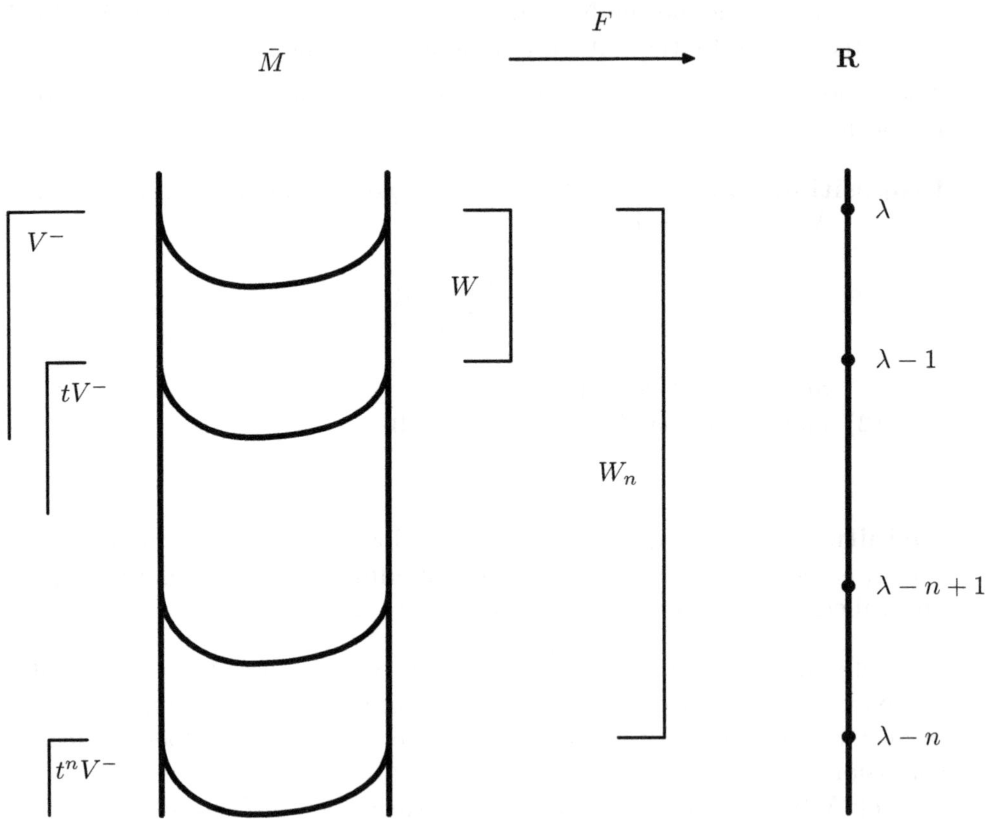

FIGURE 41.

By definition we have a basis-preserving isomorphism

$$\mathcal{N}_*(\lambda,n) \approx \mathcal{M}_*(F|W_n,\ \bar v|W_n)$$

between the truncated Novikov complex and the Morse complex of the pair $(F|W_n,\ \bar v|W_n)$. We obtain therefore homology equivalences:

$$\mathcal{N}_*(\lambda,n) \approx \mathcal{M}_*(F|W_n,\ \bar v|W_n) \xrightarrow{\sim} \mathcal{S}_*(W_n,\ \partial_0 W_n) \hookrightarrow \mathcal{S}_*(V^-, t^nV^-)$$

where the middle arrow is the canonical Morse equivalence and the right-hand arrow is induced by the inclusion of pairs. Compose the homomorphisms above to get a homology equivalence

$$\mathcal{E}' : \mathcal{N}_*(\lambda, n) \xrightarrow{\sim} \mathcal{S}_*(V^-, t^n V^-). \tag{2}$$

Observe that both chain complexes in (2) are modules over the ring

$$P_n = P/t^n P \approx \widehat{P}/t^n \widehat{P}, \quad \text{where} \quad P = \mathbf{Z}[t].$$

Our aim in this subsection is to refine the homology equivalence $\mathcal{E}'$ in order to make it compatible with this additional algebraic structure. In other words, we want it to be a chain equivalence of P_n-complexes. To explain the main idea of the construction, recall that the Morse complex for $\bar{v}|W_n$ and the corresponding Morse equivalence are constructed with the help of an auxiliary ordered Morse function. We shall see that if this ordered Morse function is in a sense adjusted to the deck transformation group of the covering, then it yields a structure of a P_n-complex in the Morse complex of $F|W_n$ and the chain equivalence (2) can be chosen so as to preserve the structure of P_n-modules.

Definition 4.1. Let $\psi : W_n \to [a, b]$ be an ordered Morse function. Let $\{Y_i\}$ denote the corresponding Morse-Smale filtration of W_n (see Definition 3.56 of Chapter 4, page 157). Put

$$X_i = t^n V^- \cup Y_i.$$

The function ψ is called *t-ordered*, if all the the spaces X_i are t-invariant, that is

$$tX_i \subset X_i \quad \text{for every} \quad i.$$

The next proposition asserts that t-ordered Morse functions exist, and moreover, we can choose a t-ordered Morse function so that the corresponding Morse-Smale filtration is as close as we like to the skeletons of the Morse stratification. The proof is technical, and it is not used in the sequel, so the reader may wish to proceed directly to the applications of the result (page 351) and return to the proof when necessary. †

Proposition 4.2. *Let*

$$D_i = D(\text{ind}{\leqslant}i\ ; \bar{v}|W_n) \cup \partial_0 W_n.$$

Let U_i be any neighbourhood of D_i in W_n. Then there is a t-ordered Morse function $\psi : W_n \to [a, b]$ adjusted to $(F, \bar{v})$, and such that the Morse-Smale filtration $\{Y_i\}$ corresponding to ψ satisfies $Y_i \subset U_i$ for every $i < \dim M$.

† Another construction of t-ordered function may be found in [**110**].

Proof. We can assume that $U_i \subset U_{i+1}$ for every i. Arrange the critical points of $F|W_n$ in a finite sequence $p_0, \dots, p_N$ in such a way that the following two conditions hold:

$$\operatorname{ind} p_i < \operatorname{ind} p_j \Rightarrow i < j;$$
$$\text{if} \quad p_i = t^r p_j \quad \text{with} \quad r > 0, \quad \text{then} \quad i < j.$$

It suffices to construct a Morse function $\chi : W_n \to \mathbf{R}$ adjusted to $(F, \bar{v})$ and a sequence of regular values $a = a_0 < a_1 < \dots < a_{N+1} < b$ of χ, such that the interval $[a_{N+1}, b]$ is regular and the subsets

$$Z_i = \chi^{-1}([a_0, a_{i+1}]) \subset W_n$$

have the following properties for $i \leqslant N$ and $\operatorname{ind} p_i \leqslant m-1$:

(A) $\quad p_i$ is the unique critical point of χ in $Z_i \setminus Z_{i-1}$.

(B) $\quad tZ_i \subset Z_{i-1} \cup t^n V^-$.

(C) $\quad Z_i \subset U_l \quad$ where $\quad l = \operatorname{ind} p_i$.

Note that the functions satisfying only the property (A) exist by Proposition 3.39 of Chapter 4 (page 149). Arguing by induction, assume that we have already constructed a function χ such that (A) holds for all i and (B) and (C) hold for all $i < s$. Let

$$Z' = \chi^{-1}\big([a_s, a_{s+1}]\big), \qquad Z'' = \chi^{-1}\big([a_{s+1}, b]\big)$$

so that $W_n = Z_{s-1} \cup Z' \cup Z''$. See Figure 42.

We shall modify our function χ in Z'. Put

$$\Delta = D(p_s, v) \cap Z'$$

(the subset Δ is shown in the figure by the fat black line). The point tp_s is in Int Z_{s-1}, therefore $t\Delta \subset \operatorname{Int} Z_{s-1} \cup t^n V^-$ and there is a neighbourhood U of Δ in Z' such that

$$tU \subset \operatorname{Int} Z_{s-1} \cup t^n V^-.$$

We can choose U in such a way that $\partial_0 Z' \subset U$. Put

$$U' = U \cap U_l \quad \text{where} \quad l = \operatorname{ind} p_s.$$

The set U' (shaded in the picture) is an open neighbourhood of $\Delta \cup \partial_0 Z'$. It suffices to consider the case $l \leqslant m-1$. Applying Theorem 3.58 of Chapter 4 (page 158) we obtain a new function $\zeta : Z' \to \mathbf{R}$ together with a regular value β of ζ, such that

(1) ζ coincides with χ nearby $\partial Z'$,

(2) $\zeta^{-1}([a_s, \beta]) \subset U'$.

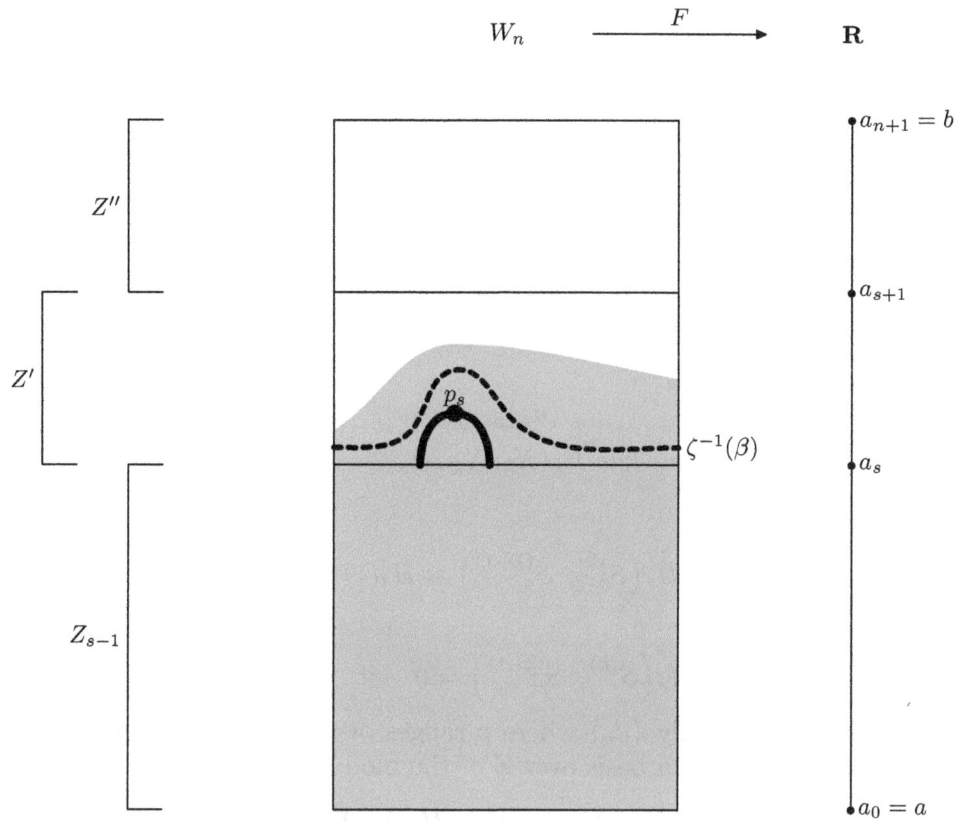

FIGURE 42.

(The regular level surface $\zeta^{-1}(\beta)$ is shown in the figure by a dashed line.) Define a function $\widetilde{\chi}$ which equals χ in $Z_{s-1} \cup Z''$, and equals ζ in Z'. Choose the sequence a'_i of regular values of $\widetilde{\chi}$ so that $a'_i = a_i$ except $i = s$, and $a'_s = \beta$. It is clear that $\widetilde{\chi}$ satisfies (B) and (C) for every $i \leqslant s$. $\square$

Now let us apply ordered Morse functions to study the homology equivalence

$$\mathcal{E}' : \mathcal{N}_*(\lambda, n) \xrightarrow{\sim} \mathcal{S}_*(V^-, t^n V^-)$$

introduced on page 349. Let $\psi : W_n \to [a, b]$ be any t-ordered Morse function such that $\bar{v}$ is a ψ-gradient. Let Y_i denote the corresponding Morse-Smale filtration of W_n, and put

$$X_i = Y_i \cup t^n V^-.$$

The filtration $(X_i, t^n V^-)$ of the pair $(V^-, t^n V^-)$ is invariant with respect to the action of t, and yields a filtration

$$\mathcal{S}_*^{(i)} = \mathcal{S}_*(X_i, t^n V^-) \subset \mathcal{S}_*(V^-, t^n V^-)$$

of the singular chain complex $\mathcal{S}_*(V^-, t^n V^-)$ by subcomplexes of P_n-modules.

For every critical point $p \in W_n$ with $\operatorname{ind} p = k$ we have the fundamental class

$$d_p = [D(p, \bar{v})] \in H_k(X_k, X_{k-1}).$$

Proposition 4.3. *The filtration* $\{\mathcal{S}_*^{(i)}\}$ *is a cellular filtration of* P_n*-complexes, and the family*

$$\{d_p\}_{p \in S_k(F|W_n)}$$

is a P_n*-basis in the module* $H_k(X_k, X_{k-1})$.

Proof. We have

$$H_*\Big(\mathcal{S}_*^{(k)},\ \mathcal{S}_*^{(k-1)}\Big) \approx H_*(Y_k, Y_{k-1})$$

therefore

$$H_*\Big(\mathcal{S}_*^{(k)},\ \mathcal{S}_*^{(k-1)}\Big) = 0 \quad \text{if} \quad * \neq k.$$

Recall that the family $\{d_p\}$ where p ranges over the family of the critical points of F in W_n is a basis over $\mathbf{Z}$ of the module

$$H_*(X_k, X_{k-1}) \approx H_*(Y_k, Y_{k-1}).$$

It is easy to see that

$$d_{t^r p} = t^r d_p \quad \text{for every} \quad r,$$

and therefore the elements d_p where $p \in S_k(F) \cap W_n$ form a P_n-basis of the module $H_k(X_k, X_{k-1})$. □

It is clear that the adjoint complex of this filtration is basis-preserving isomorphic to the truncated Novikov complex $\mathcal{N}_*(\lambda, n)$. By general theory of cellular filtrations we have a chain equivalence of P_n-complexes:

$$h(\psi) : \mathcal{N}_*(\lambda, n) \longrightarrow \mathcal{S}_*(V^-, t^n V^-)$$

which is homotopic over $\mathbf{Z}$ to the chain equivalence $\mathcal{E}'$. Similarly to the case of the usual Morse functions the chain equivalence just constructed does not depend on the choice of a t-ordered function:

Proposition 4.4. *Let* ψ_1, ψ_2 *be* t*-ordered Morse functions, such that* $\bar{v}$ *is* ψ_1*- and* ψ_2*-gradient. The chain equivalences* $h(\psi_1)$ *and* $h(\psi_2)$ *are chain homotopic.*

Proof. The proof is completely similar to the proof of Proposition 3.11 of Chapter 6 (page 217), using Proposition 4.2 instead of Theorem 3.58 of Chapter 4 (page 158). □

Definition 4.5. Let $\psi : W_n \to \mathbf{R}$ be a t-ordered Morse function such that $\bar{v}$ is a ψ-gradient. The chain equivalence $h(\psi)$ will be denoted

$$\mathcal{E}(\lambda, n) : \mathcal{N}_*(\lambda, n) \to \mathcal{S}_*(V^-, t^n V^-)$$

and called the *equivariant Morse equivalence.*

Our next aim is to compare the equivariant Morse equivalences for different values of λ and n. Let $\lambda \leqslant \mu$ be two regular values of F, and let $k \leqslant n$ be positive integers. Put

$$V_1 = F^{-1}(\lambda), \quad V_2 = F^{-1}(\mu).$$

The inclusion $V_1^- \subset V_2^-$ induces a chain map of complexes

$$J : \mathcal{S}_*(V_1^-,\ t^n V_1^-) \longrightarrow \mathcal{S}_*(V_2^-,\ t^k V_2^-). \tag{3}$$

Each of these complexes can be considered as a chain complex of $\widehat{P}$-modules, and J respects this structure. Similarly we have the map

$$I : \mathcal{N}_*(\lambda, n) \to \mathcal{N}_*(\mu, k)$$

of chain complexes over $\widehat{P}$.

Proposition 4.6. *The following diagram of chain maps of complexes of $\widehat{P}$-modules is homotopy commutative:*

$$\begin{array}{ccc} \mathcal{N}_*(\lambda, n) & \xrightarrow{\mathcal{E}(\lambda,n)} & \mathcal{S}_*(V_1^-, t^n V_1^-) \\ \Big\downarrow{\scriptstyle I} & & \Big\downarrow{\scriptstyle J} \\ \mathcal{N}_*(\mu, k) & \xrightarrow{\mathcal{E}(\mu,k)} & \mathcal{S}_*(V_2^-, t^k V_2^-) \end{array} \tag{4}$$

Proof. Pick any t-ordered Morse function

$$\phi : F^{-1}([\mu - k, \mu]) \to \mathbf{R}$$

such that $\bar{v}$ is a ϕ-gradient, and let X_i be the corresponding filtration of V_2^-. By Proposition 4.2 there is a t-ordered Morse function

$$\psi : F^{-1}([\lambda - n, \lambda]) \to \mathbf{R}$$

such that the terms $\widetilde{X}^i$ of the filtration of V_1^- corresponding to ψ satisfy $\widetilde{X}^i \subset X^i$. The inclusion map

$$(V_1^-,\ t^n V_1^-) \subset (V_2^-,\ t^k V_2^-)$$

preserves therefore the filtrations. Turning now to the diagram (4) we observe that the chain maps $J \circ \mathcal{E}(\lambda, n)$ and $\mathcal{E}(\mu, k) \circ I$ preserve filtrations and induce the same map in the adjoint complexes. Therefore the homotopy commutativity of the diagram (4) follows from Theorem 3.6 of Chapter 6 (page 213). □

Now we shall prove that the equivariant Morse equivalences are stable with respect to C^0-small perturbations of the gradient. Since we are going to consider several gradients at once, let us write

$$\mathcal{N}_*(\lambda, n; v) = \mathcal{N}_*^\lambda(f, v)/t^n \mathcal{N}_*^\lambda(f, v),$$

in order to stress the dependence of $\mathcal{N}_*$ on the choice of v.

Proposition 4.7. *There is $\delta > 0$ such that for every f-gradient w oriented similarly to v and satisfying $||w - v|| < \delta$ there is a homotopy commutative diagram*

(5)

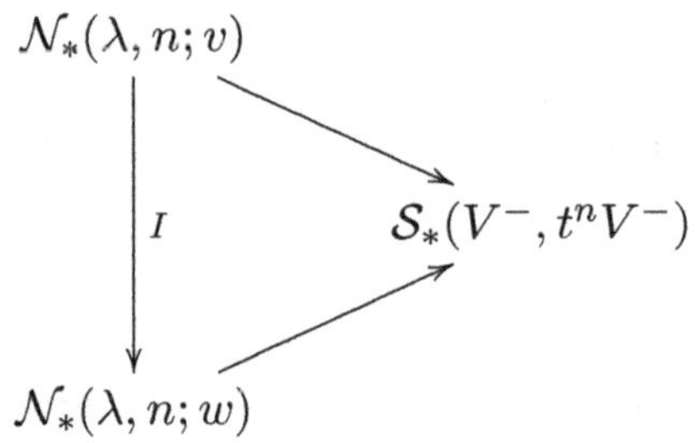

where I is a basis preserving isomorphism of chain complexes of P_n-modules, and the diagonal arrows are the equivariant Morse chain equivalences corresponding to v and respectively w.

Proof. Any t-ordered Morse function $\chi : W_n \to [a, b]$ adjusted to $(F, \bar{v})$ is also adjusted to $(F, \bar{w})$ if only $||w - v||$ is small enough. Therefore both chain complexes $\mathcal{N}(\lambda, n; v)$ and $\mathcal{N}(\lambda, n; w)$ are equal to the adjoint complex corresponding to the filtration induced by χ. Thus we obtain already the commutative diagram (5), and it remains to prove that the vertical arrow preserves the bases. This follows from Proposition 3.16 of Chapter 5 (page 184) which says that for $||w - v||$ sufficiently small the bases determined by the descending discs of w and v in the adjoint complex are equal. □

5. On the singular chain complex of the infinite cyclic covering

Let $f : M \to S^1$ be a C^∞ function, and $\bar{M} \to M$ the corresponding infinite cyclic covering. This section contains several remarks about the singular chain complex $\mathcal{S}_*(\bar{M})$ which will be useful for the sequel. Lift f to a C^∞ function $F : \bar{M} \to \mathbf{R}$. Let t be the downward generator of the structural group of the covering, that is, $F(tx) = F(x) - 1$. We shall use the abbreviation

$$\mathcal{S}_* = \mathcal{S}_*(\bar{M}).$$

Recall that $\mathcal{S}_*$ has a natural structure of a free chain complex over $L = \mathbf{Z}[t, t^{-1}]$. Let λ be a regular value of F. Put

$$V = F^{-1}(\lambda), \quad V^- = F^{-1}(]-\infty, \lambda]),$$

$$\mathcal{S}_*^\lambda = \{A \in \mathcal{S}_* \mid \operatorname{supp} A \subset V^-\};$$

then $\mathcal{S}_*^\lambda$ is obviously a graded P-module, which is a subcomplex of $\mathcal{S}_*$.

Proposition 5.1. *$\mathcal{S}_*^\lambda$ is a chain complex of free P-modules, and it is homotopically free-finite.*

Proof. The first assertion is easy: a free basis of $\mathcal{S}_*^\lambda$ is formed by all singular simplices σ in V^- such that $t^{-1}\sigma \notin V^-$. It is slightly more difficult to prove that $\mathcal{S}_*^\lambda$ is free-finite. A proof of this property in the spirit of the Morse theory will be given later (Corollary 2.3 of Chapter 13, page 391). This proof uses the results of Chapter 9 but not the results of the present chapter. For the reader who does not want to use the techniques from the theory of cellular gradients, here is another proof, which is based on the triangulation theory of smooth manifolds.†

Lemma 5.2. *There is a C^1 triangulation of M such that V is a simplicial subcomplex.*

Proof. Choose a tubular neighbourhood $N(V)$ of V diffeomorphic to $V \times [-1, 1]$. Pick a diffeomorphism

$$\Phi : N(V) \xrightarrow{\approx} V \times [-1, 1]$$
$$\text{with} \quad \Phi(V) = V \times 0, \quad \Phi(\partial N(V)) = V \times \{-1, 1\}.$$

Pick any C^1 triangulation of V. The submanifold $V \cup \partial N(V)$ is diffeomorphic to the disjoint union of three copies of V, triangulate it using this diffeomorphism. The resulting triangulation can be extended to a C^1 triangulation of $\Phi^{-1}(V \times [0, 1])$ by [**100**], Theorem 10.6. Similarly we triangulate $\Phi^{-1}(V \times [-1, 0])$ and $M \setminus \operatorname{Int} N(V)$. The union of these triangulations gives the triangulation sought. □

Now let us lift the triangulation constructed in the previous lemma to a C^1 triangulation of $\bar{M}$ which is invariant with respect to the structure group of the covering. The subspace V^- is a simplicial subcomplex. The

† A detailed exposition of the triangulation theory for C^∞ manifolds can be found in the book [**100**].

simplicial chain complex $\Delta_*(V^-)$ of the space V^- is a free finitely generated chain complex of P-modules, and there is a natural chain equivalence $\Delta_*(V^-) \xrightarrow{\sim} \mathcal{S}_*(V^-)$. This completes the proof. □

Corollary 5.3. *The $\widehat{P}$-complex*

$$\mathcal{S}_*^\lambda \underset{P}{\otimes} \widehat{P}$$

is homotopically free-finite. □

Now let us proceed to the Novikov completion of the singular chain complex. Put

$$\overline{\mathcal{S}}_*(\bar{M}) = \mathcal{S}_*(\bar{M}) \underset{L}{\otimes} \overline{L}.$$

Thus $\overline{\mathcal{S}}_*(\bar{M})$ is a chain complex of free $\overline{L}$-modules. Put

$$\overline{\mathcal{S}}_* = \overline{\mathcal{S}}_*(\bar{M}), \quad \overline{\mathcal{S}}_*^\lambda = \{A \in \overline{\mathcal{S}}_* \mid \operatorname{supp} A \subset V^-\}.$$

It is clear that the inclusion $\mathcal{S}_*^\lambda \hookrightarrow \overline{\mathcal{S}}_*^\lambda$ induces an isomorphism

$$\mathcal{S}_*^\lambda \underset{P}{\otimes} \widehat{P} \approx \overline{\mathcal{S}}_*^\lambda.$$

Proposition 5.4. *The inclusion $\mathcal{S}_*^\lambda \hookrightarrow \overline{\mathcal{S}}_*^\lambda$ induces for every $n \in \mathbf{N}$ the following isomorphisms of chain complexes of P-modules:*

$$\mathcal{S}_*^\lambda / t^n \mathcal{S}_*^\lambda \ \approx \ \overline{\mathcal{S}}_*^\lambda / t^n \overline{\mathcal{S}}_*^\lambda \ \approx \ \mathcal{S}_*(V^-, t^n V^-).$$

Proof. Obvious. □

6. The canonical chain equivalence

Now we are ready to state the theorem about the canonical chain equivalence. In this section $f : M \to S^1$ is a Morse map, v is an oriented almost transverse f-gradient. We have a commutative diagram

$$\begin{array}{ccc} \bar{M} & \xrightarrow{F} & \mathbf{R} \\ {\scriptstyle\pi}\downarrow & & \downarrow{\scriptstyle \mathrm{Exp}} \\ M & \xrightarrow{f} & S^1 \end{array}$$

where $\pi : \bar{M} \to M$ is the infinite cyclic covering induced by f and $F : \bar{M} \to \mathbf{R}$ is a lift of f to $\bar{M}$. As in the previous sections we shall use the following abbreviations:

$$\mathcal{N}_* = \mathcal{N}_*(f, v), \quad \mathcal{S}_* = \mathcal{S}_*(\bar{M}), \quad \overline{\mathcal{S}}_* = \mathcal{S}_*(\bar{M}) \underset{L}{\otimes} \overline{L}.$$

Let λ be a regular value of F. The submodule of all series in $\mathcal{N}_*(f,v)$ with support in

$$V^- = F^{-1}(]-\infty,\lambda])$$

is denoted by $\mathcal{N}_*^\lambda$; similarly $\mathcal{S}_*(V^-)$ is denoted by $\mathcal{S}_*^\lambda$.

Definition 6.1. A chain map $\Phi : \mathcal{N}_* \to \overline{\mathcal{S}}_*$ is called *compatible with the equivariant Morse equivalences* if there exists a regular value λ of F, such that

$$\Phi(\mathcal{N}_*^\lambda) \subset \overline{\mathcal{S}}_*^\lambda$$

and for every positive integer $n \in \mathbf{N}$ the quotient map of P_n-complexes

$$\mathcal{N}_*^\lambda / t^n \mathcal{N}_*^\lambda \longrightarrow \overline{\mathcal{S}}_*^\lambda / t^n \overline{\mathcal{S}}_*^\lambda \approx \mathcal{S}_*(V^-, t^n V^-)$$

is homotopic to the equivariant Morse equivalence $\mathcal{E}(\lambda, n)$.

The main aim of this chapter is the following theorem.

Theorem 6.2. *There is a homotopy unique chain equivalence*

$$\Phi : \mathcal{N}_*(f,v) \to \overline{\mathcal{S}}_*(\bar{M})$$

compatible with the equivariant Morse equivalences.

The proof will occupy the rest of this section.

6.1. Existence. Let λ be any regular value of F. Consider the following infinite diagram:

$$\begin{array}{ccccccccc}
\mathcal{N}_*^\lambda/t\mathcal{N}_*^\lambda & \leftarrow & \mathcal{N}_*^\lambda/t^2\mathcal{N}_*^\lambda & \leftarrow & \cdots & \leftarrow & \mathcal{N}_*^\lambda/t^k\mathcal{N}_*^\lambda & \leftarrow \mathcal{N}_*^\lambda/t^{k+1}\mathcal{N}_*^\lambda & \leftarrow \cdots \\
\downarrow \mathcal{E}_1 & & \downarrow \mathcal{E}_2 & & \cdots & & \downarrow \mathcal{E}_k & \downarrow \mathcal{E}_{k+1} & \\
\mathcal{S}_*^\lambda/t\mathcal{S}_*^\lambda & \leftarrow & \mathcal{S}_*^\lambda/t^2\mathcal{S}_*^\lambda & \leftarrow & \cdots & \leftarrow & \mathcal{S}_*^\lambda/t^k\mathcal{S}_*^\lambda & \leftarrow \mathcal{S}_*^\lambda/t^{k+1}\mathcal{S}_*^\lambda & \leftarrow \cdots
\end{array}$$

where $\mathcal{E}_k$ stands for the equivariant Morse equivalence $\mathcal{E}(\lambda, k)$. Each square is homotopy commutative (by Proposition 4.6, page 353). Therefore we have here a virtual homology equivalence

$$\mathcal{N}_*^\lambda \overset{\mathcal{E}}{\dashrightarrow} \mathcal{S}_*^\lambda \underset{P}{\otimes} \widehat{P}.$$

Both $\mathcal{N}_*^\lambda$ and $\mathcal{S}_*^\lambda \underset{P}{\otimes} \widehat{P}$ are homotopically free-finite $\widehat{P}$-complexes, therefore this virtual map has a homotopy unique realization

$$|\mathcal{E}| : \mathcal{N}_*^\lambda \xrightarrow{\sim} \mathcal{S}_*^\lambda \underset{P}{\otimes} \widehat{P}$$

(by Theorem 1.12 of Chapter 10, page 332). Let us state the obtained result for further use.

Proposition 6.3. *There exists a homotopy unique chain equivalence*

$$\mathcal{N}_*^\lambda \xrightarrow{\Psi} \mathcal{S}_*^\lambda \underset{P}{\otimes} \widehat{P} \tag{6}$$

such that for every $n \geqslant 0$ the following square is homotopy commutative:

$$\begin{array}{ccc} \mathcal{N}_*^\lambda & \xrightarrow{\Psi} & \mathcal{S}_*^\lambda \underset{P}{\otimes} \widehat{P} \\ \downarrow & & \downarrow \\ \mathcal{N}_*^\lambda / t^n \mathcal{N}_*^\lambda & \xrightarrow{\mathcal{E}_n} & \mathcal{S}_*^\lambda / t^n \mathcal{S}_*^\lambda \end{array}$$

where the vertical arrows are natural projections and $\mathcal{E}_n$ are the equivariant Morse equivalences. □

Since

$$\mathcal{N}_* = \mathcal{N}_*^\lambda \underset{\widehat{P}}{\otimes} \overline{L}, \qquad \text{and} \qquad \overline{\mathcal{S}}_* = (\mathcal{S}_*^\lambda \underset{P}{\otimes} \widehat{P}) \underset{L}{\otimes} \overline{L}$$

we obtain a chain equivalence

$$\Phi = \Psi \otimes \mathrm{id} : \mathcal{N}_* \to \overline{\mathcal{S}}_*$$

which satisfies the requirements of Definition 6.1.

6.2. Uniqueness. Let $\Phi' : \mathcal{N}_* \to \overline{\mathcal{S}}_*$ be another chain map compatible with Morse equivalences, and let μ be the corresponding regular value of F. Let

$$\Psi = \Phi \mid \mathcal{N}_*^\lambda, \quad \Psi' = \Phi' \mid \mathcal{N}_*^\mu.$$

Assuming that $\lambda \leqslant \mu$, consider the following diagram:

$$\begin{array}{ccc} \mathcal{N}_*^\lambda / t^n \mathcal{N}_*^\lambda & \xrightarrow{\Psi/t^n} & \mathcal{S}_*^\lambda / t^n \mathcal{S}_*^\lambda \\ \downarrow & & \downarrow \\ \mathcal{N}_*^\mu / t^n \mathcal{N}_*^\mu & \xrightarrow{\Psi'/t^n} & \mathcal{S}_*^\mu / t^n \mathcal{S}_*^\mu \end{array} \tag{7}$$

where both vertical arrows are induced by the inclusion

$$F^{-1}(]-\infty, \lambda]) \hookrightarrow F^{-1}(]-\infty, \mu]).$$

By our assumption the horizontal arrows are homotopic to the equivariant Morse equivalences, the whole diagram is identical with the diagram (4), page 353 (with $n = k$) and therefore homotopy commutative. Thus the compositions

$$\mathcal{N}_*^\lambda \xrightarrow{\Psi} \overline{\mathcal{S}}_*^\lambda \hookrightarrow \overline{\mathcal{S}}_*^\mu$$

and

$$\mathcal{N}_*^\lambda \hookrightarrow \mathcal{N}_*^\mu \xrightarrow{\Psi'} \overline{\mathcal{S}}_*^\mu$$

have the property that their t^n-quotients are homotopic for every n. Both $\mathcal{N}_*^\lambda$ and $\overline{\mathcal{S}}_*^\mu$ are homotopically free-finite over $\widehat{P}_*$, and Proposition 1.9 of Chapter 10 (page 329) implies that these compositions are homotopic. Tensoring by $\overline{L}$ we obtain $\Phi \sim \Phi'$ and Theorem 6.2 is proved. □

Definition 6.4. The chain equivalence

$$\Phi : \mathcal{N}_*(f, v) \to \overline{\mathcal{S}}_*(\bar{M})$$

compatible with the equivariant Morse equivalences will be called *the canonical chain equivalence.*

7. More about the Novikov complex

The first subsection is about the functoriality properties of the Novikov complex and the canonical chain equivalence. In the second subsection we outline the Hutchings-Lee formula for the inverse of the canonical chain equivalence.

7.1. Functoriality properties. Let

$$f_1 : M_1 \to S^1, \ f_2 : M_2 \to S^1$$

be Morse maps (where M_i are closed manifolds). Let v_i be oriented almost transverse f_i-gradients ($i = 1, 2$). Let $C_0^\infty(M_1, M_2)$ denote the set of all C^∞ maps $A : M_1 \to M_2$ such that

$$f_2 \circ A \quad \text{is homotopic to} \quad f_1.$$

This is an open subset of $C^\infty(M_1, M_2)$. Let $A : M_1 \to M_2$ be a C^∞ map in $C_0^\infty(M_1, M_2)$. Our aim in this subsection is to define a chain map in the Novikov complexes, induced by A. The construction is similar to the case of real-valued Morse functions, see Subsection 4.1 of Chapter 6 (page 218). However here we shall limit ourselves to smooth maps, leaving aside the more general framework of continuous maps. We impose the following restriction on A:

(8) $A|D(p, v_1) : D(p, v_1) \to M_2$ is transverse to $D(q, -v_2)$
for every $p \in S(f_1), q \in S(f_2)$.

Proposition 7.1. *The subset of all maps $A \in C_0^\infty(M_1, M_2)$ satisfying (8) is residual in $C_0^\infty(M_1, M_2)$.*

Proof. Similarly to the proof of Proposition 4.7 of Chapter 6 (page 222), for each $p \in S(f_1)$ cover the immersed manifold $D(p, v_1)$ by a countable family of embedded closed discs D_i of dimension $\operatorname{ind} p$. Cover each $D(q, -v_2)$ by a countable family of embedded closed discs Δ_i of dimension $\dim M_2 - \operatorname{ind} q$. By Lemma 4.9 of Chapter 6 (page 222) the set of C^∞

maps $g : M_1 \to M_2$ such that the map $g|D_i : D_i \to M_2$ is transverse to Δ_j is open and dense in $C^\infty(M_1, M_2)$, and therefore also open and dense in $C_0^\infty(M_1, M_2)$. □

Choose a lift $\bar{A} : \bar{M}_1 \to \bar{M}_2$ of the map A. Lift the maps $f_i : M_i \to S^1$ to maps $F_i : \bar{M}_i \to \mathbf{R}$ (where $i = 1, 2$). The next diagrams are commutative.

$$\begin{array}{ccccccc}
\mathbf{R} & \xleftarrow{F_1} & \bar{M}_1 & \xrightarrow{\bar{A}} & \bar{M}_2 & \xrightarrow{F_2} & \mathbf{R} \\
\downarrow & & \downarrow & & \downarrow & & \downarrow \\
S^1 & \xleftarrow{f_1} & M_1 & \xrightarrow{A} & M_2 & \xrightarrow{f_2} & S^1
\end{array}$$

Let $\dim M_1 = m_1$, $\dim M_2 = m_2$, denote by $\bar{v}_1, \bar{v}_2$ the lifts of v_1 and v_2 to $\bar{M}_1$, respectively to $\bar{M}_2$.

Proposition 7.2. *Let $p \in S_k(F_1), q \in S_{m_2-k}(F_2)$. If the map A satisfies (8), then the set*

$$T(p,q) = D(p, \bar{v}_1) \cap \bar{A}^{-1}\big(D(q, -\bar{v}_2)\big)$$

is finite.

Proof. Choose regular values α, β of F_1 and regular values μ, ν of F_2 such that

$$p \in W = F_1^{-1}([\alpha, \beta]), \quad \text{and} \quad q \in W' = F_2^{-1}([\mu, \nu]).$$

Since the function

$$F_2 \circ \bar{A} - F_1$$

is bounded, we have

$$\bar{A}\Big(F_1^{-1}(]-\infty, \beta])\Big) \subset F_2^{-1}(]-\infty, \nu]),$$

if only we choose ν sufficiently large. Furthermore, diminishing α if necessary, we can assume that

$$\bar{A}\Big(F_1^{-1}(]-\infty, \alpha])\Big) \subset F_2^{-1}(]-\infty, \mu]).$$

With this choice of α, β, μ, ν we have

$$T(p,q) = \Big(D(p, \bar{v}_1) \cap W\Big) \cap \bar{A}^{-1}\Big(D(q, -\bar{v}_2) \cap W'\Big).$$

Now to prove that the set $T(p,q)$ is finite we can apply the same argument as in the proof of Lemma 4.5 of Chapter 6 (page 221). Set

$$R = D(p, \bar{v}_1|W) \cup D(\mathrm{ind}\leqslant k-1\ ; \bar{v}_1|W),$$
$$Q = D(q, -\bar{v}_2|W') \cup D(\mathrm{ind}\leqslant m_2-k-1\ ; -\bar{v}_2|W').$$

Since v_1, v_2 are almost transverse, both R and Q are compact. The condition (8) implies that

$$T(p,q) = R \cap \bar{A}^{-1}(Q),$$

therefore $T(p,q)$ is compact. Apply once more the condition (8) to deduce that $T(p,q)$ contains only isolated points, and conclude that it is a finite set. □

For $p \in S_k(F_1)$ and $q \in S_k(F_2)$ set

$$N(p,q;\bar{A}) = \bar{A}\big(D(p,\bar{v}_1)\big) \mathrel{\#\!\!\#} D(q,-\bar{v}_2) \in \mathbf{Z}.$$

It is easy to see that $N(t^n p, t^n q; \bar{A}) = N(p,q;A)$ for every n. For $p \in S_k(F_1)$ put

$$\bar{A}_\sharp(p) = \sum_q N(p,q;\bar{A})q$$

(where the summation is over the set of all critical points of F_2 of index k). It is clear that the expression in the right-hand side of this formula is a downward series, therefore an element of $\mathcal{N}_k(f_2, v_2)$. Thus we have defined a homomorphism of graded $\overline{L}$-modules

$$\bar{A}_\sharp : \mathcal{N}_*(f_1, v_1) \to \mathcal{N}_*(f_2, v_2).$$

Theorem 7.3. *The graded homomorphism $\bar{A}_\sharp$ is a chain map, and the following diagram is chain homotopy commutative.*

$$\begin{array}{ccc} \mathcal{N}_*(f_1,v_1) & \xrightarrow{\bar{A}_\sharp} & \mathcal{N}_*(f_2,v_2) \\ \Big\downarrow{\scriptstyle \Phi_1} & & \Big\downarrow{\scriptstyle \Phi_2} \\ \overline{\mathcal{S}}_*(\bar{M}_1) & \xrightarrow{\bar{A}_*} & \overline{\mathcal{S}}_*(\bar{M}_2) \end{array}$$

(where Φ_1 and Φ_2 are the canonical chain equivalences).

Proof. Let λ be a regular value of F_1, and μ be a regular value of F_2. We have the Novikov complexes

$$\mathcal{N}_*^{(1)} = \mathcal{N}_*^{\lambda}(f_1, v_1), \quad \mathcal{N}_*^{(2)} = \mathcal{N}_*^{\mu}(f_2, v_2).$$

Put

$$V_1 = F^{-1}(\lambda), \quad V_1^- = F_1^{-1}(]-\infty, \lambda]),$$
$$V_2 = F^{-1}(\mu), \quad V_2^- = F_1^{-1}(]-\infty, \mu]).$$

Choose λ and μ in such a way, that

$$\bar{A}(V_1^-) \subset V_2^-.$$

Then

$$\bar{A}_\sharp(\mathcal{N}_*^{(1)}) \subset \mathcal{N}_*^{(2)}.$$

Consider the following diagram (where n is any positive integer):

$$(9) \qquad \begin{array}{ccc} \mathcal{N}_*^{(1)}/t^n\mathcal{N}_*^{(1)} & \xrightarrow{\bar{A}_\sharp/t^n} & \mathcal{N}_*^{(2)}/t^n\mathcal{N}_*^{(2)} \\ \downarrow \mathcal{E}_n^{(1)} & & \downarrow \mathcal{E}_n^{(2)} \\ \mathcal{S}_*(V_1^-, t^nV_1^-) & \xrightarrow{\bar{A}_*} & \mathcal{S}_*(V_2^-, t^nV_2^-) \end{array}$$

(where $\mathcal{E}_n^{(1)}, \mathcal{E}_n^{(2)}$ are the equivariant Morse equivalences). To prove our theorem it suffices to establish that the upper horizontal arrow in this diagram is a chain homomorphism, and that the diagram is chain homotopy commutative. Pick any t-ordered Morse functions on the cobordisms

$$W_{(1,n)} = F_1^{-1}([\lambda - n, \lambda]), \quad W_{(2,n)} = F_2^{-1}([\mu - n, \mu]).$$

Let

$$\{(X_k^{(1)}, t^nV_1^-)\}_{0\leqslant k\leqslant m} \quad \text{and} \quad \{(X_k^{(2)}, t^nV_2^-)\}_{0\leqslant k\leqslant m}$$

be the corresponding filtrations of the pairs $(V_1^-, t^nV_1^-)$, respectively $(V_2^-, t^nV_2^-)$. Let us first consider a particular case when the map $\bar{A}$ preserves these filtrations, that is

$$(10) \qquad \bar{A}(X_k^{(1)}) \subset (X_k^{(2)}) \quad \text{for every} \quad k.$$

The adjoint complexes of these filtrations are basis-preserving isomorphic to the chain complexes $\mathcal{N}_*^{(1)}/t^n\mathcal{N}_*^{(1)}$ and $\mathcal{N}_*^{(2)}/t^n\mathcal{N}_*^{(2)}$. The map $\bar{A}$ induces homomorphisms

$$A_\bigstar : H_k\left(X_k^{(1)}, X_{k-1}^{(1)}\right) \to H_k\left(X_k^{(2)}, X_{k-1}^{(2)}\right),$$

and the corresponding graded homomorphism

$$A_\bigstar : \mathcal{N}_*^{(1)}/t^n\mathcal{N}_*^{(1)} \to \mathcal{N}_*^{(2)}/t^n\mathcal{N}_*^{(2)}$$

is a chain map. If we replace the map $\bar{A}_\sharp/t^n$ by $A_\bigstar$ in the diagram (9) the resulting diagram will be homotopy commutative, as it follows from the functoriality of canonical chain equivalences for chain complexes, endowed with cell-like filtrations (Corollary 3.8 of Chapter 6, page 215).

Lemma 7.4. $\bar{A}_\sharp/t^n = A_\bigstar$.

Proof. For a critical point $p \in S_k(F_1)$ let d_p denote the fundamental class of $D(p, v)$ in $(X_k^{(1)}, X_{k-1}^{(1)})$. It suffices to check that

$$(\bar{A}_\sharp/t^n)(d_p) = A_\bigstar(d_p).$$

The map $A_\bigstar$ sends the class d_p to the fundamental class of the singular manifold $\bar{A}(D(p,\bar{v}_1))$ in the pair $(X_k^{(2)}, X_{k-1}^{(2)})$. The cobordism $X^{(2)} \setminus \operatorname{Int} X_{k-1}^{(2)}$ is elementary, and the assertion of the lemma follows easily from Proposition 3.13 of Chapter 5 (page 182). □

Thus we have proved the homotopy commutativity of the diagram (9) in the case when (10) holds. Let us now proceed to the general case; we are going to reduce it to the particular case considered above. Let $B : M_1 \to M_2$ be the map defined by

$$B = \Phi(T, -v_2) \circ A$$

(where $T > 0$). Lift it to the map of coverings as follows:

$$\bar{B} = \Phi(T, -\bar{v}_2) \circ \bar{A}.$$

Observe that the map B satisfies the condition (8) and that

$$\bar{A}\big(D(p,-v_1)\big) \pitchfork D(q,-v_2) = \bar{B}\big(D(p,-v_1)\big) \pitchfork D(q,-v_2).$$

Thus

$$\bar{A}_\sharp/t^n = \bar{B}_\sharp/t^n.$$

Pick any t-ordered Morse function ϕ_2 on $W_{(2,n)}$ and let

$$\{(X_k^{(2)}, t^n V_2^-)\}_{0 \leqslant k \leqslant m}$$

be the corresponding filtration of $(V_2^-, t^n V_2^-)$. The condition (8) implies that if $T > 0$ is sufficiently large, then

$$\bar{B}\big(D(p,v)\big) \subset X_k^{(2)} \cup t^n V_2^- \quad \text{if} \quad \operatorname{ind} p \leqslant k \quad \text{and} \quad p \in V_1^-.$$

Using Proposition 4.2 (page 349) we can choose a t-ordered Morse function ϕ_1 on $W_{(1,n)}$ in a such a way that

$$\bar{B}\Big(X_k^{(1)}\Big) \subset X_k^{(2)} \cup t^n V_2^-.$$

By the argument above we have

$$\bar{B}_\sharp/t^n = B_\star$$

and the proof of our theorem is now complete. □

7.2. The Hutchings-Lee formula for the inverse of the canonical chain equivalence. Let $f : M \to S^1$ be a Morse map, v a transverse f-gradient, $\bar{M} \to M$ the infinite cyclic covering induced by f from the universal covering $\mathrm{Exp} : \mathbf{R} \to S^1$. Let $F : \bar{M} \to \mathbf{R}$ be a lift of f, and $\bar{v}$ be the lift of v to $\bar{M}$. Let $\mathcal{S}_k(\bar{M}, \pitchfork v)$ denote the free abelian group generated by singular simplices $\sigma : \Delta^k \to \bar{M}$ such that σ is a C^∞ map, and the restriction of σ to every subsimplex of Δ^k is transverse to the ascending disc of every critical point p of F. The graded group $\mathcal{S}_*(\bar{M}, \pitchfork v)$ is a chain subcomplex of $\mathcal{S}_*(\bar{M})$. It is clear that for every $\sigma \in \mathcal{S}_k(\bar{M}, \pitchfork v)$ and every critical point $p \in S_k(F)$ the intersection $\sigma(\Delta^k) \cap D(p, -\bar{v})$ is a finite set. Put

$$[\sigma : p] = \sigma(\Delta^k) \,\#\, D(p, -\bar{v}).$$

Define a homomorphism

$$\mathcal{S}_k(\bar{M}, \pitchfork v) \xrightarrow{\ \eta\ } \mathcal{N}_k(f, v)$$

by the following formula

$$\eta(\sigma) = \sum_{p \in S_k(f)} [\sigma : p] p.$$

It is not difficult to prove that the expression in the right-hand side of the formula is indeed an element of $\mathcal{N}_k(f, v)$ and the homomorphism η can be extended to a homomorphism of $\overline{L}$-modules

$$\mathcal{S}_k(\bar{M}, \pitchfork v) \underset{L}{\otimes} \overline{L} \xrightarrow{\ \eta\ } \mathcal{N}_*(f, v).$$

The next theorem is a reformulation of the results implicit in the work [**64**].

Theorem 7.5 (M. Hutchings, Y. J. Lee). *The map η is a chain map and the following diagram is homotopy commutative:*

$$\begin{array}{ccc}
\mathcal{S}_*(\bar{M}, \pitchfork v) \underset{L}{\otimes} \overline{L} & \xrightarrow{\qquad \eta \qquad} & \mathcal{N}_*(f, v) \\
& \searrow \qquad \swarrow {\scriptstyle \Phi} & \\
& \mathcal{S}_*(\bar{M}) \underset{L}{\otimes} \overline{L} &
\end{array}$$

where the left diagonal arrow is the inclusion, and the right diagonal arrow is the canonical chain equivalence.

Outline of the proof. By definition of the canonical chain equivalence there is a regular value λ of F such that the map

$$\Phi | \mathcal{N}_*^\lambda : \mathcal{N}_*^\lambda \to \mathcal{S}_*^\lambda(\bar{M}) \underset{P}{\otimes} \widehat{P}$$

is a chain equivalence of $\widehat{P}$-modules. Let $\mathcal{S}_*^\lambda(\bar{M}, \pitchfork v)$ be the submodule generated by the singular simplices of $\bar{M}$ with image in $F^{-1}(]-\infty, \lambda])$. It is clear that η sends $\mathcal{S}_*^\lambda(\bar{M}, \pitchfork v)$ to $\mathcal{N}_*^\lambda$, so we have the diagram of chain equivalences of $\widehat{P}$-complexes

$$\begin{array}{ccc} \mathcal{S}_*^\lambda(\bar{M}, \pitchfork v) \underset{P}{\otimes} \widehat{P} & \xrightarrow{\quad \eta \quad} & \mathcal{N}_*^\lambda(f, v) \\ & \searrow \qquad \swarrow{\scriptstyle \Phi} & \\ & \mathcal{S}_*^\lambda(\bar{M}) \underset{P}{\otimes} \widehat{P} & \end{array}$$

and it suffices to prove that it is homotopy commutative. All the chain complexes in this diagram are homotopically free-finite, so in order to check the homotopy commutativity of the diagram above it suffices to show that for every k the truncated diagram

$$\begin{array}{ccc} \mathcal{S}_*^\lambda(\bar{M}, \pitchfork v)/t^k \mathcal{S}_*^\lambda(\bar{M}, \pitchfork v) & \xrightarrow{\quad \eta/t^k \quad} & \mathcal{N}_*^\lambda(f, v)/t^k \mathcal{N}_*^\lambda(f, v) \\ & \searrow \qquad \swarrow{\scriptstyle \Phi/t^k} & \\ & \mathcal{S}_*^\lambda(\bar{M})/t^k \mathcal{S}_*^\lambda(\bar{M}) & \end{array}$$

is homotopy commutative. This is proved similarly to Theorem 4.12 of Chapter 6 (page 224). □

Similarly to the case of real-valued Morse functions there is an analog of this formula for the simplicial framework. Choose a C^∞ triangulation of M, and assume that each simplex of this triangulation is transverse to the unstable manifold of each critical point of f. Lift the triangulation to $\bar{M}$ and obtain a $\mathbf{Z}$-invariant triangulation of $\bar{M}$. For every k-dimensional simplex σ of this triangulation and every $p \in S_k(F)$ the set $D(p, -\bar{v}) \cap \sigma$ is finite. Therefore for every oriented simplex of the triangulation the algebraic intersection number

$$[\sigma : p] = \sigma \# D(p, -\bar{v})$$

is defined. For a k-dimensional simplex σ of $\bar{M}$ set

$$\mathcal{A}_k(\sigma) = \sum_{p \in S_k(F)} [\sigma : p] p.$$

It is clear that the expression in the right hand side of the formula is a downward series, that is, an element of $\mathcal{N}_k(f, v)$, and the map $\sigma \mapsto \mathcal{A}_k(\sigma)$ extends to a graded $\overline{L}$-homomorphism

$$\mathcal{A}_* : \overline{\Delta}_*(\bar{M}) \to \mathcal{N}_*(f, v)$$

where $\overline{\Delta}_*(\bar{M})$ stands for the completion $\Delta_*(\bar{M}) \underset{L}{\otimes} \overline{L}$ of the simplicial chain complex of $\bar{M}$. The next theorem is due to D. Schütz, see [**141**].

Theorem 7.6. *The graded homomorphism $\mathcal{A}_*$ is a chain equivalence, and the following diagram is homotopy commutative.*

$$\begin{array}{ccc} \overline{\Delta}_*(\bar{M}) & \xrightarrow{\mathcal{A}_*} & \mathcal{N}_*(f, v) \\ & \searrow \quad \swarrow {\scriptstyle \Phi} & \\ & \overline{\mathcal{S}}_*(\bar{M}) & \end{array}$$

where the left diagonal arrow is the natural chain equivalence between the simplicial and singular chain complexes.

CHAPTER 12

Cellular gradients of circle-valued Morse functions and the Rationality Theorem

Let $f : M \to S^1$ be a Morse function on a closed manifold and v an f-gradient. A natural way to understand the dynamical properties of the gradient flow is to consider the *first return map* of a regular level surface $V = f^{-1}(a)$ to itself induced by the flow. To define this map, let $x \in V$; if the $(-v)$-trajectory starting at x does not converge to a critical point of f, it will reach again the manifold V and intersect it at some point, which will be denoted $\phi(x)$. The map $x \mapsto \phi(x)$ is a diffeomorphism of an open subset of V onto another open subset of V. The closed orbits of v correspond to periodic points of ϕ, and we will see shortly that the Novikov incidence coefficients can also be interpreted in terms of iterations of this map.

The map ϕ is defined not everywhere in V, so the classical theory of dynamical systems on closed manifolds can not be applied immediately. However for C^0-generic gradients this map can be in a sense regularized, and its dynamical properties are similar to those of a continuous map.

To give the definition of this class of gradients (called *cellular gradients*) we shall need another interpretation of the first return map. Consider the commutative diagram

$$\begin{array}{ccc} \bar{M} & \xrightarrow{F} & \mathbf{R} \\ {\scriptstyle\pi}\downarrow & & \downarrow{\scriptstyle\mathrm{Exp}} \\ M & \xrightarrow{f} & S^1 \end{array} \tag{1}$$

where $\pi : \bar{M} \to M$ is the infinite cyclic covering induced by f from Exp. Let t denote the downward generator of the deck transformation group of π. Lift the vector field v to a t-invariant vector field $\bar{v}$. Pick any regular value $\lambda \in \mathbf{R}$ of F. The set

$$W_\lambda = F^{-1}([\lambda - 1, \lambda])$$

is a cobordism with upper and lower boundaries

$$\partial_1 W_\lambda = F^{-1}(\lambda), \quad \partial_0 W_\lambda = F^{-1}(\lambda - 1).$$

We have a Morse function

$$F|W_\lambda : W_\lambda \to [\lambda - 1, \lambda]$$

and its gradient $\bar{v}|W_\lambda$. The first return map ϕ discussed above is now identified with the transport map

$$(-\bar{v}|W_\lambda)^{\leadsto} : U_1 \to U_0 \quad \text{with} \quad U_1 \subset \partial_1 W_\lambda, \quad U_0 \subset \partial_0 W_\lambda.$$

Endow M with a Riemannian metric, and lift it to a t-invariant Riemannian metric of $\bar{M}$. The generator t of the deck transformation group yields an isometry of $V_\lambda = F^{-1}(\lambda)$ onto $tV_\lambda = F^{-1}(\lambda - 1)$, and W_λ acquires the structure of a cyclic cobordism.

Definition 0.1. An f-gradient v is called *cellular* if there is a regular value λ of F such that the $F|W_\lambda$-gradient $\bar{v}|W_\lambda$ is a cyclic cellular gradient.[†] The set of all cellular gradients of f will be denoted by $G_C(f)$.

For a cellular f-gradient v we have a Morse-Smale filtration

$$\mathcal{F}_0 \subset \cdots \subset \mathcal{F}_k \subset \cdots$$

of V_λ such that the gradient descent from V_λ to tV_λ yields a sequence of continuous self-maps of compact spaces

$$\mathcal{F}_k/\mathcal{F}_{k-1} \to \mathcal{F}_k/\mathcal{F}_{k-1}$$

which can be iterated and form a family of dynamical systems. This can be considered as a regularization of the transport map which is only partially defined. It follows from the results of Part 3 that the subset of cellular gradients is C^0-generic in the set of all gradients (see Theorem 3.1, page 375).

The Novikov complex of a cellular gradient has remarkable properties. In general the matrix entries of the boundary operators in the Novikov complex are Laurent series in one variable. We shall see that for a cellular gradient these series are rational functions. Thus C^0-generically the Novikov complex is defined over the ring of rational functions. This confirms the so-called *Novikov exponential growth conjecture* for C^0-generic gradients. We begin with a brief introduction to this conjecture, and then proceed to the study of cellular gradients.

[†] Note that we used the same term "cellular gradient" in the context of real-valued Morse functions on cobordisms (Part 3); this can not lead to confusion.

1. Novikov's exponential growth conjecture

Let $f : M \to S^1$ be a Morse map, and v a transverse f-gradient. Recall the commutative diagram (1) on page 367. Lift the vector field v to a $\mathbf{Z}$-invariant F-gradient $\bar{v}$ on $\bar{M}$. The Novikov incidence coefficient corresponding to critical points $p, q \in S(F)$ is defined as follows:

$$N(p, q; v) = \sum_{k \in \mathbf{Z}} n(p, t^k q; \bar{v}) t^k$$

where $n(p, t^k q; \bar{v})$ is the algebraic number of flow lines of $\bar{v}$ joining $t^k q$ with p. These coefficients are the matrix entries of the boundary operators in the Novikov complex. The asymptotics of the coefficients

$$n_k(p, q; \bar{v}) = n(p, t^k q; \bar{v}) \quad \text{as} \quad k \to \infty$$

are an important invariant of the gradient flow.

Definition 1.1. Let $\{a_k\}_{k \in \mathbf{Z}}$ be a sequence of real numbers with a finite negative part (that is, for some integer N we have $a_k = 0$ if $k \leqslant N$). We say that $\{a_k\}$ has *at most exponential growth* if there are $A, B > 0$ such that

$$|a_k| < Ae^{Bk} \quad \text{for every} \quad k \in \mathbf{Z}.$$

Lemma 1.2. *A sequence $\{a_k\}_{k \in \mathbf{Z}}$ with a finite negative part has at most exponential growth if and only if for some integer $l > 0$ the series*

$$\sum_{k \in \mathbf{Z}} a_k t^{l+k} = t^l \sum_{k \in \mathbf{Z}} a_k t^k$$

has only positive powers and a strictly positive radius of convergence.

Proof. We can assume that $a_k = 0$ for $k < 0$, and then our assertion follows immediately from the standard formula for the radius of convergence of the power series $\sum a_k t^k$:

$$\rho = \frac{1}{\overline{\lim}\, |a_k|^{\frac{1}{k}}}.$$

□

Definition 1.3. We say that the *exponential growth property* holds for the pair (f, v) if for every $p, q \in S(F)$ with $\operatorname{ind} p = \operatorname{ind} q + 1$ the sequence $\{n_k(p, q; \bar{v})\}_{k \in \mathbf{Z}}$ has at most exponential growth.

Here is the initial version of the Novikov exponential growth conjecture:

> *The exponential growth property holds for every pair (f, v) where $f : M \to S^1$ is a Morse function and v is a transverse f-gradient.*

It dates back to the early 1980s. In subsequent years the conjecture was discussed and modified. At present there are several versions of the conjecture. As far as I know, the first published version appeared in the paper of V. I. Arnold "Dynamics of Intersections", 1989. This version is stated in a more general setting than the initial one: it deals with Morse 1-forms, and not only with Morse maps to the circle. Arnold writes ([**3**], p.83):

> The author is indebted to S. P. Novikov who has communicated the following conjecture, which was the starting point of the present paper. Let $p : \widetilde{M} \to M$ be a covering of a compact manifold M with fiber $\mathbf{Z}^n$, and let α be a closed 1-form on M such that $p^*\alpha = df$, where $f : \widetilde{M} \to \mathbf{R}$ is a Morse function. The Novikov conjecture states that, "generically" the number of the trajectories of the vector field $-\text{grad} f$ on $\widetilde{M}$ starting at a critical point x of the function f of index k and connecting it with the critical points y having index $k-1$ and satisfying $f(y) \geqslant f(x) - n$, grows in n more slowly than some exponential, e^{an}.

In particular this version implies that the exponential growth property holds for every generic pair (f, v) where v is a Riemannian gradient of f. Observe also that V.I. Arnold works with the *absolute number* of flow lines joining critical points rather than the algebraic number.

Here is the version of the conjecture published in 1993 by S. P. Novikov in his paper "Quasiperiodic structures in topology", [**106**], p.229:

> CONJECTURE. For any closed quantized analytic 1-form ω the boundary operator ∂ in the Morse complex has all coefficients $a_{pq}^{(i)} \in K$ with positive part convergent in some region $|t| \leqslant r_1^*(\omega) \leqslant r_1$. (If we replace S by $(-S)$ and t by t^{-1}, we have to replace r_1 by r_2^{-1}.) In particular, $r_1^* \neq 0$ and there will be coefficients $a_{pq}^{(i)}$ with radius of convergence not greater than $r_1(M, [\omega])$ if at least one jumping point really exists.

Thus the version of [**106**] says that the exponential growth property holds for every pair (f, v) where f is analytic and v is a Riemannian gradient of f.

There are also versions of the Novikov exponential growth conjecture which predict for the sequence $n_k(p, q; v)$ a stronger property than at most exponential growth.

Definition 1.4. A Laurent series $\alpha = \sum_{n=-N}^{\infty} a_n t^n \in \mathbf{R}((t))$ is called *rational* if there is a rational function

$$R(t) = \frac{P(t)}{Q(t)} \quad \text{with} \quad P, Q \in \mathbf{R}[t]$$

such that α is the Laurent expansion for $R(t)$ converging in some annulus

$$A = \{z \in \mathbf{C} \mid 0 < |z| < r\} \quad \text{with} \quad r > 0.$$

Lemma 1.5. *If* $\alpha = \sum_{n=-N}^{\infty} a_n t^n$ *is rational, then the sequence* $\{a_n\}_{n\in\mathbf{Z}}$ *has at most exponential growth.*

Proof. Follows immediately from Lemma 1.2. □

Definition 1.6. We say that the *rationality property* holds for the pair (f, v) if for every $p, q \in S(F)$ with $\operatorname{ind} p = \operatorname{ind} q + 1$ the series $N(p, q; v)$ is rational.

M. Farber and P. Vogel conjectured (unpublished) that the rationality property holds for every pair (f, v).

The results presented in this chapter show that for every given $f : M \to S^1$ the rationality property holds for every cellular f-gradient. The subset $G_C(f)$ of cellular gradients is open and dense in $G(f)$ with respect to the C^0 topology, thus the rationality property is C^0-*generic.* This was proved by the author in 1995 (see [**113**], [**112**]).

2. The boundary operators in the Novikov complex

This section contains some preliminaries necessary for the rationality theorem. We will obtain a formula expressing the Novikov boundary operators in terms of gradient descent maps. This formula in itself is only a reformulation of the definition, but has the advantage to make explicit the recursive nature of the Novikov incidence coefficients.

Let $f : M \to S^1$ be a Morse map, v be an f-gradient. Recall the diagram

$$\begin{array}{ccc} \bar{M} & \xrightarrow{F} & \mathbf{R} \\ {\scriptstyle\pi}\downarrow & & \downarrow{\scriptstyle \mathrm{Exp}} \\ M & \xrightarrow{f} & S^1. \end{array}$$

Let t denote the downward generator of the structure group of the covering π, so that $F(tx) = F(x) - 1$. Lift v to a t-invariant F-gradient $\bar{v}$. We assume that v satisfies transversality condition, then $\bar{v}$ also is transverse. Pick a regular value λ of F, and put

$$V = F^{-1}(\lambda), \quad W = F^{-1}([\lambda - 1, \lambda]).$$

Let p, q be critical points of F with $\operatorname{ind} p = \operatorname{ind} q + 1$. Recall the definition of the Novikov incidence coefficient:

$$N(p,q;\bar{v}) = \sum_{k\in\mathbf{Z}} n_k(p,q;\bar{v})t^k \quad \text{with} \quad n_k(p,q;\bar{v}) = n(p,t^kq;\bar{v}) \in \mathbf{Z}.$$

The t-invariance of the vector field $\bar{v}$ implies

$$N(t^lp, t^nq; \bar{v}) = N(p, t^{n-l}q; \bar{v}),$$

and therefore if we compute the Novikov incidence coefficients in the particular case when $p, q \in t^{-1}W$, we will deduce all of them. Let

$$\operatorname{ind} p = r+1, \quad \operatorname{ind} q = r.$$

Let

$$N = D(p,v) \cap V, \quad S = D(tq, -v) \cap V.$$

With this notation we have (for $k \geqslant 1$):

$$n_{k+1}(p,q;\bar{v}) = (-\bar{v})^{\leadsto}_{[\lambda,\lambda-k]}(N) \mathbin{\#} t^kS, \tag{2}$$

where $(-\bar{v})^{\leadsto}_{[\lambda,\lambda-k]}(N)$ and t^kS are both submanifolds of t^kV, the first is oriented, and the second cooriented, so that the intersection index in the formula is well defined. The map $(-\bar{v})^{\leadsto}_{[\lambda,\lambda-k]}$ corresponding to the gradient descent from V to t^kV is the composition of gradient descents from V to tV, then to t^2V, ... etc. up to t^kV. The vector field $\bar{v}$ is t-invariant, therefore the maps in this sequence are diffeomorphic to each other, and we can rewrite the formula (2) in another way, clarifying its recursive nature. Namely, let

$$\zeta = t^{-1} \circ (-\bar{v})^{\leadsto}_{[\lambda,\lambda-1]}$$

so that ζ is a (partially defined) map from V to itself. The gradient descent map from t^kV to $t^{k+1}V$ satisfies

$$(-\bar{v})^{\leadsto}_{[\lambda-k,\ \lambda-k-1]} = t^{k+1} \circ \zeta \circ t^{-k} : t^kV \to t^{k+1}V \quad \text{for every} \quad k$$

and a simple computation using (2) gives

$$n_{k+1}(p,q;\bar{v}) = \zeta^k(N) \mathbin{\#} S \quad \text{for every} \quad k \geqslant 0. \tag{3}$$

These constructions can be visualized in Figure 43 (where we denote $t^{k+1} \circ \zeta \circ t^{-k}$ by ζ_k).

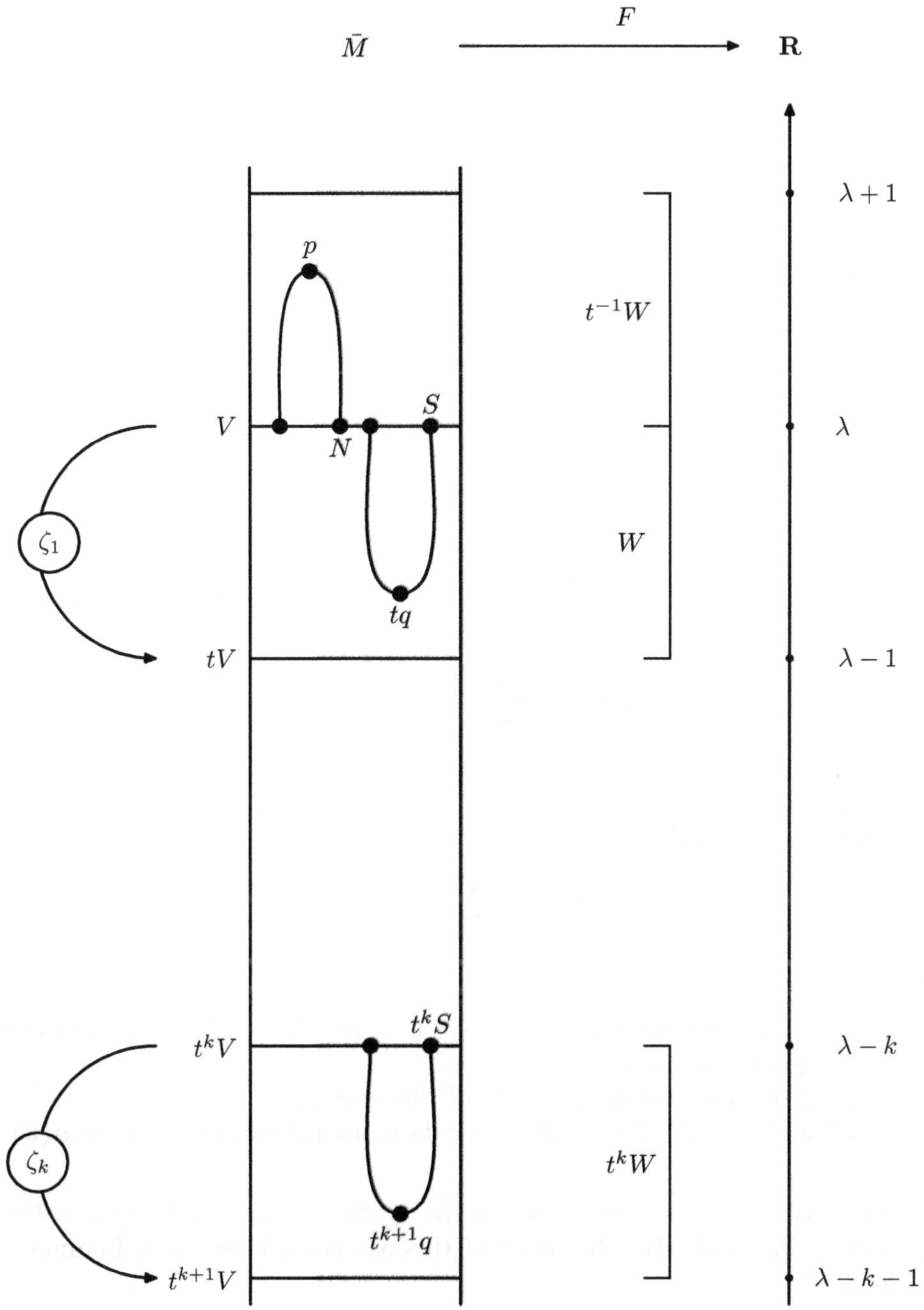
$\bar{M}$
F
$\mathbf{R}$
p
V
N
S
ζ_1
tq
tV
$t^{-1}W$
W
$\lambda+1$
λ
$\lambda-1$
t^kS
t^kV
ζ_k
$t^{k+1}q$
$t^{k+1}V$
t^kW
$\lambda-k$
$\lambda-k-1$

FIGURE 43.

If both N and S were compact, and ζ were an everywhere defined C^∞ map, the intersection indices (3) could be expressed in homological terms, and one could obtain a simple formula for the generating function of the sequence of intersection numbers. This computation, outlined below, is due to V.I.Arnold (unpublished). Let Z be a closed connected manifold, and X, Y closed submanifolds of Z, such that X is oriented, Y is cooriented and $\dim X + \dim Y = \dim Z$. Let $\phi : Z \to Z$ be a C^∞ map, such that $\phi^k(X) \pitchfork Y$ for every $k \geqslant 0$. We have then

$$\phi^k(X) \# Y = \Big\langle \omega_Y, [\phi^k(X)] \Big\rangle = \Big\langle \omega_Y, \phi^k_*([X]) \Big\rangle. \tag{4}$$

In this formula $[X] \in H_l(Z)$ is the fundamental class of X. The brackets $\langle \cdot, \cdot \rangle$ denote the canonical pairing

$$H^l(Z, Z \setminus Y) \otimes H_l(Z) \to \mathbf{Z},$$

and $\omega_Y \in H^l(Z, Z \setminus Y)$ is the Thom class of Y (see Section 4 of Chapter 5).

The power series

$$G(t) = \sum_k \Big(\phi^k(X) \# Y \Big) t^k$$

(which is also called the *generating function* of the sequence $\phi^k(X) \# Y$) can be therefore written as

$$G(t) = \sum_k \alpha(A^k \xi) t^k \tag{5}$$

where:

(1) A denotes the endomorphism $\phi_* : H_l(Z) \to H_l(Z)$ of the finitely generated abelian group $H_l(Z)$,
(2) ξ denotes the element $[X]$ of this group,
(3) and $\alpha : H_l(Z) \to \mathbf{Z}$ is the homomorphism defined by $\alpha(u) = \langle \omega_Y, u \rangle$.

It turns out that the power series (5) is the *Taylor series of a rational function.* We shall give the proof of this assertion later on in Lemma 3.8, page 379.†

† Hint:

$$\sum_{k \geqslant 0} A^k t^k = (1 - At)^{-1}.$$

Now let us return to the Novikov incidence coefficients. We would like to apply the above argument and to deduce that

$$N(p,q;v)=\sum_{k\geqslant 0} n_k(p,q;\bar v)t^k = n_0(p,q;\bar v)+\sum_{k\geqslant 0}\Big(\zeta^k(N)\pitchfork S\Big)t^{k+1}$$

is a rational function. Unfortunately this would not work since the map ζ of V to itself is not everywhere defined. Observe for example, that if $n_1(p,q;v)\neq 0$, then the manifold N is not in the domain of definition of ζ. We shall see in the next section that for cellular f-gradients there is a formula for $n_k(p,q;\bar v)$ similar to (4). For these gradients the series $N(p,q;v)$ can be computed with the help of the algebraic argument above, and this leads to the rationality theorem. The role of the homomorphism ϕ_* from the formula (4) will be played by the homological gradient descent associated with the cellular gradient.

3. Cellular gradients of circle-valued Morse functions

Cellular gradients of circle-valued Morse functions are C^0-generic as the following theorem shows.

Theorem 3.1. *Let M be a closed manifold and $f: M\to S^1$ a Morse function. The subset $G_C(f)$ of cellular f-gradients is open and dense in $G(f)$ with respect to C^0 topology.*

Moreover, let u be any almost transverse f-gradient, $\epsilon>0$, ρ a regular value of f and U a neighbourhood of $V=f^{-1}(\rho)$. Then there is $v\in G_C(f)$, such that

$$||v-u||\leqslant\epsilon \quad \textit{and} \quad \operatorname{supp}(v-u)\subset U\setminus V.$$

(As usual $||\cdot||$ stands for the C^0-norm.)

Proof. The proof is based on the results obtained in Part 3. Let us begin with the density property. Lift the function f to a Morse function $F:\bar M\to\mathbf{R}$, and lift u to an almost transverse $\mathbf{Z}$-invariant F-gradient $\bar u$. Let $\lambda\in\mathbf{R}$ be any regular value of F, with $\rho=\operatorname{Exp}(\lambda)$. Let $W=F^{-1}([\lambda-1,\lambda])$. By a certain abuse of notation the restriction of $\bar u$ to W will be denoted by u.

Theorem 2.23 of Chapter 9 (page 298) implies that for any $\epsilon>0$ and any neighbourhood U_0 of ∂W in W there is a cyclic cellular $F|W$-gradient v such that

$$||v-u||<\epsilon \quad \text{and} \quad \operatorname{supp}(v-u)\subset U_0\setminus\partial W.$$

Since the vector fields v and u are equal nearby ∂W, we can extend the field v to a $\mathbf{Z}$-invariant vector field $\bar v$ on $\bar M$, whose projection to M will be denoted by the same symbol v. Then v is a cellular f-gradient. Moreover, $||v-u||<\epsilon$ and $\operatorname{supp}(v-u)\subset U\setminus V$ if U_0 is chosen sufficiently small.

The proof of density property is completed. The openness property follows immediately from Theorem 2.23 of Chapter 9 (page 298). □

Recall that the set of all transverse f-gradients is denoted by $G_T(f)$, and the set of all almost transverse f-gradients is denoted by $G_A(f)$. It is easy to see that each cellular f-gradient is almost transverse, that is, $G_C(f) \subset G_A(f)$.

Proposition 3.2. (1) *$G_C(f)$ is open and dense in $G_A(f)$ with respect to the C^0 topology.*
(2) *$G_C(f)$ is C^0-generic in $G_T(f)$, that is, the subset $G_C(f) \cap G_T(f)$ is open and dense in $G_T(f)$ with respect to the C^0-topology.*

Proof. The first point follows immediately from Theorem 3.1. The proof of the second point is similar to Corollary 1.3 of Chapter 8 (page 243). □

Let v be a cellular f-gradient. Theorem 3.4 below says that the Novikov complex $\mathcal{N}_*(f, v)$ is defined over the ring of rational functions. Actually we shall prove a stronger statement: the Novikov incidence coefficients belong to a certain subring of the ring of all rational functions, introduced in the next definition.

Definition 3.3. Let

$$\Sigma = \{1 + tP(t) \mid P \in \mathbf{Z}[t]\}.$$

The localization of the ring $L = \mathbf{Z}[t, t^{-1}]$ with respect to the multiplicative subset Σ will be denoted $\widetilde{L}$.

Thus $\widetilde{L}$ is a subring of the ring $\mathbf{Q}(t)$ of rational functions. Since every element of Σ is invertible in $\overline{L} = \mathbf{Z}((t))$ we have a natural injective ring homomorphism

$$\widetilde{L} = \Sigma^{-1} L \xrightarrow{J} \mathbf{Z}((t)).$$

For an element $R = \frac{P}{Q} \in \widetilde{L}$, where $P, Q \in \mathbf{Z}[t]$, the element $J(R)$ can be considered as the Laurent expansion of the rational function R in a neighbourhood of 0. We shall identify each element of $\widetilde{L}$ with its J-image in $\mathbf{Z}((t))$.

Theorem 3.4. *Let M be a closed manifold and $f : M \to S^1$ a Morse map. Let $\bar{M} \to M$ be the infinite cyclic covering corresponding to f, and $F : \bar{M} \to \mathbf{R}$ a lift of f to $\bar{M}$. Let v be an oriented cellular f-gradient. Then:*

(1) *For every pair p, q of critical points of F with $\operatorname{ind} p = \operatorname{ind} q + 1$ the Novikov incidence coefficient $N(p, q; v)$ is in $\widetilde{L}$.*

(2) *There is* $\delta > 0$ *such that for every cellular* f*-gradient* w *with* $||w - v|| < \delta$ *and every pair* p, q *of critical points of* F *with* $\operatorname{ind} p = \operatorname{ind} q + 1$ *we have*

$$N(p, q; v) = N(p, q; w),$$

if w *is oriented similarly to* v.

Proof. Let $v \in G_C(f)$, and $p, q \in S(F)$ with $\operatorname{ind} q = r$, $\operatorname{ind} p = r + 1$. The main idea of the proof is to obtain a formula for the coefficients $N(p, q; v)$ in terms of the homological gradient descent. Since v is a cellular gradient, there is a regular value λ of F such that the restriction of $\bar{v}$ to $W = F^{-1}([\lambda - 1, \lambda])$ is a cyclic cellular gradient for the function $F|W$. We can assume that $p, q \in t^{-1}W$. Put

$$N = D(p, \bar{v}) \cap \partial_1 W.$$

Put

$$\zeta = t^{-1} \circ (-\bar{v})^{\leadsto}_{[\lambda, \lambda-1]},$$

so that ζ is a diffeomorphism of an open subset of $\partial_1 W$ onto another open subset of $\partial_1 W$. Let

$$N_k = \zeta^k(N) \subset \partial_1 W.$$

We begin by studying the manifold N_k and its homological properties. Let $\partial_1 W^r$ and $\partial_0 W^r$ denote the Morse-Smale filtrations of $\partial_1 W$, respectively, $\partial_0 W$ from the definition of cellular gradient (Definition 2.22 of Chapter 9, page 297).

Lemma 3.5. *For every* $k \geqslant 1$ *the submanifold* N_k *is in* $\partial_1 W^r$, *and* $N_k \setminus \operatorname{Int} \partial_1 W^{r-1}$ *is compact.*

Proof. The lemma follows by induction on k via a direct application of the first part of Proposition 1.4 of Chapter 9 (page 283). □

In particular the manifold N_k has its fundamental class

$$[N_k] \in H_r(\partial_1 W^r, \partial_1 W^{r-1}).$$

Recall from Subsection 2.1 of Chapter 7 (page 239) that the gradient descent for the cellular gradient v yields a continuous map of quotients

$$(-v)^{\twoheadrightarrow} : \partial_1 W^r / \partial_1 W^{r-1} \to \partial_0 W^r / \partial_0 W^{r-1}.$$

We obtain therefore a continuous self-map

$$t^{-1} \circ (-v)^{\twoheadrightarrow} : \partial_1 W^r / \partial_1 W^{r-1} \to \partial_1 W^r / \partial_1 W^{r-1}.$$

The homomorphism induced by this map in homology will be denoted by

$$\overline{\mathcal{H}}_r(-v) : H_r(\partial_1 W^r, \partial_1 W^{r-1}) \to H_r(\partial_1 W^r, \partial_1 W^{r-1}).$$

We have a recursive formula for the fundamental class of N_k:

Lemma 3.6.

$$[N_k] = \left(\overline{\mathcal{H}}_r(-v)\right)^k\left([N]\right).$$

Proof. The proof follows immediately from Proposition 1.4 of Chapter 9 (page 283). □

Since the manifold N_k is in $\partial_1 W^r$ the multiplicity $m([N_k], q; \bar{v})$ of the homology class $[N_k]$ with respect to the critical point tq is defined (see Subsection 1.4 of Chapter 9, page 286 for the definition of multiplicities).

Proposition 3.7. *For every $k \geqslant 0$ we have:*

$$n_{k+1}(p, q; \bar{v}) = m\big([N_k], tq; \bar{v}\big). \tag{6}$$

Proof. Let us first consider the case when v is transverse. Let

$$S = D(tq, -\bar{v}) \cap \partial_1 W.$$

We have

$$n_{k+1}(p, q; \bar{v}) = N_k \mathbin{\#} S = \zeta^k(N) \mathbin{\#} S,$$

see the formula (3), page 372. Applying Proposition 1.9 of Chapter 9 (page 286) we deduce

$$N_k \mathbin{\#} S = m([N_k], tq; \bar{v}).$$

This completes the proof for the case of transverse gradients. The general case, when v is assumed to be only almost transverse is reduced to the transverse case by a perturbation argument. We will show that both sides of (6) are stable with respect to C^0-small perturbations of v. Since we are going to deal with an arbitrary f-gradient w, sufficiently C^0-close to v, we will modify our notation for the end of this proof, and write

$$N(v) = D(p, \bar{v}) \ \cap\ \partial_1 W, \quad N(w) = D(p, \bar{w}) \ \cap\ \partial_1 W.$$

Similarly, we define a diffeomorphism $\zeta(w)$ and a manifold $N_k(w) = \zeta^k(w)(N(w))$. Applying Proposition 1.10 of Chapter 9 (page 287), Corollary 1.3 of Chapter 9 (page 282), and Proposition 1.7 of Chapter 9 (page 285), we deduce that there is $\delta > 0$ such that for every f-gradient w with $||w - v|| < \delta$ we have

$$m([N_k(v)],\ tq; \bar{v}) = m([N_k(w)],\ tq; \bar{w})$$

if v and w are similarly oriented. Therefore the right-hand part of (6) is C^0-stable. As for the left-hand side, apply Theorem 2.6 of Chapter 6 (page 212) to the cobordism $F^{-1}\big([\lambda - k - 1, \lambda]\big)$ and deduce that there is $\delta > 0$ such that for every f-gradient w with $||w - v|| < \delta$ we have

$$n_{k+1}(p, q; \bar{v}) = n_{k+1}(p, q; \bar{w})$$

if v and w are similarly oriented.

By Corollary 2.10 of Chapter 4 (page 126) there is a transverse f-gradient w in an arbitrary small neighbourhood of v. For the gradient w the formula (6) holds as we have proved above. □

Therefore we obtain a recursive formula which expresses the Novikov incidence coefficients in terms of the homological gradient descent:

$$n_{k+1}(p,q;\bar v) = m\Big(\big(\overline{\mathcal{H}}_r(-v)\big)^k([N]),\ tq;\ \bar v\Big). \tag{7}$$

Let us write this formula somewhat shorter:

$$n_{k+1}(p,q;v) = \lambda\big(A^k(\xi)\big)$$

where the entries in the right-hand side are as follows:

- $\xi = [N] \in H_r(\partial_1 W^r, \partial_1 W^{r-1})$;
- $A = \overline{\mathcal{H}}_r(-v)$ is an endomorphism of the group $G = H_r(\partial_1 W^r, \partial_1 W^{r-1})$;
- $\lambda : G \to \mathbf{Z}$ is a homomorphism given by the formula $\lambda(x) = m(x, tq;\ \bar v)$.

Thus we have the following formula:

$$N(p,q;v) = n_0(p,q;\bar v) + \sum_{k\geqslant 0} \lambda\big(A^k(\xi)\big)t^{k+1}$$

and the proof of the first point of our theorem is completed by the next lemma.

Lemma 3.8. *Let G be a finitely generated abelian group, $A : G \to G$ an endomorphism and $\lambda : G \to \mathbf{Z}$ a homomorphism. Then for every $\xi \in G$ we have*

$$\sum_{k\geqslant 0} \lambda(A^k\xi)t^k = \frac{P(t)}{Q(t)} \quad \textit{where} \quad P, Q \in \mathbf{Z}[t] \quad \textit{and} \quad Q(0) = 1.$$

Proof. It is sufficient to prove the lemma in the case when G is a free finitely generated abelian group. Consider a free finitely generated $\mathbf{Z}[[t]]$-module $R = G[[t]]$ and a homomorphism $\phi : R \to R$, given by $\phi(x) = (1 - At)x$. Then ϕ is invertible, and

$$\phi^{-1}(x) = \Big(\sum_{k\geqslant 0} A^k t^k\Big)(x).$$

On the other hand the inverse of ϕ is given by the Cramer formula, which implies that each matrix entry of ϕ^{-1} is the ratio of two polynomials with integral coefficients. Moreover the denominator $Q(t)$ of this ratio equals $\det(1 - At)$, therefore $Q(0) = 1$. □

It remains to prove the invariance of $N(p,q;v)$ with respect to C^0-small perturbations of v, as required in the second part of our theorem. Let us return to the formula (7). By Proposition 1.10 of Chapter 9 (page 287), Corollary 1.3 of Chapter 9 (page 282), and Proposition 1.7 of Chapter 9, page 285 the right-hand side of (6) is stable with respect to C^0-small perturbations of the vector field v, and the proof is over. □

4. Gradient-like vector fields and Riemannian gradients

In this section we establish analogs of Theorem 3.1 for special classes of gradients. Recall that the space of all gradient-like vector fields for a circle-valued Morse function f is denoted $GL(f)$. The subset of all almost transverse gradient-like vector fields is denoted by $GL_A(f)$, and the subset of all transverse gradient-like vector fields is denoted by $GL_T(f)$.

Proposition 4.1. *Let $f : M \to S^1$ be a Morse function on a closed manifold M.*

(1) *The space $G_C(f) \cap GL(f)$ is open and dense in $GL(f)$ with respect to C^0 topology.*
(2) *The space $G_C(f) \cap GL_A(f)$ is open and dense in $GL_A(f)$ with respect to C^0 topology.*
(3) *The space $G_C(f) \cap GL_T(f)$ is open and dense in $GL_T(f)$ with respect to C^0 topology.*

Proof. The subset $G_C(f)$ is C^0-open in $G(f)$, therefore all the assertions of the proposition concerning openness properties follow immediately. Let us move on to the density properties.

(1) Let v be a gradient-like vector field for f. We can assume that v is almost transverse. Let $\epsilon > 0$, and λ be a regular value of f. Put $V = f^{-1}(\lambda)$. Applying Theorem 3.1 we obtain an f-gradient $u \in G_C(f)$ with $||u - v|| \leqslant \epsilon$ and supp $(u - v) \subset U \setminus V$, where U is a neighbourhood of V which we can choose as we like. Observe that if U is sufficiently small, then u is necessarily a gradient-like vector field for f, and this completes the proof of point (1).
(2) This point follows immediately since $G_C(f) \subset G_A(f)$.
(3) Let $v \in GL_T(f)$. Let $\delta > 0$; pick a cellular gradient-like vector field w for f with $||w - v|| < \delta/2$. The set $G_C(f)$ being C^0-open, there is $\epsilon > 0$ such that every f-gradient u with $||w - u|| < \epsilon$ is in $G_C(f)$. Now pick a transverse gradient-like vector field u with $||w-u|| < \min(\epsilon, \delta/2)$ (this is possible by Corollary 2.17 of Chapter 4, page 128). Then $u \in GL_T(f) \cap G_C(f)$ and $||u - v|| < \delta$. □

Now let us proceed to the version of Theorem 3.1 for Riemannian gradients. We denote the space of all Riemannian gradients for f by $GR(f)$, the subset of almost transverse Riemannian gradients by $GR_A(f)$, and the subset of transverse Riemannian gradients by $GR_T(f)$.

Proposition 4.2. *Let $f : M \to S^1$ be a Morse function on a closed manifold M.*

(1) *The space $G_C(f) \cap GR(f)$ is open and dense in $GR(f)$ with respect to C^0 topology.*
(2) *The space $G_C(f) \cap GR_A(f)$ is open and dense in $GR_A(f)$ with respect to C^0 topology.*
(3) *The space $G_C(f) \cap GR_T(f)$ is open and dense in $GR_T(f)$ with respect to C^0 topology.*

Proof. As in the proof of Proposition 4.1 it suffices to prove the density properties.

(1) Let v be an almost transverse Riemannian gradient for f. Choose any regular value λ of f, and put $V = f^{-1}(\lambda)$. Let $\epsilon > 0$. Applying Theorem 3.1 we obtain an f-gradient $u \in G_C(f)$ such that $||u - v|| < \epsilon$, and supp $(u - v) \subset U \setminus V$, where U is a neighbourhood of V which can be chosen as we like. By Corollary 2.21 of Chapter 4 (page 131) the vector field u is a Riemannian gradient for f if U is chosen sufficiently small.
(2) This point follows immediately since $G_C(f) \subset G_A(f)$.
(3) The proof of this point is completely similar to the proof of the point (3) of Proposition 4.1 with the only difference that we use Corollary 2.26 of Chapter 4 (page 132) instead of Corollary 2.17 of Chapter 4 (page 128). □

There is an alternative version of the preceding proposition, dealing with the space of Riemannian metrics rather than with the space of gradients.

Definition 4.3. Let M be a closed manifold. The space of all Riemannian metrics on M is denoted by $\mathcal{R}(M)$. For a circle-valued Morse function f on M let

$$\mathcal{R}_C(M, f), \quad \mathcal{R}_T(M, f), \quad \mathcal{R}_A(M, f)$$

denote the subsets of all Riemannian metrics R on M such that the Riemannian gradient $\text{grad}_R(f)$ is, respectively, cellular, transverse, almost transverse.

Proposition 4.4. *Let $f : M \to S^1$ be a Morse function.*

(1) *The subset $\mathcal{R}_C(M, f)$ is open and dense in $\mathcal{R}(M)$ with respect to C^0 topology.*
(2) *The subset $\mathcal{R}_C(M, f)$ is open and dense in $\mathcal{R}_A(M, f)$ with respect to C^0 topology.*

(3) *The subset* $\mathcal{R}_C(M, f) \cap \mathcal{R}_T(M, f)$ *is open and dense in* $\mathcal{R}_T(M, f)$ *with respect to* C^0 *topology.*

Proof. The map $\mathcal{R}(M) \to G(f)$ sending each Riemannian metric to the corresponding f-gradient is continuous in both C^∞ and C^0 topologies. Therefore all the assertions of our proposition concerning openness properties are immediate. Let us proceed to the density properties and prove the point (1) of our proposition. Let R be any Riemannian metric on M and $v = \text{grad}_R f$. Let $\mathcal{U}$ be a neighbourhood of R in $\mathcal{R}(M)$ with respect to C^0 topology. Choose any regular value λ of f, and let $\delta > 0$ be small enough so that the interval $I_\delta = [\lambda - \delta, \lambda + \delta]$ is regular. Let $K = f^{-1}(I_\delta)$. Pick any $\epsilon > 0$ and let w be a cellular f-gradient such that $||w - v|| < \epsilon$ and supp $(w - v) \subset K$. Applying Corollary 2.21 of Chapter 4 (page 131) we deduce that if ϵ is sufficiently small, then there is a Riemannian metric $R' \subset \mathcal{U}$ such that w is the Riemannian gradient of f with respect to R'. The proof of the first point is now complete. The proofs of the second and third point are similar to the proofs of the corresponding points of Proposition 4.2 and will be omitted. □

CHAPTER 13

Counting closed orbits of the gradient flow

In Sections 1 and 2 we introduce and study the *Lefschetz zeta function* of the gradient flow of a circle-valued Morse function. This function is a power series which contains information about the closed orbits of the flow. The main result of this chapter (Theorem 4.3) states that the Lefschetz zeta function can be expressed in terms of Whitehead torsion invariants, associated with the Novikov complex. We discuss the Whitehead torsion and related topics in Section 3.

1. Lefschetz zeta functions of the gradient flows

Definition 1.1. Let w be a C^1 vector field on a manifold M. A *closed trajectory* of w is a non-constant integral curve $\gamma : [a,b] \to M$ of w, such that $\gamma(a) = \gamma(b)$. We say that two closed trajectories

$$\gamma : [a,b] \to M, \quad \widetilde{\gamma} : [a',b'] \to M$$

are *equivalent*, if there is C such that $a' = a + C$, $b' = b + C$ and $\widetilde{\gamma}(t + C) = \gamma(t)$ for every t. A *closed orbit* of w is a class of equivalence of closed trajectories. The set of all closed orbits of w is denoted by $Cl(w)$. We shall identify a closed trajectory with the corresponding closed orbit, when no confusion is possible.

For a closed orbit $\gamma : [a,b] \to M$ let $c \in [a,b]$ be the smallest number $> a$ such that $\gamma(a) = \gamma(c)$. It is easy to see that $b - a = n(c - a)$ where n is a natural number, called the *multiplicity* of γ, and denoted $mult(\gamma)$. If $mult(\gamma) = 1$, the orbit γ is called a *prime closed orbit*. The set of all prime closed orbits of w is denoted by $ClPr(w)$.

Let $\gamma : [a,b] \to M$ be a closed trajectory. Let Σ be any submanifold of M with $\dim \Sigma = \dim M - 1$, containing $p = \gamma(a)$ and such that w is transverse to Σ at p (that is, $w(p) \notin T_p\Sigma$). The classical construction of the *Poincaré return map* associates with these data a diffeomorphism $R : U \to V$, where U, V are neighbourhoods of p in Σ. Here is a brief description of the construction (we refer to the book of J. Palis and W. de Melo [**122**] for more details): for each point $x \in \Sigma$ in a small neighbourhood of p the w-trajectory starting at x will intersect Σ again at some point y at some moment t_0 close to b; we set by definition $R(x) = y$. If p is an isolated fixed point of R, then its index is called the *index of* γ, and denoted by $i(\gamma)$. The number

$$i_F(\gamma) = \frac{i(\gamma)}{m(\gamma)}$$

is called *the Fuller index* of γ. We say that γ is *a hyperbolic closed orbit* if p is a hyperbolic fixed point of the Poincaré return map.

Now we will concentrate on the case of gradients of circle-valued Morse functions. In the case when $w = -v$ with v a gradient of a Morse map $f : M \to S^1$ there is one more natural invariant for closed orbits of w. Let $\gamma : I \to M$ be a closed orbit of $(-v)$, defined on some closed interval $I \subset \mathbf{R}$. The composition $f \circ \gamma : I \to S^1$ sends the endpoints of the segment to the same point, and we obtain a C^1 map from the circle $I/\partial I$ to S^1. The derivative of this map is strictly negative, and the degree of this map is a strictly negative integer; the opposite number will be called the *winding number of* γ and denoted by $n(\gamma) > 0$.

As usual we lift the function $f : M \to S^1$ to a real-valued Morse function $F : \bar{M} \to \mathbf{R}$ on the infinite cyclic covering $\pi : \bar{M} \to M$. Let $\bar{v}$ denote the lift of v to $\bar{M}$. Let t be the downward generator of the structure group of the covering. Pick a regular value λ of F, and put

$$V = F^{-1}(\lambda), \quad W = F^{-1}([\lambda - 1, \lambda]).$$

We have the partially defined map

$$\Theta = t^{-1} \circ (-\bar{v})^{\leadsto}_{[\lambda, \lambda-1]}$$

from an open subset of V to V. There is a close relation between the periodic points of Θ and closed orbits of $(-v)$. Namely, let $\sigma \in \mathrm{Per}(\Theta)$, and $l(\sigma)$ be the least period of σ. It is clear that the $(-v)$-trajectory starting at the point $\pi(\sigma)$ determines a prime closed orbit of $(-v)$ with winding number $l(\sigma)$; we shall denote this closed orbit by $c(\sigma)$. Also for $k \in \mathbf{N}$ the k-th iteration of the closed orbit $c(\sigma)$ is a closed orbit $c^k(\sigma)$ of multiplicity k. We obtain a natural surjection

$$(\sigma, k) \mapsto c^k(\sigma), \quad \mathrm{Per}(\Theta) \times \mathbf{N} \xrightarrow{\mathcal{C}} Cl(-v).$$

Observe that

$$n\big(\mathcal{C}(\sigma, k)\big) = l(\sigma) \cdot k, \quad m\big(\mathcal{C}(\sigma, k)\big) = k.$$

The closed orbit γ is hyperbolic if and only if every $\sigma \in \mathcal{C}^{-1}(\gamma)$ is a hyperbolic fixed point of Θ, and in this case all the points in $\mathcal{C}^{-1}(\gamma)$ have the same index equal to the index of γ.

Lemma 1.2. *Assume that v is almost transverse and all closed orbits of v are hyperbolic. Then for each $k \in \mathbf{N}$ there is only a finite number of closed orbits γ with $n(\gamma) = k$.*

Proof. It suffices to prove that for every k the set of fixed points of Θ^k is finite. Consider the cobordism $W_k = F^{-1}([\lambda - k, \lambda])$. The deck transformation t^{-k} is an isometry $\partial_0 W_k \to \partial_1 W_k$. Thus W_k is a cyclic cobordism, and the vector field v is a CAT-gradient for the Morse function $F|W_k$ (see Definition 2.14 of Chapter 9, page 293 for definition of CAT condition). The fixed point set of

$$\Theta^k = t^{-k} \circ (-\bar{v})^{\leadsto}_{[\lambda,\lambda-k]}$$

is therefore compact (see page 294), and since every periodic point of Θ is hyperbolic, this set is finite. □

Now we will define counting functions for the closed orbits of the gradient flow.

Definition 1.3. Let v be an almost transverse f-gradient such that all closed orbits of $(-v)$ are hyperbolic. Set

$$\eta_L(-v) = \sum_{\gamma \in Cl(-v)} i_F(\gamma) t^{n(\gamma)}, \quad \zeta_L(-v) = \exp\big(\eta_L(-v)\big).$$

The formal series $\eta_L(-v)$ and $\zeta_L(-v)$ are called the *Lefschetz eta function* and the *Lefschetz zeta function* of $(-v)$.

Lemma 1.2 implies that the expression defining $\eta_L(-v)$ is indeed an element of $\mathbf{Q}[[t]]$, and since the constant term of $\eta_L(-v)$ is equal to zero, the exponent of this series is also a power series with rational coefficients.

Lemma 1.4. $\eta_L(-v) = \eta_L(\Theta), \quad \zeta_L(-v) = \zeta_L(\Theta).$

Proof. To prove the first formula, it suffices to show that for every positive integer k,

$$\sum_{\gamma | n(\gamma) = k} \frac{i(\gamma)}{m(\gamma)} = \sum_{a \in \mathrm{Fix}(\Theta^k)} \frac{I(\Theta^k, a)}{k}.$$

This is obvious: every closed orbit γ with $n(\gamma) = k$ and $m(\gamma) = m$ gives rise to k/m fixed points of Θ^k; these points have the same indices equal to the index of γ. □

The map $\mathcal{C}$ induces a bijective correspondence between the set of all prime closed orbits of $(-v)$ and the set of all finite orbits of Θ. This observation together with Proposition 2.12 of Chapter 9 (page 292) leads to the factorization formula for the Lefschetz zeta function:

Proposition 1.5. *Let v be an almost transverse f-gradient, such that all closed orbits of v are hyperbolic. Then*

$$\zeta_L(-v) = \prod_{\gamma \in ClPr(-v)} \left(1 - \varepsilon(\gamma) t^{n(\gamma)}\right)^{-\nu(\gamma)}$$

where $\varepsilon(\gamma), \nu(\gamma) \in \{-1, 1\}$. In particular, the power series $\zeta_L(-v)$ has integral coefficients. □

Now we shall drop the hyperbolicity assumption and generalize the above constructions to the case of any almost transverse f-gradient v. In this case the closed orbits are not necessarily isolated, and the definition of the counting functions above does not apply.

Proposition 2.18 of Chapter 9 (page 294) implies that for every $k \geqslant 0$ the set of the fixed points of

$$\Theta^k = t^{-k} \circ (-\bar{v})^{\leadsto}_{[\lambda, \lambda - k]}$$

is compact and the index $L(\Theta^k) \in \mathbf{Z}$ is defined.

Lemma 1.6. *The integer $L(\Theta^k)$ does not depend on the particular choice of the regular value λ.*

Proof. The map

$$\Theta = t^{-1} \circ (-\bar{v})^{\leadsto}_{[\lambda, \lambda - 1]}$$

depends obviously on λ and during this proof we shall write $\Theta = \Theta(\lambda)$ in order to emphasize this dependence. Let μ be another regular value, assume for example that $\mu < \lambda$. The gradient descent map $(-\bar{v})^{\leadsto}_{[\lambda, \mu]}$ is defined on a neighbourhood U of Fix $\Theta^k(\lambda)$ and maps it diffeomorphically onto a neighbourhood U' of Fix $\Theta^k(\mu)$. Denote this diffeomorphism by ξ, then the two maps

$$\Theta^k(\lambda)|U, \quad \Theta^k(\mu)|U'$$

are conjugate via ξ, and the result follows. □

Definition 1.7. For an almost transverse f-gradient v put

$$\eta_L(-v) = \eta_L(\Theta) = \sum_{k \geqslant 1} \frac{L(\Theta^k)}{k} t^k, \quad \zeta_L(-v) = \zeta_L(\Theta) = \exp\left(\eta_L(-v)\right).$$

The series $\eta_L(-v)$ and $\zeta_L(-v)$ are called the *Lefschetz eta function* and the *Lefschetz zeta function* of $(-v)$.

Thus by definition $\eta_L(-v)$, $\zeta_L(-v)$ are the power series in one variable with rational coefficients. We shall see later on that the coefficients of $\zeta_L(-v)$ are integral.

Let $\eta_k(-v)$ be the k-truncation of the series $\eta_L(-v)$, that is, the image of $\eta_L(-v)$ in the ring P/t^kP. Similarly, we denote by $z_k(-v)$ the k-truncation of $\zeta_L(-v)$. These polynomials are stable with respect to C^0-small perturbations of the vector field v, as the following proposition shows.

Proposition 1.8. *Let v be an almost transverse f-gradient. Let $k \in \mathbf{N}$. There is $\delta > 0$ such that for every f-gradient w with $||w - v|| < \delta$ we have:*

$$\eta_k(-v) = \eta_k(-w), \quad z_k(-v) = z_k(-w).$$

Proof. In order to emphasize the dependence of the diffeomorphism Θ on the vector field v, let us write

$$\Theta_v = t^{-1} \circ (-\bar{v})^{\leadsto}_{[\lambda, \lambda-1]}.$$

Let r be any natural number. Applying Proposition 2.19 of Chapter 9, page 295 to the cyclic cobordism $W_r = F^{-1}\big([\lambda - r, \lambda]\big)$ we deduce that

$$L(\Theta_v^r) = L(\Theta_w^r)$$

for every w sufficiently C^0-close to v. The stability of $\eta_k(-v)$ follows and implies the stability of $z_k(-v)$. □

2. Homological and dynamical properties of the cellular gradients

For every almost transverse gradient of a circle-valued Morse function the Novikov complex associated to this data computes the completed homology of the corresponding infinite cyclic covering. In this section we will see that the Morse-theoretic data associated to a *cellular* gradient allows us to obtain more, namely the homology of the cyclic covering itself. Our constructions are based on the techniques developed in Part 3. In Section 4 we discuss the relation between the zeta function of the gradient flow and the Whitehead torsion of the canonical equivalence.

2.1. A filtration of $\bar{M}$ associated with a cellular gradient. Let $f : M \to S^1$ be a Morse map. Recall the commutative diagram

$$\begin{array}{ccc} \bar{M} & \xrightarrow{F} & \mathbf{R} \\ {\scriptstyle \pi}\downarrow & & \downarrow{\scriptstyle \mathrm{Exp}} \\ M & \xrightarrow{f} & S^1. \end{array}$$

Assume that M is endowed with a Riemannian metric; then $\bar{M}$ inherits the Riemannian structure, and the downward generator t of the structure group of the covering is an isometry of $\bar{M}$. Let v be a cellular f-gradient;

denote by $\bar{v}$ the lift of v to $\bar{M}$. There is a regular value λ of F such that the restriction of $\bar{v}$ to the cobordism

$$W = F^{-1}([\lambda - 1, \lambda])$$

is a cellular $F|W$-gradient, and W acquires the corresponding handle-like filtration $\{W^{\langle k\rangle}\}$ (Section 3 of Chapter 9, page 301). Define a filtration $\{V^-_{\langle k\rangle}\}$ of

$$V^- = F^{-1}(]-\infty, \lambda])$$

by

$$V^-_{\langle k\rangle} = \bigcup_{s\geqslant 0} t^s W^{\langle k\rangle}.$$

Our first aim is to describe the homology of the pair $\big(V^-_{\langle k\rangle}, V^-_{\langle k-1\rangle}\big)$. Since $V^-_{\langle k\rangle}, V^-_{\langle k-1\rangle}$ are invariant with respect to the action of t, the homology of this pair acquires the structure of a module over the ring

$$P = \mathbf{Z}[t].$$

We are going to prove that this homology vanishes in all degrees except k, and the module

$$\mathcal{F}_k = H_k\Big(V^-_{\langle k\rangle}, V^-_{\langle k-1\rangle}\Big)$$

is a free P-module. We have the inclusion of pairs

$$(W^{\langle k\rangle}, W^{\langle k-1\rangle}) \xhookrightarrow{I} (V^-_{\langle k\rangle}, V^-_{\langle k-1\rangle}).$$

Recall from Theorem 3.1 of Chapter 9 (page 306) that the homology of the pair $(W^{\langle k\rangle}, W^{\langle k-1\rangle})$ is equal to zero in all degrees except k, and the group

$$E_k = H_k(W^{\langle k\rangle}, W^{\langle k-1\rangle})$$

is the direct sum of the images of the four monomorphisms:

$$\begin{array}{ccccc} & & \mathcal{M}_k & & \\ & & \downarrow{\scriptstyle \mu} & & \\ \mathcal{M}_k^{(1)} & \xrightarrow{\lambda_1} & E_k & \xleftarrow{\lambda_0} & \mathcal{M}_k^{(0)} \\ & & \uparrow{\scriptstyle \tau} & & \\ & & \mathcal{M}_{k-1}^{(1)} & & \end{array}$$

where $\mathcal{M}_*^{(0)}, \mathcal{M}_*^{(1)}$ are the Morse complexes associated with ordered Morse functions on $\partial_0 W$, respectively $\partial_1 W$, and $\mathcal{M}_*$ is the Morse complex

$\mathcal{M}_*(F|W, \bar{v}|W)$. Along with the I_*-images of the four submodules above the P-module $\mathcal{F}_*$ contains also the t^n-shifts of these images for $n \geqslant 0$. Let

$$\mathcal{R}_* = \mathcal{M}_*^{(1)} \underset{\mathbf{Z}}{\otimes} P, \qquad \Pi_* = \mathcal{M}_* \underset{\mathbf{Z}}{\otimes} P.$$

Since $\mathcal{M}_*^{(1)}$ and $\mathcal{M}_*$ are free abelian groups, $\mathcal{R}_*$ and Π_* are free P-modules. The free generators of $\mathcal{R}_*$ are the critical points of the ordered Morse function on $\partial_1 W$, and the free generators of $\mathcal{M}_*$ are the critical points of $F|W$. Let

$$\Lambda_k : \mathcal{R}_k \to \mathcal{F}_k, \qquad \varpi_k : \Pi_k \to \mathcal{F}_k, \qquad \mathscr{T}_k : \mathcal{R}_{k-1} \to \mathcal{F}_k$$

be the corresponding extensions of the homomorphisms $I_* \circ \lambda_1$, $I_* \circ \mu$, $I_* \circ \tau$.

Theorem 2.1. *The subspaces $V^-_{\langle k \rangle}$ form a cellular filtration of V^- and*

$$\mathfrak{L}_k = (\Lambda_k, \varpi_k, \mathscr{T}_k) : \mathcal{R}_k \oplus \Pi_k \oplus \mathcal{R}_{k-1} \to \mathcal{F}_k = H_k\big(V^-_{\langle k \rangle}, V^-_{\langle k-1 \rangle}\big)$$

is an isomorphism of P-modules.

Proof. We shall use the techniques developed in Section 3.1 of Chapter 9 (page 302). Recall the space $Q_k \subset W^{\langle k-1 \rangle}$ from (6), page 307. Let

$$U_k = \bigcup_{n \in \mathbf{N}} t^n Q_k.$$

Lemma 2.2. *The inclusion $U_k \hookrightarrow V^-_{\langle k \rangle}$ is a homology equivalence.*

Proof. By excision

$$H_*(V^-_{\langle k \rangle}, U_k) \approx \bigoplus_{n \in \mathbf{N}} H_*(t^n W^{\langle k \rangle}, t^n Q_k).$$

The homology of each pair $(t^n W^{\langle k \rangle}, t^n Q_k)$ vanishes by Lemma 3.2 of Chapter 9 (page 308). □

Thus it suffices to compute the homology $H_*(U_k, V^-_{\langle k-1 \rangle})$. Consider the quotient space

$$U_k / V^-_{\langle k-1 \rangle} = \bigcup_{n \in \mathbf{N}} t^n Q'_k \quad \text{where} \quad Q'_k = Q_k / W^{\langle k-1 \rangle}.$$

This space acquires from $\bar{M}$ the action of the generator t of the deck transformation group. The map t has the unique fixed point ω corresponding to the collapsed subspace $W^{\langle k-1 \rangle}$. Two copies $t^n Q'_k$, $t^r Q'_k$ of the space Q'_k intersect by the point ω if $|r - n| > 1$, and

$$t^n(Q'_k) \cap t^{n+1}(Q'_k) = t^{n+1}\Big(\partial_1 W^k / \partial_1 W^{k-1}\Big).$$

The homeomorphism type of the space $Q_k/W^{\langle k-1\rangle}$ was studied in Subsection 3.1 of Chapter 9 (page 302). We proved there that this space is homeomorphic to the wedge of four spaces:

$$Q_k/W^{\langle k-1\rangle} \approx \partial_1 W^k/\partial_1 W^{k-1} \vee \widehat{\mathcal{H}} \vee \widehat{T}'_k \vee \widehat{Z}'_k.$$

Referring to page 308 for the definition of the summands of the wedge we only recall here that the space $\widehat{\mathcal{H}}$ is homotopy equivalent to the wedge of $m_k(f)$ spheres of dimension k, the space $\widehat{Z}'_k$ is homotopy equivalent to $\partial_0 W^k/\partial_0 W^{k-1}$ and the space $\widehat{T}'_k$ has the homotopy type of the space $\partial_1 W^{k-1}/\partial_1 W^{k-2}$. Thus we have a splitting

$$U_k/V^-_{\langle k-1\rangle} \approx \partial_1 W^k/\partial_1 W^{k-1} \vee \bigvee_{n\in\mathbf{N}} t^n\Big(\widehat{\mathcal{H}} \vee \widehat{T}'_k \vee \widehat{Z}'_k\Big)$$

and in view of the equality

$$t(\partial_1 W^k/\partial_1 W^{k-1}) = \partial_0 W^k/\partial_0 W^{k-1}$$

we deduce that the inclusion

$$(1) \qquad \bigvee_{n\in\mathbf{N}} t^n\Big(\partial_1 W^k/\partial_1 W^{k-1} \vee \widehat{\mathcal{H}} \vee \widehat{T}'_k\Big) \xrightarrow{J} U_k/V^-_{\langle k-1\rangle}$$

is a homology equivalence. Since the reduced homology of each of the three spaces $\partial_1 W^k/\partial_1 W^{k-1}$, $\widehat{\mathcal{H}}$, $\widehat{T}'_k$ is equal to zero in all dimensions except k, this is also the case for the reduced homology of $U_k/V^-_{\langle k-1\rangle}$.

To complete the proof of our theorem it remains to observe that the homology of the space on the left-hand side of (1) is isomorphic to

$$\mathcal{R}_k \oplus \Pi_k \oplus \mathcal{R}_{k-1},$$

the inclusion

$$U_k/V^-_{\langle k-1\rangle} \hookrightarrow V^-_{\langle k\rangle}/V^-_{\langle k-1\rangle}$$

is a homology equivalence, and the resulting isomorphism

$$\mathcal{R}_k \oplus \Pi_k \oplus \mathcal{R}_{k-1} \longrightarrow H_k(V^-_{\langle k\rangle}/V^-_{\langle k-1\rangle})$$

is equal to $\mathfrak{L}_k$. □

The filtration $\{V^-_{\langle k\rangle}\}$ of the space

$$V^- = F^{-1}(]-\infty,\lambda])$$

induces a cellular filtration in the singular chain complex

$$\mathcal{S}^\lambda_* = \mathcal{S}_*(V^-).$$

The P-modules

$$\mathcal{F}_k = H_k(V^-_{\langle k\rangle}, V^-_{\langle k-1\rangle})$$

form the adjoint complex, and by Corollary 3.8 of Chapter 6 (page 215) there is a chain equivalence

$$\nu : \mathcal{F}_* \to \mathcal{S}_*^\lambda = \mathcal{S}_*(V^-)$$

of P-complexes. We obtain in particular the following finiteness property for the complex $\mathcal{S}_*^\lambda$ (another proof using the triangulation theory of manifolds was given in Section 6 of Chapter 11, page 356).

Corollary 2.3. *The P-complex $\mathcal{S}_*^\lambda$ is homotopically free-finite.* □

Having identified the modules $\mathcal{F}_k$ of the adjoint complex, our next aim is to compute the matrices of the boundary operators $d_k : \mathcal{F}_k \to \mathcal{F}_{k-1}$. This computation follows immediately from results of Subsection 3.3 of Chapter 9 (page 314) once we introduce a convenient terminology. Extend the boundary operators in the Morse complexes $\mathcal{M}_*^{(1)}$ and $\mathcal{M}_*$ to the operators in the extended graded modules $\mathcal{R}_*, \Pi_*$; we shall keep the same notation $\partial_*^{(1)}, \partial_*$ for the extended operators. These operators form the diagonal entries of the matrix. To describe the other entries consider the map

$$t^{-1} \circ (-\bar{v})^{\twoheadrightarrow} : \partial_1 W^k / \partial_1 W^{k-1} \to \partial_1 W^k / \partial_1 W^{k-1}.$$

Let

$$\overline{\mathcal{H}}_k(-v) : \mathcal{M}_k^{(1)} \to \mathcal{M}_k^{(1)}$$

denote the homomorphism induced by this map in homology, and extend it to an endomorphism of the P-module $\mathcal{R}_k$. Similarly, the homomorphism $\sigma_k : \mathcal{M}_k \to \mathcal{M}_{k-1}^{(0)}$ yields a homomorphism of abelian groups

$$\bar{\sigma}_k = t^{-1} \circ \sigma_k : \mathcal{M}_k \to \mathcal{M}_{k-1}^{(1)},$$

which is extended to a homomorphism of P-modules $\Pi_k \to \mathcal{R}_{k-1}$. Finally extend the homomorphism η_k (page 315) to a homomorphism $\mathcal{R}_k \to \Pi_k$ of P-modules. The matrix of the boundary operator in $\mathcal{F}_*$ is now deduced immediately from Theorem 3.8 of Chapter 9 (page 315):

Proposition 2.4. *The matrix of the boundary operator $d_{k+1} : \mathcal{F}_{k+1} \to \mathcal{F}_k$ with respect to the direct sum decomposition from Theorem 2.1 is*

$$\begin{pmatrix} \partial_{k+1}^{(1)} & t\bar{\sigma}_{k+1} & \mathrm{Id} - t\overline{\mathcal{H}}_k(-v) \\ 0 & \partial_{k+1} & -\eta_k \\ 0 & 0 & -\partial_k^{(1)} \end{pmatrix}. \tag{2}$$

□

2.2. Chain complex $\mathcal{F}_*$ and the Novikov complex. Tensoring by $\widehat{P}$ the chain equivalence

$$\nu : \mathcal{F}_* \xrightarrow{\sim} \mathcal{S}_*^\lambda = \mathcal{S}_*(V^-)$$

constructed in the previous subsection, we obtain a chain equivalence

$$\widehat{\nu} : \widehat{\mathcal{F}}_* = \mathcal{F}_* \underset{P}{\otimes} \widehat{P} \xrightarrow{\sim} \mathcal{S}_*^\lambda \underset{P}{\otimes} \widehat{P}.$$

Composing the inverse of this chain equivalence with the canonical chain equivalence of Proposition 6.3 of Chapter 11 (page 358) we obtain a chain equivalence

$$\mathcal{N}_*^\lambda(f,v) \to \widehat{\mathcal{F}}_*.$$

The aim of this subsection is to give an explicit formula for this chain equivalence. Here is an outline of the construction. A homomorphism $\mathcal{N}_k^\lambda \to \widehat{\mathcal{F}}_k$ is determined by its values on the generators of $\mathcal{N}_k^\lambda$, that is, on the critical points of $F|V^-$ of index k. Let p be such a point. Since v is cellular, the descending disc $D(p,\bar{v})$ is in $V^-_{\langle k\rangle}$. The set

$$\Delta_p = D(p,\bar{v}) \setminus \operatorname{Int} V^-_{\langle k-1\rangle}$$

is not compact in general, and the manifold $D(p,\bar{v})$ does not have a fundamental class in $H_k\big(V^-_{\langle k\rangle}, V^-_{\langle k-1\rangle}\big)$. However the intersection of the set Δ_p with each cobordism $W_n = F^{-1}\big([\lambda - n, \lambda]\big)$ is compact, therefore Δ_p is a *downward infinite* disjoint union of compact sets. This allows us to define the fundamental class of $D(p,\bar{v})$ as an element of the completion of the module $\mathcal{F}_k = H_k(V^-_{\langle k\rangle}, V^-_{\langle k-1\rangle})$. The homomorphism $\mathcal{N}_k \to \widehat{\mathcal{F}}_k$ which we are about to construct sends the point p to this fundamental class.

To write down an explicit formula, let p be a critical point of F of index k with $F(p) \leqslant \lambda$. Recall from the previous subsection the homomorphism

$$\varpi_k : \Pi_k = \mathcal{M}_k \underset{\mathbf{Z}}{\otimes} P \to \mathcal{F}_k$$

and define an element $\xi(p) \in \widehat{\mathcal{F}}_k$ by the following formula:

$$\xi(p) = \varpi(p) - \sum_{r=0}^{\infty} t^{r+1}\, \mathscr{T}\Big(\overline{\mathcal{H}}^r\big(\bar{\sigma}(p)\big)\Big). \tag{3}$$

To simplify the notation we omit in this formula the indices showing the dimension of the corresponding objects; also we abbreviate $\overline{\mathcal{H}}_{k-1}(-v)$ to $\overline{\mathcal{H}}$. Thus $\bar{\sigma}(p)$ is here an element of

$$\mathcal{R}_{k-1} = \mathcal{M}_{k-1}^{(1)} \underset{\mathbf{Z}}{\otimes} P$$

and $\overline{\mathcal{H}}$ is an endomorphism of this module, while $\mathscr{T}$ is a homomorphism $\mathcal{R}_{k-1} \to \mathcal{F}_k$, see the preceding subsection for the definitions. To relate this formula to the discussion above, observe that if in the right-hand side of (3) we omit the terms with $r \geqslant n$, we obtain exactly the fundamental class of $D(p, \bar{v}) \cap W_{n+1}$ modulo $V^-_{\langle k-1\rangle}$.

Extend the map $p \mapsto \xi(p)$ to a homomorphism $\xi : \mathcal{N}^\lambda_* \to \widehat{\mathcal{F}}_*$ of graded $\widehat{P}$-modules.

Theorem 2.5. *The graded homomorphism ξ is a chain homotopy equivalence and the next diagram is homotopy commutative:*

$$\begin{array}{ccc} \mathcal{N}^\lambda_* & \xrightarrow{\ \xi\ } & \widehat{\mathcal{F}}_* \\ & {\scriptstyle\Psi}\searrow \quad \swarrow{\scriptstyle\widehat{\nu}} & \\ & \mathcal{S}^\lambda_* \underset{P}{\otimes} \widehat{P} & \end{array}$$

(where Ψ is the homotopy equivalence from Proposition 6.3 of Chapter 11, page 358).

Proof. Both $\mathcal{N}^\lambda_*$ and $\widehat{\mathcal{F}}_*$ are free finitely generated complexes, and $\mathcal{S}^\lambda_* \underset{P}{\otimes} \widehat{P}$ is homotopically free-finite. Therefore to prove our theorem it suffices to establish the following proposition.

Proposition 2.6. *Let n be a positive integer. In the t^n-quotient of the above diagram*

$$\begin{array}{ccc} \mathcal{N}^\lambda_*/t^n\mathcal{N}^\lambda_* & \xrightarrow{\ \xi_n\ } & \mathcal{F}_*/t^n\mathcal{F}_* \\ & {\scriptstyle\Psi_n}\searrow \quad \swarrow{\scriptstyle\nu_n} & \\ & \mathcal{S}^\lambda_*/t^n\mathcal{S}^\lambda_* & \end{array} \tag{4}$$

the graded map ξ_n is a chain homomorphism, and $\nu_n \circ \xi_n \sim \Psi_n$.

Proof. The main idea of the proof is that both Ψ_n and ν_n are canonical chain equivalences associated with certain cellular filtrations of the chain complex $\mathcal{S}^\lambda_*/t^n\mathcal{S}^\lambda_*$ and can be compared using the techniques of Section 3.1 of Chapter 6 (page 212).

Recall from Section 4 of Chapter 11 (page 348) the cellular filtration of the chain complex

$$\mathcal{S}^\lambda_*/t^n\mathcal{S}^\lambda_* = \mathcal{S}_*(V^-)/\mathcal{S}_*(t^nV^-)$$

associated to a t-ordered Morse function on $W_n = F^{-1}\big([\lambda - n, \lambda]\big)$. The map Ψ_n is the canonical chain equivalence associated with this filtration.

This filtration is not quite convenient for our present purposes and we shall modify it. Let $\psi : W_{n+1} \to [a,b]$ be a t-ordered Morse function on the cobordism W_{n+1}. Let $\{X_k\}$ be the Morse-Smale filtration of W_{n+1} associated with ψ; put $Y_k = X_k \cup t^n V^-$. The pairs $(Y_k, t^n V^-)$ form a t-invariant filtration of $(V^-, t^n V^-)$, and we obtain a filtration

$$\mathcal{A}_*^{(k)} = \mathcal{S}_*(Y_k, t^n V^-) \subset \mathcal{S}_*(V^-, t^n V^-) = \mathcal{S}_*^\lambda / t^n \mathcal{S}_*^\lambda$$

by P-modules. The homology $H_k(Y_k, Y_{k-1})$ has the natural basis over $P_n = \mathbf{Z}[t]/t^n$ formed by the fundamental classes of the descending discs $D(p, \bar{v})$ (where p ranges over the set $S_k(F) \cap W$). The adjoint complex is obviously identified with the truncated Novikov complex $\mathcal{N}_*^\lambda / t^n \mathcal{N}_*^\lambda$ and the canonical chain equivalence

$$\mathcal{N}_*^\lambda / t^n \mathcal{N}_*^\lambda \to \mathcal{S}_*(V^-, t^n V^-)$$

is homotopic to Ψ_n. (These assertions are easily checked by comparison of the filtration above with the filtration induced by any t-ordered function on the cobordism W_n. See a similar argument in the proof of Proposition 4.6 of Chapter 11, page 353.)

Define another filtration of $\mathcal{S}_*(V^-, t^n V^-)$ as follows:

$$\mathcal{B}_*^{(k)} = \mathcal{S}_*(L_k, t^n V^-), \qquad \text{where} \quad L_k = V^-_{\langle k \rangle} \cup t^n V^-.$$

Observe that we have the inclusion of pairs

$$\left(V^-_{\langle k \rangle}, V^-_{\langle k-1 \rangle}\right) \xrightarrow{\ j\ } (L_k, L_{k-1})$$

which induces a homomorphism of P_n-modules

$$j_* : H_*(V^-_{\langle k \rangle}, V^-_{\langle k-1 \rangle}) \underset{P}{\otimes} P_n \to H_*(L_k, L_{k-1}).$$

Lemma 2.7. *The filtration $\mathcal{B}_*^{(k)}$ is cellular, j_* is an isomorphism and the next diagram is homotopy commutative:*

$$(5) \qquad \begin{array}{ccc} \mathcal{B}_*^{gr} & \xleftarrow{\quad j_* \quad} & \mathcal{F}_*/t^n\mathcal{F}_* \\ & {\scriptstyle \mu_n} \searrow \quad \swarrow {\scriptstyle \nu_n} & \\ & \mathcal{S}_*^\lambda / t^n \mathcal{S}_*^\lambda & \end{array}$$

where $\mathcal{B}_*^{gr}$ stands for the adjoint complex of the filtration $\mathcal{B}_*^{(k)}$ and μ_n is the canonical chain equivalence.

Proof. We have inclusions of pairs

$$\left(V^-_{\langle k \rangle}, V^-_{\langle k-1 \rangle}\right) \hookrightarrow \left(V^-_{\langle k \rangle}, t^n V^-_{\langle k \rangle} \cup V^-_{\langle k-1 \rangle}\right) \hookrightarrow (L_k, L_{k-1}).$$

The second inclusion induces an isomorphism in homology (by excision). It is not difficult to deduce from the exact sequence of the triple

$$\Big(V^-_{\langle k\rangle},\ t^nV^-_{\langle k\rangle}\ \cup\ V^-_{\langle k-1\rangle},\ V^-_{\langle k-1\rangle}\Big)$$

that the first inclusion induces an isomorphism

$$H_*\big(V^-_{\langle k\rangle}, V^-_{\langle k-1\rangle}\big) \underset{P}{\otimes} P_n \xrightarrow{\ \approx\ } H_*\Big(V^-_{\langle k\rangle},\ t^nV^-_{\langle k\rangle}\ \cup\ V^-_{\langle k-1\rangle}\Big).$$

The other assertions of the lemma are obvious. □

Now let us compare the filtrations $\mathcal{B}^{(k)}_*$ and $\mathcal{A}^{(k)}_*$. Since v is cellular, we have

$$\partial_0 W_{n+1} \cup D(\mathrm{ind}{\leqslant}k\ ; \bar v | W_{n+1}) \ \subset\ \mathrm{Int}\ L_k.$$

By Proposition 4.2 of Chapter 11 (page 349) we can choose a t-ordered function ψ on W_{n+1} in such a way that

$$X_k \subset \mathrm{Int}\ L_k \quad \text{for every} \quad k.$$

Then we have $Y_k \subset L_k$, hence

$$\mathcal{A}^{(k)}_* \subset \mathcal{B}^{(k)}_* \quad \text{for every} \quad k.$$

Thus the identity map of the pair (V^-, t^nV^-) respects the filtrations if the source is endowed with the filtration $\mathcal{A}^{(k)}_*$ and the target with $\mathcal{B}^{(k)}_*$. Therefore the identity map induces a chain homomorphism $\widetilde{\xi}_n$ such that the next diagram is homotopy commutative:

$$\begin{array}{ccccc}
\mathcal{N}^\lambda_*/t^n\mathcal{N}^\lambda_* & \xrightarrow{\ \widetilde{\xi}_n\ } & \mathcal{B}^{gr}_* & \xleftarrow{\ j_*\ } & \mathcal{F}_*/t^n\mathcal{F}_* \\
 & {\scriptstyle\Psi_n}\searrow & \downarrow{\scriptstyle\mu_n} & \swarrow{\scriptstyle\nu_n} & \\
 & & \mathcal{S}^\lambda_*/t^n\mathcal{S}^\lambda_* & &
\end{array}$$

The proof of our proposition will be complete if we show that $j_*^{-1}\circ\widetilde{\xi}_n \sim \xi_n$ (recall that ξ_n is the reduction modulo t^n of the graded homomorphism ξ defined in (3), page 392). For this it suffices to prove that for every critical point $p \in W$ of F of index k the fundamental class

$$[D(p,\bar v)] \in H_k(L_k, L_{k-1}) \approx \mathcal{F}_k/t^n\mathcal{F}_k$$

is equal to the n-truncation of the power series in the right-hand side of (3). Let us have a closer look at the manifold $D(p,\bar v)\cap W_n$. Let $\phi : W \to [a,b]$ be the ordered Morse function on W used for construction of the filtrations $W^{\langle k\rangle}$ and $V^-_{\langle k-1\rangle}$. Let $a = a_0 < a_1 < \cdots < a_{m+1} = b$ be the ordering sequence for ϕ (where $m = \dim M$). Let

$$V' = \phi^{-1}(a_k).$$

The condition ($\mathfrak{C}$) implies that the sole of $D(p,\bar{v}) \cap W$ is in $\partial_0 W^{k-1}$, therefore the part of $D(p,\bar{v})$ between V' and tV is in $V^{-}_{\langle k-1\rangle}$. Let

$$W' = \phi^{-1}\big([a,a_k]\big), \quad W'' = \phi^{-1}\big([a_k,b]\big).$$

The condition ($\mathfrak{C}$) applied to the cobordism tW implies that the part of $D(p,\bar{v})$ between tV and tV' is in

$$tT_k = tT\big(\partial_1 W^{k-1}; -\bar{v}|W''\big)$$

and the part of $D(p,\bar{v})$ between tV' and t^2V is in $V^{-}_{\langle k-1\rangle}$. In Figure 44 the set $V^{-}_{\langle k-1\rangle}$ is shaded dark grey, and the sets T_k, tT_k are shaded light grey. The thick curve depicts the descending disc $D(p,\bar{v})$ and its dotted part depicts the part of $D(p,\bar{v})$ which is in the union of the t-shifts of W'.

It is not difficult to show by induction that for every $j \geqslant 1$ we have

$$D(p,\bar{v}) \cap t^jW'' \subset t^jT_k, \quad D(p,\bar{v}) \cap t^jW' \subset t^jV^{-}_{\langle k-1\rangle}.$$

Put

$$\Sigma_j = D(p,\bar{v}) \cap t^jV, \quad D_j = D(p,\bar{v}) \cap t^jW''.$$

Then Σ_j is a submanifold of t^jV. The manifold $\partial D_0 = D_0 \cap V'$ is diffeomorphic to a sphere and for $j \geqslant 1$ we have

$$D_j = T(\Sigma_j, -\bar{v}|t^jW'').$$

Every $(-\bar{v})$-trajectory starting at a point of Σ_j reaches the manifold t^jV', and therefore D_j is a manifold with boundary diffeomorphic to the product $\Sigma_j \times [a_k,b]$ and

$$\partial D_j = D_j \cap t^jV' \bigsqcup D_j \cap t^jV.$$

Applying Lemma 4.7 of Chapter 5 (page 191) we deduce that the fundamental class

$$[D(p,\bar{v})] \in H_k\Big(V^{-}_{\langle k\rangle}, V^{-}_{\langle k-1\rangle} \cup t^nV^{-}_{\langle k\rangle}\Big)$$

is equal to the sum of the fundamental classes of the manifolds D_j for $0 \leqslant j \leqslant n-1$. By definition

$$[D_0] = [\varpi(p)].$$

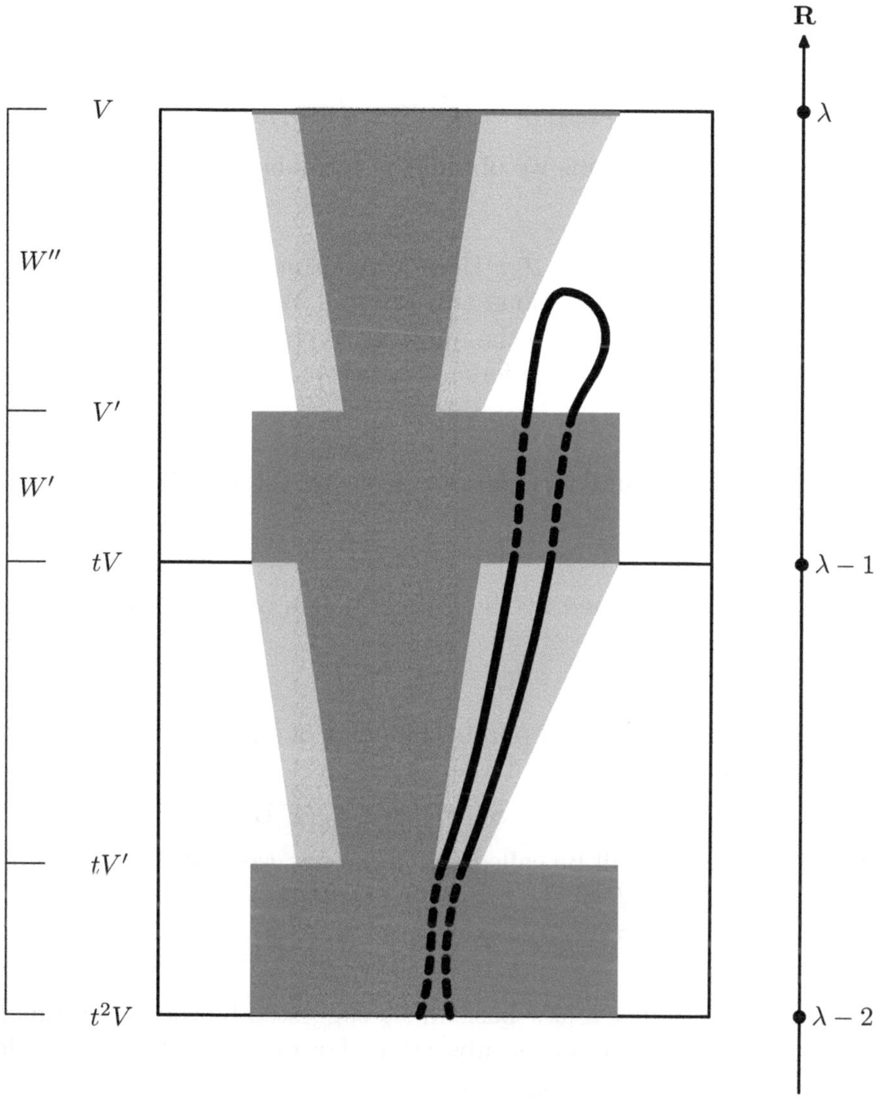

FIGURE 44.

As for $j \geqslant 1$, the fundamental class of D_j in the pair $(t^jW^{\langle k\rangle}, t^jW^{\langle k-1\rangle})$ equals $-\mathscr{T}_k([\Sigma_j])$, where

$$[\Sigma_j] \in H_{k-1}\Big(t^j\partial_1 W^{k-1}, t^j\partial_1 W^{k-2}\Big).$$

By one of the basic properties of homological gradient descent (see Proposition 1.4 of Chapter 9, page 283) we have:

$$[\Sigma_j] = t^j\, \overline{\mathcal{H}}^{j-1}(-v)\big(\bar\sigma(p)\big),$$

and adding up these terms we obtain the formula $\xi_n(p) = \widetilde{\xi}_n(p)$ which we had to prove. □

2.3. Change of basis in $\widehat{\mathcal{F}}_*$. Here we continue our study of the chain map ξ, defined in the previous subsection. We show that it is a split monomorphism, and describe the quotient $\widehat{\mathcal{F}}_*/\operatorname{Im}\xi$. By a certain abuse of notation we shall make no difference between the chain complex $\mathcal{F}_*$ and the chain complex

$$\mathcal{R}_* \oplus \Pi_* \oplus \mathcal{R}_{*-1}$$

endowed with the boundary operator given by the matrix (2) (page 391). The P-module

$$\Pi_k = \mathcal{M}_k \underset{\mathbf{Z}}{\otimes} P$$

has a natural basis formed by the critical points of index k of $F|W$. Similarly, the module

$$\mathcal{R}_k = \mathcal{M}_k^{(1)} \underset{\mathbf{Z}}{\otimes} P$$

has a basis formed by critical points of index k of an ordered Morse function $\phi_1 : \partial_1 W \to \mathbf{R}$. The resulting basis

$$\mathcal{B} = \{q, p, r \mid \text{ where } q \in S_k(\phi_1),\ p \in S_k(F|W),\ r \in S_{k-1}(\phi_1)\}$$

of the P-module $\mathcal{F}_k$ will be called the *geometric basis of* $\mathcal{F}_k$. The family $\mathcal{B}$ is also the basis over $\widehat{P}$ of the completed module

$$\widehat{\mathcal{F}}_k = \mathcal{F}_k \underset{P}{\otimes} \widehat{P} = \widehat{\mathcal{R}}_k \oplus \widehat{\Pi}_k \oplus \widehat{\mathcal{R}}_{k-1}.$$

Now we shall introduce a new basis in $\widehat{\mathcal{F}}_k$ suggested by the definition of the chain map ξ from the previous subsection. For every $p \in S_k(F) \cap W$ define an element $\widehat{p} \in \widehat{\mathcal{F}}_k$ by the following formula:

$$\widehat{p} = p - \sum_{r=0}^{\infty} t^{r+1}\, \mathcal{T}\Big(\overline{\mathcal{H}}^r\big(\bar\sigma(p)\big)\Big).$$

The difference between $\widehat{p}$ and p is in the submodule $\widehat{\mathcal{R}}_{k-1}$, and we obtain the next lemma.

Lemma 2.8. *The family*

$$\mathcal{B}' = \{q, \widehat{p}, r \mid \textit{ where } q \in S_k(\phi_1),\ p \in S_k(F|W),\ r \in S_{k-1}(\phi_1)\}$$

is a $\widehat{P}$-basis for the module $\widehat{\mathcal{F}}_k$. □

By the definition of the chain map ξ we have

$$\xi(p) = \widehat{p}, \qquad \text{for every} \qquad p \in S_k(F),$$

so that ξ sends the elements of the geometric basis of $\mathcal{N}_*^\lambda$ to the elements of some basis of the module $\widehat{\mathcal{F}}_*$, and therefore ξ is a split monomorphism. For the quotient complex $\mathcal{G}_*$ we have an isomorphism

$$\mathcal{G}_k \approx \widehat{\mathcal{R}}_k \oplus \widehat{\mathcal{R}}_{k-1}. \tag{6}$$

The projections to $\mathcal{G}_k$ of the elements q, r (where $q \in S_k(\phi_1)$ and $r \in S_{k-1}(\phi_1)$) form a $\widehat{P}$-basis of $\mathcal{G}_k$ which will be also called geometric.

Recall that two bases of a free module are called *elementary equivalent* if the transition matrix from one to another is elementary (i.e. it differs from the identity matrix in at most one term, which is non-diagonal). Two bases β, γ are called *equivalent* if there is a finite sequence of bases $\beta = \beta_0, \beta_1, \ldots, \beta_N = \gamma$ such that for every i the bases β_i and β_{i+1} are elementary equivalent. It is clear that the bases $\mathcal{B}$ and $\mathcal{B}'$ are equivalent.

Proposition 2.9. *We have a short exact sequence of free chain complexes over $\widehat{P}$:*

$$0 \longrightarrow \mathcal{N}_*^\lambda \xrightarrow{\ \xi\ } \widehat{\mathcal{F}}_* \longrightarrow \mathcal{G}_* \longrightarrow 0$$

where the matrix of the boundary operator $\partial_{k+1} : \mathcal{G}_{k+1} \to \mathcal{G}_k$ with respect to the direct sum decomposition (6) equals

$$\begin{pmatrix} \partial_{k+1}^{(1)} & \mathrm{Id} - t\overline{\mathcal{H}}_k \\ 0 & * \end{pmatrix}.$$

The basis in $\widehat{\mathcal{F}}_*$ induced by the geometric bases in $\mathcal{G}_*$ and $\mathcal{N}_*^\lambda$ is equivalent to the geometric basis of $\widehat{\mathcal{F}}_*$.† □

3. Whitehead groups and Whitehead torsion

In this section we give a brief survey of the algebraic K_1-functor, and its geometric applications.

3.1. Whitehead groups. Let R be a ring with a unit. As usual we denote by $\mathrm{GL}(n, R)$ the group of all invertible $n \times n$-matrices over R. Consider the group $\mathrm{GL}(R) = \varinjlim \mathrm{GL}(n, R)$ and set

$$K_1(R) = \mathrm{GL}(R)/[\mathrm{GL}(R), \mathrm{GL}(R)].$$

There is another description of this group in terms of elementary matrices. A theorem of J. H. C. Whitehead (see for example [**93**], §3) says that the

† The terms denoted $*$ are not important for our present purposes.

commutator group $[\mathrm{GL}(R), \mathrm{GL}(R)]$ is equal to the normal subgroup generated by the elementary matrices (recall that a matrix is called *elementary* if it differs from the identity matrix in only one non-diagonal term). If the ring A is commutative, then the determinant homomorphism

$$A \mapsto \det(A), \quad \mathrm{GL}(R) \to R^\bullet$$

(where $R^\bullet$ denotes the multiplicative group of all invertible elements of the ring R) factors through the natural projection

$$A \mapsto [A], \quad \mathrm{GL}(R) \to K_1(R),$$

so that we have a commutative diagram

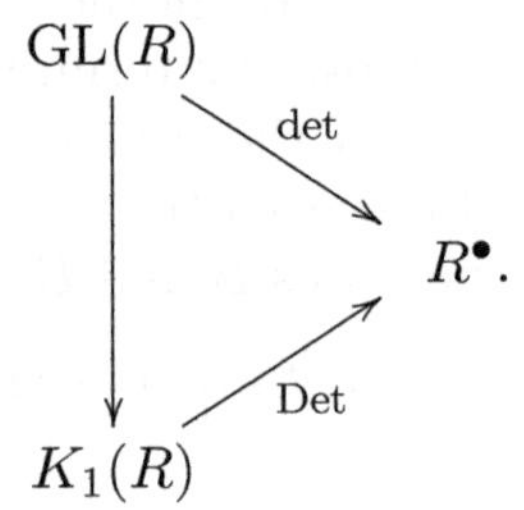

The next proposition is classical, the proof can be found for example in the book of J. Rosenberg [**139**].

Proposition 3.1. *If R is a commutative Euclidean ring, then the homomorphism* $\mathrm{Det} : K_1(R) \to R^\bullet$ *is an isomorphism.*

In particular, $K_1(\mathbf{Z}) \approx \mathbf{Z}/2\mathbf{Z}$; the only non-trivial element of this group is the class $[-1]$ of the 1×1 matrix (-1).

Theorem 3.2. *The homomorphism*

$$\mathrm{Det} : K_1(R) \to R^\bullet$$

is an isomorphism if R is one of the following rings:

$$P = \mathbf{Z}[t], \quad L = \mathbf{Z}[t, t^{-1}], \quad P_m = \mathbf{Z}[t]/t^m\mathbf{Z}[t], \quad \widehat{P} = \mathbf{Z}[[t]], \quad \overline{L} = \mathbf{Z}((t)).$$

Proof. The first two cases are classical: see the work of Bass-Heller-Swan [**12**] for the case of the ring $\mathbf{Z}[t]$, and the work of Higman [**59**] for the case of $\mathbf{Z}[t, t^{-1}]$. The K_1-groups here are particularly simple:

$$K_1\big(\mathbf{Z}[t]\big) \approx \mathbf{Z}/2\mathbf{Z}; \quad K_1\big(\mathbf{Z}[t, t^{-1}]\big) \approx \mathbf{Z} \oplus \mathbf{Z}/2\mathbf{Z}.$$

The case of $\mathbf{Z}((t))$ is easy, since the Novikov ring is Euclidean by Theorem 2.4 of Chapter 10 (page 333). Let us consider the ring $\widehat{P}$ (the case of P_m is similar). It suffices to show that the determinant homomorphism is injective. Let B be an invertible square matrix with entries in $\mathbf{Z}[[t]]$, such that $\det B = 1$. Let $\overline{B}$ be the image of B in $\mathrm{GL}(\mathbf{Z})$ via the natural projection $\mathbf{Z}[[t]] \to \mathbf{Z}$. The class of $\overline{B}$ in $K_1(\mathbf{Z})$ equals zero, therefore there is a matrix $C \in [\mathrm{GL}(\mathbf{Z}), \mathrm{GL}(\mathbf{Z})]$ such that $\overline{B} \cdot C = \mathrm{Id}$, hence the matrix $B \cdot C \in \mathrm{GL}(\mathbf{Z}[[t]])$ equals $\mathrm{Id} + tD$, where D is a matrix with entries in $\mathbf{Z}[[t]]$. The diagonal entries of BC are therefore of the form $1 + t\lambda$ with $\lambda \in \mathbf{Z}[[t]]$. Such elements are invertible in $\mathbf{Z}[[t]]$, and the matrix is easily reduced to a diagonal one by elementary transformations, so that there is a finite sequence $A_1, \ldots, A_k$ of elementary matrices such that BC, multiplied by the product $A_1 \cdot \ldots \cdot A_k$ is a diagonal matrix Δ. The product of the diagonal entries of Δ equals 1, since $\det B = 1$. Thus the class of Δ in $K_1(\mathbf{Z}[[t]])$ vanishes, and it is now easy to deduce that the class of B is also equal to zero. □

Let us describe the group of units $\widehat{P}^\bullet$. We have a subgroup

$$\mathcal{W} = \Big\{1 + \sum_{k \geqslant 1} a_k t^k \mid a_k \in \mathbf{Z}\Big\} \subset \widehat{P}^\bullet,$$

the elements of this subgroup are called *Witt vectors*. The whole group $\widehat{P}^\bullet$ is isomorphic to the direct sum of $\mathcal{W}$ and the group $\{1, -1\}$ of order 2.

As for the rings L and $\overline{L}$, their groups of units contain a subgroup

$$\mathcal{T} = \big\{(-1)^m t^k \mid k \in \mathbf{Z},\ m = 0, 1\big\} \approx (\mathbf{Z}/2\mathbf{Z}) \oplus \mathbf{Z},$$

and it is easy to prove that

$$L^\bullet = \mathcal{T}, \qquad \overline{L}^\bullet = \mathcal{W} \oplus \mathcal{T}.$$

Definition 3.3. The group $K_1(\overline{L})/\mathcal{T}$ is called the *Whitehead group of the Novikov ring* $\overline{L}$ and denoted $\mathrm{Wh}(\overline{L})$.

It is clear that the determinant homomorphism $\mathrm{Det} : K_1(\overline{L}) \to \overline{L}^\bullet$ yields an isomorphism

$$\overline{\mathrm{Det}} : \mathrm{Wh}(\overline{L}) \to \mathcal{W},$$

such that the following diagram is commutative:

$$\begin{array}{ccc} \overline{K}_1(\overline{L}) & \xrightarrow[\approx]{\mathrm{Det}} & \overline{L}^\bullet \\ \big\downarrow & & \big\downarrow \\ \mathrm{Wh}(\overline{L}) & \xrightarrow[\approx]{\overline{\mathrm{Det}}} & \mathcal{W} \end{array}$$

where the vertical arrows are natural projections and the horizontal arrows are isomorphisms.

3.2. Whitehead torsion. Applications of the functor K_1 in algebraic topology are mostly related to secondary invariants (torsions) of acyclic chain complexes. The first invariant of this kind was defined by Reidemeister ([**136**]). We shall work mainly with the *Whitehead torsion* introduced by J. H. C. Whitehead in [**162**].

Definition 3.4. Let R be a ring with a unit. The *reduced K_1-group* of R is by definition the quotient of $K_1(R)$ by its subgroup $\{0, [-1]\}$; it is denoted $\overline{K}_1(R)$.

A *free-finite based module* over R (*fb-module* for short) is a left free R-module endowed with a finite basis over R.

A *free-finite based complex* over R (*fb-complex* for short) is a finite chain complex of fb-modules.

An isomorphism $\phi : C_* \to D_*$ of two free-finite based complexes over R is called *simple* if for every k the image of the matrix of $\phi_k : C_k \to D_k$ in $\overline{K}_1(R)$ equals 0.

A classical construction due to K. Reidemeister and J. H. C. Whitehead associates to any acyclic fb-complex over R its *torsion* $\tau(C_*) \in \overline{K}_1(R)$. Without reproducing the definition (the reader can find it for example in the works [**23**], [**162**]) let us indicate three basic properties of the torsion.

(A1) If there is a simple isomorphism between C_*, D_*, then $\tau(C_*) = \tau(D_*)$.

(A2) $$\tau\Big(0 \longleftarrow C_i \xleftarrow{\ \alpha\ } C_{i+1} \longleftarrow 0\Big) = (-1)^i[\alpha].$$

(In this formula α is an isomorphism of two fb-modules, and $[\alpha]$ is the image in $\overline{K}_1(R)$ of the matrix of α with respect to the given bases.)

(A3) $$\tau(C_* \oplus D_*) = \tau(C_*) + \tau(D_*).$$

These three properties determine a unique function

$$C_* \mapsto \tau(C_*) \in \overline{K}_1(R)$$

defined on the category of acyclic fb-complexes over R (for the proof we refer to the book of M. Cohen [**23**]).

Here is one more useful property of the torsion invariant:

(A4) Let

$$0 \longrightarrow C_* \longrightarrow D_* \longrightarrow E_* \longrightarrow 0$$

be an exact sequence of acyclic fb-complexes. Let c_i, d_i, e_i denote the bases in the modules C_i, D_i, E_i. Then

(7) $$\tau(D_*) = \tau(C_*) + \tau_*(E_*) + \sum_i (-1)^i [c_i e_i / d_i]$$

where c_ie_i is the base in D_i obtained as the direct sum of the image of c_i and the pullback of e_i, and c_ie_i/d_i is the transition matrix. (We refer to **[23]**, 17.2, for the proof of (A4); it is based on the fact that each short exact sequence of acyclic fb-complexes splits.)

Remark 3.5. In the definition of the matrix $M(\phi)$ associated to a linear map $\phi : L \to K$ of two based vector spaces, one has two possibilities: either the coordinates of the ϕ-images of the basis elements of L constitute the rows of the matrix, or the columns. One has a similar choice for the homomorphisms of free based R-modules. The first definition leads to the formula $M(\phi \circ \psi) = M(\psi)M(\phi)$, and the second implies $M(\phi \circ \psi) = M(\phi)M(\psi)$. The first convention is used in the seminal paper by J. H. C. Whitehead **[162]** (where the Whitehead torsion was introduced) and in many other sources, for example in the paper **[95]** and in M.Cohen's book **[23]**. These works are the basic references about Whitehead torsion. We use the second convention which is more common in linear algebra. Observe that in our setting the two choices lead essentially to the same formulas, since for the rings R which we consider (see Theorem 3.2) any element of $\overline{K}_1(R)$ can be represented by a 1×1 matrix, and therefore the class in $\overline{K}_1(R)$ of any invertible matrix equals the class of its conjugate matrix.

Proceeding to the definition of the Whitehead torsion of a chain equivalence of two fb-complexes, let us recall the definition of the cone of a chain map.

Definition 3.6. Let $\phi : C_* \to D_*$ be a chain map. The *chain cone* $Cone_*(\phi)$ of ϕ is a chain complex defined as follows:

(1) $Cone_k(\phi) = D_k \oplus C_{k-1}$.
(2) The boundary operator is given by the following formula:

$$\partial_k : Cone_k(\phi) \to Cone_{k-1}(\phi);$$

$$\partial_k(x_k, y_{k-1}) = \Big(d_k(x_k) + \phi_{k-1}(y_{k-1}),\ -d'_{k-1}(y_{k-1})\Big),$$

where $x_k \in D_k, y_{k-1} \in C_{k-1}$ and d_k, d'_k are the boundary operators in C_*, respectively D_*.

If ϕ is a homotopy equivalence, then $Cone_*(\phi)$ is acyclic.

Definition 3.7. Let $\phi : C_* \to D_*$ be a homotopy equivalence of two free-finite based complexes over R. The *Whitehead torsion* $\tau(\phi) \in \overline{K}_1(R)$ is by definition the torsion of the chain cone $Cone_*(\phi)$. If $\tau(\phi) = 0$ we say that ϕ is a *simple homotopy equivalence.*

Now we will describe the properties of the Whitehead torsion, outlining some of the proofs since this subject is less developed in the classical sources.

(B1) $\tau(\phi) = \tau(\psi)$ if $\phi, \psi : C_* \to D_*$ are chain homotopic.

Proof: A chain homotopy between ϕ and ψ yields a simple isomorphism between $Cone_*(\phi)$ and $Cone_*(\psi)$. □

(B2) The Whitehead torsion of the identity map of any fb-complex is equal to 0.
Proof: It is easy to show that for any fb-complex C_* the cone of the identity map is simply isomorphic to the direct sum of the chain complexes $0 \leftarrow C_i \xleftarrow{\text{Id}} C_i \leftarrow 0$, and such complexes have zero torsion. □

(B3) If $\phi : C_* \to D_*$ is an isomorphism, then

$$\tau(\phi) = \sum_i (-1)^i [\phi_i]$$

where $[\phi_i]$ is the class in $K_1(R)$ of the matrix of the isomorphism $\phi_i : C_i \to D_i$.
Proof: The assertion is easily reduced to the case when the map ϕ is basis-preserving, and then proved by applying (B2). □

(B4) For a short exact sequence

$$0 \longrightarrow C_* \xrightarrow{\phi} D_* \longrightarrow F_* \longrightarrow 0$$

where C_*, D_*, F_* are fb-complexes and ϕ is a chain equivalence, we have

$$\tau(\phi) = \tau(F_*),$$

if the basis in D_* is formed as the direct sum of the ϕ-image of the basis in C_* and the pullback of the basis in F_*.
Proof: Since F_* is acyclic, the exact sequence splits as follows:

$$0 \longrightarrow C_* \xrightarrow{\phi \oplus 0} D_* \oplus F_* \xrightarrow{(0,\text{Id})} F_* \longrightarrow 0$$

where ϕ preserves the bases. It remains to observe that the torsion of the chain equivalence

$$C_* \xrightarrow{\phi \oplus 0} D_* \oplus F_*$$

equals the torsion of F_*. □

We finish this subsection with two lemmas to be used later. Let C_*, D_* be two fb-complexes, denote by

$$d_k : C_k \to C_{k-1}, \quad d'_k : D_k \to D_{k-1}$$

the boundary operators. Let $A_* : C_* \to D_*$ be a graded homomorphism. Put

$$E_k = D_k \oplus C_{k-1}$$

and define a homomorphism $\partial_k : E_k \to E_{k-1}$ by its matrix with respect to this direct sum decomposition:

$$\partial_k : E_k \to E_{k-1}; \quad M(\partial_k) = \begin{pmatrix} d'_k & A_{k-1} \\ 0 & -d_{k-1} \end{pmatrix}.$$

Lemma 3.8. *The graded map ∂_* is a boundary operator in E_* if and only if A_* is a chain homomorphism.*

Proof. An obvious matrix computation. □

Lemma 3.9. *Assume that ∂_* is a boundary operator, and that A_* is an isomorphism. Then the fb-complex E_* is acyclic, and*

$$\tau(E_*) = \sum_i (-1)^i [A_i].$$

Proof. The chain complex E_* is the cone of the chain isomorphism A_*. Our formula follows now from (B3). □

3.3. Whitehead torsion of Morse equivalences. Let G be a group. The *Whitehead group of* G is by definition the quotient of $K_1(\mathbf{Z}G)$ by the subgroup of the images of all the elements of the form $\pm g$ with $g \in G$. The functor $G \mapsto \mathrm{Wh}(G)$ is very important for the geometric topology, and there is an enormous literature dedicated to this functor and its ramifications. In this subsection we outline the relation between this functor and the Morse theory. Let M be a closed connected manifold; put $G = \pi_1(M)$, and let $\widetilde{M} \to M$ denote the universal covering of M. Choose any C^1 triangulation of M, lift it to $\widetilde{M}$ and obtain a G-invariant C^1 triangulation of $\widetilde{M}$. The simplicial chain complex $\Delta_*(\widetilde{M})$ is a free-finite complex of $\mathbf{Z}G$-modules; any lift to $\widetilde{M}$ of the family of all simplices of M yields a finite basis of $\Delta_*(\widetilde{M})$ over $\mathbf{Z}G$. We have a natural chain equivalence

$$\Delta_*(\widetilde{M}) \xrightarrow{\mathcal{L}} \mathcal{S}_*(\widetilde{M})$$

of $\mathbf{Z}G$-complexes. Another choice of C^1 triangulation of M leads to another simplicial chain complex $\Delta'_*(\widetilde{M})$, and we have two natural chain equivalences

$$\Delta'_*(\widetilde{M}) \xrightarrow{\mathcal{L}'} \mathcal{S}_*(\widetilde{M}) \xleftarrow{\mathcal{L}} \Delta_*(\widetilde{M}).$$

The composition

$$\Delta'_*(\widetilde{M}) \xrightarrow{\mathcal{L}^{-1} \circ \mathcal{L}'} \Delta_*(\widetilde{M})$$

is therefore a chain equivalence of free-based chain complexes over $\mathbf{Z}G$. Its torsion

$$\tau(\mathcal{L}^{-1}\circ\mathcal{L}')\in\overline{K}_1(\mathbf{Z}G),$$

depends on the choice of lifts of simplices of the two triangulations to $\widetilde{M}$. It is not difficult to show that the image

$$\bar{\tau}(\mathcal{L}^{-1}\circ\mathcal{L}')$$

of the element $\tau(\mathcal{L}^{-1}\circ\mathcal{L}')$ in the group $\mathrm{Wh}(G)$ does not depend on the choice of the lifts. It turns out that

$$\bar{\tau}(\mathcal{L}^{-1}\circ\mathcal{L}')=0.$$

The proof of this important result relies on the fact that for every two C^1 triangulations of M there are subdivisions of these triangulations which yield isomorphic simplicial chain complexes (see for example the book of J. Munkres [**100**], Theorem 10.5). A much more difficult and deep theorem due to T. A. Chapman [**22**] says that for any two CW decompositions of a closed manifold M there is a chain homotopy equivalence of the corresponding cellular chain complexes

$$\mathcal{C}_*(\widetilde{M})\xrightarrow{\ \phi\ }\mathcal{C}'_*(\widetilde{M})$$

such that $\bar{\tau}(\phi)=0$.

Let us now return to Morse equivalences. Let M be a closed connected manifold, $f:M\to\mathbf{R}$ a Morse function, v an oriented transverse f-gradient. Let $\widetilde{\mathcal{M}}_*(f,v)$ denote the universal Morse complex (see Subsection 4.3 of Chapter 6, page 225). Recall from Theorem 4.15 of Chapter 6 (page 226) the chain equivalence

$$\widetilde{\mathcal{E}}:\widetilde{\mathcal{M}}_*(f,v)\xrightarrow{\ \sim\ }\mathcal{S}_*(\widetilde{M}).$$

Choose a C^1 triangulation of M and let $\Delta_*(\widetilde{M})$ denote the corresponding simplicial chain complex of $\widetilde{M}$. Composing $\widetilde{\mathcal{E}}$ with the chain equivalence

$$\mathcal{L}^{-1}:\mathcal{S}_*(\widetilde{M})\longrightarrow\Delta_*(\widetilde{M})$$

we obtain a chain equivalence

$$\mathcal{E}_\Delta:\mathcal{M}_*(f,v)\xrightarrow{\ \sim\ }\Delta_*(\widetilde{M}).$$

Both chain complexes in the previous formula can be endowed with free bases over $\mathbf{Z}G$ (using lifts of critical points and simplices to $\widetilde{M}$), thus $\mathcal{E}_\Delta$ is a chain equivalence of two fb-complexes and it is easy to see that the image

$$\bar{\tau}(\mathcal{E}_\Delta)\in\mathrm{Wh}(G)$$

of its torsion

$$\tau(\mathcal{E}_\Delta) \in \overline{K}_1(\mathbf{Z}G)$$

does not depend on the choices of lifts of critical points or simplices to $\widetilde{M}$.

Theorem 3.10.

$$\bar{\tau}(\mathcal{E}_\Delta) = 0.$$

A Morse stratification can be considered as a generalized CW decomposition of M, and Theorem 3.10 is in a sense an analog of T. A. Chapman's theorem cited above. This theorem has been known for a long time, see for example [**95**]. For a detailed proof in the style of the present exposition see [**110**] (Appendix).

4. The Whitehead torsion of the canonical chain equivalence

Now let us proceed to the circle-valued Morse theory. Let M be a closed manifold, $f : M \to S^1$ a Morse map, and v an oriented almost transverse f-gradient. The Novikov complex $\mathcal{N}_*(f, v)$ is a free-finite chain complex of $\overline{L}$-modules, and we have the canonical chain equivalence

$$\Phi : \mathcal{N}_*(f, v) \to \mathcal{S}_*(\bar{M}) \underset{L}{\otimes} \overline{L}$$

(see Section 6 of Chapter 11, page 356). Similarly to the real-valued Morse theory we can derive from this chain equivalence a torsion invariant, but contrary to the case of real-valued Morse maps this invariant does not vanish in general, and its value is related to the dynamical properties of the gradient flow.

The relevant algebraic tool here is the Whitehead group $\mathrm{Wh}(\overline{L})$ of the Novikov ring. Recall that

$$\mathrm{Wh}(\overline{L}) = K_1(\overline{L})/\mathcal{T},$$
$$\text{where} \quad \mathcal{T} = \{(-1)^m t^k \mid k \in \mathbf{Z},\ m = 0, 1\} \subset \overline{L}^\bullet.$$

It follows from Theorem 3.2 (page 400) that the inclusion $\mathcal{W} \hookrightarrow \overline{L}^\bullet$ (where $\mathcal{W}$ is the group of the Witt vectors) induces an isomorphism

$$\mathcal{W} \xrightarrow{\approx} \mathrm{Wh}(\overline{L}).$$

We are going to derive from the chain equivalence Φ a torsion invariant in $\mathrm{Wh}(\overline{L})$. To do this we have to replace the singular chain complex by a free-finite complex over $\overline{L}$. Observe that the L-complex $\mathcal{S}_*(\bar{M})$ is homotopy equivalent to a free-finite based complex. (Indeed, such a homotopy equivalence is easy to construct from a homotopy equivalence $K \to M$ with K a finite CW complex.) Choose any such chain equivalence

$$h : \mathcal{S}_*(\bar{M}) \to \mathcal{C}_*$$

and form a tensor product of h with $\overline{L}$ to obtain a chain equivalence of $\overline{L}$-complexes:

$$\mathcal{S}_*(\bar{M}) \underset{L}{\otimes} \overline{L} \xrightarrow{\overline{h}} \mathcal{C}_* \underset{L}{\otimes} \overline{L}.$$

We have then a chain equivalence

$$\Phi_h = \overline{h} \circ \Phi : \mathcal{N}_*(f, v) \to \mathcal{C}_* \underset{L}{\otimes} \overline{L}$$

of free-finite based $\overline{L}$-complexes (here as usual we endow the Novikov complex with a geometric basis). Consider the image

$$\bar{\tau}(\Phi_h) \in \mathrm{Wh}(\overline{L})$$

of the torsion $\tau(\Phi_h) \in K_1(\overline{L})$.

Lemma 4.1. *The element $\bar{\tau}(\Phi_h) \in Wh(\overline{L})$ does not depend on the choice of bases in $\mathcal{C}_*$ and $\mathcal{N}_*(f, v)$ and does not depend on the choice of the chain equivalence h.*

Proof. Any two lifts of a given point in $S(f)$ to $\bar{M}$ differ by the action of t^n with some $n \in \mathbf{Z}$. Thus, replacing a lift of $S(f)$ to $\bar{M}$ by another lift leads to adding to $\tau(\Phi_h)$ of a term of the form $n \cdot [t]$ (where $n \in \mathbf{Z}$ and $[t]$ is the class in $K_1(\overline{L})$ of the unit $t \in L^\bullet$), and the image of $\tau(\Phi_h)$ in the Whitehead group $\mathrm{Wh}(\overline{L}) = K_1(\overline{L})/\mathcal{T}$ does not change. A similar argument proves that $\bar{\tau}(\Phi_h)$ does not depend on the choice of a basis in $\mathcal{C}_*$.

Further, let $h' : \mathcal{S}_*(\bar{M}) \xrightarrow{\sim} \mathcal{C}'_*$ be another chain equivalence where $\mathcal{C}'_*$ is a free-finite based L-complex. The torsion of the composition $h'^{-1} \circ h$ vanishes in the group

$$\mathrm{Wh}(L) = K_1(L)/\mathcal{T},$$

since this group is trivial. Therefore the torsions of the chain equivalences Φ_h, $\Phi_{h'}$ corresponding to h, and h' become equal after projection to $\mathrm{Wh}(\overline{L})$. □

Definition 4.2. The Witt vector

$$\overline{\mathrm{Det}}\,\big(\bar{\tau}(\Phi_h)\big) \in \mathcal{W}$$

is called the *Witt invariant* of the pair (f, v) and denoted $w(f, v)$.

Theorem 4.3. *Let M be a closed manifold, $f : M \to S^1$ a Morse map, and v an oriented almost transverse f-gradient. We have*

$$w(f, v) = \big(\zeta_L(-v)\big)^{-1}. \tag{8}$$

Proof. Both sides of (8) are elements of $\widehat{P} = \mathbf{Z}[[t]]$. To prove that they are equal it suffices to check that for every natural n their images in

$$P/t^n P = \mathbf{Z}[t]/t^n\mathbf{Z}[t]$$

are equal. Let

$$w_n(v),\ z_n(v) \in P/t^n P$$

denote these images. First we shall show that for every n both $w_n(v)$ and $z_n(v)$ are invariant with respect to C^0-small perturbations of v, and then we shall prove that for every cellular f-gradient v the equality (8) holds. Then the theorem follows, since the cellular gradients are C^0-dense in the set of all gradients.

The stability of the truncated zeta function with respect to C^0-small perturbations is already proved in Proposition 1.8, page 387. Let us proceed to the truncated Witt invariant.

Proposition 4.4. *Let v be an almost transverse f-gradient. Let $n \in \mathbf{N}$. There is $\delta > 0$ such that*

$$w_n(v) = w_n(u)$$

for every almost transverse f-gradient u with $||u - v|| < \delta$.

Let λ be a regular value of F; as usual we denote

$$V = F^{-1}(\lambda), \quad V^- = F^{-1}(]-\infty, \lambda]), \quad \mathcal{S}_*^\lambda = \mathcal{S}_*(V^-).$$

The chain complex $\mathcal{S}_*^\lambda$ is then homotopically free-finite over P (see Corollary 2.3, page 391). Let

$$g : \mathcal{S}_*^\lambda \longrightarrow \mathcal{D}_*$$

be a chain equivalence where $\mathcal{D}_*$ is a finitely generated free chain complex over P. Recall the chain equivalence

$$\Psi : \mathcal{N}_*^\lambda(f, v) \to \mathcal{S}_*^\lambda \underset{P}{\otimes} \widehat{P}$$

from Proposition 6.3 of Chapter 11 (page 358). The composition

$$\Psi_g = (g \otimes \mathrm{id}) \circ \Psi : \mathcal{N}_*^\lambda(f, v) \longrightarrow \mathcal{D}_* \underset{P}{\otimes} \widehat{P} \tag{9}$$

is a chain equivalence. Put $\mathcal{C}_* = \mathcal{D}_* \underset{P}{\otimes} L$ and let

$$h = (g \otimes \mathrm{id}) : \mathcal{S}_*(\bar{M}) = \mathcal{S}_*^\lambda \underset{P}{\otimes} L \to \mathcal{D}_* \underset{P}{\otimes} L = \mathcal{C}_*.$$

Then the map

$$\Phi_h : \mathcal{N}_*(f, v) \to \mathcal{D}_* \underset{P}{\otimes} \overline{L}$$

is homotopic to the tensor product of Ψ_g by the identity map $\mathrm{id} : \overline{L} \to \overline{L}$. To compute the truncated Witt invariant $w_n(f, v)$ we can therefore consider the tensor product of the map (9) by P_n over $\widehat{P}$, and compute the torsion of the resulting chain equivalence of P_n-complexes. The truncated map $\Psi/t^n\Psi$ is homotopic to the equivariant Morse equivalence

$$\mathcal{E}_*(\lambda, n; v) : \mathcal{N}_*(\lambda, n; v) = \mathcal{N}_*^\lambda(f, v)/t^n\mathcal{N}_*^\lambda(f, v) \longrightarrow \mathcal{S}_*^\lambda/t^n\mathcal{S}_*^\lambda.$$

Now our assertion follows immediately from Proposition 4.7 of Chapter 11 (page 354). □

Therefore it remains to consider the case of cellular gradients, and the proof of the theorem will be completed with the next proposition.

Proposition 4.5. *Let v be a cellular f-gradient. Then*

$$w(f, v) = (\zeta_L(-v))^{-1}.$$

Proof. Recall that in Theorem 2.5, page 393 we have constructed a homotopy commutative diagram of chain equivalences

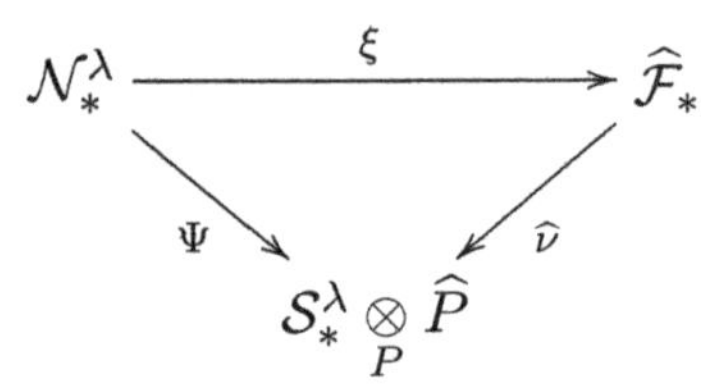

Here $\mathcal{F}_*$ is the free finitely generated based P-complex, associated with the filtration of V^- provided by the cellular structure of v. Therefore the Whitehead torsion invariant $\overline{\tau}(\Phi_h)$ is equal to the image in the group $\mathrm{Wh}(\overline{L})$ of the element $\tau(\xi) \in \overline{K}_1(\widehat{P})$. The chain equivalence ξ is a split monomorphism, as we have proved in Subsection 2.3. The quotient complex is computed in Proposition 2.9 (page 399), and combining this Proposition with Lemma 3.9, page 405 we deduce that the torsion of ξ (with respect to the bases chosen) satisfies

$$\tau(\xi) = \sum_i (-1)^i \ [\mathrm{Id} - t\overline{\mathcal{H}}_i(-v)] \in \overline{K}_1(\widehat{P}).$$

Therefore we obtain the following formula for the determinant of $\tau(\xi)$:

$$\overline{\mathrm{Det}}\ \tau(\xi) = \prod_i \det\Big(\mathrm{Id} - t\overline{\mathcal{H}}_i(-v)\Big)^{(-1)^i} \in \mathcal{W}.$$

Recall that by definition (Definition 1.7, page 386) the element $\zeta_L(-v) \in \widehat{P}$ is the Lefschetz zeta function of the (partially defined) map

$$\Theta = t^{-1} \circ (-\overline{v})^{\leadsto} : V \to V.$$

Apply now Theorem 2.29 of Chapter 9 (page 301) which asserts that

$$\zeta_L(\Theta) = \prod_i \det\left(\mathrm{Id} - t \cdot \overline{\mathcal{H}}_i(-v)\right)^{(-1)^{i+1}}$$

and the proof is complete. $\square$

CHAPTER 14

Selected topics in the Morse-Novikov theory

1. Homology with local coefficients and the de Rham framework for the Morse-Novikov theory

For a real-valued Morse function $f : M \to \mathbf{R}$ on a closed manifold M the Morse inequalities in their simplest form read as

$$m_k(f) \geqslant b_k(M)$$

where $m_k(f)$ is the number of critical points of f of index k and $b_k(M)$ is the k-th Betti number of M, a common algebro-topological invariant: $b_k(M) = \dim H_k(M, \mathbf{Q})$.

The simplest lower bound for the number of critical points of a circle-valued Morse function is as follows:

$$m_k(f) \geqslant b_k(M, \xi)$$

where the Novikov Betti number $b_k(M, \xi)$ is the rank over $\mathbf{Z}((t))$ of the completed module

$$H_k(\bar{M}) \underset{\mathbf{Z}[t,t^{-1}]}{\otimes} \mathbf{Z}((t)).$$

This rank is not so easy to calculate, since it involves the $\mathbf{Z}[t, t^{-1}]$-module structure of the homology of the cyclic covering $\bar{M}$. It turns out however that there is a simpler way to compute the Novikov Betti number. Namely $b_k(M, \xi)$ is equal to $\dim(H_k(M, \mathcal{L}))$ where $H_k(M, \mathcal{L})$ stands for the homology of M with coefficients in a *generic local system* associated with ξ. In the next subsection we explain what it means and how to obtain the corresponding inequalities without using the Novikov complex (the *Witten deformation method*).

1.1. Homology with local coefficients. Let X be a finite connected CW complex, G its fundamental group and

$$\rho : G \to \mathrm{GL}(1, k) \approx k^*$$

be a homomorphism, where k is a field. Consider the universal covering $\widetilde{X} \to X$. The cellular chain complex $C_*(\widetilde{X})$ is a complex of free finitely generated left $\mathbf{Z}G$-modules. Consider the chain complex

$$C_*(X,\rho) = k \underset{\mathbf{Z}G}{\otimes} C_*(\widetilde{X})$$

where k is endowed with the structure of $\mathbf{Z}G$-module via the representation ρ. The homology of this complex is called the *homology with local coefficients with respect to* ρ and denoted $H_*(X,\rho)$. In a similar way one defines the cohomology with local coefficients. If ρ is the trivial representation, then $H_*(X,\rho)$ is isomorphic to the ordinary homology $H_*(X,k)$.

Now let $\xi \in H^1(X,\mathbf{C})$ be a cohomology class, and $t \in \mathbf{C}$. Consider a homomorphism

$$\rho_t : \pi_1(X) = G \to \mathbf{C}^*; \quad \rho_t(\gamma) = e^{t\langle \xi, \bar\gamma\rangle},$$

where $\bar\gamma$ is the image of γ in $H_1(X)$, and $\langle\cdot,\cdot\rangle$ stands for the Kronecker pairing between homology and cohomology. Let us write

$$B_k(\xi,t) = \dim H_k(X,\rho_t), \quad \beta_k(\xi) = \min_{t\in\mathbf{C}} B_k(\xi,t).$$

Proposition 1.1. *Let* $\xi \in H^1(X,\mathbf{C})$. *There is a discrete subset* $\Delta \in \mathbf{C}$ *such that for every* $t \in \mathbf{C}\setminus\Delta$ *we have*

$$B_k(\xi,t) = \beta_k(\xi).$$

Proof. It follows from the next lemma, which is easily deduced from the principle of isolated zeros for holomorphic functions.

Lemma 1.2. *Let*

$$\mathcal{M} : \mathbf{C} \to Mat(n\times m,\mathbf{C})$$

be a matrix function, such that every matrix coefficient of $\mathcal{M}$ *is an entire holomorphic function. Then the function* $t \mapsto \operatorname{rk}\mathcal{M}(t)$ *is constant on the complement to some discrete subset* $\Delta\subset\mathbf{C}$, *and takes on* $\mathbf{C}\setminus\Delta$ *its maximal value.* □

In the case when $\xi \in H^1(X,\mathbf{Z})$ we can say more about $B_k(\xi,t)$ and $\beta_k(\xi)$.

Proposition 1.3 ([**107**], Lemma 2, see also [**103**]). *Let* $\xi \in H^1(X,\mathbf{Z})$. *Then*

(1) *There is a finite subset* $\Gamma\subset\mathbf{C}$ *whose elements are algebraic numbers, and such that* $B_k(\xi,t) = \beta_k(\xi)$ *for every* t *with* $e^t \in \mathbf{C}\setminus\Gamma$.
(2) $\beta_k(\xi)$ *is equal to the Novikov Betti number* $b_k(X,\xi)$.

Now we shall give an interpretation of the cohomology $H^*(X,\rho_t)$ in the de Rham framework. Let M be a closed manifold. Recall that the cohomology $H^*(M,\mathbf{C})$ is isomorphic to the *de Rham cohomology*, defined as follows. Consider the vector space $\Omega^k(M)$ of all C^∞ complex differential forms of degree k. The exterior differential

$$d : \Omega^k(M) \to \Omega^{k+1}(M)$$

satisfies $d^2 = 0$, and thus the sequence

$$\Omega^*(M,d) = \{ \cdots \longrightarrow \Omega^k(M) \xrightarrow{d} \Omega^{k+1}(M) \longrightarrow \cdots \}$$

is a cochain complex, whose cohomology is called the *de Rham cohomology* of M. Let ω be any closed form on M and $t \in \mathbf{C}$. Define a map

$$d_t : \Omega^k(M) \to \Omega^{k+1}(M)$$

as follows:

$$d_t(\theta) = d\theta + t\omega \wedge \theta. \tag{1}$$

It is easy to check that $d_t^2 = 0$, and we obtain an analog of the de Rham complex:

$$\Omega^*(M,d_t) = \{ \cdots \longrightarrow \Omega^k(M) \xrightarrow{d_t} \Omega^{k+1}(M) \longrightarrow \cdots \}.$$

For $t = 0$ this complex is equal to the ordinary de Rham complex $\Omega^*(M,d)$ described above, so $\Omega^*(M,d_t)$ can be considered as *a deformation* of $\Omega^*(M,d)$. It is not difficult to prove that

$$H^*\big(\Omega^*(M,d_t)\big) \approx H^*(M,\rho_t),$$

where ρ_t is the representation $\pi_1(M) \to GL(1,\mathbf{C})$ given by the formula

$$\rho_t(\gamma) = e^{t\int_\gamma \omega}.$$

The formula (1) for the perturbed de Rham differential d_t contains explicitly a closed 1-form ω representing the given cohomology class ξ, and if ω is a Morse form one may expect to find a direct relationship between the deformed de Rham cohomology $H^*\big(\Omega^*(M,d_t)\big)$ and the zeros of ω. In the next section we describe the *Witten deformation method* realizing this program.

1.2. The Witten deformation method applied to Morse forms. The Witten method provides a lower bound for the number of zeros of a Morse form independently of the construction of the Novikov complex:

Theorem 1.4 ([**107**], Theorem 1). *Let ω be a Morse form on a closed manifold M, denote by ξ its cohomology class. Then for every $t \in \mathbf{R}$ sufficiently large we have:*

$$m_k(\omega) \geqslant B_k(\xi,t).$$

For the case when ω is cohomologous to zero, this theorem is exactly the Witten theorem from [**163**]. In this particular case the local system is trivial, and the cohomology $H^*(M,\rho_t)$ is equal to the ordinary de Rham cohomology of M for any t. Witten's proof of his theorem generalizes to the case of arbitrary Morse forms without any changes. Here is an outline of the proof. It is well known that the cohomology $H^k(\Omega^*(M,\rho_t))$ is isomorphic to the space of harmonic forms of degree k with respect to the Laplacian corresponding to the differential d_t. This Laplacian Δ_t can be computed explicitly (we assume here that $t\in\mathbf{R}$):

$$(\mathscr{L})\qquad \Delta_t\alpha = \Delta\alpha + t^2|\omega|^2\alpha + tG_\omega(\alpha).$$

Here Δ is the Laplacian corresponding to the ordinary de Rham differential, and the linear operator $G_\omega : \Omega^k(M)\to\Omega^k(M)$ is a tensor field and it does not depend on t.

This formula resembles the definition of the Schrödinger operator. The function $t^2|\omega|^2$ plays the role of the potential function, its points of minimum coincide with the zeros of ω. For any $x\notin S(\omega)$ we have

$$t^2|\omega|^2(x)\to\infty \quad \text{as} \quad t\to\infty$$

and therefore the eigenforms of the Laplacian Δ_t concentrate in a neighbourhood of the set of zeros of ω. To get a better idea of what goes on in the neighbourhood of zeros of ω, consider an example: $M=\mathbf{R}^1$ (endowed with the Euclidean metric), $\omega = d(x^2) = 2xdx$. An explicit computation shows:

$$\Delta_t f = -\frac{d^2 f}{dx^2} + t^2x^2 f - tf$$

(where f is any C^∞ function on $\mathbf{R}$). Recall that the Schrödinger operator corresponding to the 1-dimensional system with potential t^2x^2 has the form

$$H_t = -\frac{d^2 f}{dx^2} + t^2x^2 f.$$

It is well known that the eigenvalues of H_t are of the form $t(2n+1)$ with $n\in\mathbf{N}$. The first eigenvalue of $\Delta_t = H_t - t$ is therefore 0, it corresponds to the eigenfunction $\exp\big(-\frac{tx^2}{2}\big)$. The other eigenvalues converge to infinity as $t\to\infty$.

It turns out that this pattern is reproduced in the case of the Laplacian $(\mathscr{L})$ above. Namely, fix some $A>0$. As $t\to\infty$ for every zero of ω of index p there is exactly one p-eigenform of Δ_t with eigenvalue less than A. One can prove that the space of all such forms is a subcomplex $A^*(M)$ of the deformed de Rham complex $\Omega^*(M,d_t)$. Moreover, $A^*(M)$ has the same cohomology as the complex itself. Thus

$$m_p(\omega) = \dim A^p(M) \geqslant \dim H^p_t(M) = B_p([\omega],t).$$

2. The universal Novikov complex

The construction of the Novikov complex uses a procedure of counting the flow lines joining critical points of the circle-valued Morse function. There is a refinement of this construction where one takes into account the homotopy classes of these flow lines. This construction (similar to the universal Morse complex from Subsection 4.3 of Chapter 6, page 225) can be called *the universal Novikov complex* or *the Novikov-Sikorav complex*, since it was first introduced by J.-C. Sikorav in his thesis [**147**].

Let us first describe the necessary algebraic notions. Let G be a group and $\xi : G \to \mathbf{R}$ a group homomorphism. Put $\Lambda = \mathbf{Z}[G]$ and let Λ' denote the abelian group of all series of the form $\lambda = \sum_{g\in G} n_g g$ where $n_g \in \mathbf{Z}$. For $\lambda \in \Lambda'$ and $C \in \mathbf{R}$ put

$$\mathrm{supp}\,(\lambda, C) = \{g \in G \mid n_g \neq 0 \quad \& \quad \xi(g) > C\}.$$

The Novikov ring is defined as follows:

$$\widehat{\Lambda}_\xi = \{\lambda \in \Lambda' \mid \mathrm{supp}\,(\lambda, C) \quad \text{is finite for every} \quad C \in \mathbf{R}\}.$$

It is easy to show that the subset $\widehat{\Lambda}_\xi \subset \Lambda'$ has indeed a natural structure of a ring, containing Λ as a subring.

Let $f : M \to S^1$ be a Morse function on a connected closed manifold M, and v a transverse oriented f-gradient. Let $G = \pi_1(M)$. Consider the universal covering

$$\Pi : \widetilde{M} \to M$$

of M. We have a free left action of G on $\widetilde{M}$, and the quotient space is M. Pick any lift $\mathcal{F} : \widetilde{M} \to \mathbf{R}$ of the function f. Then $\mathcal{F}$ is a Morse function, and the lift $\widetilde{v}$ of the vector field v to $\widetilde{M}$ is a transverse oriented $\mathcal{F}$-gradient. The universal Novikov complex which we are about to construct is a free finitely generated chain complex over $\widehat{\Lambda}_\xi$, where $\Lambda = \mathbf{Z}G$, and

$$\xi = f_\# : G = \pi_1(M) \to \mathbf{Z} = \pi_1(S^1)$$

is the homomorphism induced by f.

Let C_k be a free $\mathbf{Z}$-module, generated by the critical points of $\mathcal{F}$ of index k. Let $\widehat{C}_k$ be the abelian group, consisting of all series

$$\lambda = \sum_{p\in S_k(\mathcal{F})} n_p p \quad \text{with} \quad n_p \in \mathbf{Z}$$

such that above any given level surface $\mathcal{F}^{-1}(C)$ the series λ contains only a finite number of critical points $p \in S_k(\mathcal{F})$ (in other words, the set

$$\mathrm{supp}\,(\lambda, C) = \{p \mid n_p \neq 0 \;\&\; \mathcal{F}(p) > C\}$$

is finite for every C). It is clear that $\widehat{C}_k$ is a free left module over $\widehat{\Lambda}_\xi$, any lift to $\widetilde{M}$ of the set $S_k(f)$ being a basis of $\widehat{C}_k$ over $\widehat{\Lambda}_\xi$.

Let p, q be two critical points of $\mathcal{F}$ with $\operatorname{ind} p = \operatorname{ind} q + 1$, and let γ be any flow line of $(-v)$, joining p with q. The orientation of v allows us to attribute a sign $\varepsilon(\gamma) \in \{1, -1\}$ to γ. The following results were proved in [**110**].

Lemma 2.1. *Let p, q be critical points of $\mathcal{F}$, $\operatorname{ind}\ p = \operatorname{ind}\ q + 1$. Then:*

(1) *The set of flow lines of $(-v)$ joining p with q is finite. The sum of the signs $\varepsilon(\gamma)$ over all these flow lines will be denoted $n(p, q; v)$.*

(2) *For every $p \in S_k(\mathcal{F})$ the series*

$$\partial_k p = \sum_{q \in S_{k-1}(\mathcal{F})} n(p, q; v) q$$

is in $\widehat{C}_{k-1}$. Extending ∂_k to $\widehat{C}_k$ by linearity we obtain a homomorphism $\partial_k : \widehat{C}_k \to \widehat{C}_{k-1}$.

Theorem 2.2. (1) *$\partial_k \circ \partial_{k+1} = 0$ for every k.*

(2) *The resulting free-based chain complex $\widetilde{\mathcal{N}}_*(f, v)$ is homotopy equivalent to the completed singular chain complex of the universal covering:*

$$\widetilde{\mathcal{N}}_*(f, v) \sim \widehat{\Lambda}_\xi \underset{\Lambda}{\otimes} \mathcal{S}_*(\widetilde{M}).^\dagger$$

As in the real-valued case one can strengthen the previous theorem, and obtain a *simple homotopy equivalence* between the Novikov complex and the completed simplicial chain complex of the universal covering. Let us first introduce the corresponding K-theoretic notion.

Definition 2.3. A unit of the ring $\widehat{\Lambda}_\xi$ will be called *trivial* if it is of the form $\pm g + u$, where $g \in G$ and the power series u has its support in the subset $\{g \in G \mid \xi(g) < 0\}$. The group of all trivial units will be denoted by U_ξ, and the image of this group in $K_1(\widehat{\Lambda}_\xi)$ will be denoted by $\bar{U}_\xi$. The quotient group $K_1(\widehat{\Lambda}_\xi)/\bar{U}_\xi$ will be denoted by $\widehat{Wh}(G, \xi)$. Two free finitely generated based chain complexes over $\widehat{\Lambda}_\xi$ will be called *simply homotopy equivalent* if there is a homotopy equivalence between them such that the image of its torsion in $\widehat{Wh}(G, \xi)$ vanishes.

† The ring Λ being non-commutative in general, we need to distinguish between right and left modules. In this formula for example we have the tensor product of the *right* Λ-module $\widehat{\Lambda}_\xi$ by the *left* Λ-module $\mathcal{S}_*(\widetilde{M})$.

Pick any C^1 triangulation of M, and lift it to a G-invariant triangulation of $\widetilde{M}$. The simplicial chain complex corresponding to this triangulation will be denoted by $\Delta_*(\widetilde{M})$, this is a finite chain complex of finitely generated free $\widehat{\Lambda}_\xi$-modules. Both $\widetilde{\mathcal{N}}_*(f,v)$ and $\Delta_*(\widetilde{M})$ have natural free bases. Namely, any lift to $\widetilde{M}$ of the set of all critical points of f forms a family of free generators of $\widetilde{\mathcal{N}}_*(f,v)$. Similarly, if we choose a lift to $\widetilde{M}$ for every simplex of the triangulation of M we obtain a free base of $\Delta_*(\widetilde{M})$ over Λ. The following theorem was proved in [**110**]:

Theorem 2.4. *The chain complexes $\widetilde{\mathcal{N}}_*(f,v)$ and $\widehat{\Lambda}_\xi \underset{\Lambda}{\otimes} \Delta_*(\widetilde{M})$ are simply homotopy equivalent.*

Corollary 2.5. *If a homotopy class $\xi \in [M,S^1] = \mathrm{Hom}(G,\mathbf{Z}) = H^1(M,\mathbf{Z})$ contains a C^∞ fibration over a circle, then the completed chain complex $\widehat{\Lambda}_\xi \underset{\Lambda}{\otimes} \Delta_*(\widetilde{M})$ is simply homotopy equivalent to zero.*

It turns out that the necessary condition provided by this corollary is also sufficient under some restrictions ([**111**], [**82**]):

Theorem 2.6. *Let M be a closed connected manifold, $\dim M \geqslant 6$. Let $\xi : \pi_1(M) \to \mathbf{Z}$ be an epimorphism such that* Ker ξ *is a finitely presented group.*

Then $\xi = f_\#$ for a C^∞ fibration $f : M \to S^1$ if and only if the chain complex $\widehat{\Lambda}_\xi \underset{\Lambda}{\otimes} \Delta_*(\widetilde{M})$ is simply homotopy equivalent to zero.

Thus there is a natural relation of the theory of the Novikov complex to the problem of fibring manifolds over a circle. In the next section we discuss this subject in detail.

3. The Morse-Novikov theory and fibring obstructions

The theory of fibrations of manifolds over a circle was developed in the 1960s and early 1970s. The story began with a theorem of J. Stallings [**154**] which says that if M is an irreducible compact 3-dimensional manifold and $\xi : \pi_1(M) \to \mathbf{Z}$ an epimorphism such that Ker ξ is a finitely generated group not isomorphic to $\mathbf{Z}/2\mathbf{Z}$, then there is a fibration of M over S^1 whose fiber is a 2-dimensional compact manifold.

The next step was done by W. Browder and J. Levine [**17**] who considered the case $\dim M \geqslant 6$, and solved the problem completely with the additional assumption $\pi_1(M) \approx \mathbf{Z}$. Here is their necessary and sufficient condition for fibring: the homotopy groups of M must be finitely generated (or equivalently, the universal covering $\widetilde{M}$ is homotopy equivalent to a finite CW complex).

If we drop the restriction $\pi_1(M) \approx \mathbf{Z}$ the statement and the proof of the corresponding result are much more complicated. A necessary and sufficient condition for fibring in the general case was obtained by T. Farrell [**33**], [**34**]. His answer to the fibring problem is as follows. Let M be a closed connected manifold of dimension $\geqslant 6$, put $G = \pi_1(M)$ and let $\xi : G \to \mathbf{Z}$ be an epimorphism. If ξ contains a C^∞ fibration $f : M \to S^1$, then the infinite cyclic covering $\bar{M}_\xi$ is homotopy equivalent to a finite CW complex (namely, to the fiber of f). This provides the first necessary condition for fibring. This condition granted, there is an obstruction $c(\xi)$ for fibring which is an element of a certain K-theoretic group $C(\mathbf{Z}G, \alpha)$ introduced and studied by T. Farrell. (In the case when the epimorphism $\xi : G \to \mathbf{Z}$ splits, this group can be described as the Grothendiek group of the category of nilpotent matrices over $\mathbf{Z}H$, where $H = \operatorname{Ker} \xi$.) If $c(\xi) = 0$, then there arises a second obstruction $\tau(\xi)$ which is an element of a certain quotient of the Whitehead group $\operatorname{Wh}(G)$. Finally, if $\tau(\xi) = 0$, then the class ξ contains a C^∞ fibration over S^1.

In Farrell's thesis the obstruction $c(\xi)$ was defined under the assumption that $\bar{M}_\xi$ is *finitely dominated*, which is weaker than the finiteness condition. Recall that a CW complex X is called *finitely dominated* if there exists a finite CW complex Y and continuous maps $f : X \to Y$ and $g : Y \to X$ such that $g \circ f$ is homotopic to the identity map.

Another approach to the theory of fibring manifolds over a circle was developed by L. Siebenmann in [**144**]. Let M be a manifold with $\dim M \geqslant 6$, and let $\xi : \pi_1(M) \to \mathbf{Z}$ be a homomorphism. Assuming that the infinite cyclic covering $\bar{M}_\xi$ is finitely dominated, L. Siebenmann constructs an element $\sigma(\xi) \in \operatorname{Wh}(\pi_1(M))$ such that $\sigma(\xi)$ is equal to zero if and only if the class ξ contains a C^∞ fibration.

Observe that Corollary 2.5 and Theorem 2.6 of the previous section provide yet another solution to the problem of fibring manifolds over a circle. Namely, assuming that $\operatorname{Ker} \xi$ is a finitely presented group, the class ξ contains a C^∞ fibration if and only if the completed chain complex $\widehat{\Lambda}_\xi \underset{\Lambda}{\otimes} \Delta_*(\widetilde{M})$ is simple homotopy equivalent to zero (here $\Lambda = \mathbf{Z}G$).

These approaches to the fibring problem can be identified purely algebraically. This was done by A. Ranicki [**134**], [**132**]. His results express the finite domination condition in terms of the Novikov homology without using the geometric construction of the Novikov complex. Here is the main theorem of [**134**]:

Theorem 3.1. *Let X be a finite CW complex with universal cover $\widetilde{X}$ and fundamental group $\pi_1(X) = \pi \times \mathbf{Z}$, so that $\mathbf{Z}[\pi_1(X)] = \mathbf{Z}[\pi][z, z^{-1}]$. The infinite cyclic cover $\overline{X} = \widetilde{X}/\pi$ is finitely dominated if and only if X is $\mathbf{Z}[\pi]((z))$- and $\mathbf{Z}[\pi]((z^{-1}))$- acyclic*

$$H_*(X, \mathbf{Z}[\pi]((z))) = H_*(X, \mathbf{Z}[\pi]((z^{-1}))) = 0.$$

Another theorem of [**134**] says that if the finite domination condition holds, then the fibring obstruction provided by the Morse-Novikov theory is equivalent to the Farrell-Siebenmann obstruction:

Theorem 3.2. *Let M be a compact n-dimensional manifold with $\pi_1(M) = \pi \times \mathbf{Z}$. The infinite cyclic cover $\overline{M} = \widetilde{M}/\pi$ of M is finitely dominated if and only if $H_*(M, \mathbf{Z}[\pi]((z))) = 0$, in which case the natural map $Wh(\pi \times \mathbf{Z}) \to Wh(\mathbf{Z}[\pi]((z)))$ sends the Farrell-Siebenmann fibring obstruction $\Phi(M) \in Wh(\pi \times \mathbf{Z})$ to the Pajitnov fibring obstruction $\tau(M, \mathbf{Z}[\pi]((z))) \in Wh(\mathbf{Z}[\pi]((z)))$ and $\Phi(M) = 0$ if and only if $\tau(M, \mathbf{Z}[\pi]((z))) = 0$. Thus if $\overline{M}$ is finitely dominated $\tau(M, \mathbf{Z}[\pi]((z))) = 0$ if (and for $n \geqslant 6$ only if) M fibers over S^1, as proved in* [**111**] *using S^1-valued Morse theory.*

See the book [**132**] for a detailed exposition of these results.

4. Exactness theorems and localization constructions

Let $f : M \to S^1$ be a Morse map. Recall the Novikov inequalities

$$(\mathcal{N}) \qquad m_k(f) \geqslant b_k(M, \xi) + q_k(M, \xi) + q_{k-1}(M, \xi)$$

for the numbers of critical points of f (here $\xi \in H^1(M, \mathbf{Z}) \approx [M, S^1]$ is the homotopy class of f).

Definition 4.1. The inequalities $(\mathcal{N})$ are called *exact* for the given M and $\xi \in H^1(M, \mathbf{Z})$ if there is a Morse map $f : M \to S^1$ such that

$$m_k(f) = b_k(M, \xi) + q_k(M, \xi) + q_{k-1}(M, \xi) \quad \text{for every} \quad k.$$

In 1984 M. Farber proved the following theorem:

Theorem 4.2 ([**30**]). *Let M be a connected closed manifold with* $\dim M \geqslant 6$ *and $\pi_1(M) \approx \mathbf{Z}$. Let $\xi \in H^1(M, \mathbf{Z})$ be a non-zero cohomology class. Then the inequalities $(\mathcal{N})$ are exact for M and ξ.*

The algebraic technique of this paper is different from the initial Novikov setting. Instead of the Novikov completion $\overline{L} = \mathbf{Z}((t))$ of the ring $L = \mathbf{Z}[t, t^{-1}]$ M. Farber works with the localization

$$\widetilde{L} = S^{-1}L, \quad \text{where} \quad S = \{P(t) \mid n \in \mathbf{Z}, P(t) \in \mathbf{Z}[t], P(0) = 1\}.$$

This ring is principal, as it is proved in a paper of D. Quillen [**131**]. The ring $\widetilde{L}$ can be considered as a subring of $\overline{L} = \mathbf{Z}((t))$ which is a faithfully flat module over $\widetilde{L}$. Therefore the algebraic properties of the Novikov homology $H_*(\bar{M}_\xi)\underset{L}{\otimes}\overline{L}$ are very close to the algebraic properties of the localized module

$$S^{-1}H_*(\bar{M}_\xi) = H_*(\bar{M}_\xi) \underset{L}{\otimes} \widetilde{L}.$$

In particular we have

$$b_k(M,\xi) = \operatorname{rk}_{\widetilde{L}} S^{-1}H_k(\bar{M}_\xi); \quad q_k(M,\xi) = \text{t.n.}_{\widetilde{L}} S^{-1}H_k(\bar{M}_\xi),$$

and therefore

$$S^{-1}H_*(\bar{M}_\xi) = 0 \;\Leftrightarrow\; H_*(\bar{M}_\xi) \underset{L}{\otimes} \widetilde{L} = 0. \tag{2}$$

The Novikov inequalities are exact also for manifolds with any free abelian fundamental group, although in this case the cohomology class ξ must be in a general position. The following theorem was proved by the author:

Theorem 4.3 ([**108**], [**109**], [**111**]). *Let M be a connected closed manifold with $\dim M \geqslant 6$ and $\pi_1(M) \approx \mathbf{Z}^n$. Then there is a finite set of integral hyperplanes $\{\Delta_i\}_{i\in I}$ in $H^1(M,\mathbf{Z})$ such that for any non-zero $\xi \notin \cup_i\Delta_i$ the inequalities $(\mathcal{N})$ are exact. The numbers $b_k(M,\xi)$, $q_k(M,\xi)$ do not depend on ξ while ξ varies in any connected component of $H^1(M,\mathbf{R}) \setminus \bigcup_i \Delta_i \otimes \mathbf{R}$.*

5. The Morse-Novikov theory of closed 1-forms

Most notions and results of the Morse theory of circle-valued functions can be generalized to Morse forms. In this section we will briefly describe the basic results in this direction, referring to the recent book "Topology of closed 1-forms" by M. Farber [**32**] for a detailed discussion. Let M be a closed connected manifold, ω be a Morse form on M. We shall identify the de Rham cohomology class $[\omega]$ with the period homomorphism $\pi_1(M) \to \mathbf{R}$ defined by the formula

$$\gamma \mapsto \int_\gamma \omega.$$

Let $H = \operatorname{Im}[\omega]$ and $H' = \operatorname{Ker}[\omega]$. The number rk H is called the *degree of irrationality* of ω. Observe that the degree of irrationality of ω is $\leqslant 1$ if and only if $\omega = \lambda df$ for some $\lambda \in \mathbf{R}$ and some Morse map $f : M \to S^1$. We have a commutative diagram

$$\begin{array}{ccc} \pi_1(M) & \longrightarrow & H \approx \mathbf{Z}^k \\ & {\scriptstyle [\omega]} \searrow \quad \swarrow {\scriptstyle \xi} & \\ & \mathbf{R} & \end{array}$$

where ξ is a monomorphism, and the horizontal arrow is an epimorphism.

Let $\Lambda = \mathbf{Z}H$. Recall that the Novikov ring $\widehat{\Lambda}_\xi$ is by definition the set of all series of the form

$$\lambda = \sum_{g \in H} n_g g \quad \text{where} \quad n_g \in \mathbf{Z}$$

such that for every $C > 0$ the set of $g \in H$ with $\xi(g) > C$ and $n_g \neq 0$ is finite. In our present case $\widehat{\Lambda}_\xi$ is a commutative ring and $\Lambda \subset \widehat{\Lambda}_\xi$ is its subring. A theorem of J.-C. Sikorav says that $\widehat{\Lambda}_\xi$ is principal (see [**108**]).

Let $\overline{M} \to M$ denote the covering corresponding to the subgroup $H' \subset \pi_1(M)$; this is a regular covering with the structural group H. Put

$$\widehat{H}_*(M, [\omega]) = H_k(\overline{M}) \underset{\Lambda}{\otimes} \widehat{\Lambda}_\xi$$

and

$$b_k(M, [\omega]) = \operatorname{rk}_{\widehat{\Lambda}_\xi} \widehat{H}_k(M, [\omega]), \quad q_k(M, [\omega]) = \text{t.n.}_{\widehat{\Lambda}_\xi} \widehat{H}_k(M, [\omega]).$$

We have the following generalization of Theorem 1.8 of Chapter 11 (page 339):

Theorem 5.1. *Let ω be a Morse form on a closed connected manifold M. There is a chain complex $\mathcal{N}_*(\omega, v)$ of $\widehat{\Lambda}_\xi$-modules, such that:*

(1) *$\mathcal{N}_k(\omega, v)$ is the free $\widehat{\Lambda}_\xi$-module generated by $S_k(\omega)$.*

(2) *$H_k(\mathcal{N}_*(\omega, v)) \approx \widehat{H}_*(M, [\omega])$.*

(3) *$m_k(\omega) \geqslant b_k(M, [\omega]) + q_k(M, [\omega]) + q_{k-1}(M, [\omega])$.*

The construction of the Novikov complex $\mathcal{N}_k(\omega, v)$ is similar to the case of the Morse forms of irrationality degree 1, explained in Chapter 11. See the papers of S. P. Novikov [**102**], F. Latour [**82**], and the book of M. Farber [**32**] for details and generalizations. One can generalize the construction of $\mathcal{N}_k(\omega, v)$ yet further so as to obtain a refined version of the Novikov complex, defined over the Novikov completion of the group ring $\mathbf{Z}\pi_1(M)$ of the fundamental group of M (see [**82**] and [**140**]). The C^0-generic rationality of the boundary operators in the Novikov complex holds also in the case of Morse forms of irrationality degree > 1 (see [**112**]). Finally let us mention that the relation of closed orbits of the gradient flow to the torsion invariants of the Novikov complex discussed in Chapter 13 for the case of circle-valued Morse functions holds also for Morse forms of arbitrary irrationality degree (see [**140**], [**141**]).

6. Circle-valued Morse theory for knots and links

In this section we give a brief survey of recent developments in the circle-valued Morse theory of knots and links in S^3. Let $\mathscr{L} \subset S^3$ be an oriented link, that is, an oriented compact one-dimensional C^∞ submanifold of S^3. Let us call a *framing* of $\mathscr{L}$ any orientation preserving diffeomorphism $\Phi : \mathscr{L} \times B \longrightarrow U$ where $B = B^2(0, \rho)$ is the 2-dimensional open disc of radius ρ, and U is a neighbourhood of $\mathscr{L}$ in S^3, such that the following diagram is commutative:

$$\begin{array}{ccc} \mathscr{L} \times B & \xrightarrow{\Phi} & U \\ \uparrow & & \uparrow \\ \mathscr{L} \times 0 & \xrightarrow{\text{id}} & \mathscr{L}. \end{array}$$

For any $x \in \mathscr{L}$ and $r < \rho$ the image $\Phi(x \times S^1(0, r))$ is called a *meridian* of $\mathscr{L}$ corresponding to the framing Φ. Any meridian is an oriented circle in $S^3 \setminus \mathscr{L}$. The complement $C_{\mathscr{L}} = S^3 \setminus \mathscr{L}$ has a 1-dimensional cohomology class $\xi \in H^1(S^3 \setminus \mathscr{L})$ characterized by the following property: for any meridian μ of $\mathscr{L}$ the restriction of ξ to μ is the positive generator of the group $H^1(\mu) \approx \mathbf{Z}$. We shall study Morse maps $S^3 \setminus \mathscr{L} \to S^1$ such that their homotopy class equals $\xi \in [S^3 \setminus \mathscr{L}, S^1] \approx H^1(S^3 \setminus \mathscr{L})$. The manifold $S^3 \setminus \mathscr{L}$ is not compact and to develop a reasonable Morse theory it is natural to impose a restriction on the behavior of the Morse map in a neighbourhood of $\mathscr{L}$.

Definition 6.1. Let $\mathscr{L} \subset S^3$ be a link. A Morse map $f : C_{\mathscr{L}} \to S^1$ is said to be *regular* if there is a framing

$$\Phi : \mathscr{L} \times B^2(0, \rho) \to U$$

for $\mathscr{L}$ such that

$$(f \circ \Phi)(x, z) = \frac{z}{|z|}; \quad f \circ \Phi : \mathscr{L} \times \Big(B^2(0, \rho) \setminus \{0\}\Big) \to S^1.$$

It is not difficult to prove that the set of critical points of a regular Morse map f is finite.

Definition 6.2. The minimal possible number of critical points of a regular Morse map $f : S^3 \setminus \mathscr{L} \to S^1$ is called *the Morse-Novikov number of the link* $\mathscr{L}$ and denoted by $\mathcal{MN}(\mathscr{L})$.

The invariant $\mathcal{MN}(\mathscr{L})$ was introduced and studied in [**119**]. In the present section we give a survey of this work and other developments in this domain.

6.1. Background. The first result directly related to the circle-valued Morse theory of knots and links is the Stallings theorem, which we already discussed in Section 3. The theorem gives a necessary and sufficient condition for a link $\mathscr{L}$ to be fibred (the reader will find an exposition of the Stallings theorem for the case of complements to knots in the book [**18**]).

The universal Novikov homology of 3-manifolds was studied by J.-C. Sikorav (see his thesis [**147**]). He proves in particular that the universal Novikov homology of $S^3 \setminus \mathscr{L}$ vanishes if and only if $\mathscr{L}$ is fibred.

In the papers [**47**] and [**48**] H. Goda defined an important numerical invariant of a link $\mathscr{L}$ endowed with a Seifert surface $\mathscr{R}$. (Recall that a C^∞ connected oriented 2-submanifold $\mathscr{R}$ is called a *Seifert surface* for $\mathscr{L}$ if $\partial\mathscr{R} = \mathscr{L}$.) H. Goda's definitions and results are formulated in the language of handlebodies and sutured manifolds; we will translate them to the language of Morse functions and critical points.

Definition 6.3. Let $\mathscr{L}$ be an oriented link and $\mathscr{R}$ a Seifert surface for $\mathscr{L}$. The *handle number* $h(\mathscr{R})$ of the surface $\mathscr{R}$ is by definition one half of the minimal possible number of critical points of a regular Morse function $f : S^3 \setminus \mathscr{L} \to S^1$, such that $\mathscr{R} \setminus \mathscr{L}$ is a regular level surface of f (it is not difficult to show that the number of critical points of any regular Morse function is even).

It is clear that $2h(\mathscr{R}) \geqslant \mathcal{MN}(\mathscr{L})$ and

$$\mathcal{MN}(\mathscr{L}) = 2 \min_{\mathscr{R}} h(\mathscr{R}).$$

In the cited papers H. Goda develops efficient methods for constructing Morse functions with a given regular surface $\mathscr{R}$, thus providing computable upper bounds for the number $\mathcal{MN}(\mathscr{L})$. In particular, it follows from his results that $\mathcal{MN}(A_n) = 2$ where A_n is the boundary of the n-twisted unknotted annulus (see the figure on page 426). Another computation of $\mathcal{MN}(A_n)$ can be found in the paper of M. Hirasawa and Lee Rudolph [**60**].

Another theorem due to H. Goda [**48**] says that the handle number of the Murasugi sum $\mathscr{R}_1 \star \mathscr{R}_2$ of two Seifert surfaces $\mathscr{R}_1$, $\mathscr{R}_2$ satisfies the inequality

$$h(\mathscr{R}_1 \star \mathscr{R}_2) \leqslant h(\mathscr{R}_1) + h(\mathscr{R}_2).$$

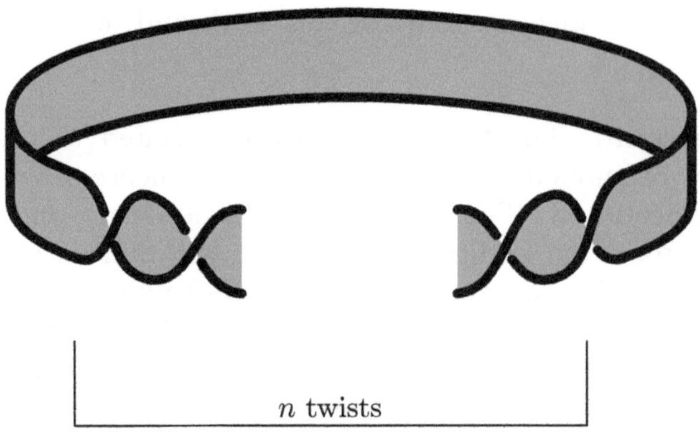

In the paper [**87**] A. Lazarev studied the homology with local coefficients of the complement to a knot $S^3 \setminus K$. He indicated a relationship between knot polynomials and the Novikov numbers of $S^3 \setminus K$. The homology with local coefficients for the complements of the links with several components was studied by A. Alania [**2**].

In the paper [**71**] Akio Kawauchi found a relation between the Novikov homology of the complement to a link in S^3, and the so-called $\times$-distance between the link and a fibred link. To formulate his results we need some more definitions. For a link $\mathscr{L}$ in S^3 we denote by $\sharp\mathscr{L}$ the number of its components. Let $\mathscr{L}_0, \mathscr{L}_1$ be two links with $\sharp\mathscr{L}_0 = \sharp\mathscr{L}_1$. We say that $d^\times(\mathscr{L}_0, \mathscr{L}_1) \leqslant 1$ if $\mathscr{L}_0$, $\mathscr{L}_1$ are isotopic to links $\mathscr{L}_0'$, respectively $\mathscr{L}_1'$ having both a regular projection onto some plane in $\mathbf{R}^3$ such that the diagram of $\mathscr{L}_0'$ with respect to this projection is obtained from the diagram of $\mathscr{L}_1'$ by at most one crossing move.

We say that $d^\times(\mathscr{L}_0, \mathscr{L}_1) \leqslant k$ if there is a sequence $\mathscr{L}_0', \dots, \mathscr{L}_k'$ of links such that $\mathscr{L}_0 = \mathscr{L}_0'$, $\mathscr{L}_1 = \mathscr{L}_k'$ and for every i we have $d^\times(\mathscr{L}_i', \mathscr{L}_{i+1}') \leqslant 1$. We define $d^\times(\mathscr{L}_0, \mathscr{L}_1)$ as the minimum of all natural numbers k with $d^\times(\mathscr{L}_0, \mathscr{L}_1) \leqslant k$. The following result is proved in [**71**]:

Theorem 6.4. *Let $\mathscr{L}$ be any link, and $\mathscr{L}_0$ any fibred link with $\sharp\mathscr{L}_0 = \sharp\mathscr{L}$. Then*

$$d^\times(\mathscr{L}, \mathscr{L}_0) \geqslant \widehat{b}_1(S^3 \setminus \mathscr{L}, \xi) + \widehat{q}_1(S^3 \setminus \mathscr{L}, \xi).$$

This theorem suggests a possible relation between the notion of the $\times$-distance and the Morse-Novikov numbers; this domain is still unexplored.

6.2. The Morse-Novikov inequalities for $\mathcal{MN}(\mathscr{L})$. Let $\mathscr{L}$ be an oriented link, and $f : S^3 \setminus \mathscr{L} \to S^1$ be a regular Morse function. The manifold $S^3 \setminus \mathscr{L}$ is not compact, and Proposition 2.4 of Chapter 11 (page 341) can not be applied directly. However one can show using the regularity condition that the proof of this proposition is still valid in the present context, with only minor modifications. Thus we obtain the next result:

Theorem 6.5. *If* $f : C_{\mathscr{L}} \to S^1$ *is a regular Morse function, then*

$$m_i(f) \geqslant \widehat{b}_i(S^3 \setminus \mathscr{L}, \xi) + \widehat{q}_i(S^3 \setminus \mathscr{L}, \xi) + \widehat{q}_{i-1}(S^3 \setminus \mathscr{L}, \xi),$$

where as usual $m_i(f)$ *stands for the number of critical points of* f *of index* i.

Since the cohomology class ξ is determined by the orientation of the link, we shall omit it in the notation and write $\widehat{b}_i(S^3 \setminus \mathscr{L})$ and $\widehat{q}_i(S^3 \setminus \mathscr{L})$. The numbers $\widehat{b}_i(S^3 \setminus \mathscr{L})$, $\widehat{q}_i(S^3 \setminus \mathscr{L})$ satisfy certain relations due to the Poincaré duality. Namely one can prove that

$$\widehat{b}_i(S^3 \setminus \mathscr{L}) = \widehat{q}_i(S^3 \setminus \mathscr{L}) = 0 \quad \text{for} \quad i \geqslant 3 \quad \text{and} \quad i = 0,$$

and

$$\widehat{b}_1(S^3 \setminus \mathscr{L}) = \widehat{b}_2(S^3 \setminus \mathscr{L}), \quad \text{and} \quad \widehat{q}_2(S^3 \setminus \mathscr{L}) = 0.$$

Thus we arrive at the following inequalities:

$$(3) \qquad m_2(f) \geqslant \widehat{b}_1(S^3 \setminus \mathscr{L}) + \widehat{q}_1(S^3 \setminus \mathscr{L}) \leqslant m_1(f).$$

The structure of the Novikov homology module $\widehat{H}_*(S^3 \setminus \mathscr{L})$ is related to the classical polynomials of the knot. To explain this relation let us first recall some definitions from elementary commutative algebra. Let T be a finitely generated L-module. Consider a free resolution of T:

$$0 \longleftarrow T \longleftarrow F_0 \overset{D}{\longleftarrow} F_1 \longleftarrow \cdots,$$

with $F_0 \approx L^m$ and $F_1 \approx L^n$, and $m \leqslant n$. The ideals $E_s(D)$ generated by the $(m-s)\times(m-s)$-subdeterminants of D are invariants of T (that is, they do not depend on the particular choice of the resolution). We denote by $\alpha_s(T)$ the greatest common divisor of all the elements of the ideal $E_s(D)$. In general, $\alpha_s(T) \notin E_s(D)$.

Theorem 6.6 ([**119**]). *Let* $T = H_1(\overline{S^3 \setminus \mathscr{L}})$, *and let* $\alpha_s = \alpha_s(T)$.
1. *The module* $\widehat{T} = T \underset{L}{\otimes} \overline{L}$ *is isomorphic to the following sum of cyclic modules:*†

$$\widehat{T} \approx \bigoplus_{s=0}^{m-1} \overline{L}/\gamma_s\overline{L}, \quad \text{where} \quad \gamma_s = \alpha_s/\alpha_{s+1}.$$

2. *The Novikov Betti number* $\widehat{b}_1(S^3 \setminus \mathscr{L})$ *is equal to the number of those* α_s *that are equal to zero.*

† With the conventions $0/0 = 0$ and $\alpha_m = 1$.

3. *The Novikov torsion number $\widehat{q}_1(S^3 \setminus \mathscr{L})$ is equal to the number of those γ_s that are non-zero and non-monic.*

A regular Morse function $f : S^3 \setminus \mathscr{L} \to S^1$ is called *minimal* if for every regular function $g : S^3 \setminus \mathscr{L} \to S^1$ we have

$$m_k(f) \leqslant m_k(g) \quad \text{for every} \quad k.$$

Proposition 6.7 ([**119**]). *Let $\mathscr{L}$ be a link. There exists a minimal regular Morse function $f : S^3 \setminus \mathscr{L} \to S^1$ such that:*

(1) $m_0(f) = m_3(f) = 0$.
(2) *One of the regular level surfaces of f is connected.*
(3) $m_1(f) = m_2(f)$.

Let us mention a challenging problem due to M. Boileau and C. Weber. It is not difficult to prove that the Morse-Novikov number of the connected sum $\mathscr{K}_1 \# \mathscr{K}_2$ of two knots $\mathscr{K}_1, \mathscr{K}_2$ satisfies the inequality

$$\mathcal{MN}(\mathscr{K}_1 \# \mathscr{K}_2) \leqslant \mathcal{MN}(\mathscr{K}_1) + \mathcal{MN}(\mathscr{K}_2).$$

Thus it is natural to ask:

$$\text{Is it true that} \quad \mathcal{MN}(K_1 \# K_2) = \mathcal{MN}(K_1) + \mathcal{MN}(K_2)?$$

6.3. Twisted Novikov homology. The lower bounds for Morse-Novikov numbers considered in the previous subsection are in general not sufficient to compute the Morse-Novikov number of a link. To get an idea about the strength of these inequalities, let us consider the problem of detecting fibred knots. If a knot is fibred, then its Novikov homology is equal to zero. It is not difficult to prove via a simple homological algebra that the Novikov homology of a knot vanishes if and only if its Alexander polynomial is monic (that is, the leading term of this polynomial equals ± 1). This condition is sufficient to detect all the fibred knots with $\leqslant 10$ crossings (due to a theorem of T. Kanenobu [**70**]). The first examples of non-fibred knots with monic Alexander polynomial have 11 crossings; they are presented in Figures 45 and 46.

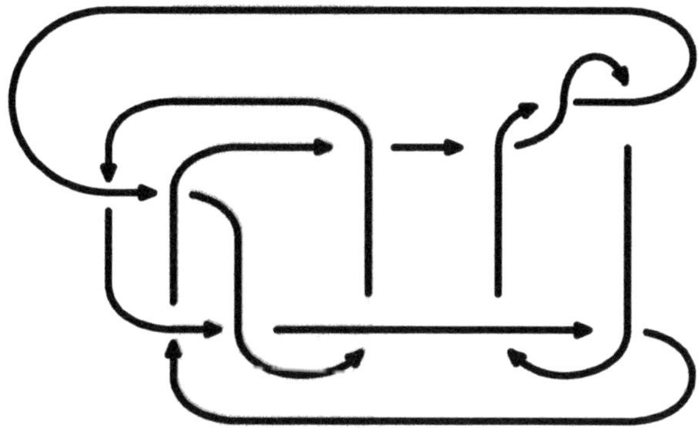

FIGURE 45.

The Conway knot

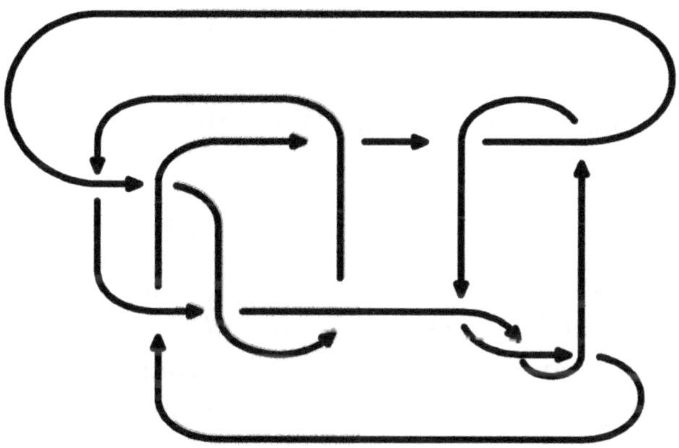

FIGURE 46.

The Kinoshita-Terasaka knot

It is natural to expect that a more sophisticated version of Novikov homology would be sufficient to prove that these knots are not fibred and obtain lower bounds for their Morse-Novikov numbers. This is indeed so, and the corresponding version of the Novikov homology was introduced by H. Goda and the author in the paper [**49**]. We will describe it now. Let X be a connected CW complex; let $G = \pi_1 X$, and let $\xi : G \to \mathbf{Z}$ be a homomorphism. Let $\rho : G \to GL(n, \mathbf{Z})$ be a map such that $\rho(g_1 g_2) = \rho(g_2)\rho(g_1)$

for every $g_1, g_2 \in G$. Such a map will be called a *right representation* of G. The group homomorphism $\xi : G \to \mathbf{Z}$ extends to a ring homomorphism $\mathbf{Z}[G] \to L$, which will be denoted by the same symbol ξ. The tensor product $\rho \otimes \xi$ (where ξ is considered as a representation $G \to GL(1, L)$) induces a right representation $\rho_\xi : G \to GL(n, L)$. The composition of this right representation with the natural inclusion $L \hookrightarrow \overline{L}$ gives a right representation

$$\widehat{\rho}_\xi : G \to GL(n, \overline{L}).$$

Let us form a chain complex

$$\widehat{C}_*(\widetilde{X}; \xi, \rho) = \overline{L}^n \underset{\widehat{\rho}_\xi}{\otimes} C_*(\widetilde{X}). \tag{4}$$

Here $\widetilde{X}$ is the universal cover of X, $C_*(\widetilde{X})$ is the cellular chain complex of $\widetilde{X}$ (it is a left module over $\mathbf{Z}[G]$), and $\overline{L}^n$ is a right $\mathbf{Z}G$-module via the right representation $\widehat{\rho}_\xi$. Then (4) is a chain complex of free left modules over $\overline{L}$, and the same is true for its homology. The modules

$$\widehat{H}_*(X; \xi, \rho) = H_*\big(\widehat{C}_*(\widetilde{X}; \xi, \rho)\big)$$

will be called the *ρ-twisted Novikov homology* or simply the *twisted Novikov homology* if no confusion is possible. When these modules are finitely generated we set

$$\widehat{b}_i(X; \xi, \rho) = \operatorname{rk}_{\overline{L}}\big(\widehat{H}_i(X; \xi, \rho)\big), \quad \widehat{q}_i(X; \xi, \rho) = \text{t.n.}_{\overline{L}}\big(\widehat{H}_i(X; \xi, \rho)\big),$$

where t.n. stands for the *torsion number* of the $\overline{L}$-module, that is, its minimal possible number of generators over $\overline{L}$.

The numbers $\widehat{b}_i$ and $\widehat{q}_i$ can be recovered from the canonical decomposition of $\widehat{H}_i(X; \xi, \rho)$ into a direct sum of cyclic modules. Namely, let

$$\widehat{H}_i(X; \xi, \rho) = \overline{L}^{\alpha_i} \oplus \Big(\bigoplus_{j=1}^{\beta_i} \overline{L}/\lambda_j^{(i)} \overline{L} \Big)$$

where $\lambda_j^{(i)}$ are non-zero non-invertible elements of $\overline{L}$ and

$$\lambda_{j+1}^{(i)} \mid \lambda_j^{(i)} \quad \text{for every} \quad j.$$

(Such a decomposition exists since $\overline{L}$ is a principal ideal domain.) Then

$$\alpha_i = \widehat{b}_i(X; \xi, \rho) \quad \text{and} \quad \beta_i = \widehat{q}_i(X; \xi, \rho).$$

It is not difficult to show that we can always choose $\lambda_j^{(i)}$ to be elements of L.

When ρ is a trivial 1-dimensional representation, we obtain the usual Novikov homology, which can be also calculated from the infinite cyclic covering $\bar{X}_\xi$ associated to ξ, namely

$$\widehat{H}_*(X;\xi,\rho) = \overline{L} \underset{L}{\otimes} H_*(\bar{X}_\xi) \quad \text{for} \quad \rho = 1 : G \to GL(1,\mathbf{Z}).$$

6.4. The twisted version of the Novikov inequalities. Let $\mathscr{L} \subset S^3$ be an oriented link; put

$$C_{\mathscr{L}} = S^3 \setminus \mathscr{L}, \quad G = \pi_1(C_{\mathscr{L}}).$$

Recall the canonical element $\xi \in H^1(C_{\mathscr{L}})$ characterized by the property that the restriction of ξ to every meridian μ is the positive generator of the group $H^1(\mu) \approx \mathbf{Z}$. We shall identify the cohomology class ξ with the corresponding homomorphism $G \to \mathbf{Z}$.

Let $\rho : G \to GL(n,\mathbf{Z})$ be any right representation of G. The proof of the next theorem is similar to the proof of the main theorem in [**110**] (with the necessary modifications due to the non-compactness of the complement to $\mathscr{L}$).

Theorem 6.8. *Let $f : C_{\mathscr{L}} \to S^1$ be a regular Morse map. There is a chain complex $\mathcal{N}_*$ of free $\overline{L}$-modules such that*

(1) *for every i the number of free generators of $\mathcal{N}_i$ equals $n \times m_i(f)$;*

(2) $H_*(\mathcal{N}_*) \approx \widehat{H}_*(C_{\mathscr{L}};\xi,\rho)$.

We shall denote $\widehat{H}_*(C_{\mathscr{L}},\xi,\rho)$ by $\widehat{H}_*(C_{\mathscr{L}},\rho)$. The numbers $\widehat{b}_i(C_{\mathscr{L}},\xi,\rho)$ and $\widehat{q}_i(C_{\mathscr{L}},\xi,\rho)$ will be denoted by $\widehat{b}_i(C_{\mathscr{L}},\rho)$ and $\widehat{q}_i(C_{\mathscr{L}},\rho)$. (We omit the cohomology class ξ in the notation since it is determined by the orientation of the link.)

Corollary 6.9. *Let $f : C_{\mathscr{L}} \to S^1$ be any regular map. Then*

$$m_i(f) \geqslant \frac{1}{n}\Big(\widehat{b}_i(C_{\mathscr{L}},\rho) + \widehat{q}_i(C_{\mathscr{L}},\rho) + \widehat{q}_{i-1}(C_{\mathscr{L}},\rho)\Big) \tag{5}$$

for every i. □

Corollary 6.10. *If $\mathscr{L}$ is fibred, then*

$$\widehat{H}_*(C_{\mathscr{L}},\rho) = 0 \quad \textit{and} \quad \widehat{b}_i(C_{\mathscr{L}},\rho) = \widehat{q}_i(C_{\mathscr{L}},\rho) = 0$$

for every representation ρ and every i.

Proposition 6.11. *The twisted Novikov numbers satisfy the following relations:*

$$\widehat{b}_i(C_{\mathscr{L}},\rho) = \widehat{q}_i(C_{\mathscr{L}},\rho) = \widehat{q}_2(C_{\mathscr{L}},\rho) = 0 \quad \textit{for} \quad i = 0, i \geqslant 3,$$
$$\widehat{b}_1(C_{\mathscr{L}},\rho) = \widehat{b}_2(C_{\mathscr{L}},\rho).$$

Proof. Pick a regular Morse map $f : C_{\mathscr{L}} \to S^1$ satisfying the conditions (1) – (3) of Proposition 6.7. The first assertion of our proposition follows since f has only critical points of indices 1 and 2. Since $m_1(f) = m_2(f)$, the Euler characteristics of the chain complex $\mathcal{N}_*$ is equal to 0 and we deduce the second assertion. □

Thus the non-trivial part of the Novikov inequalities is as follows:

$$m_1(f) \geqslant \frac{1}{n}\big(\widehat{b}_1(\mathscr{L},\rho) + \widehat{q}_1(\mathscr{L},\rho)\big) \leqslant m_2(f).$$

Now we shall investigate the twisted Novikov homology for the connected sum of knots. Let $\mathscr{K}$ be the connected sum of two oriented knots $\mathscr{K}_1, \mathscr{K}_2$. We have:

$$\pi_1(S^3 \setminus \mathscr{K}) = \pi_1(S^3 \setminus \mathscr{K}_1) \underset{Z}{*} \pi_1(S^3 \setminus \mathscr{K}_2),$$

where Z is the infinite cyclic group generated by a meridian μ of $\mathscr{K}$ (see [**18**], Ch.7, Prop. 7.10). In particular the groups $\pi_1(S^3 \setminus \mathscr{K}_1)$, $\pi_1(S^3 \setminus \mathscr{K}_2)$ are naturally embedded into $\pi_1(S^3 \setminus \mathscr{K})$, and some meridian element $\mu \in \pi_1(S^3 \setminus \mathscr{K})$ is the image of some meridian elements

$$\mu_1 \in \pi_1(S^3 \setminus \mathscr{K}_1), \quad \mu_2 \in \pi_1(S^3 \setminus \mathscr{K}_2).$$

Now let

$$\rho_1 : \pi_1(S^3 \setminus \mathscr{K}_1) \to GL(n, \mathbf{Z}), \; \rho_2 : \pi_1(S^3 \setminus \mathscr{K}_2) \to GL(n, \mathbf{Z})$$

be two right representations. Assume that $\rho_1(\mu_1) = \rho_2(\mu_2)$. Form the product representation

$$\rho_1 * \rho_2 : \pi_1(S^3 \setminus \mathscr{K}) \to GL(n, \mathbf{Z}).$$

Theorem 6.12. $\widehat{H}_*(\mathscr{K}, \rho_1 * \rho_2) \approx \widehat{H}_*(\mathscr{K}_1, \rho_1) \oplus \widehat{H}_*(\mathscr{K}_2, \rho_2)$.

Proof. The complement $C_{\mathscr{K}}$ is the union of two subspaces C_1, C_2 with C_i having the homotopy type of $C_{\mathscr{K}_i}$ (for $i = 1, 2$). The intersection $C_1 \cap C_2$ is homeomorphic to the twice punctured sphere $\Delta' = S^2 \setminus \{*, *\}$. The universal covering of $C_{\mathscr{K}}$ is therefore the union of two subspaces, which have the Novikov homology respectively equal to $\widehat{H}_*(\mathscr{K}_1, \rho_1)$ and $\widehat{H}_*(\mathscr{K}_2, \rho_2)$. The intersection of these two subspaces has the same Novikov homology as Δ', and this module vanishes. Then a standard application of the Mayer-Vietoris sequence proves the result sought. □

Corollary 6.13. *Denote by $m\mathscr{K}$ the connected sum of m copies of the knot $\mathscr{K}$. Let $\rho : \pi_1(S^3 \setminus \mathscr{K}) \to GL(n, \mathbf{Z})$ be a representation. Let $\rho^m : \pi_1(S^3 \setminus m\mathscr{K}) \to GL(n, \mathbf{Z})$ be the product of m copies of the representation ρ. Then*

$$\widehat{q}_1(m\mathscr{K}, \rho^m) = m \cdot \widehat{q}_1(\mathscr{K}, \rho).$$

Proof. This follows from the purely algebraic equality:

$$\text{t.n.}(mN) = m \cdot \big(\text{t.n.}(N)\big)$$

where N is any finitely generated module over a principal ideal domain, and mN stands for the direct sum of m copies of N. □

6.5. The Kinoshita-Terasaka and Conway knots. The Kinoshita-Terasaka knot $\mathfrak{KT}$ was introduced in the paper [**74**], and the Conway knot $\mathfrak{C}$ was discovered by J. Conway much later [**25**]. These two knots are very much alike (see the figures in Subsection 6.3), and many classical invariants take the same value on them. Still these knots are different, as was proved by R. Riley in [**137**], and they can be distinguished by the twisted Alexander polynomials (as shown by M. Wada, see [**159**]).

These knots are not fibred. Indeed, for a fibred knot the degree of its Alexander polynomial is equal to twice the genus of the knot, and the Alexander polynomial is trivial for both knots. In the paper [**49**], we proved the following result.

Theorem 6.14. *There are right representations*

$$\rho_1 : \pi_1(S^3 \setminus \mathfrak{C}) \to SL(5, \mathbf{Z}), \quad \rho_2 : \pi_1(S^3 \setminus \mathfrak{KT}) \to SL(5, \mathbf{Z}),$$

such that $\widehat{q}_1(\mathfrak{C}, \rho_1) \neq 0$, $\widehat{q}_1(\mathfrak{KT}, \rho_2) \neq 0$.

The representations ρ_1, ρ_2 were found with the help of K. Kodama's KNOT program [**78**]. By Corollary 6.13, this theorem implies

$$\mathcal{MN}(m\mathfrak{C}) \geqslant 2m/5, \quad \mathcal{MN}(m\mathfrak{KT}) \geqslant 2m/5 \quad \text{for every} \quad m.$$

Applying H. Goda's methods one can construct regular Morse functions on the complements of $\mathfrak{KT}$ and $\mathfrak{C}$ with exactly two critical points (see [**49**] for more details). Therefore

$$\mathcal{MN}(m\mathfrak{KT}) \leqslant 2m, \quad \mathcal{MN}(m\mathfrak{C}) \leqslant 2m.$$

6.6. The asymptotic Morse-Novikov number of a knot. Let $\mathscr{K} \subset S^3$ be an oriented knot. Let $m\mathscr{K}$ denote the connected sum of m copies of $\mathscr{K}$. Observe that

$$\mathcal{MN}(m_1\mathscr{K}) + \mathcal{MN}(m_2\mathscr{K}) \geqslant \mathcal{MN}\big((m_1 + m_2)\mathscr{K}\big),$$

therefore the sequence

$$\mu_m(\mathscr{K}) = \frac{\mathcal{MN}(m\mathscr{K})}{m}$$

converges (see [**126**], B.1, Ex. 98); its limit will be called the *asymptotic Morse-Novikov number of* $\mathscr{K}$ and denoted by $\mu(\mathscr{K})$. If the Morse-Novikov number is additive (cf. the Boileau-Weber problem, page 428), then obviously $\mu(\mathscr{K}) = \mathcal{MN}(\mathscr{K})$.

Corollary 6.15. *The asymptotic Morse-Novikov numbers of Kinoshita-Terasaka knot* $\mathfrak{KT}$ *and of the Conway knot* $\mathfrak{C}$ *satisfy*

$$\tfrac{2}{5} \leqslant \mu(\mathfrak{KT}), \mu(\mathfrak{C}) \leqslant 2. \qquad \square$$

History and Sources

The whole field as it exists now was initiated by S. P. Novikov. He was motivated by his joint work with I. Schmeltser about applications of the variational methods to the Kirchhof equations (see [**101**], [**104**]). The problem in theoretical mechanics studied in these papers turned out to be equivalent to a problem of minimial action paths with respect to a certain Lagrangian. The only difference with the classical setting was that the Lagrangian was not a single-valued function but a multi-valued one, in other words the differential of this Lagrangian was a closed non-exact one-form. In the two cited papers the critical point theory for such multi-valued functions was developed already sufficiently far to obtain important physical applications. In particular the authors have shown the existence of many periodic trajectories in the Kirchhof problem.

The resulting topological theory is the subject of S. P. Novikov's article [**102**]. This remarkable paper opens with the following challenge, which stimulated the development of the theory for several subsequent years:

Let M be a finite or infinite dimensional manifold and ω a closed 1-form, $d\omega = 0$. Integrating ω over paths in M defines a "multivalued function" S which becomes single-valued on some covering $\widehat{M} \xrightarrow{\pi} M$ with a free abelian monodromy group: $dS = \pi^*\omega$. The number of generators of the monodromy group is equal to the number of rationally independent integrals of the 1-form ω over integral cycles of M.

PROBLEM. To construct an analogue of Morse theory for the multi-valued functions S. That is, to find a relationship between the stationary points $dS = 0$ of different index and the topology of the manifold M.

S. P. Novikov outlined in this paper an approach to this problem, based on the construction of the *Novikov complex* (he focuses in this paper mainly on the case of circle-valued Morse functions). The paper continues with the Morse-type inequalities which are derived from the existence of the Novikov complex (ibidem, p.225):

$$(\mathcal{N}) \qquad m_k(f) \geqslant b_k(M,\xi) + q_k(M,\xi) + q_{k-1}(M,\xi).$$

In the article [**105**] S. P. Novikov asked whether these inequalities are exact for the case $\pi_1(M) \approx \mathbf{Z}$. This problem was solved by M. Farber [**30**]. It is interesting to observe that the notion of the Novikov complex or any other chain complex generated by the critical points of the Morse function in question does not appear in this work. One by-product of these results is an alternative proof of the Novikov inequalities (which was actually the first rigorous proof of these inequalities).

The first proof of the properties of the Novikov complex of the circle-valued Morse function was given by the author in a 1991 preprint [**110**] (published in 1995). The case of Morse forms of arbitrary degree of irrationality was considered by F. Latour [**82**].

In the beginning of the 1980s S. P. Novikov formulated the *Exponential Growth Conjecture* which says that generically the boundary operators in the Novikov complex have a non-zero convergence radius, or equivalently, the incidence coefficients $n_k(p, q; v)$ have at most exponential growth in k. In 1995 the author proved that for a C^0-generic gradient of any circle-valued Morse function the boundary operators are rational functions ([**112**], [**113**]). This confirms the exponential growth conjecture for the C^0-generic case (see Chapter 12 for a detailed discussion of this subject). The dynamical aspects of this problem are discussed in the papers of V. I. Arnold [**3**], [**4**]. The analytical approach to this conjecture was developed in the works of D. Burghelea and S. Haller [**19**], [**20**].

In 1996 M. Hutchings and Y. J. Lee proved that the torsion of the Novikov complex associated to a circle-valued Morse function is equal to the zeta function of its gradient flow (see [**64**], [**65**]). Their result was generalized by the author to the case when the Novikov complex is not acyclic, and defined over the completion of the group ring of the fundamental group ([**115**], [**117**], [**121**]). The methods used in these articles are different from the Hutchings-Lee original techniques; they are based on the theory of cellular gradients (see Part 3 and Chapter 12).

Bibliography

[1] R. Abraham, J. Robbin, *Transversal mappings and flows*, Benjamin, New York, 1967.

[2] L. Alaniya (Л. Алания), *Многообразия типа Александера*, Успехи Мат. Наук, т. **46** no. 1 (1991). English translation: *Manifolds of Alexander type*, Russian Math. Surveys **46** no. 1 (1991).

[3] V. I. Arnold, *Dynamics of intersections*, Proceedings of a Conference in Honour of J. Moser, edited by P. Rabinowitz and R. Hill, Academic Press, 1990, pp. 77–84.

[4] V. I. Arnold, *Dynamics of Complexity of Intersections*, Boletim Soc. Brasil. Mat. (N. S.) **21** (1) (1990), pp. 1–10.

[5] V. I. Arnold, *Bounds for Milnor numbers of intersections in holomorphic dynamical systems*, In: Topological Methods in Modern Mathematics, Publish or Perish, 1993, pp. 379–390.

[6] V. I. Arnold, *Ordinary Differential Equations*, Moscow, Nauka, 1975 (in Russian).

[7] V. I. Arnold, *Sur quelques problèmes de la théorie des systèmes dynamiques*, Journal of the Julius Schauder center **4** (1994), pp. 209–225.

[8] M. Artin, B. Mazur, *On periodic points*, Annals of Math. **102** (1965), pp. 82–99.

[9] M. F. Atiyah, I. G. Macdonald, *Introduction to commutative algebra*, Addison-Wesley, 1969.

[10] V. Baladi, *Periodic orbits and dynamical spectra*, Ergodic theory and dynamical systems, **18** (1998), pp. 255–292.

[11] H. Bass, *Algebraic K-theory*, Benjamin, 1968.

[12] H. Bass, A. Heller, R. G. Swan, *The Whitehead group of a polynomial extension*, Inst. Hautes Etudes Sci. Publ. Math. **22** (1964), 61–79.

[13] G. Birkhoff, G-C. Rota, *Ordinary differential equations*, Blaisdell Publishing Company, 1962.

[14] N. Bourbaki, *Groupes de Lie, Algèbres de Lie*, Hermann, 1971.

[15] M. Braverman, M. Farber, *Novikov type inequalities for differential forms with non-isolated zeros*, e-print: dg-ga/9508006, Journal publication: Math. Proc. Cambridge Philos. Soc. **122** (1997), pp. 357–375.

[16] M. Braverman, M. Farber, *The Novikov-Bott inequalities*, C. R. Acad. Sci. Paris **321** (1995), pp. 897–902.

[17] W. Browder, J. Levine, *Fibering manifolds over the circle*, Comment. Math. Helv. **40** (1966), pp. 153–160.

[18] G. Burde, H. Zieschang, *Knots*, Walter de Gruyter, 1985.

[19] D. Burghelea, S. Haller, *On the topology and analysis of a closed 1-form (Novikov's theory revisited)* , E-print: dg-ga/0101043, Journal publication: Essays on geometry and related topics, Vol. 1,2, pp. 133–175, Monogr. Enseign. Math., 38, Enseignement Math., Geneva, 2001.

[20] D. Burghelea, S. Haller, *Laplace transform, dynamics and spectral geometry*, E-print: math. DG/0405037.

[21] H. Cartan, *Cours de Calcul Différentiel*, Hermann, 1977.

[22] T. A. Chapman, *Topological invariance of Whitehead torsion*, Amer. J. Math. **96** (1974), pp. 488–497.

[23] M. M. Cohen, *A course in Simple-Homotopy theory*, Springer, 1972.

[24] P. M. Cohn, *Free rings and their relations*, Academic Press, 1971.

[25] J. H. Conway, *An enumeration of knots and links, and some of their algebraic properties*, Computational Problems in Abstract Algebra, Pergamon Press, New York, 1970, pp. 329–358.

[26] O. Cornea, A. Ranicki, *Rigidity and glueing for Morse and Novikov complexes*, e-print: AT. 0107221. Journal publication: Journal of the European Mathematical Society 5 (2003), pp. 343–394.

[27] J. Dieudonné, *Eléments d'analyse, Vol. III*, Gauthier-Villars, 1970.

[28] J. Dieudonné, *Foundations of modern analysis*, Academic Press, 1960.

[29] A. Dold, *Lectures on Algebraic Topology*, Springer, 1972.

[30] M. Farber (M. Фарбер), *Точность неравенств Новикова,* Функциональный анализ и его приложения **19**, 1985. English translation: *Exactness of Novikov inequalities*, Functional Analysis and Applications **19**, 1985.

[31] M. Farber, A. Ranicki, *The Morse-Novikov theory of circle-valued functions and noncommutative localization,* E-print: dg-ga/9812122, Journal Publication: Proc. 1998 Moscow Conference for the 60th Birthday of S. P. Novikov, Tr. Mat. Inst. Steklova **225** (1999), pp. 381–388.

[32] M. Farber *Topology of closed 1-forms*, Mathematical Surveys and Monographs, 108. American Mathematical Society, Providence, RI, 2004.

[33] F. T. Farrell, *The obstruction to fibering a manifold over a circle*, Bull. Amer. Math. Soc. **73** (1967), pp. 737–740.

[34] F. T. Farrell, *The obstruction to fibering a manifold over a circle*, Indiana Univ. J. **21** (1971), pp. 315–346.

[35] F. T. Farrell, W. -C. Hsiang, *A formula for* $K_1R_\alpha[T]$, Proc. Symp. Pure Math., Vol. **17** (1968), pp. 192–218.

[36] A. Fel'shtyn, *Dynamical zeta functions, Nielsen Theory and Reidemeister torsion,* preprint: ESI 539 of The Erwin Schrödinger International Institute for Mathematical Physics, Journal Publication: Mem. Amer. Math. Soc. 147 (2000), no. 699.

[37] J. Franks *Homology and dynamical systems*, CBMS Reg. Conf. vol. 49, AMS, Providence 1982.

[38] D. Fried, *Homological Identities for closed orbits*, Inv. Math. **71** (1983), pp. 419–442.

[39] D. Fried, *Periodic points and twisted coefficients*, Lect. Notes in Math. **1007** (1983), pp. 261–293.

[40] D. Fried, *Flow equivalence, hyperbolic systems and a new zeta function for flows*, Comm. Math. Helv. **57** (1982), pp. 237–259.

[41] F. B. Fuller, *An index of fixed point type for periodic orbits*, Amer. J. Math. **89** (1967), 133–148.

[42] D. Gabai, *The Murasugi sum is a natural geometric operation*, Contemp. Math., **20** (1983), pp. 131–143.

[43] S. Gallot, D. Hulin, J. Lafontaine, *Riemannian Geometry*, Universitext. Springer, 2004.

[44] R. Geoghegan, A. Nicas, *Trace and torsion in the theory of flows*, Topology **33** (1994), pp. 683–719.

[45] R. Geoghegan, A. Nicas, *Higher Euler characteristics 1*, L'Enseignement Mathématique **41** (1995), pp. 3–62.

[46] H. Goda, T. Kitano, T. Morifugi, *Reidemeister torsion, twisted Alexander polynomial and fibered knots*, preprint, 2002; Journal Publication: Comment. Math. Helv. **80** (2005), pp. 51–61.

[47] H. Goda, *Heegaard splitting for sutured manifolds and Murasugi sum*, Osaka J. Math. **29** (1992), pp. 21–40.

[48] H. Goda, *On handle number of Seifert surfaces in S^3*, Osaka J. Math. **30** (1993), pp. 63–80.

[49] H. Goda, A. Pajitnov, *Twisted Novikov homology and circle-valued Morse theory for knots and links*, e-print: math. GT/0312374, Journal Publication: Osaka Journal of Mathematics, v. **42** $N°$ 3, 2005.

[50] D. M. Grobman (Д. М. Гробман), *Гомеоморфизм систем дифференциальных уравнений*, ДАН СССР **128** (1959), pp. 880–881. English translation: Homeomorphisms of systems of differential equations, Sov. Math. Dokl. **128** (1959), pp. 880–881.

[51] D. M. Grobman, *Топологическая классификация окрестности особой точки в n-мерном пространстве,* Мат. Сборник **56** (1962), pp. 77–94. English translation: *Topological classification of the neighbourhood of a singular point in n-dimensional space*, Matem. Sbornik **56** (1962), pp. 77–94.

[52] J. Hadamard, *Sur l'itération et les solutions asymptotiques des équations différentielles*, Bull. Soc. Math. France **29** (1901), pp. 224–228.

[53] R. S. Hamilton, *On the inverse function theorem of Nash and Moser*, Bull. AMS **7** (1982), pp. 65–229.

[54] P. Hartman, *Ordinary Differential Equations*, Wiley, New York, 1964.

[55] R. Harvey, B. Lawson, *Morse theory and finite volume flows*, Ann. Math **153** (2001), pp. 1–25.

[56] R. Harvey, B. Lawson, *Morse theory and Stokes theorem*, Surveys in Differential Geometry **7**, pp. 65–229.

[57] R. Harvey, G. Minervini, *Morse Novikov theory and cohomology with finite supports*, e-print: math. DG/0212295.

[58] A. Hatcher, *Algebraic Topology*, Cambridge University Press, 2002.

[59] G. Higman, *Units in group rings*, Proc. London Math. Soc. **46** (1940), pp. 231–248.

[60] M. Hirasawa, L. Rudolph, *Constructions of Morse maps for knots and links, and upper bounds on the Morse-Novikov number*, e-prints mathGT/0108006 (2001), math. GT/0311134 (2003).

[61] M. Hirsch, *Differential Topology*, Springer, 1976.

[62] W. Huebsch, M. Morse, *The bowl theorem and a model non-degenerate function*, Proc. Nat. Acad. Sci. U. S. A. **51** (1964), pp. 49–51.

[63] D. Husemoller, *Fibre Bundles*, McGraw-Hill, 1966.

[64] M. Hutchings, Y. -J. Lee, *Circle-valued Morse theory, Reidemeister torsion and Seiberg-Witten invariants of 3-manifolds*, E-print: dg-ga/9612004, 3. Dec. 1996, journal publication: Topology **38** (1999), pp. 861–888.

[65] M. Hutchings, Y-J. Lee, *Circle-valued Morse theory and Reidemeister torsion,* Geometry and Topology **3** (1999), pp. 369–396.

[66] M. Irwin, *On the Stable Manifold Theorem*, Bull. London Math. Soc. **2** (1970), pp. 196–198.

[67] M. C. Irwin, *Smooth Dynamical Systems*, Academic Press, 1980.

[68] B. Jiang, S. Wang, *Twisted topological invariants associated with representations*, in: Topics in Knot Theory, Kluwer, 1993, pp. 211–227.

[69] B. Jiang, *Estimation of the number of periodic orbits,* Preprint: Universität Heidelberg, Mathematisches Institut, Heft 65, Mai 1993, Journal Publication: Pac. J. Math. **172**, no. 1 (1996), pp. 151–185.

[70] T. Kanenobu, *The augmentation subgroup of a pretzel link,* Mathematics Seminar Notes of Kobe University, v. 7 (1979), pp. 363–384.

[71] A. Kawauchi, *Distance between links by zero-linking twists,* Kobe Journal of Mathematics **13**, no. 2 (1996), pp. 183–190.

[72] J. L. Kelley, *General Topology*, van Nostrand, 1957.

[73] Kiang Tsai-han, *The Theory of Fixed Point Classes*, Springer, 1989.

[74] S. Kinoshita, H. Terasaka, *On unions of knots*, Osaka Math. J. **9** (1957), pp. 131–153.

[75] P. Kirk, Ch. Livingston, *Twisted Alexander invariants, Reidemeister torsion, and Casson-Gordon invariants,* Topology **38** (1999), pp. 635–661.

[76] T. Kitano, *Twisted Alexander polynomial and Reidemeister torsion,* Pacific J. Mat. **174** (1996), pp. 431–442.

[77] W. Klingenberg, *Lectures on Closed Geodesics*, Springer, 1978.

[78] K. Kodama, *KNOT program*, http://www.math.kobe-u.ac.jp/~kodama/knot.html.

[79] O. S. Kozlovskii (О. С. Козловский), *Динамика пересечений аналитических многообразий*, Доклады АН СССР **323** (1992). English translation: *The dynamics of intersections of analytic manifolds*, Sov. Math. Dokl. **45** (1992), pp. 425–427.

[80] I. Kupka, *Contribution à la théorie des champs génériques*, Contributions to Differential Equations **2** (1963), pp. 457–484, **3** (1964), pp. 411–420.

[81] S. Lang, *Algebra,* Addison-Wesley (1965).

[82] F. Latour, *Existence de 1-formes fermées non singulières dans une classe de cohomologie de de Rham*, Publ. IHES **80** (1995).

[83] J. Latschev, *A generalization of the Morse complex*, PhD thesis, SUNY, 1998.

[84] J. Latschev, *Gradient flows of Morse-Bott functions*, Mathematische Annalen, **318** no. 4 (2000), pp. 731–759.

[85] F. Laudenbach, *On the Thom-Smale complex*, In the book *An extension of a theorem by Cheeger and Müller*, Asterisque **205** (1992), pp. 219–233.

[86] F. Laudenbach, J. -C. Sikorav, *Persistance d'intersection avec la section nulle au cours d'une isotopic hamiltonienne dans un fibre cotangent*, Invent. Math. **82** (1985), pp. 349–357.

[87] A. Lazarev (А. Лазарев), *Гомологии Новикова в теории узлов*, Мат. Заметки, т. 51 $N°$ 3, 1992. English translation: *Novikov homologies in knot theory*, Mathematical Notes **51** (1992), pp. 259–261.

[88] A. M. Lyapunov (А. М. Ляпунов). *Общая проблема устойчивости движения*, 1892. French Translation: *Problème général de la stabilité du mouvement*, Ann. Fac. Sci. Toulouse **9** (1907), pp. 203–474. English translation: *The General Problem of the Stability of Motion*, Taylor & Francis, 1992.

[89] W. Lück, *The universal functorial Lefschetz invariant*, Fundam. Math. **161** no. 1–2 (1999), pp. 167–215.

[90] W. Massey, *Homology and Cohomology Theory*, Marcel Dekker, 1978.

[91] J. Milnor, *Infinite cyclic coverings*, In the book *Conference on the topology of manifolds*, 1968.

[92] J. Milnor, *Lectures on the h-Cobordism Theorem*, Princeton University Press, 1965.

[93] J. Milnor, *Introduction to Algebraic K-Theory*, Princeton University Press, 1971.

[94] J. Milnor, *Morse theory*, Princeton University Press, 1963.

[95] J. Milnor, *Whitehead Torsion*, Bull. Amer. Math. Soc. **72** (1966), pp. 358–426.

[96] J. Milnor and J. Stasheff, *Characteristic Classes*, Princeton University Press, 1974.

[97] G. Minervini, *A Current Approach to Morse and Novikov theories*, PhD thesis, Università La Sapienza, Roma, 2003.

[98] M. Morse, *Relations between the critical points of a real function of n independent variables*, Trans. Amer. Math. Soc. **15** (1925), pp. 345–396.

[99] M. Morse, *Calculus of Variations in the Large*, American Mathematical Society Colloquium Publications, Vol. 18, 1934.

[100] J. R. Munkres, *Elementary Differential Topology*, Annals of Math. Studies, Vol. 54, Princeton 1963.

[101] S. P. Novikov (С. П. Новиков), I. Shmel'tser (И. Шмельцер), *Периодические решения уравнений типа Кирхгофа для свободного движения твёрдого тела и расширенная теория Люстерника-Шнирельмана-Морса (ЛШМ) 1*, Функциональный анализ и его приложения **15** (1981), 54–66. English translation: *Periodic solutions of the Kirchhof equations for the free motion of a rigid body in an ideal fluid and the extended Lyusternik-Shnirel'man-Morse theory, 1*, Funktional analysis and applications **15** (1981), pp. 197–207.

[102] S. P. Novikov, *Многозначные функции и функционалы. Аналог теории Морса*, Доклады АН СССР **260** (1981), pp. 31–35. English translation: *Many-valued functions and functionals. An analogue of Morse theory*, Sov. Math. Dokl. **24** (1981), pp. 222–226.

[103] S. P. Novikov, *Блоховские гомологии и критические точки функций и 1-форм*, ДАН СССР **287** no. 6 (1982), English translation: *Bloch homology, critical points of functions and closed 1-forms*, Soviet Math. Dokl. **287** (1986), no. 2.

[104] S. P. Novikov, *Вариационные методы и периодические решения уравнений типа Кирхгофа, 2* Функциональный анализ и его приложения **15** (1981), English translation: *Variational methods and periodic solutions of equations of Kirchhof type, 2*, Funkstionalnyi analiz i ego prilozheniya, Functional analysis and applications **15** (1981), pp. 263–274.

[105] S. P. Novikov, *Гамильтонов формализм и многозначный аналог теории Морса*, Успехи Мат. Наук **37** (1982), pp. 3–49, English translation: *The hamiltonian formalism and a multivalued analogue of Morse theory*, Russ. Math. Surveys **37** (1982), pp. 1–56.

[106] S. P. Novikov, *Quasiperiodic Structures in topology*, in the book: Topological Methods in Modern Mathematics, Publish or Perish, 1993, pp. 223–235.

[107] A. Pajitnov (А. В. Пажитнов), *Аналитическое доказательство вещественной части неравенств Новикова*, Доклады АН СССР **293** no. 6. (1987)
English translation: *An analytic proof of the real part of the Novikov inequalities* Soviet Math. Dokl. **35** no. 2 (1987).

[108] A. Pajitnov *О точности неравенств типа Новикова для многообразий со свободной абелевой фундаментальной группой*, Математический Сборник, N° 11 (1989). English translation: *On the sharpness of Novikov-type inequalities for manifolds with free abelian fundamental group* , Math. USSR Sbornik **68** (1991), pp. 351–389.

[109] A. V. Pajitnov, *О модулях над некоторыми локализациями кольца Лорановских полиномов,* Математические заметки **46** no. 5 (1989). English translation: *Modules over some localizations of the ring of Laurent polynomials*, Mathematical Notes **46** no. 5 (1989).

[110] A. V. Pajitnov, *On the Novikov complex for rational Morse forms,* preprint: Institut for Matematik og datalogi, Odense Universitet Preprints, 1991, no. 12, Oct. 1991; Journal publication: Annales de la Faculté de Sciences de Toulouse **4** (1995), pp. 297–338.

[111] A. V. Pajitnov, *Surgery on the Novikov Complex*, Preprint: Rapport de Recherche CNRS URA 758, Nantes, 1993; Journal publication: K-theory **10** (1996), pp. 323-412.

[112] A. V. Pajitnov, *Rationality and exponential growth properties of the boundary operators in the Novikov Complex*, Mathematical Research Letters **3** (1996), pp. 541–548.

[113] A. V. Pajitnov, *The incidence coefficients in the Novikov complex are generically rational functions,* E-print: dg-ga/9603006 14 March 96, Journal publication: *Рациональность граничных операторов в комплексе Новикова в случае общего положения,* Алгебра и анализ **9** no. 5 (1997), pp. 92–139. English translation: Sankt-Petersbourg Mathematical Journal **9** no. 5 (1998), pp. 969–1006.

[114] A. V. Pajitnov, *On the asymptotics of Morse numbers of finite covers of manifold*, E-print: math. DG/9810136, Journal publication: Topology **38** no. 3 (1999), pp. 529–541 .

[115] A. V. Pajitnov, *Simple homotopy type of Novikov complex and ζ-function of the gradient flow,* E-print: dg-ga/970614 9 July 1997; journal article: *Простой гомотопический тип комплекса Новикова и ζ функция градиентного потока,* Успехи Мат. Наук **54** (1999), pp. 117–170. English translation: Russian Mathematical Surveys **54** (1999).

[116] A. V. Pajitnov, *C^0-generic properties of boundary operators in the Novikov complex,* E-print: math. DG/9812157, 29 Dec 1998, journal article: Advances in Mathematical Sciences, **197** (1999), pp. 29–117.

[117] A. V. Pajitnov, *Closed orbits of gradient flows and logarithms of non-abelian Witt vectors,* E-print: math. DG/9908010, 2. Aug. 1999, Journal publication: K-theory **21** no. 4, 2000.

[118] A. V. Pajitnov, A. Ranicki, *The Whitehead group of the Novikov ring,* E-print: math. AT/0012031, 5. dec. 2000, Journal publication: K-theory **21** no. 4, 2000.

[119] A. V. Pajitnov, C. Weber, L. Rudolph, *Morse-Novikov number for knots and links*, Algebra i Analiz **13** no. 3 (2001), (in Russian), English translation: Sankt-Petersbourg Mathematical Journal **13** no. 3 (2002), pp. 417–426.

[120] A. V. Pajitnov, *C^0-topology in Morse theory,* E-print: math. DG/0303195, 5. dec. 2000.

[121] A. V. Pajitnov, *Counting closed orbits of gradients of circle-valued maps*, E-print: math. DG/0104273 28 Apr. 2001, journal publication: *О замкнутых орбитах градиентных потоков отображений в окружность,* Алгебра и анализ **14** no. (2002), pp. 92–139. English translation: Sankt-Petersbourg Mathematical Journal **14** no. 3 (2003).

[122] J. Palis, Jr. , W. de Melo, *Geometric theory of dynamical systems*, Springer, 1982.

[123] M. Peixoto, *On an approximation theorem of Kupka and Smale*, J. Diff. Eq. **3** (1966), pp. 423–430.

[124] E. Pitcher, *Critical points of a map to a circle*, Proc. Nat. Acad. Sci. USA **25** (1939), pp. 428–431.

[125] E. Pitcher, *Inequalities of critical point theory*, Bull. Amer. Math. Soc. **64** (1958), pp. 1–30.

[126] G. Polya, G. Szegö, *Aufgaben und Lehrsätze aus der Analysis,* Springer, 1964.

[127] M. M. Postnikov (M. M. Постников), *Introduction to Morse Theory*, Moscow, Nauka, 1971, in Russian.

[128] M. Pozniak, *Triangulation of compact smooth manifolds and Morse theory*, preprint: University of Warwick preprint series, 11/1990.

[129] M. Pozniak, *The Morse complex, Novikov Cohomology and Fredholm Theory*, preprint: University of Warwick preprint series, 08/1991.

[130] M. Pozniak, *Floer homology, Novikov rings and clean intersections*, In: Northern California Symplectic Geometry Seminar, AMS Translations **196**, AMS, 1999.

[131] D. Quillen, *Projective modules over polynomial rings*, Inventiones Math. **36** (1976), pp. 167–171.

[132] A. Ranicki, *High-dimensional Knot Theory,* Springer, 1998.

[133] A. Ranicki, *The algebraic theory of torsion I,* Lecture Notes in Mathematics **1126** (1985), pp. 199–237.

[134] A. Ranicki, *Finite domination and Novikov rings*, preprint: 1993, Edinburgh University, journal publication: Topology **34** (1995), pp. 619–632.

[135] A. Ranicki, *The algebraic construction of the Novikov complex of a circle-valued Morse function*, E-print: math. AT/9903090, journal publication: Mathematische Annalen **322** (2002), pp. 745–785.

[136] K. Reidemeister, *Homotopieringe und Linsenräume*, Hamburger Abhandl. **11** (1938), pp. 102–109.

[137] R. Riley, *Homomorphisms of knots on finite groups*, Mathematics of Computation, v. 25, **115** (1971), pp. 603–619.

[138] D. Rolfsen, *Knots and Links*, Publish or Perish (1976, 1990).

[139] J. Rosenberg, *Algebraic K-theory and its applications*, Springer, 1994.

[140] D. Schütz, *Gradient flows of closed 1-forms and their closed orbits*, e-print: math. DG/0009055, journal article: Forum Math. **14** (2002), pp. 509–537.

[141] D. Schütz, *One-parameter fixed point theory and gradient flows of closed 1-forms*, e-print: math. DG/0104245, journal article: K-theory 25 (2002), pp. 59–97.

[142] V. V. Sharko, *Funktsii na mnogoobraziyah (algebraicheskie i topologicheskie aspekty)*, Kiev, Naukova dumka (1990).

[143] D. Sheiham, *Non-commutative Characteristic Polynomials and Cohn Localization*, e-print: math. RA/0104158, journal publication: Journal of the London Mathematical Society (2) **64** (2001) no. 1, pp. 13–28.

[144] L. Siebenmann, *A total Whitehead torsion obstruction to fibering over the circle*, Comment. Math. Helv. **45** (1970), pp. 1–48.

[145] J. -Cl. Sikorav, *Homologie associée a une fonctionnelle (d'après Floer)*, Asterisque 201 - 202 - 203, 1991, Séminaire Bourbaki **733**, 1990–1991, November 1991.

[146] J. -Cl. Sikorav, *Un probleme de disjonction par isotopie symplectique dans un fibré cotangent*, Ann. Scient. Ecole Norm. Sup. **19** (1986), pp. 543–552.

[147] J. -Cl. Sikorav, *Points fixes de difféomorphismes symplectiques, intersections de sous-variétés lagrangiennes, et singularités de un-formes fermées,* Thése de Doctorat d'Etat Es Sciences Mathématiques, Université Paris-Sud, Centre d'Orsay, 1987.

[148] S. Smale, *On the structure of manifolds*, Am. J. Math. **84** (1962), pp. 387–399.

[149] S. Smale, *Generalized Poincaré conjecture in dimensions greater than four*, Ann. Math. **74** (1961), pp. 391–406.
[150] S. Smale, *On gradient dynamical systems*, Am. J. Math. **74** (1961), pp. 199–206.
[151] S. Smale, *Differential dynamical systems*, Bull. Amer. Math. Soc. **73** (1967), pp. 747–817.
[152] S. Smale, *Stable manifolds for differential equations and diffeomorphisms*, Ann. Scuola Norm. Superiore Pisa **18** (1963), pp. 97–16.
[153] E. H. Spanier, *Algebraic Topology*, McGraw-Hill, (1966).
[154] J. Stallings, *On fibering certain 3-manifolds*, Proc. 1961 Georgia conference on the Topology of 3-manifolds, Prentice-Hall, 1962, pp. 95–100.
[155] R. Stong, *Notes on Cobordism Theory*, Princeton, New Jersey, 1968.
[156] R. M. Switzer, *Algebraic Topology – Homotopy and Homology*, Springer, 1975.
[157] R. Thom, *Sur une partition en cellules associée à une fonction sur une variété*, Comptes Rendus de l'Académie de Sciences **228** (1949), pp. 973–975.
[158] B. L. van der Waerden, *Algebra 1*, Springer, 1971.
[159] M. Wada, *Twisted Alexander polynomial for finitely presentable groups*, Topology **33** (1994), pp. 241–256.
[160] C. A. Weibel, *An Introduction to Homological Algebra*, Cambridge University Press, (1997).
[161] R. O. Wells, *Differential Analysis on Complex Manifolds*, Prentice-Hall, 1973.
[162] J. H. C. Whitehead, *Simple homotopy types*, Amer. J. Math. **72** (1952), pp. 1–57.
[163] E. Witten, *Supersymmetry and Morse theory*, Journal of Diff. Geom. **17** (1982), pp. 661–692.

Selected Symbols and Abbreviations

Subject Index

A

B

C

D

E

F

G

N

O

P

Q

R

S

T

U

V

W

Z

Bei Fragen zur Produktsicherheit wenden Sie sich bitte an:
If you have any questions regarding product safety,
please contact:

Walter de Gruyter GmbH
Genthiner Straße 13
10785 Berlin
productsafety@degruyterbrill.com